UNITEXT – La Matematica per il 3+2

Volume 65

For further volumes:
http://www.springer.com/series/5418

Sandro Salsa • Federico M.G. Vegni
Anna Zaretti • Paolo Zunino

A Primer on PDEs

Models, Methods, Simulations

 Springer

Sandro Salsa
Department of Mathematics
Politecnico di Milano

Federico M.G. Vegni
Department of Mathematics
Politecnico di Milano

Anna Zaretti
Department of Mathematics
Politecnico di Milano

Paolo Zunino
MOX – Department of Mathematics
Politecnico di Milano, and
Department of Mechanical Engineering and
Materials Science
University of Pittsburgh (USA)

Translated and extended version of the original Italian edition:
S. Salsa, F.M.G. Vegni, A. Zaretti, P. Zunino: Invito alle equazioni a derivate parziali,
© Springer-Verlag Italia 2009

UNITEXT – La Matematica per il 3+2
ISSN 2038-5722   ISSN 2038-5757 (electronic)

ISBN 978-88-470-2861-6   ISBN 978-88-470-2862-3 (eBook)
DOI 10.1007/978-88-470-2862-3

Library of Congress Control Number: 2012949463

Springer Milan Heidelberg New York Dordrecht London

© Springer-Verlag Italia 2013

9 8 7 6 5 4 3 2 1

Cover-Design: Beatrice ẞ, Milano

Typesetting with LaTeX: PTP-Berlin, Protago TeX-Production GmbH, Germany
(www.ptp-berlin.eu)
Printing and Binding: Grafiche Porpora, Segrate (MI)

Springer-Verlag Italia S.r.l., Via Decembrio 28, I-20137 Milano
Springer is a part of Springer Science+Business Media (www.springer.com)

Preface

This book is designed as an advanced undergraduate or a first-year graduate course for students from various disciplines like applied mathematics, physics, engineering. It has evolved while teaching courses on partial differential equations during the last decade at the Politecnico di Milano.

The main purpose of these courses was twofold: on the one hand, to train the students to appreciate the interplay between theory and modelling in problems arising in the applied sciences and on the other hand to give them a solid background for numerical methods, such as finite differences and finite elements, also through numerical simulations for selected problems. Accordingly, this textbook is divided into two parts.

The **first one**, Chapters 2 to 6, has a rather elementary character with the goal of developing and studying basic problems from the macro-areas of *diffusion, propagation and transport, waves and vibrations*. A knowledge of advanced calculus and ordinary differential equations is required to this part. Also, the repeated use of the method of separation of variables assumes some basic results from the theory of Fourier series. All this background material is summarized in the introductory Chapter 1 and in the Appendices.

Chapter 2 is devoted to first order equations and in particular to first order scalar conservation laws. Simple models from traffic dynamics are used to introduce concepts as characteristics lines, rarefaction and shock waves.

Chapters 3 and 5 deal with diffusion/reaction diffusion models, respectively. The heat and the Fisher-Kolmogoroff equations constitutes the reference models to illustrate the qualitative properties of the solutions and the asymptotic behavior towards equilibria.

In Chapter 4, the main properties of solutions to the Laplace/Poisson equation, Maximum principle, mean value properties, Green's function and Newtonian potential are the main topics.

In Chapter 6 the fundamental aspects of waves propagation are examined, leading to the classical formulas of d'Alembert, Kirchhoff and Poisson.

The **second part**, Chapters 7,8 and 9, develops the Hilbert spaces methods for the *variational formulation* and the analysis of *linear boundary* and *initial-boundary value problems*.

The understanding of these topics requires some basic knowledge of Lebesgue measure and integration, summarized in Chapter 7. This chapter contains tools from functional analysis in Hilbert spaces. The main theme is the solvability of abstract variational problems, leading to the Lax-Milgram Theorem. Then, we present a brief introduction to the theory of distributions of L. Schwarz and the most common Sobolev spaces, necessary for a correct variational formulation of the most common boundary value problems.

Chapter 8 is devoted to the variational formulation of elliptic boundary value problems and their solvability. The development starts with one-dimensional problems, continues with Poisson's equation and ends with general second order equations in divergence form. The last section contains an application to a simple control problem, with both distributed observation and control.

The issue in Chapter 9 is the variational formulation of initial-boundary value problems for second order parabolic operators in divergence form.

At the end of each chapter, a brief account of numerical methods is included, with a discussion of some particular case study, to complete a *model-theory-simulation* path.

Also a number of exercises is presented. Some of them can be solved by a routine application of the theory or of the methods developed in the text. Other problems are intended as a completion of some arguments or proofs in the text. Also, there are problems in which the student is required to be more autonomous. Most problems are supplied with answers or hints at the end of the volume.

In the first part the exposition if flexible enough to allow substantial changes in the order of presentation of the material, without compromising the comprehension. All chapters are in practice mutually independent, with the exception of Chapter 5, which presumes the knowledge of Chapters 3 and 4.

In the second part, which, in principle, may be presented independently of the first one, more attention has to be paid to the order of the arguments.

A huge number of books on partial differential equation has been written. At the end of this volume we have indicated some of the most popular ones, to which the reader can refer for a more advanced comprehension of the subject.

Acknowledgments. While writing this book we benefitted from comments, suggestions and criticisms of many colleagues. In particular, we express our gratitude to Cristina Cerutti, Michele Di Cristo, Maurizio Grasselli, Alessandro Veneziani and Gianmaria A. Verzini.

Milan, September 2012

Sandro Salsa
Federico M.G. Vegni
Anna Zaretti
Paolo Zunino

Contents

Part II Functional Analysis Techniques for Differential Problems

Part III Solutions

10 Solutions of selected exercises

Part IV Appendices

1

Introduction

1.1 Mathematical Modelling

Mathematical modelling plays a big role in the description of a large part of phenomena in the applied sciences and in several aspects of technical and industrial activity.

By a "mathematical model" we mean a set of equations and/or other mathematical relations capable of capturing the essential features of a complex natural or artificial system, in order to describe, forecast and control its evolution. The applied sciences are not confined to the classical ones; in addition to *physics* and *chemistry*, the practice of mathematical modelling heavily affects disciplines like *finance, biology, ecology, medicine, sociology*.

In the industrial activity (e.g. for aerospace or naval projects, nuclear reactors, combustion problems, production and distribution of electricity, traffic control, etc...) the mathematical modelling, involving first the analysis and the numerical simulation and followed by experimental tests, has become a common procedure, necessary for innovation, and also motivated by economic factors. It is clear that all of this is made possible by the enormous computational power now available.

In general, the construction of a mathematical model is based on two main ingredients: *general laws* and *constitutive relations*. In this book we shall deal with general laws coming from continuum mechanics and appearing as conservation or balance laws (e.g. of mass, energy, linear momentum, etc...).

The constitutive relations are of an experimental nature and strongly depend on the features of the phenomena under examination. Examples are the Fourier law of heat conduction, the Fick law for the diffusion of a substance or the way the speed of a driver depends on the density of cars ahead.

The outcome of the combination of the two ingredients is usually a *partial differential equation or a system of them*.

Salsa S., Vegni F.M.G., Zaretti A., Zunino P.: *A Primer on PDEs. Models, Methods, Simulations.*
Unitext – La Matematica per il 3+2 65.
DOI 10.1007/978-88-470-2862-3_1, © Springer-Verlag Italia 2013

1.2 Partial Differential Equations

A partial differential equation is a relation of the following type:

$$F(x_1, ..., x_n, u, u_{x_1}, ..., u_{x_n}, u_{x_1 x_1}, u_{x_1 x_2} ..., u_{x_n x_n}, u_{x_1 x_1 x_1}, ...) = 0 \qquad (1.1)$$

where the unknown $u = u(x_1, ..., x_n)$ is a function of n variables and $u_{x_j}, ...,$ $u_{x_i x_j}, ...$ are its partial derivatives. The highest order of differentiation occurring in the equation is the *order of the equation*.

A first important distinction is between *linear* and *nonlinear* equations.

Equation (1.1) is *linear* if F is linear with respect to u and all its derivatives, otherwise it is *nonlinear*.

A second distinction concerns the types of nonlinearity. We distinguish:

- *Semilinear* equations where F is nonlinear only with respect to u but is linear with respect to all its derivatives.
- *Quasi-linear* equations where F is linear with respect to the highest order derivatives of u.
- *Fully nonlinear equations where* F is nonlinear with respect to the highest order derivatives of u.

The theory of linear equations can be considered sufficiently well developed and consolidated, at least for what concerns the most important questions. On the contrary, the non linearities present such a rich variety of aspects and complications that a general theory does not appear to be conceivable. The existing results and the new investigations focus on more or less specific cases, especially interesting in the applied sciences.

To give the reader an idea of the wide range of applications we present a series of examples, suggesting one of the possible interpretations. Most of them are considered at various level of deepness in this book. In the examples, **x** represents a space variable (usually in dimension $n = 1, 2, 3$) and t is a time variable.

We start with **linear equations**. In particular, equations (1.2)–(1.5) are fundamental and their theory constitutes a starting point for many other equations.

1. *Transport equation* (first order):

$$u_t + \mathbf{v} \cdot \nabla u = 0. \qquad (1.2)$$

    It describes for instance the transport of a solid polluting substance along a channel; here u is the concentration of the substance and **v** is the stream speed. We consider the one-dimensional version of (1.2) in Section 2.2.

2. *Diffusion* or *heat equation* (second order):

$$u_t - D\Delta u = 0, \qquad (1.3)$$

    where $\Delta = \partial_{x_1 x_1} + \partial_{x_2 x_2} + ... + \partial_{x_n x_n}$ is the *Laplace operator*. It describes the conduction of heat through a homogeneous and isotropic medium; u

is the temperature and D encodes the thermal properties of the material. Chapter 3 is devoted to the heat equation and its variants.

3.   *Wave equation* (second order):

$$u_{tt} - c^2 \Delta u = 0. \qquad (1.4)$$

It describes for instance the propagation of transversal waves of small amplitude in a perfectly elastic chord (e.g. of a violin) if $n = 1$, or membrane (e.g. of a drum) if $n = 2$. If $n = 3$ it governs the propagation of electromagnetic waves in vacuum or of small amplitude sound waves (Section 6.6.4). Here u may represent the wave amplitude and c is the propagation speed.

4.   *Laplace's or potential equation* (second order):

$$\Delta u = 0, \qquad (1.5)$$

where $u = u(\mathbf{x})$. The diffusion and the wave equations model evolution phenomena. The Laplace equation describes the corresponding *steady state*, in which the solution does not depend on time anymore. Together with its nonhomogeneous version

$$\Delta u = f,$$

called *Poisson's* equation, it plays an important role in electrostatics as well. Chapter 4 is devoted to these equations.

5.   *Black-Scholes equation* (second order):

$$u_t + \frac{1}{2}\sigma^2 x^2 u_{xx} + rx u_x - ru = 0.$$

Here $u = u(x,t)$, $x \geq 0$, $t \geq 0$. Fundamental in mathematical finance, this equation governs the evolution of the price u of a so called *derivative* (e.g. an *European option*), based on an underlying asset (a stock, a currency, etc.) whose price is x.

6.   *Vibrating plate* (fourth order):

$$u_{tt} - \Delta^2 u = 0,$$

where $\mathbf{x} \in \mathbb{R}^2$ and

$$\Delta^2 u = \Delta(\Delta u) = \frac{\partial^4 u}{\partial x_1^4} + 2\frac{\partial^4 u}{\partial x_1^2 \partial x_2^2} + \frac{\partial^4 u}{\partial x_2^4}$$

is the *biharmonic operator*. In the theory of linear elasticity, it models the transversal waves of small amplitude of a homogeneous isotropic plate.

7. *Schrödinger equation* (second order):

$$-iu_t = \Delta u + V\left(\mathbf{x}\right)u$$

where i is the complex unit. This equation is fundamental in quantum mechanics and governs the evolution of a particle subject to a potential V. The function $|u|^2$ represents a *probability density*.

Let us list now some examples of **nonlinear equations**.

8. *Burgers' equation* (quasilinear, first order):

$$u_t + cuu_x = 0 \qquad \left(x \in \mathbb{R}\right).$$

It governs a one-dimensional flux of a non viscous fluid but it is used to model traffic dynamics as well. Its viscous variant

$$u_t + cuu_x = \varepsilon u_{xx} \qquad \left(\varepsilon > 0\right)$$

constitutes a basic example of competition between *dissipation* (due to the term εu_{xx}) and *steepening* (shock formation due to the term cuu_x). We will discuss these topics in Sections 2.5 and 2.6.1.

9. *Fisher's equation* (semilinear, second order):

$$u_t - D\Delta u = ru\left(M - u\right)$$

It governs the evolution of a population of density u, subject to diffusion and logistic growth (represented by the right hand side).

10. *Porous medium equation* (quasilinear, second order):

$$u_t = k \operatorname{div}\left(u^\gamma \nabla u\right)$$

where $k > 0, \gamma > 1$ are constant. This equation appears in the description of filtration phenomena, e.g. of the motion of water through the ground.

11. *Minimal surface equation* (quasilinear, second order):

$$\operatorname{div}\left(\frac{\nabla u}{\sqrt{1 + |\nabla u|^2}}\right) = 0 \qquad \left(\mathbf{x} \in \mathbb{R}^2\right)$$

The graph of a solution u minimizes the area among all surfaces $z = v\left(x_1, x_2\right)$ whose boundary is a given curve. For instance, soap balls are minimal surfaces. We will not examine this equation (for deeper insights see [9]).

12. *Eikonal equation* (fully nonlinear, first order):

$$|\nabla u| = c\left(\mathbf{x}\right).$$

It appears in geometrical optics: if u is a solution, its level surfaces $u\left(\mathbf{x}\right) = t$ describe the position of a light wave front at time t.

Let us now give some examples of **systems**.

13. *Navier's equation of linear elasticity* (three scalar equations of second order):

$$\varrho\mathbf{u}_{tt} = \mu\Delta\mathbf{u} + (\mu + \lambda)\text{grad div }\mathbf{u}$$

where $\mathbf{u} = (u_1(\mathbf{x},t), u_2(\mathbf{x},t), u_3(\mathbf{x},t))$, $\mathbf{x} \in \mathbb{R}^3$. The vector $\mathbf{u}$ represents the displacement from equilibrium of a deformable continuum body of (constant) density ϱ (see e.g. *Dautray and Lions* [22], Vol. 1,6).

14. *Maxwell's equations in vacuum* (six scalar linear equations of first order):

$$\mathbf{E}_t - \text{curl }\mathbf{B} = \mathbf{0}, \qquad \mathbf{B}_t + \text{curl }\mathbf{E} = \mathbf{0} \qquad \text{(Ampère and Faraday laws)}$$

$$\text{div }\mathbf{E} = 0 \qquad \text{div }\mathbf{B} = 0 \qquad \text{(Gauss' law)}$$

where $\mathbf{E}$ is the electric field and $\mathbf{B}$ is the magnetic induction field. The unit measures are the "natural" ones, i.e. the light speed is $c = 1$ and the magnetic permeability is $\mu_0 = 1$ (see e.g. *Dautray and Lions* [22], Vol. 1).

15. *Navier-Stokes equations* (three quasilinear scalar equations of second order and one linear equation of first order):

$$\begin{cases} \mathbf{u}_t + (\mathbf{u}\cdot\nabla)\,\mathbf{u} = -\frac{1}{\rho}\nabla p + \nu\Delta\mathbf{u} \\ \text{div }\mathbf{u} = 0 \end{cases}$$

where $\mathbf{u} = (u_1(\mathbf{x},t), u_2(\mathbf{x},t), u_3(\mathbf{x},t))$, $p = p(\mathbf{x},t)$, $\mathbf{x} \in \mathbb{R}^3$. These equations governs the motion of a viscous, homogeneous and incompressible fluid. Here $\mathbf{u}$ is the fluid speed, p its pressure, ρ its density (constant) and ν is the kinematic viscosity, given by the ratio between the fluid viscosity and its density. The term $(\mathbf{u}\cdot\nabla)\,\mathbf{u}$ represents the inertial acceleration due to fluid transport.

1.3 Well Posed Problems

Usually, in the construction of a mathematical model, only some of the general laws of continuum mechanics are relevant, while the others are eliminated through the constitutive laws or suitably simplified according to the current situation. In general, additional information is necessary to select or to predict the existence of a unique solution. This information is commonly supplied in the form of *initial and/or boundary data*, although other forms are possible. For instance, typical boundary conditions prescribe the value of the solution or of its normal derivative, or a combination of the two. A main goal of a theory is to establish suitable conditions on the data in order to have a problem with the following features:

a) *there exists at least one solution;*
b) *there exists at most one solution;*
c) *the solution depends continuously on the data.*

This last condition requires some explanation. Roughly speaking, property c) states that the correspondence

$$data \to solution \tag{1.6}$$

is *continuous* or, in other words, that a *small error on the data entails a small error on the solution.*

This property is extremely important and may be expressed as a **local stability of the solution with respect to the data**. Think for instance of using a computer to find an approximate solution: the insertion of the data and the computation algorithms entail approximation errors of various type. A significant sensitivity of the solution on small variations of the data would produce an unacceptable result.

The notion of continuity and the error measurements, both in the data and in the solution, are made precise by introducing a suitable notion of *distance*. In dealing with a numerical or a finite-dimensional set of data, an appropriate distance may be the usual *euclidean distance:* if $\mathbf{x} = (x_1, x_2, ..., x_n)$, $\mathbf{y} = (y_1, y_2, ..., y_n)$ then

$$\text{dist}(\mathbf{x}, \mathbf{y}) = \|\mathbf{x} - \mathbf{y}\| = \sqrt{\sum_{k=1}^{n} (x_k - y_k)^2}.$$

When dealing for instance with real functions, defined on a set A, common distances are:

$$\text{dist}(f, g) = \max_{\mathbf{x} \in A} |f(\mathbf{x}) - g(\mathbf{x})|$$

which measures the maximum difference between f and g over A, or

$$\text{dist}(f, g) = \sqrt{\int_A (f - g)^2}$$

which is the so called *least square distance between f and g.*

Once the notion of distance has been chosen, the continuity of the correspondence (1.6) is easy to understand: *if the distance of the data tends to zero then the distance of the corresponding solutions tends to zero.*

When a problem possesses the properties a), b) c) above it is said to be **well posed**. When using a mathematical model, it is extremely useful, sometimes essential, to deal with well posed problems: existence of the solution indicates that the model is coherent, uniqueness and stability increase the possibility of providing accurate numerical approximations.

As one can imagine, complex models lead to complicated problems which require rather sophisticated techniques of theoretical analysis. Often, these problems become well posed and efficiently treatable by numerical methods if suitably reformulated in the abstract framework of Functional Analysis, as we will see in Chapter 7.

On the other hand, not only well posed problems are interesting for the applications. There are problems that are intrinsically *ill posed* because of the lack of uniqueness or of stability, but still of great interest for the modern technology. We only mention an important class of ill posed problems, given by the so called **inverse problems**, closely related to *control theory* (an example of control problem is in Section 8.7).

1.4 Basic Notations and Facts

We specify some of the symbols we will constantly use throughout the book and recall some basic notions about sets, topology and functions.

Sets and Topology

We denote by: $\mathbb{N}$, $\mathbb{Z}$, $\mathbb{Q}$, $\mathbb{R}$, $\mathbb{C}$ the sets of natural numbers, integers, rational, real and complex numbers, respectively. $\mathbb{R}^n$ is the n-dimensional vector space of the n-uples of real numbers. We denote by $\mathbf{e}^1, \dots, \mathbf{e}^n$ the unit vectors in the canonical base in $\mathbb{R}^n$. In $\mathbb{R}^2$ and $\mathbb{R}^3$ we may denote them by $\mathbf{i}$, $\mathbf{j}$ and $\mathbf{k}$.

The symbol $B_r(\mathbf{x})$ denotes the *open* ball in $\mathbb{R}^n$, with radius r and center at $\mathbf{x}$, that is

$$B_r(\mathbf{x}) = \{\mathbf{y} \in \mathbb{R}^n; \ |\mathbf{x} - \mathbf{y}| < r\}.$$

If there is no need to specify the radius, we write simply $B(\mathbf{x})$. The volume of $B_r(\mathbf{x})$ and the area of $\partial B_r(\mathbf{x})$ are given by

$$|B_r| = \frac{\omega_n}{n} r^n \quad \text{and} \quad |\partial B_r| = \omega_n r^{n-1}$$

where ω_n is the surface area of the unit sphere[1] ∂B_1 in $\mathbb{R}^n$; in particular $\omega_2 = 2\pi$ and $\omega_3 = 4\pi$.

Let $A \subseteq \mathbb{R}^n$. A point $\mathbf{x} \in A$ is:

- an *interior point* if there exists a ball $B_r(\mathbf{x}) \subset A$;
- a *boundary point* if any ball $B_r(\mathbf{x})$ contains points of A **and** of its complement $\mathbb{R}^n \backslash A$. The set of boundary points of A, the *boundary of A*, is denoted by ∂A;
- a *limit point* of A if there exists a sequence $\{\mathbf{x}_k\}_{k \geq 1} \subset A$ such that $\mathbf{x}_k \to \mathbf{x}$.

A is *open* if every point in A is an interior point; the set $\overline{A} = A \cup \partial A$ is the *closure of A*; A is *closed* if $A = \overline{A}$. A set is closed if and only if it contains all of its limit points.

---

[1] In general, $\omega_n = n\pi^{n/2}/\Gamma\left(\frac{1}{2}n + 1\right)$ where $\Gamma(s) = \int_0^{+\infty} t^{s-1} e^{-t} dt$ is the *Euler gamma function*.

An open set is *connected* if for every couple of points $\mathbf{x}, \mathbf{y} \in A$ there exists a regular curve joining them. By a *domain* we mean an *open connected* set. Domains are usually denoted by the letter Ω.

If $U \subset A$, we say that U is *dense in* A if $\overline{U} = A$. This means that any point $\mathbf{x} \in A$ is a limit point of U. For instance, $\mathbb{Q}$ is dense in $\mathbb{R}$.

A is *bounded* if it is contained in some ball $B_r(\mathbf{0})$; it is *compact* if it is *closed and bounded*. If $\overline{A}_0$ is compact and contained in A, we write $A_0 \subset\subset A$ and we say that A_0 is *compactly contained* in A.

Infimum and supremum of a set of real numbers

A set $A \subset \mathbb{R}$ is *bounded from below* if there exists a number K such that

$$K \leq x \quad \text{for every } x \in A. \tag{1.7}$$

The greatest among the numbers K with the property (1.7) is called the *infimum* or *the greatest lower bound* of A and denoted by $\inf A$.

More precisely, we say that $\lambda = \inf A$ if $\lambda \leq x$ for every $x \in A$ and if, for every $\varepsilon > 0$, we can find $\bar{x} \in A$ such that $\bar{x} < \lambda + \varepsilon$. If $\inf A \in A$, then $\inf A$ is actually called the *minimum of* A, and may be denoted by $\min A$.

Similarly, $A \subset \mathbb{R}$ is *bounded from above* if there exists a number K such that

$$x \leq K \quad \text{for every } x \in A. \tag{1.8}$$

The smallest among the numbers K with the property (1.8) is called the *supremum* or the *lowest upper bound of* A and denoted by $\sup A$.

Precisely, we say that $\Lambda = \sup A$ if $\Lambda \geq x$ for every $x \in A$ and if, for every $\varepsilon > 0$, we can find $\bar{x} \in A$ such that $\bar{x} > \Lambda - \varepsilon$. If $\sup A \in A$, then $\sup A$ is actually called the *maximum of* A, and may be denoted by $\max A$.

Functions

Let $A \subseteq \mathbb{R}$ and $u : A \to \mathbb{R}$ be a real valued function defined in A. We say that u is *continuous* at $\mathbf{x} \in A$ if $u(\mathbf{y}) \to u(\mathbf{x})$ as $\mathbf{y} \to \mathbf{x}$. If u is continuous at any point of A we say that u is continuous in A. The set of such functions is denoted by $C(A)$.

The **support** of a continuous function is the *closure of the set where it is different from zero*. A continuous function is *compactly supported* in A if it vanishes outside a compact set contained in A.

We say that u is *bounded from below* (resp. *above*) in A if the image

$$u(A) = \{y \in \mathbb{R}, \, y = u(\mathbf{x}) \text{ for some } \mathbf{x} \in A\}$$

is *bounded by below* (resp. *above*). The infimum (supremum) of $u(A)$ is called the *infimum* (*supremum*) *of* u and is denoted by

$$\inf_{\mathbf{x} \in A} u(\mathbf{x}) \quad (\text{resp. } \sup_{\mathbf{x} \in A} u(\mathbf{x})).$$

We will denote by χ_A the *characteristic function of* A: $\chi_A = 1$ on A and $\chi_A = 0$ in $\mathbb{R}^n \backslash A$.

We use one of the symbols u_{x_j}, $\partial_{x_j} u$, $\frac{\partial u}{\partial x_j}$ for the first partial derivatives of u, and ∇u or grad u for the *gradient* of u. Accordingly, for the higher order derivatives we use the notations $u_{x_j x_k}$, $\partial_{x_j x_k} u$, $\frac{\partial^2 u}{\partial x_j \partial x_k}$ and so on.

We say that u is of class $C^k(\Omega)$, $k \geq 1$, or that it is a C^k–function, if u has continuous partials up to the order k (included) in the domain Ω. The class of continuously differentiable functions of any order in Ω, is denoted by $C^\infty(\Omega)$.

If $u \in C^1(\Omega)$ then u is differentiable in Ω and we can write, for $\mathbf{x} \in \Omega$ and $\mathbf{h} \in \mathbb{R}^n$ small:

$$u(\mathbf{x} + \mathbf{h}) - u(\mathbf{x}) = \nabla u(\mathbf{x}) \cdot \mathbf{h} + o(\mathbf{h})$$

where the symbol $o(\mathbf{h})$, "*little o of* $\mathbf{h}$", denotes a quantity such that $o(\mathbf{h})/|\mathbf{h}| \rightarrow 0$ as $|\mathbf{h}| \rightarrow 0$.

The symbol $C^k(\overline{\Omega})$ will denote the set of functions in $C^k(\Omega)$ whose derivatives up to the order k included, can be extended continuously up to $\partial \Omega$.

Integrals

Up to Chapter 6 included, the integrals can be considered in the Riemann sense (proper or improper). A brief introduction to Lebesgue measure and integral is provided in Section 7.1. Let $1 \leq p < \infty$ and $q = p/(p-1)$, the *conjugate exponent of* p. The following Hölder's inequality holds

$$\left| \int_\Omega uv \right| \leq \left(\int_\Omega |u|^p \right)^{1/p} \left(\int_\Omega |v|^q \right)^{1/q}. \tag{1.9}$$

The case $p = q = 2$ is known as the Schwarz inequality.

Uniform convergence

A series $\sum_{m=1}^\infty u_m$, where $u_m : \Omega \subseteq \mathbb{R}^n \rightarrow \mathbb{R}$, is said to be *uniformly convergent in* Ω, with *sum* u if, setting $S_N = \sum_{m=1}^N u_m$, we have

$$\sup_{\mathbf{x} \in \Omega} |S_N(\mathbf{x}) - u(\mathbf{x})| \rightarrow 0 \text{ as } N \rightarrow \infty.$$

Weierstrass test. Let $|u_m(\mathbf{x})| \leq a_m$, for every $m \geq 1$ and $\mathbf{x} \in \Omega$. If the numerical series $\sum_{m=1}^\infty a_m$ is convergent, then $\sum_{m=1}^\infty u_m$ converges absolutely and uniformly in Ω.

Limit and series. Let $\sum_{m=1}^\infty u_m$ be uniformly convergent in Ω. If u_m is continuous at $\mathbf{x}_0$ for every $m \geq 1$, then u is continuous at $\mathbf{x}_0$ and

$$\lim_{\mathbf{x} \rightarrow \mathbf{x}_0} \sum_{m=1}^\infty u_m(\mathbf{x}) = \sum_{m=1}^\infty u_m(\mathbf{x}_0).$$

Term by term integration. Let $\sum_{m=1}^{\infty} u_m$ be uniformly convergent in Ω. If Ω is bounded and u_m is integrable in Ω for every $m \geq 1$, then:

$$\int_{\Omega} \sum_{m=1}^{\infty} u_m = \sum_{m=1}^{\infty} \int_{\Omega} u_m.$$

Term by term differentiation. Let Ω be bounded and $u_m \in C^1\left(\overline{\Omega}\right)$ for every $m \geq 0$. If the series $\sum_{m=1}^{\infty} u_m\left(\mathbf{x}_0\right)$ is convergent at some $\mathbf{x}_0 \in A$ and the series $\sum_{m=1}^{\infty} \partial_{x_j} u_m$ are uniformly convergent in $\overline{\Omega}$ for every $j = 1, ..., n$, then $\sum_{m=1}^{\infty} u_m$ converges uniformly in $\overline{\Omega}$, with sum in $C^1\left(\overline{\Omega}\right)$ and

$$\partial_{x_j} \sum_{m=1}^{\infty} u_m\left(\mathbf{x}\right) = \sum_{m=1}^{\infty} \partial_{x_j} u_m\left(\mathbf{x}\right) \qquad (j = 1, ..., n).$$

1.5 Integration by Parts Formulas

Let $\Omega \subset \mathbb{R}^n$, be a C^1-domain. For vector fields

$$\mathbf{F} = \left(F_1, F_2, ..., F_n\right) : \Omega \to \mathbb{R}^n$$

with $\mathbf{F} \in C^1\left(\overline{\Omega}\right)$, the **Gauss divergence formula** holds:

$$\int_{\Omega} \mathrm{div}\mathbf{F} \, d\mathbf{x} = \int_{\partial \Omega} \mathbf{F} \cdot \boldsymbol{\nu} \, d\sigma \qquad (1.10)$$

where $\mathrm{div}\mathbf{F} = \sum_{j=1}^{n} \partial_{x_j} F_j$, $\boldsymbol{\nu}$ denotes the *outward normal* unit vector to $\partial \Omega$, and $d\sigma$ is the "surface" measure on $\partial \Omega$, locally given in terms of local charts by

$$d\sigma = \sqrt{1 + |\nabla \varphi\left(\mathbf{y}'\right)|} d\mathbf{y}'.$$

A number of useful identities can be derived from (1.10). Applying (1.10) to $v\mathbf{F}$, with $v \in C^1\left(\overline{\Omega}\right)$, and recalling the identity

$$\mathrm{div}(v\mathbf{F}) = v\,\mathrm{div}\mathbf{F} + \nabla v \cdot \mathbf{F}$$

we obtain the following **integration by parts** formula:

$$\int_{\Omega} v \,\mathrm{div}\mathbf{F} \, d\mathbf{x} = \int_{\partial \Omega} v\mathbf{F} \cdot \boldsymbol{\nu} \, d\sigma - \int_{\Omega} \nabla v \cdot \mathbf{F} \, d\mathbf{x}. \qquad (1.11)$$

Choosing $\mathbf{F} = \nabla u$, $u \in C^2\left(\Omega\right) \cap C^1\left(\overline{\Omega}\right)$, since $\mathrm{div}\nabla u = \Delta u$ and $\nabla u \cdot \boldsymbol{\nu} = \partial_{\boldsymbol{\nu}} u$, the following **Green's identity** follows:

$$\int_{\Omega} v\Delta u \, d\mathbf{x} = \int_{\partial \Omega} v\partial_{\boldsymbol{\nu}} u \, d\sigma - \int_{\Omega} \nabla v \cdot \nabla u \, d\mathbf{x}. \qquad (1.12)$$

In particular, the choice $v \equiv 1$ yields

$$\int_\Omega \Delta u \; d\mathbf{x} = \int_{\partial\Omega} \partial_\nu u \; d\sigma. \tag{1.13}$$

If also $v \in C^2(\Omega) \cap C^1(\overline{\Omega})$, interchanging the roles of u and v in (1.12) and subtracting, we derive a second **Green's identity**:

$$\int_\Omega (v\Delta u - u\Delta v) \; d\mathbf{x} = \int_{\partial\Omega} (v\partial_\nu u - u\partial_\nu v) \; d\sigma. \tag{1.14}$$

1.6 Abstract Methods and Variational Formulation

The abstract methods that we present and use in the second part of the book combine either analytical and geometrical aspects. These techniques are the core of the branch of Mathematics called Functional Analysis. In order to have a rough understanding of the main ideas, it could be useful to examine in an informal way how they come out, working on a specific example.

We consider the equilibrium position of a stretched membrane having the shape of a square Ω, subject to an external load f (force per unit mass) and kept at level zero on the boundary $\partial\Omega$.

Since there is no time evolution, the position of the membrane may be described by a function $u = u(\mathbf{x})$, solution of the (Dirichlet) problem

$$\begin{cases} -\Delta u = f & \text{in } \Omega \\ u = 0 & \text{on } \partial\Omega. \end{cases} \tag{1.15}$$

Suppose that we want to reformulate the problem (1.15) in order to have also solution that are less "classical", namely solutions which are not $C^2(\Omega)$. Proceeding formally, we multiply the equation $-\Delta u = f$ by a smooth function (that is called test function) vanishing on $\partial\Omega$, and we integrate over Ω. Using the Gauss' formula, we obtain

$$\int_\Omega \nabla u \cdot \nabla v \; d\mathbf{x} = \int_\Omega f v \; d\mathbf{x} \qquad \forall v \text{ test} \tag{1.16}$$

which is the *variational formulation* of the problem (1.15). Now, this equation has an interesting physical interpretation. The integral in the left hand side represents the work done by the internal elastic forces, due to a *virtual displacement* v. On the other hand, $\int_\Omega f v \, d\mathbf{x}$ expresses the work done by the external forces. The solution of (1.16) is the so called *variational solution* of the problem (1.15).

Thus, the variational formulation (1.16) states that these two works balance, which constitutes a version of the *principle of virtual work*.

There is more, if we bring into play the energy. In fact, the *total potential energy* is proportional to

$$E\left(v\right) = \quad \underbrace{\frac{1}{2}\int_{\Omega}\left|\nabla v\right|^{2}dx}_{\text{internal elastic energy}} - \underbrace{\int_{\Omega}fv\,dx}_{\text{external potential energy}}. \qquad (1.17)$$

Since nature likes to save energy, the equilibrium position u corresponds to the minimizer of (1.17) among all the *admissible* configurations v. This fact is closely connected with the principle of virtual work and, actually, it is equivalent to it (see Section 7.4.2).

Thus, changing point of view, instead of looking for a variational solution of (1.16) we may, equivalently, look for a minimizer of (1.17).

We remark that whenever you seek for the minimizer of a functional it is important to "choose wisely" the set where you look for it. For instance, if we are looking for the minimizer of the function

$$f\left(x\right) = \left(x - \pi\right)^{2}$$

among the rational numbers, it is obvious that such a minimizer does not exist, and that a wiser choice would have been to search for it among real numbers. Analogously, for the functional (1.17), we see that it is natural to require that the gradient of u is *square integrable*. The minimizer of (1.17) belongs to the so called *Sobolev space* $H_{0}^{1}\left(\Omega\right)$, whose elements are exactly the square integrable functions with square integrable first derivatives, vanishing on $\partial\Omega$. In view of the physical meaning of $E\left(v\right)$, representing energy, we could call them functions of finite energy!

Furthermore, even theoretically the space $H_{0}^{1}\left(\Omega\right)$ is special and the conjunction between geometrical and analytical aspects comes here into play. In fact, for instance, although it is an infinite-dimensional vector space, we may endow $H_{0}^{1}\left(\Omega\right)$ with a structure which reflects as much as possible the structure of a finite-dimensional vector space like $\mathbb{R}^{n}$, where life is obviously easier: in particular an inner product is introduced (See Section 7.2).

These concepts of Functional Analysis are presented in the Chapter 7, while the variational formulations of several problems are in the Chapters 8 and 9.

1.7 Numerical approximation methods

Each chapter ends with a brief introduction to numerical approximation techniques for the specific problem at hand, with the double purpose of complementing the theory and extending the range of applications that can be addressed.

From the theoretical standpoint, numerical simulations help to put into action and visualize the theoretical properties of the models that will be analysed. The presentation of numerical methods is limited to the simplest classes of methods and the description of the schemes is complemented with a brief discussion of their fundamental properties, such as stability and convergence.

We believe that this minimal knowledge on numerical approximation schemes represents an useful tool for training on model applications. Some case studies are proposed at the end of each chapter, complementing the solution of theoretical exercises. We mention for instance the study of traffic flow on Chapter 2, of mass transfer on Chapter 3, of elastic deformation of membranes in Chapter 4 and of pattern formation in biology in Chapter 5.

Among the many different techniques for the approximation of partial differential equations, we will focus on finite difference and finite element methods. Finite difference schemes are more appropriate for the approximation of problems that are formulated within the classical theory of partial differential equations, which focuses on the properties of classical solutions. Indeed, this family of schemes will be applied to the discretization of problems addressed in the first six chapters. Along Chapters 7, 8, and 9 we will introduce the Galerkin method for the numerical approximation of variational problems and in particular we will address the finite element method for the approximation of second order (elliptic) problems. Anyway, we point out that this material is not sufficient for a complete course on numerical approximation of partial differential equations, because fundamental topics such as a rigorous approach to approximation theory, error analysis (partially addressed for finite difference methods in Appendix C) and efficient algorithms for the solution of large systems of algebraic equations are not properly developed here. For further studies, we refer the interested reader to the selected bibliography.

Part I
Differential Models

2

Scalar Conservation Laws

2.1 Introduction

In this chapter we consider equations of the form

$$u_t + q(u)_x = 0, \qquad x \in \mathbb{R},\, t > 0. \tag{2.1}$$

In general, $u = u(x,t)$ represents the *density* or the *concentration* of a physical quantity Q and $q(u)$ is its *flux function*[1]. Equation (2.1) constitutes a *link* between density and flux and expresses a **(scalar) conservation law** for the following reason. If we consider a control interval $[x_1, x_2]$, the integral

$$\int_{x_1}^{x_2} u(x,t)\, dx$$

gives the amount of Q between x_1 and x_2 at time t. A *conservation law* states that, without sources or sinks, the rate of change of Q in the interior of $[x_1, x_2]$ is determined by the net flux through the end points of the interval. If the flux is modeled by a function $q = q(u)$, the law translates into the equation

$$\frac{d}{dt} \int_{x_1}^{x_2} u(x,t)\, dx = -q(u(x_2,t)) + q(u(x_1,t)), \tag{2.2}$$

where we assume that $q > 0$ ($q < 0$) for a flux along the positive (negative) direction of the x axes. If u and q are smooth functions, equation (2.2) can be rewritten in the form

$$\int_{x_1}^{x_2} [u_t(x,t) + q(u(x,t))_x]\, dx = 0$$

which implies (2.1), due to the arbitrariness of the interval $[x_1, x_2]$.

---

[1] The dimensions of q are $[mass] \times [time]^{-1}$.

Salsa S., Vegni F.M.G., Zaretti A., Zunino P.: *A Primer on PDEs. Models, Methods, Simulations.*
Unitext – La Matematica per il 3+2 65.
DOI 10.1007/978-88-470-2862-3_2, © Springer-Verlag Italia 2013

The conservation law (2.1) occurs, for instance, in one-dimensional fluid dynamics where it often describes the formation and propagation of the so called *shock waves*. Along a shock curve a solution undergoes a *jump discontinuity* and an important question is how to reinterpret the differential equation (2.1) in order to admit discontinuous solutions.

A typical problem associated with equation (2.1) is the *initial value problem*:

$$\begin{cases} u_t + q\,(u)_x = 0 \\ u\,(x,0) = g\,(x) \end{cases} \tag{2.3}$$

where $x \in \mathbb{R}$. Sometimes x varies in a half-line or in a finite interval; in these cases some other conditions have to be added to obtain a well posed problem.

To proceed into the analysis of the model we must decide which type of flux function we are dealing with, or, in other words, we have to establish a *constitutive relation for q*.

Let us use introduce a simple example.

Pollution in a channel

We examine a convection-diffusion model of a pollutant on the surface of a narrow channel. A water stream of constant speed v transports the pollutant along the positive direction of the x axis. We can neglect the depth of the water (thinking to a floating pollutant) and the transverse dimension (thinking of a very narrow channel).

Our purpose is to derive a mathematical model capable of describing the evolution of the concentration[2] $c = c\,(x,t)$ of the pollutant. Accordingly, the integral

$$\int_x^{x+\Delta x} c\,(y,t)\,dy \tag{2.4}$$

gives the mass inside the interval $[x, x + \Delta x]$ at time t (Fig. 2.1). In the present case there are neither sources nor sinks of pollutant, therefore to construct a model we use the **law of mass conservation**: *the growth rate of the mass*

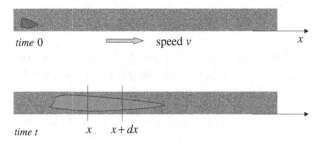

time 0   ⟹   speed *v*   *x*

time t   *x*   *x + dx*

Fig. 2.1. Pollution in a narrow channel

---

[2] $[c] = [mass] \times [length]^{-1}$.

contained in an interval $[x, x + \Delta x]$ equals the net mass flux into $[x, x + \Delta x]$ through the end points.

From (2.4), the growth rate of the mass contained in an interval $[x, x + \Delta x]$ is given by [3]

$$\frac{d}{dt} \int_x^{x+\Delta x} c(y, t)\, dy = \int_x^{x+\Delta x} c_t(y, t)\, dy. \tag{2.5}$$

Denote by $q = q(x, t)$ the mass flux[4] *entering* the interval $[x, x + \Delta x]$, through the point x at time t. The net mass flux into $[x, x + \Delta x]$ through the end points is

$$q(x, t) - q(x + \Delta x, t). \tag{2.6}$$

Equating (2.5) and (2.6), the law of mass conservation reads

$$\int_x^{x+\Delta x} c_t(y, t)\, dy = q(x, t) - q(x + \Delta x, t).$$

Dividing by Δx and letting $\Delta x \to 0$, we find the basic law

$$c_t = -q_x. \tag{2.7}$$

At this point we have to decide a *constitutive relation for* q. There are several possibilities, for instance:

Convection (drift). The flux is determined by the water stream only. This case corresponds to a bulk of pollutant that is driven by the stream, without deformation or expansion. Translating into mathematical terms we find

$$q(x, t) = vc(x, t)$$

where, we recall, v denotes the stream speed.

Diffusion. The pollutant expands from higher concentration regions to lower ones. Here we can adopt the so called *Fick's law* which reads

$$q(x, t) = -Dc_x(x, t)$$

where the constant D depends on the pollutant and has physical dimensions ($[D] = [length]^2 \times [time]^{-1}$).

In our case, convection and diffusion are both present and therefore we superpose the two effects, by writing

$$q(x, t) = vc(x, t) - Dc_x(x, t).$$

From (2.7) we deduce

$$c_t = Dc_{xx} - vc_x \tag{2.8}$$

which constitutes our mathematical model.

---

[3] Assuming we can take the derivative inside the integral.
[4] $[q] = [mass] \times [time]^{-1}$.

Remark 2.1. Notice that if v and D were non constant, we would get an equation of the form

$$c_t = (Dc_x)_x - (vc)_x.$$

2.2 Linear transport equation

We discuss here the case of the *pure transport* equation

$$c_t + vc_x = 0 \tag{2.9}$$

that is when $D = 0$ in (2.8). We want to determine the evolution of the concentration c, by knowing its initial profile

$$c(x,0) = g(x). \tag{2.10}$$

Introducing the vector

$$\mathbf{v} = v\mathbf{i} + \mathbf{j}$$

equation (2.9) can be written in the form

$$vc_x + c_t = \nabla c \cdot \mathbf{v} = 0,$$

pointing out the orthogonality of ∇c and $\mathbf{v}$. But ∇c is orthogonal to the level lines of c, along which c is constant. Therefore the level lines of c are the straight lines parallel to $\mathbf{v}$ with equation

$$x = vt + x_0.$$

These straight lines are called **characteristics**.

To compute the solution of (2.9), (2.10) at a point $(\bar{x}, \bar{t})$, $t > 0$, is now very simple. Let $x = vt + x_0$ be the equation of the characteristic passing through $(\bar{x}, \bar{t})$. and go back in time along this characteristic from $(\bar{x}, \bar{t})$ until the point $(x_0, 0)$, of intersection with the $x-$axes (see Fig. 2.2).

Since c is constant along the characteristic and $c(x_0, 0) = g(x_0)$, it must be

$$c(\bar{x}, \bar{t}) = g(x_0) = g(\bar{x} - v\bar{t}).$$

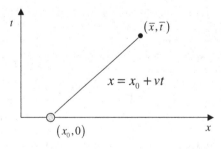

Fig. 2.2. Characteristic line for the linear transport problem

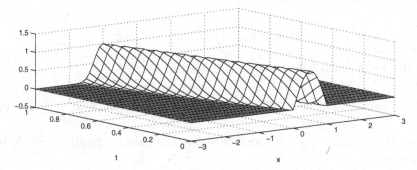

Fig. 2.3. Travelling wave solution of the linear transport equation

Thus, if $g \in C^1(\mathbb{R})$, the solution of the initial value problem (2.9), (2.10) is given by

$$c(x,t) = g(x - vt).\tag{2.11}$$

The solution (2.11) represents a *travelling wave*, moving with speed v in the positive x−direction. In Fig. 2.3 an initial profile $g(x) = \sin(\pi x)\chi_{[0,1]}(x)$ is *transported* in the plane (x,t) along the straight-lines $x - t = $ constant, i.e. with speed $v = 1$.

2.2.1 Distributed source

Suppose now we take into account the effect of an external distributed *source* of pollutant along the channel, of intensity $f = f(x,t)$, measured in concentration per unit time. Instead of equation (2.2) we have

$$\frac{d}{dt}\int_{x_1}^{x_2} c(x,t)\,dx = -q\left(c(x_2,t)\right) + q\left(c(x_1,t)\right) + \int_{x_1}^{x_2} f(x,t)\,dx,\tag{2.12}$$

which leads to the nonhomogeneous differential equation

$$c_t + vc_x = f(x,t),\tag{2.13}$$

since $q = vc$, with the initial condition

$$c(x,0) = g(x).\tag{2.14}$$

Again, to compute the value of the solution u at a point $(\bar{x}, \bar{t})$ is not difficult. Let $x = x_0 + vt$ be the characteristic passing through $(\bar{x}, \bar{t})$ and compute u along this characteristic, setting $w(t) = c(x_0 + vt, t)$. From (2.13), w satisfies the *ordinary differential equation*

$$\frac{dw}{dt} = vc_x(x_0 + vt, t) + c_t(x_0 + vt, t) = f(x_0 + vt, t)$$

with the initial condition

$$w(0) = g(x_0).$$

Thus

$$w\left(t\right) = g\left(x_0\right) + \int_0^t f\left(x_0 + vs, s\right) ds.$$

Letting $t = \bar{t}$ and recalling that $x_0 = \bar{x} - v\bar{t}$, we get

$$c\left(\bar{x}, \bar{t}\right) = w\left(\bar{t}\right) = g\left(\bar{x} - v\bar{t}\right) + \int_0^t f\left(\bar{x} - v(\bar{t} - s), s\right) ds. \qquad (2.15)$$

Since $(\bar{x}, \bar{t})$ is arbitrary, if g and f are reasonably smooth functions, (2.15) is our solution.

Proposition 2.1. *Let* $g \in C^1\left(\mathbb{R}\right)$ *and* $f, f_x \in C\left(\mathbb{R} \times \mathbb{R}_+\right)$. *The solution of the initial value problem*

$$\begin{cases} c_t + vc_x = f\left(x, t\right) & x \in \mathbb{R}, \, t > 0 \\ c(x, 0) = g\left(x\right) & x \in \mathbb{R} \end{cases}$$

is given by the formula

$$c\left(x, t\right) = g\left(x - vt\right) + \int_0^t f\left(x - v(t - s), s\right) ds. \qquad (2.16)$$

Example 2.1. The solution of the problem

$$\begin{cases} c_t + vc_x = e^{-t} \sin x & x \in \mathbb{R}, \, t > 0 \\ c\left(x, 0\right) = 0 & x \in \mathbb{R} \end{cases}$$

is given by

$$\begin{aligned} c\left(x, t\right) &= \int_0^t e^{-s} \sin\left(x - v(t - s)\right) ds \\ &= \frac{1}{1 + v^2} \left\{ -e^{-t}\left(\sin x + v \cos x\right) + \left[\sin(x - vt) + v\cos(x - vt)\right] \right\}. \end{aligned}$$

2.2.2 Extinction and localized source

Suppose that, due to *biological decomposition*, the pollutant decays at the rate

$$r\left(x, t\right) = -\gamma c\left(x, t\right) \qquad \gamma > 0.$$

Without external sources and diffusion, the mathematical model is

$$c_t + vc_x = -\gamma c,$$

with the initial condition

$$c\left(x, 0\right) = g\left(x\right).$$

Setting

$$u(x,t) = c(x,t)\, e^{\frac{\gamma}{v}x}, \tag{2.17}$$

we have

$$u_x = \left(c_x + \frac{\gamma}{v}c\right) e^{\frac{\gamma}{v}x} \quad \text{and} \quad u_t = c_t e^{\frac{\gamma}{v}x}$$

and therefore the equation for u is

$$u_t + vu_x = 0$$

with the initial condition

$$u(x,0) = g(x)\, e^{\frac{\gamma}{v}x}.$$

From Proposition 2.1, we get

$$u(x,t) = g(x - vt)\, e^{\frac{\gamma}{v}(x-vt)}$$

and from (2.17)

$$c(x,t) = g(x - vt)\, e^{-\gamma t}$$

which is a *damped travelling wave*.

We now examine the effect of a source of pollutant placed at a certain point of the channel, e.g. at $x = 0$. Typically, one can think of waste material from industrial machineries. Before the machines start working, for instance before time $t = 0$, we assume that the channel is clean. We want to determine the pollutant concentration, supposing that at $x = 0$ it is kept at a constant level $\beta > 0$, for $t > 0$.

To model this source we introduce the Heaviside function

$$\mathcal{H}(t) = \begin{cases} 1 & t \geq 0 \\ 0 & t < 0, \end{cases}$$

and we consider the *boundary condition*

$$c(0,t) = \beta \mathcal{H}(t) \tag{2.18}$$

and the initial condition

$$c(x,0) = 0 \qquad \text{for } x > 0. \tag{2.19}$$

As before, let $u(x,t) = c(x,t)\, e^{\frac{\gamma}{v}x}$, which is a solution of $u_t + vu_x = 0$. Then:

$$u(x,0) = c(x,0)\, e^{\frac{\gamma}{v}x} = 0 \qquad x > 0$$
$$u(0,t) = c(0,t) = \beta \mathcal{H}(t).$$

Since u is constant along the characteristics it must be of the form

$$u(x,t) = u_0(x - vt) \tag{2.20}$$

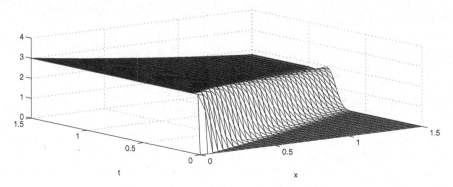

Fig. 2.4. Propagation of a discontinuity

where u_0 is to be determined from the boundary condition (2.18) and the initial condition (2.19).

To compute u for $x < vt$, observe that a characteristic leaving the $t-$axis from a point $(0,t)$ carries the data $\beta \mathcal{H}(t)$. Therefore, we must have

$$u_0(-vt) = \beta \mathcal{H}(t).$$

Letting $s = -vt$ we get

$$u_0(s) = \beta \mathcal{H}\left(-\frac{s}{v}\right)$$

and from (2.20),

$$u(x,t) = \beta \mathcal{H}\left(t - \frac{x}{v}\right).$$

This formula gives the solution also in the sector

$$x > vt, \quad t > 0,$$

since the characteristics leaving the $x-$axis carry *zero data* and hence we deduce $u = c = 0$ there. This means that the pollutant has not yet reached the point x at time t, if $x > vt$.

Finally, recalling (2.17), we find

$$c(x,t) = \beta \mathcal{H}\left(t - \frac{x}{v}\right) e^{-\frac{\gamma}{v}x}.$$

Observe that in $(0,0)$ there is a *jump discontinuity which is transported along the characteristic $x = vt$.* The Fig. 2.4 shows the solution for $\beta = 3$, $\gamma = 0.7$, $v = 2$.

2.2.3 Inflow and outflow characteristics. A stability estimate

The domain in the localized source problem is the quadrant $x > 0, t > 0$. To uniquely determine the solution we have used the initial data on the $x-$axis,

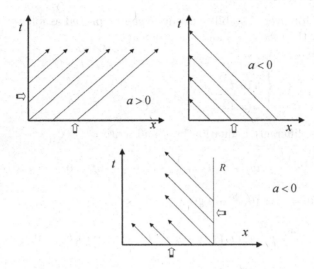

Fig. 2.5. The arrows indicate where the data should be assigned

$x > 0$, and the boundary data on the $t-$axis, $t > 0$. The problem is therefore well posed. This is due to the fact that, since $v > 0$, when time increases, *all* the characteristics carry the information (the data) *towards the interior* of the quadrant $x > 0, t > 0$. In other words the characteristics are **inflow characteristics**.

More generally, consider the equation

$$u_t + au_x = f(x,t)$$

in the domain $x > 0$, $t > 0$, where a is a constant $(a \neq 0)$. The characteristics are the lines

$$x - at = \text{constant}$$

as shown in Fig. 2.5. If $a > 0$, we are in the case of the pollutant model: all the characteristics **are inflow** and **the data must be assigned on both semi-axes.**

If $a < 0$, the characteristics leaving the $x-$axis are **inflow**, while those leaving the $t-$axis are **outflow**. In this case the initial data alone are sufficient to uniquely determine the solution, while **no data has to be assigned on the semi-axis** $x = 0$, $t > 0$.

Coherently, a problem in the half-strip $0 < x < R$, $t > 0$, besides the initial data, requires a data assignment on the inflow boundary, namely

$$\begin{cases} u(0,t) = h_0(t) & \text{if } a > 0 \\ u(R,t) = h_R(t) & \text{if } a < 0. \end{cases}$$

The resulting initial-boundary value problem is well posed, since the solution is uniquely determined at every point in the strip by its values along the characteristics. Moreover, a stability estimate can be proved as follows. Consider, for instance, the case $a > 0$ and the problem

$$\begin{cases} u_t + au_x = 0 & 0 < x < R, t > 0 \\ u(0,t) = h(t) & t > 0 \\ u(x,0) = g(x) & 0 < x < R. \end{cases} \tag{2.21}$$

Multiply the differential equation by u and write

$$uu_t + auu_x = \frac{1}{2}\frac{d}{dt}u^2 + \frac{a}{2}\frac{d}{dx}u^2 = 0.$$

Integrating in x over $(0, R)$ we get:

$$\frac{d}{dt}\int_0^R u^2(x,t)\,dx + a\left[u^2(R,t) - u^2(0,t)\right] = 0.$$

Now use the data $u(0,t) = h(t)$ and the positivity of a to obtain

$$\frac{d}{dt}\int_0^R u^2(x,t)\,dx \le ah^2(t).$$

Integrating in t we have, using the initial condition $u(x,0) = g(x)$,

$$\int_0^R u^2(x,t)\,dx \le \int_0^R g^2(x)\,dx + a\int_0^t h^2(s)\,ds. \tag{2.22}$$

Now, let u_1 and u_2 be solutions of problem (2.21) with initial data g_1, g_2 and boundary data h_1, h_2 on $x = 0$. Then, by linearity, $w = u_1 - u_2$ is a solution of problem (2.21) with initial data $g_1 - g_2$ and boundary data $h_1 - h_2$ on $x = 0$. Applying the inequality (2.22) to w we have

$$\int_0^R [u_1(x,t) - u_2(x,t)]^2 dx \le \int_0^R [g_1(x) - g_2(x)]^2 dx + a\int_0^t [h_1(s) - h_2(s)]^2 ds.$$

Thus, a least-squares approximation of the data controls a least-squares approximation of the corresponding solutions. In this sense, the solution of problem (2.21) depends continuously on the initial data and on the boundary data on $x = 0$. We point out that the values of u on $x = R$ do not appear in (2.22).

2.3 Traffic Dynamics

2.3.1 A macroscopic model

From far away, an intense traffic on a highway can be considered as a fluid flow and described by means of macroscopic variables such as the *density* of cars[5]

---

[5] Number of cars per unit length.

ρ, their *average speed* v and their *flux*[6] q. The three (more or less regular) functions ρ, v and q are linked by the simple convection relation

$$q = v\rho.$$

To construct a model for the evolution of ρ we assume the following hypotheses.

1. *There is only one lane and overtaking is not allowed.* This is realistic for instance for traffic in a tunnel. Multi-lanes models with overtaking are beyond the scope of this introduction. However the model we will present is often in agreement with observations also in this case.

2. *No car "sources" or "sinks".* We consider a road section without exit/ entrance gates.

3. *The average speed is not constant and depends on the density alone*, that is

$$v = v\left(\rho\right).$$

This rather controversial assumption means that at a certain density the speed is uniquely determined and that a density change causes an immediate speed variation. Clearly

$$v'\left(\rho\right) = \frac{dv}{d\rho} \leq 0$$

since we expect the speed to decrease as the density increases.

As in Section 1.1, from hypotheses 2 and 3 we derive the conservation law:

$$\rho_t + q(\rho)_x = 0 \tag{2.23}$$

where

$$q(\rho) = v\left(\rho\right)\rho.$$

We need a constitutive relation for $v = v\left(\rho\right)$. When ρ is small, it is reasonable to assume that the average speed v is more or less equal to the maximal velocity v_m, given by the speed limit. When ρ increases, traffic slows down and stops at the maximum density ρ_m (bumper-to-bumper traffic). We adopt the simplest model consistent with the above considerations, namely

$$v\left(\rho\right) = v_m\left(1 - \frac{\rho}{\rho_m}\right),$$

so that

$$q\left(\rho\right) = v_m\rho\left(1 - \frac{\rho}{\rho_m}\right). \tag{2.24}$$

Since

$$q(\rho)_x = q'\left(\rho\right)\rho_x = v_m\left(1 - \frac{2\rho}{\rho_m}\right)\varrho_x$$

---

[6] Cars per unit time.

equation (2.23) becomes

$$\rho_t + \underbrace{v_m \left(1 - \frac{2\rho}{\rho_m} \right)}_{q'(\rho)} \rho_x = 0. \tag{2.25}$$

According to the terminology in Section 1.1, this is *quasilinear* equation. We also point out that

$$q''(\rho) = -\frac{2v_m}{\rho_m} < 0$$

so that q is strictly *concave*. We couple the equation (2.25) with the initial condition

$$\rho(x, 0) = g(x). \tag{2.26}$$

2.3.2 The method of characteristics

We want to solve the initial value problem (2.25), (2.26). To compute the density ρ at a point (x, t) we follow the idea we used in the homogeneous linear transport case: *to connect the point (x, t) with a point $(x_0, 0)$ on the x−axis, through a curve along which ρ is constant* (Fig. 2.6).

Clearly, if we manage to find such a curve, that we call **characteristic based at** $(x_0, 0)$, the value of ρ at (x, t) is given by $\rho(x_0, 0) = g(x_0)$. Moreover, if this procedure can be repeated for every point (x, t), $x \in \mathbb{R}$, $t > 0$, then we can compute ρ at every point and the problem is completely solved. This is the *method of characteristics*.

Adopting a slightly different point of view, we can implement the above idea as follows: assume that $x = x(t)$ is the equation of the characteristic based at the point $(x_0, 0)$; along $x = x(t)$ we *observe always the same initial density $g(x_0)$*. In other words

$$\rho(x(t), t) = g(x_0) \tag{2.27}$$

for every $t > 0$. If we differentiate the identity (2.27), we get

$$\frac{d}{dt}\rho(x(t), t) = \rho_x(x(t), t)x'(t) + \rho_t(x(t), t) = 0 \qquad (t > 0). \tag{2.28}$$

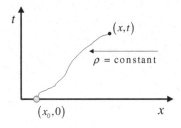

Fig. 2.6. Characteristic curve

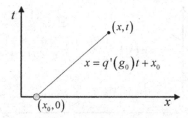

Fig. 2.7. Characteristic straight line $(g_0 = g(x_0))$

On the other hand, (2.25) yields

$$\rho_t\left(x\left(t\right),t\right) + q'\left(g_0\right)\rho_x\left(x\left(t\right),t\right) = 0 \tag{2.29}$$

so that, subtracting (2.29) from (2.28), we obtain

$$\rho_x\left(x\left(t\right),t\right)\left[\frac{dx}{dt} - q'\left(g\left(x_0\right)\right)\right] = 0.$$

Assuming $\rho_x\left(x\left(t\right),t\right) \neq 0$, we deduce

$$\frac{dx}{dt} = q'\left(g\left(x_0\right)\right).$$

Since $x\left(0\right) = x_0$ we find

$$x\left(t\right) = q'\left(g\left(x_0\right)\right)t + x_0. \tag{2.30}$$

Thus, the characteristics are **straight lines** with slope $q'\left(g\left(x_0\right)\right)$ (Fig. 2.7). Different values of x_0 give, in general, different values of the slope.

We can now derive a formula for ρ. To compute $\rho\left(x,t\right)$, $t > 0$, we go back in time along the characteristic through (x,t) until we reach its base point $(x_0, 0)$. Then $\rho\left(x,t\right) = g\left(x_0\right)$. From (2.30) we have, since $x\left(t\right) = x$,

$$x_0 = x - q'\left(g\left(x_0\right)\right)t$$

and finally

$$\rho\left(x,t\right) = g\left(x - q'\left(g\left(x_0\right)\right)t\right). \tag{2.31}$$

Formula (2.31) represents **a travelling wave propagating with speed** $q'\left(g\left(x_0\right)\right)$ along the positive $x-$direction.

We emphasize that $q'\left(g\left(x_0\right)\right)$ is the *local wave speed* and it must not be confused with the traffic velocity. In fact, in general,

$$\frac{dq}{d\rho} = \frac{d\left(\rho v\right)}{d\rho} = v + \rho\frac{dv}{d\rho} \leq v$$

since $\rho \geq 0$ and $\frac{dv}{d\rho} \leq 0$.

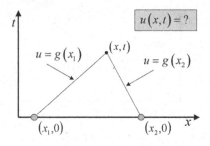

Fig. 2.8. Intersection of characteristics

The different nature of the two speeds becomes more evident if we observe that the wave speed *may be negative* as well. This means that, while the traffic advances along the positive x–direction, the disturbance given by the travelling wave may propagate in the opposite direction. Indeed, in our model (2.24), $\frac{dq}{d\rho} < 0$ when $\rho > \frac{\rho_m}{2}$.

Formula (2.31) seems to be rather satisfactory, since, apparently, it gives the solution of the initial value problem (2.25), (2.26) at every point. Actually, a more accurate analysis shows that, even if the initial data g are smooth, the solution may develop a singularity in finite time (e.g. a jump discontinuity). When this occurs, the method of characteristics does not work anymore and formula (2.31) is not effective. A typical case is described in Fig. 2.8: two characteristics based at different points $(x_1, 0)$ e $(x_2, 0)$ intersect at the point (x, t) and the value $u(x, t)$ is not uniquely determined as soon as $g(x_1) \neq g(x_2)$.

In this case we have to weaken the concept of solution and the computation technique. We will come back on these questions later. For the moment, we analyze the method of characteristics in some particularly significant cases.

2.3.3 The green light problem. Rarefaction waves

Suppose that bumper-to-bumper traffic is standing at a red light, placed at $x = 0$, while the road ahead is empty. Accordingly, the initial density profile is

$$g(x) = \begin{cases} \rho_m & \text{for } x \leq 0 \\ 0 & \text{for } x > 0. \end{cases} \tag{2.32}$$

At time $t = 0$ the traffic light turns green and we want to describe the car flow evolution for $t > 0$. At the beginning, only the cars nearer to the light start moving while most remain standing.

Since $q'(\rho) = v_m \left(1 - \frac{2\rho}{\rho_m}\right)$, the local wave speed is given by

$$q'(g(x_0)) = \begin{cases} -v_m & \text{for } x_0 \leq 0 \\ v_m & \text{for } x_0 > 0 \end{cases}$$

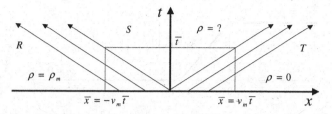

Fig. 2.9. Characteristic for the green light problem

and the characteristics are the straight lines

$$x = -v_m t + x_0 \qquad \text{if } x_0 < 0$$
$$x = v_m t + x_0 \qquad \text{if } x_0 > 0.$$

The lines $x = v_m t$ and $x = -v_m t$ partition the upper half-plane in the three regions R, S and T, shown in Fig. 2.9.

Inside R we have $\rho(x,t) = \rho_m$, while inside T we have $\rho(x,t) = 0$. Consider the points on the horizontal line $t = \bar{t}$. At the points $(x, \bar{t}) \in T$ the density is zero: the traffic has not yet arrived in x at time $t = \bar{t}$. The front car is located at the point

$$\bar{x} = v_m \bar{t}$$

which moves at the maximum speed, since ahead the road is empty.

The cars placed at the points $(x, \bar{t}) \in R$ are still standing. The first car that starts moving at time $t = \bar{t}$ is at the point

$$\bar{x} = -v_m \bar{t}.$$

In particular, it follows that *the green light signal propagates back through the traffic at the speed* v_m.

What is the value of the density inside the sector S? No characteristic extends into S due to the discontinuity of the initial data at the origin, and the method as it stands does not give any information on the value of ρ inside S.

A strategy that may give a reasonable answer is the following:

a) approximate the initial data by a continuous function g_ε, which converges to g as $\varepsilon \to 0$ at every point x, except 0;

b) construct the solution ρ_ε of the ε−problem by the method of characteristics;

c) let $\varepsilon \to 0$ and check that the limit of ρ_ε is a solution of the original problem.

Clearly we run the risk of constructing many solutions, each one depending on the way we regularize the initial data, but for the moment we are satisfied if we construct *at least one* solution.

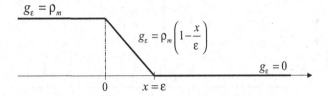

Fig. 2.10. Smoothing of the initial data in the green light problem

a) Let us choose as g_ε the function (Fig. 2.10)

$$g_\varepsilon(x) = \begin{cases} \rho_m & x \leq 0 \\ \rho_m(1 - \dfrac{x}{\varepsilon}) & 0 < x < \varepsilon \\ 0 & x \geq \varepsilon. \end{cases}$$

When $\varepsilon \to 0$, $g_\varepsilon(x) \to g(x)$ for every $x \neq 0$.

b) The characteristics for the ε−problem are:

$$x = -v_m t + x_0 \qquad\qquad \text{if } x_0 < 0$$
$$x = -v_m \left(1 - 2\frac{x_0}{\varepsilon}\right) t + x_0 \qquad \text{if } 0 \leq x_0 < \varepsilon$$
$$x = v_m t + x_0 \qquad\qquad \text{if } x_0 \geq \varepsilon$$

since, for $0 \leq x_0 < \varepsilon$,

$$q'(g_\varepsilon(x_0)) = v_m \left(1 - \frac{2g_\varepsilon(x_0)}{\rho_m}\right) = -v_m \left(1 - 2\frac{x_0}{\varepsilon}\right).$$

The characteristics in the region $-v_m t < x < v_m t + \varepsilon$ form a *rarefaction fan* (Fig. 2.11). Clearly, $\rho_\varepsilon(x,t) = 0$ for $x \geq v_m t + \varepsilon$ and $\rho_\varepsilon(x,t) = \rho_m$ for $x \leq -v_m t$. Let now (x,t) belong to the region

$$-v_m t < x < v_m t + \varepsilon.$$

Solving for x_0 in the equation of the characteristic $x = -v_m \left(1 - 2\frac{x_0}{\varepsilon}\right) t + x_0$, we find

$$x_0 = \varepsilon \frac{x + v_m t}{2v_m t + \varepsilon}.$$

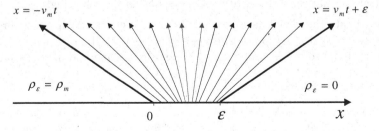

Fig. 2.11. Fanlike characteristics

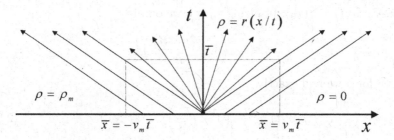

Fig. 2.12. Characteristics in a rarefaction wave

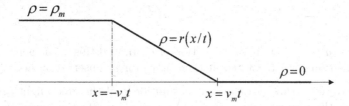

Fig. 2.13. Profile of a rarefaction wave at time t

Then

$$\rho_\varepsilon\,(x,t) = g_\varepsilon\,(x_0) = \rho_m(1 - \frac{x_0}{\varepsilon}) = \rho_m\left(1 - \frac{x + v_m t}{2v_m t + \varepsilon}\right). \qquad (2.33)$$

c) Letting $\varepsilon \to 0$ in (2.33) we obtain

$$\rho\,(x,t) = \begin{cases} \rho_m & \text{for } x \leq -v_m t \\ \dfrac{\rho_m}{2}\left(1 - \dfrac{x}{v_m t}\right) & \text{for } -v_m t < x < v_m t \\ 0 & \text{for } x \geq v_m t \end{cases} \qquad (2.34)$$

It is easy to check that ρ is a solution of the equation (2.25) in the regions R, S, T. For fixed t, the function ρ decreases linearly from ρ_m to 0 as x varies from $-v_m t$ to $v_m t$. Moreover, ρ is constant on the fan of straight lines

$$x = ht \qquad -v_m < h < v_m.$$

These type of solutions are called **rarefaction** or **simple waves** (centered at the origin).

The formula for $\rho\,(x,t)$ in the sector S can be obtained, a posteriori, by a formal procedure that emphasizes its structure. The equation of the characteristics can be written in the form

$$x = v_m\left(1 - \frac{2g\,(x_0)}{\rho_m}\right)t + x_0 = v_m\left(1 - \frac{2\rho\,(x,t)}{\rho_m}\right)t + x_0.$$

because $\rho(x,t) = g(x_0)$. Inserting $x_0 = 0$ we obtain

$$x = v_m \left(1 - \frac{2\rho(x,t)}{\rho_m}\right) t.$$

Solving for ρ we find exactly

$$\rho(x,t) = \frac{\rho_m}{2}\left(1 - \frac{x}{v_m t}\right) \qquad (t > 0). \qquad (2.35)$$

Since $v_m\left(1 - \frac{2\rho}{\rho_m}\right) = q'(\rho)$, we see that (2.35) is equivalent to

$$\rho(x,t) = r\left(\frac{x}{t}\right)$$

where $r = (q')^{-1}$ is the inverse function of q'. Indeed this is the general form of a rarefaction wave (centered at the origin) for a conservation law.

We have constructed a continuous solution ρ of the green light problem, connecting the two constant states ρ_m and 0 by a rarefaction wave. However, it is not clear in which sense ρ is a solution across the lines $x = \pm v_m t$, since, there, its derivatives undergo a jump discontinuity. Also, it is not clear whether or not (2.34) is the only solution. We will return later on these important points.

2.3.4 Traffic jam ahead. Shock waves. Rankine–Hugoniot condition

Suppose that the initial density profile is

$$g(x) = \begin{cases} \frac{1}{8}\rho_m & \text{for } x < 0 \\ \rho_m & \text{for } x > 0. \end{cases}$$

For $x > 0$, the density is maximal and therefore the traffic is bumper-to-bumper. The cars on the left move with speed $v = \frac{7}{8}v_m$ so that we expect congestion propagating back into the traffic. We have

$$q'(g(x_0)) = \begin{cases} \frac{3}{4}v_m & \text{if } x_0 < 0 \\ -v_m & \text{if } x_0 > 0 \end{cases}$$

and therefore the characteristics are

$$x = \frac{3}{4}v_m t + x_0 \qquad \text{if } x_0 < 0$$
$$x = -v_m t + x_0 \qquad \text{if } x_0 > 0.$$

The characteristics configuration (Fig. 2.14) shows that the latter intersect somewhere in finite time and the theory predicts that ρ becomes a "multivalued" function of the position. In other words, ρ should assume two different

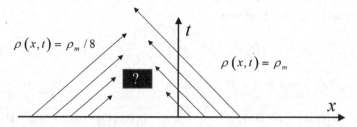

Fig. 2.14. Expecting a shock

values at the same point, which clearly makes no sense in our situation. Therefore we have to admit solutions with jump discontinuities (**shocks**), but then we have to reexamine the derivation of the conservation law, because the smoothness assumption for ρ does not hold anymore.

Thus, let us go back to the conservation of cars in integral form (see (2.2)):

$$\frac{d}{dt} \int_{x_1}^{x_2} \rho(x,t)\ dx = -q\left(\rho\left(x_2,t\right)\right) + q\left(\rho\left(x_1,t\right)\right), \qquad (2.36)$$

valid in any control interval $[x_1, x_2]$. Suppose now that ρ is a smooth function except along a curve

$$x = s(t) \qquad t \in [t_1, t_2],$$

that we call **shock curve**, on which ρ undergoes a *jump discontinuity*.

For fixed t, let $[x_1, x_2]$ be an interval containing the discontinuity point

$$x = s(t).$$

From (2.36) we have

$$\frac{d}{dt} \left\{ \int_{x_1}^{s(t)} \rho(y,t)\,dy + \int_{s(t)}^{x_2} \rho(y,t)\,dy \right\} + q\left[\rho\left(x_2,t\right)\right] - q\left[\rho\left(x_1,t\right)\right] = 0. \quad (2.37)$$

The fundamental theorem of calculus gives

$$\frac{d}{dt} \int_{x_1}^{s(t)} \rho(y,t)\,dy = \int_{x_1}^{s(t)} \rho_t(y,t)\,dy + \rho^- \left(s\left(t\right),t\right) \frac{ds}{dt}$$

and

$$\frac{d}{dt} \int_{s(t)}^{x_2} \rho(y,t)\,dy = \int_{s(t)}^{x_2} \rho_t(y,t)\,dy - \rho^+ \left(s\left(t\right),t\right) \frac{ds}{dt},$$

where

$$\rho^- \left(s\left(t\right),t\right) = \lim_{y \uparrow s(t)} \rho(y,t), \qquad \rho^+ \left(s\left(t\right),t\right) = \lim_{y \downarrow s(t)} \rho(y,t).$$

Hence, equation (2.37) becomes

$$\int_{x_1}^{x_2} \cdot \rho_t\,(y,t)\,dy + \left[\rho^-\,(s\,(t)\,,t) - \rho^+\,(s\,(t)\,,t)\right] \frac{ds}{dt} = q\left[\rho\,(x_1,t)\right] - q\left[\rho\,(x_2,t)\right].$$

Letting $x_2 \downarrow s\,(t)$ and $x_1 \uparrow s\,(t)$ we obtain

$$\left[\rho^-\,(s\,(t)\,,t) - \rho^+\,(s\,(t)\,,t)\right] \frac{ds}{dt} = q\left[\rho^-\,(s\,(t)\,,t)\right] - q\left[\rho^+\,(s\,(t)\,,t)\right]$$

that is:

$$\frac{ds}{dt} = \frac{q\left[\rho^+\,(s,t)\right] - q\left[\rho^-\,(s,t)\right]}{\rho^+\,(s,t) - \rho^-\,(s,t)} \tag{2.38}$$

which is often written in the concise form

$$\frac{ds}{dt} = \frac{[q\,(\rho)]_-^+}{[\rho]_-^+}$$

where $[z]_-^+$ denotes the jump of z from left to right.

The relation (2.38) is an ordinary differential equation for s and it is known as **Rankine-Hugoniot condition**. The discontinuity propagating along the shock curve is called **shock wave**.

The Rankine-Hugoniot condition gives the *shock speed* $\frac{ds}{dt}$ as the quotient of the flux jump over the density jump. To determine the shock curve we need to know *its initial point* and the values of ρ from both sides of the curve.

Let us apply the above considerations to our traffic problem[7]. We have

$$\rho^+ = \rho_m, \qquad \rho^- = \frac{\rho_m}{8}$$

while

$$q\left[\rho^+\right] = 0 \qquad q\left[\rho^-\right] = \frac{7}{64}v_m\rho_m$$

and (2.38) gives

$$\frac{ds}{dt} = \frac{q\left[\rho^+\right] - q\left[\rho^-\right]}{\rho^+ - \rho^-} = -\frac{1}{8}v_m.$$

Since clearly $s\,(0) = 0$, the shock curve is the straight line

$$x = -\frac{1}{8}v_m t.$$

Note that *the slope is negative: the shock propagates back with speed* $-\frac{1}{8}v_m$, as it is revealed by the braking of the cars, slowing down because of a traffic jam ahead.

---

[7] In the present case the following simple formula holds:

$$\frac{q\,(w) - q\,(z)}{w - z} = v_m\left(1 - \frac{w + z}{\rho_m}\right).$$

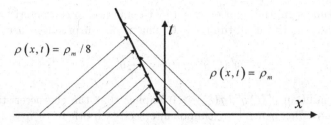

Fig. 2.15. Shock wave

As a consequence, the solution of our problem is given by the following formula (Fig. 2.15)

$$\rho(x,t) = \begin{cases} \frac{1}{8}\rho_m & x < -\frac{1}{8}v_m t \\ \rho_m & x > -\frac{1}{8}v_m t. \end{cases}$$

The two constant values $\frac{1}{8}\rho_m$ and ρ_m are connected by a **shock wave**.

2.4 The method of characteristics revisited

The method of characteristics applied to the problem

$$\begin{cases} u_t + q(u)_x = 0 \\ u(x,0) = g(x) \end{cases} \tag{2.39}$$

gives the travelling wave (see (2.31) with $x_0 = \xi$)

$$u(x,t) = g\left[x - q'(g(\xi))t\right] \quad \left(q' = \frac{dq}{du}\right) \tag{2.40}$$

with local speed $q'(g(\xi))$, in the positive x−direction.

Since $u(x,t) \equiv g(\xi)$ along the characteristic $x = q'(g(\xi))t + \xi$, based at $(\xi,0)$, from (2.40) we obtain that u is implicitly defined by the equation

$$G(x,t,u) \equiv u - g\left[x - q'(u)t\right] = 0. \tag{2.41}$$

If g and q' are smooth, the *Implicit Function Theorem*, implies that equation (2.41) defines u as a function of (x,t), as long as the condition

$$G_u(x,t,u) = 1 + tq''(u)g'\left[x - q'(u)t\right] \neq 0 \tag{2.42}$$

holds. An immediate consequence is that if

$$q''(u) = q''(g(\xi)) \quad \text{and} \quad g'\left[x - q'(u)t\right] = g'(\xi)$$

have the same sign, the solution given by the method of characteristics is well defined and smooth for all times $t \geq 0$. This is not surprising, since

$$g'\left(\xi\right)q''\left(g\left(\xi\right)\right) = \frac{d}{d\xi}g'\left(\xi\right)$$

and the condition $g'\left(\xi\right)q''\left(g\left(\xi\right)\right) \geq 0$ implies that the characteristic slopes are *nondecreasing*, hence they cannot intersect each other.

Precisely, we have:

Proposition 2.2. *Suppose that $q \in C^2\left(\mathbb{R}\right)$, $g \in C^1\left(\mathbb{R}\right)$ and $g'\left(\xi\right)q''\left(g\left(\xi\right)\right) \geq 0$ in $\mathbb{R}$. Then formula (2.41) defines the unique solution u of problem (2.39) in the half-plane $t \geq 0$. Moreover, $u\left(x,t\right) \in C^1\left(\mathbb{R} \times [0,\infty)\right)$.*

Thus, if $q'' \circ g$ and g' have the same sign, the characteristics do not intersect. Note that in the ε-approximation of the green light problem, q is concave and g_ε is decreasing. Although g_ε is not smooth, the characteristics do not intersect and ρ_ε is well defined for all times $t > 0$. In the limit as $\varepsilon \to 0$, the discontinuity of g reappears and the fan of characteristics produces a rarefaction wave.

What happens if $q''\left(g\left(\xi\right)\right)$ and $g'\left(\xi\right)$ have a different sign in some interval $[a,b]$? Proposition 2.2 still holds for small times, since $G_u \sim 1$ if $t \sim 0$, but when time goes on we expect the formation of a shock. Indeed, suppose, for instance, that q is concave and g is increasing. The family of characteristics based on a point in the interval $[a,b]$ is

$$x = q'\left(g\left(\xi\right)\right)t + \xi \qquad \xi \in [a,b]. \tag{2.43}$$

When ξ increases, g increases as well, while $q'\left(g\left(\xi\right)\right)$ decreases so that we expect intersection of characteristics along a shock curve. The main question is to find the positive time t_s (*breaking time*) and the location x_s of **first appearance of the shock**.

According to the above discussion, the *breaking time* must coincide with the first time t at which the expression

$$G_u\left(x,t,u\right) = 1 + tq''(u)g'\left[x - q'\left(u\right)t\right]$$

becomes zero. Computing G_u along the characteristic (2.43), we have $u = g\left(\xi\right)$ and

$$G_u\left(x,t,u\right) = 1 + tq''(g\left(\xi\right))g'(\xi).$$

Assume that the nonnegative function

$$z\left(\xi\right) = -q''(g(\xi))g'(\xi)$$

attains its *positive* maximum $z\left(\xi_M\right)$ only at the point $\xi_M \in [a,b]$. Then

$$t_s = \min_{\xi \in [a,b]} \frac{1}{z\left(\xi\right)} = \frac{1}{z\left(\xi_M\right)}. \tag{2.44}$$

Since x_s belongs to the characteristics $x = q'\left(g\left(\xi_M\right)\right)t + \xi_M$, we find

$$x_s = \frac{q'\left(g\left(\xi_M\right)\right)}{z\left(\xi_M\right)} + \xi_M. \tag{2.45}$$

Remark 2.2. The point (x_s, t_s) has an interesting geometrical meaning. In fact, it turns out that if $q''(g(\xi))g'(\xi) < 0$, the family of characteristics (2.43) admits an *envelope* and (x_s, t_s) is the point on the envelope with minimum time coordinate. To find the envelope[8] it is enough to eliminate the parameter ξ from equations (2.43) and

$$0 = q''(g(\xi))g'(\xi)t + 1,$$

obtained by differentiation of (2.43) with respect to ξ. Clearly, the envelope has not to be confused with the shock curve.

Example 2.2. Consider the initial value problem

$$\begin{cases} u_t + (1 - 2u)u_x = 0 \\ u\left(x, 0\right) = \arctan x. \end{cases} \tag{2.46}$$

We have $q\left(u\right) = u - u^2$, $q'\left(u\right) = 1 - 2u$, $q''\left(u\right) = -2$, and $g\left(\xi\right) = \arctan\xi$, $g'\left(\xi\right) = 1/\left(1 + \xi^2\right)$. Therefore, the function

$$z\left(\xi\right) = -q''(g(\xi))g'(\xi) = \frac{2}{\left(1 + \xi^2\right)}$$

has a maximum at $\xi_M = 0$ and $z\left(0\right) = 2$. The breaking-time is $t_S = 1/2$ and $x_S = 1/2$. Thus, the shock curve starts from $(1/2, 1/2)$. For $0 \leq t < 1/2$ the solution u is smooth and implicitly defined by the equation

$$u - \arctan\left[x - (1 - 2u)t\right] = 0. \tag{2.47}$$

After $t = 1/2$, equation (2.47) defines u as a multivalued function of (x, t) and does not define a solution anymore. Fig. 2.16 shows what happens for $t = 1/4$, $1/2$ and 1. Note that the common point of intersection is $(1/2, \tan 1/2)$ which is not the first shock point.

How does the solution evolve after $t = 1/2$? We have to insert a shock wave into the multivalued graph in Fig. 2.16 in such a way the conservation law is

---

[8] Recall that the *envelope* of a family of curves $\phi(x, t, \xi) = 0$, depending on the parameter ξ, is a curve $\psi(x, t) = 0$ *tangent at each one of its points to a curve of the family*. If the family of curves $\phi(x, t, \xi) = 0$ has an envelope, its parametric equations are obtained by solving the system

$$\begin{cases} \phi(x, t, \xi) = 0 \\ \phi_\xi(x, t, \xi) = 0 \end{cases}$$

with respect to x and t.

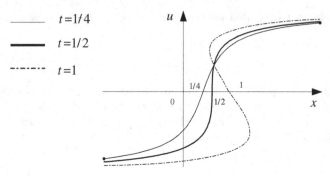

Fig. 2.16. Breaking time for problem (2.46)

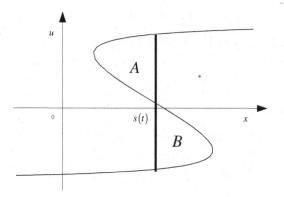

Fig. 2.17. Inserting a shock wave by the Whitham *equal-area rule*

preserved. The correct insertion point is prescribed exactly by the Rankine–Hugoniot condition. It turns out that this corresponds to cutting off from the multivalued profile two **equal area** lobes A and B as described in Fig. 2.17 (*G.B. Whitham equal area rule*[9]).

2.5 Generalized solutions. Uniqueness and entropy condition

We have seen that the method of characteristics is not sufficient, in general, to determine the solution of an initial value problem for all times $t > 0$. In the green light problem a rarefaction wave was used to construct the solution in a region not covered by characteristics. In the traffic jam case the solution undergoes a shock, propagating according to the Rankine-Hugoniot condition.

[9] The *equal-area* rule holds for a general conservation law (see [27]).

We will call rarefaction waves and shock waves obeying the Rankine-Hugoniot condition **generalized solutions**.

Some questions arise naturally.

- *Is there always a unique generalized solution?*
- *If uniqueness fails, is there a criterion to select the "physically correct" solution?*

These questions require a deeper analysis as the following example shows.

Example 2.3 (Non uniqueness). Imagine a flux of particles along the $x-$axis, each one moving with constant speed. Suppose that $u = u(x,t)$ represents *the velocity field*, which gives the speed of the particle located at x at time t. If $x = x(t)$ is the *path* of a particle, its velocity at time t is given by

$$\dot{x}(t) = u(x(t),t) \equiv \text{constant}.$$

Thus, we have

$$0 = \frac{d}{dt}u(x(t),t) = u_t(x(t),t) + u_x(x(t),t)\dot{x}(t)$$
$$= u_t(x(t),t) + u_x(x(t),t)\ u(x(t),t).$$

Therefore $u = u(x,t)$ satisfies *Burgers' equation*

$$u_t + uu_x = u_t + \left(\frac{u^2}{2}\right)_x = 0 \qquad (2.48)$$

which is a conservation law with $q(u) = u^2/2$. Note that q is strictly convex: $q'(u) = u$ and $q''(u) = 1$. We couple (2.48) with the initial condition $u(x,0) = g(x)$, where

$$g(x) = \begin{cases} 0 & x < 0 \\ 1 & x > 0. \end{cases}$$

The characteristics are the straight lines

$$x = g(x_0)t + x_0. \qquad (2.49)$$

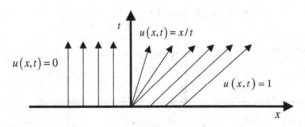

Fig. 2.18. The rarefaction wave of Example 2.3

Therefore, $u = 0$ if $x < 0$ and $u = 1$ if $x > t$. The region $S = \{0 < x < t\}$ is not covered by characteristics. As in the green light problem, we connect the states 0 and 1 through a *rarefaction wave*. Since $q'(u) = u$, we have $r(s) = (q')^{-1}(s) = s$, so that we construct the weak solution.

$$u(x,t) = \begin{cases} 0 & x \le 0 \\ \dfrac{x}{t} & 0 < x < t \\ 1 & x \ge t. \end{cases} \tag{2.50}$$

However, u **is not the unique weak solution!** There exists also a *shock wave* solution. In fact, since

$$u_- = 0, \ u_+ = 1, \ q(u_-) = 0, \ q(u_+) = \frac{1}{2},$$

the Rankine–Hugoniot condition yields

$$\dot{s}(t) = \frac{q(u_+) - q(u_-)}{u_+ - u_-} = \frac{1}{2}.$$

Given the discontinuity at $x = 0$ of the initial data, the shock curve starts at $s(0) = 0$ and it is the straight line

$$x = \frac{t}{2}.$$

Hence, the function

$$w(x,t) = \begin{cases} 0 & x < \frac{t}{2} \\ 1 & x > \frac{t}{2} \end{cases}$$

is another weak solution (Fig. 2.19). As we shall see, this shock wave has to be considered not physically acceptable.

The above example shows that the answer to the first question is *negative*. Thus, we need a criterion to establish which one is the physically correct solution.

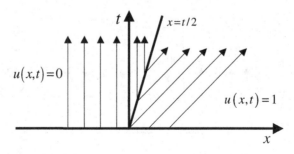

Fig. 2.19. A *non physical* shock

The answer is not elementary and comes from an analogy with gas dynamics, according to which the entropy increases across a shock curve. In fact, there are a few selection criteria, called *entropy criteria*, that in some generalized sense translate the same idea. A rarefaction wave, being continuous satisfies automatically these criteria. For a shock wave the following condition is known as *Lax entropy condition*.

Definition 2.1. *Let u be a shock wave, generalized solution of the conservation law $u_t + q(u)_x = 0$, with q convex or concave. We say that u satisfies the Lax entropy condition if, across the shock curve:*

$$q'(u^+) < \frac{ds}{dt} < q'(u^-). \qquad (2.51)$$

A shock wave satisfying (2.51) is said to be an **entropy solution**.

The geometrical meaning of (2.51) is remarkable: *the slope of a shock curve is less than the slope of the left-characteristics and greater than the slope of the right-characteristics.* Roughly, the characteristics **hit forward** in time the shock line, so that it is not possible to go back in time along characteristics and hit a shock line, expressing a sort of irreversibility after a shock.

The above considerations lead us to select the entropy solutions as the only physically meaningful shocks. On the other hand, if the characteristics hit a shock curve backward in time, the shock wave is to be considered *non-physical*.

Thus, in the non-uniqueness Example 2.3, the solution w represents a non-physical shock since it does not satisfy the entropy condition. The correct solution is therefore the simple wave (2.50). The following important result holds (see e.g. *Smoller* [16]).

Theorem 2.1. *If $q \in C^2(\mathbb{R})$ is convex (or concave) and g is bounded, there exists a unique entropy solution of the problem*

$$\begin{cases} u_t + q(u)_x = 0 & x \in \mathbb{R},\, t > 0 \\ u(x,0) = g(x) & x \in \mathbb{R}. \end{cases} \qquad (2.52)$$

Example 2.4. We apply Theorem 2.1 to solve explicitly problem (2.52) with initial data

$$g(x) = \begin{cases} u_+ & x > 0 \\ u_- & x < 0, \end{cases} \qquad (2.53)$$

where u_+ and u_- are constants, $u_+ \neq u_-$ and $q \in C^2(\mathbb{R})$, and $q'' \geq h > 0$.

This problem is known as **Riemann problem**, and it is particularly important for the numerical approximation of more complex problems.

Now, we claim and prove the following:

a) If $u_+ < u_-$, the unique entropy solution is the shock wave

$$u(x,t) = \begin{cases} u_+ & \frac{x}{t} > \frac{ds}{dt} \\ u_- & \frac{x}{t} < \frac{ds}{dt} \end{cases} \qquad (2.54)$$

where

$$\frac{ds}{dt} = \frac{q(u_+) - q(u_-)}{u_+ - u_-}.$$

b) If $u_+ > u_-$, the unique entropy solution is the rarefaction wave

$$u\,(x,t) = \begin{cases} u_- & \frac{x}{t} < q'\,(u_-) \\ r\left(\frac{x}{t}\right) & q'\,(u_-) < \frac{x}{t} < q'\,(u_+) \\ u_+ & \frac{x}{t} > q'\,(u_+) \end{cases}$$

where $r = (q')^{-1}$, is the inverse function of q'.

Proof (a). The shock wave (2.54) satisfies the Rankine Hugoniot condition and therefore it is clearly a generalized solution. Moreover, since $u_+ < u_-$ the Lax entropy condition holds as well, and u is the unique entropy solution of problem (2.53) by Theorem 2.1.

Proof (b). Since

$$r\,(q'\,(u_+)) = u_+ \quad \text{and} \quad r\,(q'\,(u_-)) = u_-,$$

u is continuous in the half-plane $t > 0$ and we have only to check that u satisfies the equation $u_t + q\,(u)_x = 0$ in the region

$$S = \left\{ (x,t) : q'\,(u_-) < \frac{x}{t} < q'\,(u_+) \right\}.$$

Let $u\,(x,t) = r\left(\frac{x}{t}\right)$. We have:

$$u_t + q\,(u)_x = -r'\left(\frac{x}{t}\right)\frac{x}{t^2} + q'\,(r)\,r'\left(\frac{x}{t}\right)\frac{1}{t} = r'\left(\frac{x}{t}\right)\frac{1}{t}\left[q'\,(r) - \frac{x}{t}\right] \equiv 0.$$

Thus, u is a generalized solution in the upper half-plane. □

2.6 The Vanishing Viscosity Method

There is another instructive and perhaps more natural way to construct discontinuous solutions of the conservation law

$$u_t + q\,(u)_x = 0, \tag{2.55}$$

the so called *vanishing viscosity method*. This method consists in viewing equation (2.55) as the limit for $\varepsilon \to 0^+$ of the equation

$$u_t + q\,(u)_x = \varepsilon u_{xx}, \tag{2.56}$$

that corresponds to choosing the flux function

$$\tilde{q}\,(u, u_x) = q\,(u) - \varepsilon u_x, \tag{2.57}$$

where ε is a *small positive* number. Although we recognize εu_{xx} as a diffusion term, this kind of model arises mostly in fluid dynamics where u is the fluid velocity and ε its *viscosity*, from which comes the name of the method.

There are several good reasons in favor of this approach. First of all, a small amount of diffusion or viscosity makes the mathematical model more realistic in most applications. Note that εu_{xx} becomes relevant only when u_{xx} is large, that is in a region where u_x changes rapidly and a shock occurs. For instance in our model of traffic dynamics, it is natural to assume that drivers would slow down when they see increased (relative) density ahead. Thus, an appropriate model for their velocity is

$$\tilde{v}\left(\rho\right) = v\left(\rho\right) - \varepsilon\frac{\rho_x}{\rho}$$

which corresponds to $\tilde{q}\left(\rho\right) = \rho v\left(\rho\right) - \varepsilon\rho_x$ for the flow-rate of cars.

Another reason comes from the fact that shocks constructed by the vanishing viscosity method are *physical shocks*, since they satisfy the entropy inequality.

As for the heat equation, in principle we expect to obtain smooth solutions even with discontinuous initial data. On the other hand, the nonlinear term may force the evolution towards a shock wave.

Here we are interested in solutions of (2.56) connecting two constant states u_L and u_R, that is, satisfying the conditions

$$\lim_{x\to-\infty} u\left(x,t\right) = u_L, \qquad \lim_{x\to+\infty} u\left(x,t\right) = u_R. \tag{2.58}$$

Since we are looking for shock waves, it is reasonable to seek a solution depending only on a coordinate $\xi = x - vt$ moving with the (unknown) shock speed v. Thus, let us look for *bounded travelling waves* solution of (2.56) of the form

$$u\left(x,t\right) = U\left(x - vt\right) \equiv U\left(\xi\right)$$

with

$$U\left(-\infty\right) = u_L \quad \text{and} \quad U\left(+\infty\right) = u_R \tag{2.59}$$

and $u_L \neq u_R$. We have

$$u_t = -v\frac{dU}{d\xi}, \qquad u_x = \frac{dU}{d\xi}, \qquad u_{xx} = \frac{d^2U}{d\xi^2}$$

so that we obtain for U the ordinary differential equation

$$\left(q'\left(U\right) - v\right)\frac{dU}{d\xi} = \varepsilon\frac{d^2U}{d\xi^2}$$

which can be integrated to yield

$$q\left(U\right) - vU + A = \varepsilon\frac{dU}{d\xi}$$

where A is an arbitrary constant. Assuming that $\dfrac{dU}{d\xi} \to 0$ as $\xi \to \pm\infty$ and using (2.59) we get

$$q\left(u_L\right) - vu_L + A = 0 \quad \text{and} \quad q\left(u_R\right) - vu_R + A = 0. \tag{2.60}$$

Subtracting these two equations we find

$$v = \frac{q\left(u_R\right) - q\left(u_L\right)}{u_R - u_L} \equiv \bar{v}. \tag{2.61}$$

and then $A = \dfrac{-q\left(u_R\right)u_L + q\left(u_L\right)u_R}{u_R - u_L} \equiv \bar{A}.$

Thus, if there exists a travelling wave solution satisfying conditions (2.58), it moves with a speed $\bar{v}$ predicted by the Rankine–Hugoniot formula. Still it is not clear whether such travelling wave solution exists. In order to verify this fact, examine the equation

$$\varepsilon\frac{dU}{d\xi} = q\left(U\right) - \bar{v}U + \bar{A}. \tag{2.62}$$

From (2.60), equation (2.62) has the two equilibria $U = u_R$ and $U = u_L$. A bounded travelling wave connecting u_R and u_L corresponds to a solution of (2.62) starting from a point ξ_0 between u_R and u_L. On the other hand, conditions (2.59) require u_R to be *asymptotically stable* and u_L *unstable*. At this point, we need to have information on the shape of q.

Assume $q'' < 0$. Then the phase diagram for equation (2.62) is described in Fig. (2.20) for the two cases $u_L > u_R$ and $u_L < u_R$. Between u_L and u_R, $q\left(U\right) - \bar{v}U + \bar{A} > 0$ and, as the arrows indicate, U is *increasing*. We see that only the case $u_L < u_R$ is compatible with conditions (2.59) and this corresponds precisely to a shock formation for the non diffusive conservation law. Thus,

$$q'\left(u_L\right) - \bar{v} > 0 \quad \text{and} \quad q'\left(u_R\right) - \bar{v} < 0$$

or

$$q'\left(u_R\right) < \bar{v} < q'\left(u_L\right) \tag{2.63}$$

which is *the entropy inequality.*

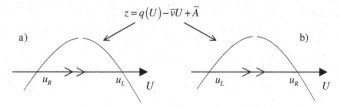

Fig. 2.20. Case b) only is compatible with conditions (2.58)

Similarly, if $q'' > 0$, a travelling wave solution connecting the two states u_R and u_L exists only if $u_L > u_R$ and (2.63) holds.

Let us see what happens when $\varepsilon \to 0$. Assume $q'' < 0$. For ε small, we expect that our travelling wave increases abruptly from a value $U(\xi_1)$ close to u_L to a value $U(\xi_2)$ close to u_R within a narrow region called the *transition layer*. For instance we may choose ξ_1 and ξ_2 such that

$$U(\xi_2) - U(\xi_1) \geq (1 - \beta)(u_R - u_L)$$

with a positive β, very close to 0. We call the number $\varkappa = \xi_2 - \xi_1$ *thickness* of the transition layer. To compute it, we separate the variables U and ξ in (2.62) and integrate over (ξ_1, ξ_2); this yields

$$\xi_2 - \xi_1 = \varepsilon \int_{U(\xi_1)}^{U(\xi_2)} \frac{ds}{q(s) - vs + \bar{A}}.$$

Thus, the thickness of the transition layer is proportional to ε. As $\varepsilon \to 0$, the transition region becomes more and more narrow and eventually a shock wave that satisfies the entropy inequality is obtained.

This phenomenon is clearly seen in the important case of *viscous Burgers'* equation that we examine in more details in the next subsection.

Example 2.5. *Burgers' shock solution.* Let us determine a travelling wave solution of the viscous Burgers' equation

$$u_t + uu_x = \varepsilon u_{xx} \tag{2.64}$$

connecting the states $u_L = 1$ and $u_R = 0$. Note that $q(u) = u^2/2$ is convex. Then $\bar{v} = 1/2$ and $\bar{A} = 0$. Equation (2.62) becomes

$$2\varepsilon \frac{dU}{d\xi} = U^2 - U$$

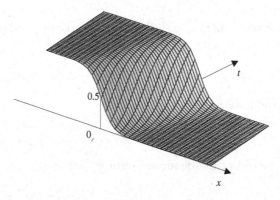

Fig. 2.21. The travelling wave in Example 2.5

that can be easily integrated to give

$$U\left(\xi\right) = \frac{1}{1 + \exp\left(\dfrac{\xi}{2\varepsilon}\right)}.$$

Thus the travelling wave is given by

$$u\left(x,t\right) = U\left(x - \frac{t}{2}\right) = \frac{1}{1 + \exp\left(\dfrac{2x - t}{4\varepsilon}\right)}. \tag{2.65}$$

When $\varepsilon \to 0$,

$$u\left(x,t\right) \to w\left(x,t\right) = \begin{cases} 0 & x > t/2 \\ 1 & x < t/2 \end{cases}$$

which is the entropy shock solution for the non viscous Burgers' equation with initial data 1 if $x < 0$ and 0 if $x > 0$.

2.6.1 The viscous Burgers' equation

The viscous Burgers' equation is one of the most celebrated examples of non-linear diffusion equation. It arose (Burger, 1948) as a simplified form of the Navier-Stokes equation, in an attempt to study some aspects of turbulence. It appears also in gas dynamics, in the theory of sound waves and in traffic flow modelling and it constitutes a basic example of competition between *dissipation* (due to linear diffusion) and *steepening* (shock formation due to the nonlinear transport term uu_x).

The success of Burgers' equation is also due to the rather surprising fact that the initial value problem can be solved analytically. In fact, via the so called *Hopf-Cole transformation*, Burgers' equation is converted into the heat equation. Let us see how this can be done. Write the equation in the form

$$\frac{\partial u}{\partial t} + \frac{\partial}{\partial x}\left(\frac{1}{2}u^2 - \varepsilon u_x\right) = 0.$$

Then, the planar vector field $(-u, \frac{1}{2}u^2 - \varepsilon u_x)$ is curl-free and therefore there exists a potential $\psi = \psi\left(x,t\right)$ such that

$$\psi_x = -u \quad \text{and} \quad \psi_t = \frac{1}{2}u^2 - \varepsilon u_x.$$

Thus, ψ solves the equation

$$\psi_t = \frac{1}{2}\psi_x^2 + \varepsilon\psi_{xx}. \tag{2.66}$$

Now we try to get rid of the quadratic term letting $\psi = g\left(\varphi\right)$, with g to be chosen. We have

$$\psi_t = g'\left(\varphi\right)\varphi_t, \quad \psi_x = g'\left(\varphi\right)\varphi_x, \quad \psi_{xx} = g''\left(\varphi\right)\left(\varphi_x\right)^2 + g'\left(\varphi\right)\varphi_{xx}.$$

Substituting into (2.66) we find

$$g'(\varphi)\left[\varphi_t - \varepsilon\varphi_{xx}\right] = \left[\frac{1}{2}(g'(\varphi))^2 + \varepsilon g''(\varphi)\right](\varphi_x)^2.$$

Hence, if we choose $g(s) = 2\varepsilon \log s$, then the right hand side vanishes and we are left with

$$\varphi_t - \varepsilon\varphi_{xx} = 0. \tag{2.67}$$

Thus

$$\psi = 2\varepsilon \log \varphi$$

and from $u = -\psi_x$ we obtain

$$u = -2\varepsilon\frac{\varphi_x}{\varphi} \tag{2.68}$$

which is the *Hopf-Cole transformation*. An initial data

$$u(x,0) = u_0(x) \tag{2.69}$$

is transformed into an initial data of the form[10]

$$\varphi_0(x) = \exp\left\{-\int_a^x \frac{u_0(z)}{2\varepsilon}dz\right\} \qquad (a \in \mathbb{R}). \tag{2.70}$$

As we will see in the next chapter, using formula (3.72), if

$$\frac{1}{x^2}\int_a^x u_0(z)\,dz \to 0 \quad \text{as } |x| \to \infty,$$

the initial value problem (2.67), (2.70) has a unique smooth solution in the half-plane $t > 0$, given by formula

$$\varphi(x,t) = \frac{1}{\sqrt{4\pi\varepsilon t}}\int_{-\infty}^{+\infty}\varphi_0(y)\exp\left(-\frac{(x-y)^2}{4\varepsilon t}\right)dy.$$

This solution is continuous with its x-derivative up to $t = 0$ at any continuity point of u_0. Consequently, using (2.68), problem (2.64) has a unique smooth solution in the half-plane $t > 0$, continuous up to $t = 0$ at any continuity point of u_0, given by

$$u(x,t) = \frac{\int_{-\infty}^{+\infty}\frac{x-y}{t}\varphi_0(y)\exp\left(-\frac{(x-y)^2}{4\varepsilon t}\right)dy}{\int_{-\infty}^{+\infty}\varphi_0(y)\exp\left(-\frac{(x-y)^2}{4\varepsilon t}\right)dy}. \tag{2.71}$$

We use formula (2.71) to solve an initial pulse problem.

---

[10] The choice of a is arbitrary and does not affect the value of u.

Example 2.6. *Initial pulse.* Consider problem (2.64) with the initial condition

$$u_0(x) = M\delta(x)$$

where δ denotes the Dirac density at the origin. We have, choosing $a = 1$,

$$\varphi_0(x) = \exp\left\{-\int_1^x \frac{u_0(y)}{2\varepsilon}dy\right\} = \begin{cases} 1 & x > 0 \\ \exp\left(\dfrac{M}{2\varepsilon}\right) & x < 0. \end{cases}$$

Formula (2.71), gives, after some routine calculations,

$$u(x,t) = \sqrt{\frac{4\varepsilon}{\pi t}} \; \frac{\exp\left(-\dfrac{x^2}{4\varepsilon t}\right)}{\dfrac{2}{\exp(M/2\varepsilon)-1} + \dfrac{\sqrt{\pi}}{2}\left[1 - \mathrm{erf}\left(\dfrac{x}{\sqrt{4\varepsilon t}}\right)\right]}$$

where

$$\mathrm{erf}(x) = \int_0^x e^{-z^2}dz$$

is the *error function.*

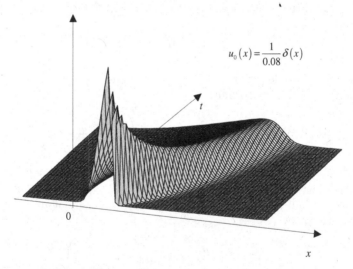

Fig. 2.22. Evolution of an initial pulse for the viscous Burgers' equation ($M = 1, \varepsilon = 0.04$)

2.7 Numerical methods

2.7.1 Finite difference approximation of scalar conservation laws

Let us consider equation (2.21) on a domain $x \in (0, R)$, $t > 0$. The basic idea of the finite difference method consists in building up an approximation of the problem by replacing the derivatives in the differential equation by a difference quotient. For the scalar conservation law $u_t + a u_x = 0$ this approximation has to be carried out for both time and space derivatives.

First, let us define a discretization of the physical domain into computational nodes or cells. For the particular case at hand, it is more convenient to look for an approximation of $u(x, t)$ in the nodes (x_i, t^n) defined as (see also Fig. 2.23)

$$x_i = i\,h \text{ with } h = \frac{R}{N} \text{ and } i, N \in \mathbb{N}, \quad t^n = n\,\tau \text{ with } n \in \mathbb{N}$$

corresponding to a uniform partition of the time and space domains. The collection of the nodes is called computational grid or mesh.

Second, let us use the computational nodes to define suitable difference quotients for the approximation of time and space derivatives of u, respectively,

$$u_t(x_i, t^n) = \frac{1}{\tau}\big(u(x_i, t^{n+1}) - u(x_i, t^n)\big) + \mathcal{O}(\tau) \tag{2.72}$$

$$u_x(x_i, t^n) = \begin{cases} \frac{1}{h}\big(u(x_i, t^n) - u(x_{i-1}, t^n)\big) + \mathcal{O}(h) \\ \frac{1}{h}\big(u(x_{i+1}, t^n) - u(x_i, t^n)\big) + \mathcal{O}(h). \end{cases} \tag{2.73}$$

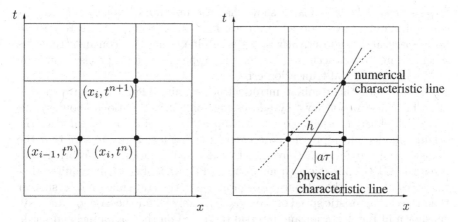

Fig. 2.23. On the left we show the computational grid for the approximation of $u_t + a u_x = 0$, where the nodes involved to build up the upwind scheme with $a > 0$ are highlighted. On the right we provide a graphical interpretation of the CFL condition

Let u_i^n be an approximation of $u(x_i, t^n)$, solution of problem (2.21). Rewriting the above difference quotients for u_i^n and replacing the corresponding expressions into $u_t + au_x = 0$, we obtain the following approximation scheme for problem (2.21) in the particular case $a > 0$,

$$\frac{1}{\tau}\left(u_i^{n+1} - u_i^n\right) + \frac{a}{h}\left(u_i^n - u_{i-1}^n\right) = 0$$

$$u_i^{n+1} - u_i^n + a\lambda\left(u_i^n - u_{i-1}^n\right) = 0, \quad \text{where } \lambda = \frac{\tau}{h}.$$

According to the difference quotients adopted in (2.72), we have obtained a *one-sided* scheme referring to space approximation, because the scheme only involves the nodes x_i and x_{i-1}, *forward* in time, because the time discretization is performed moving forward with respect to the reference time level t^n. Similarly to the continuum model, by this way the numerical scheme propagates the information in the same direction of the characteristic lines, when $a > 0$. Then, the scheme has to be modified to maintain this good property when $a < 0$. To this purpose, we take the following difference quotient for the space approximation,

$$au_x(x_i, t^n) \simeq \begin{cases} \frac{a}{h}\left(u_i^n - u_{i-1}^n\right) & \text{if } a > 0 \\ \frac{a}{h}\left(u_{i+1}^n - u_i^n\right) & \text{if } a < 0 \end{cases}$$

$$\simeq \frac{1}{2}\frac{a}{h}\left(u_{i+1}^n - u_{i-1}^n\right) - \frac{1}{2}\frac{|a|}{h}\left(u_{i+1}^n - 2u_i^n + u_{i-1}^n\right),$$

such that the so called **upwind** scheme is obtained,

$$u_i^{n+1} = u_i^n - \frac{1}{2}a\lambda\left(u_{i+1}^n - u_{i-1}^n\right) + \frac{1}{2}|a|\lambda\left(u_{i+1}^n - 2u_i^n + u_{i-1}^n\right). \qquad (2.74)$$

Expressions (2.72) and (2.73) show that the upwind scheme is obtained by combining difference quotients that are first order accurate with respect to both space and time derivatives. As a result of that, we conclude that the scheme is first order accurate. We refer to Appendix C for a precise definition of accuracy and local truncation error.

We conclude this minimal introduction to finite difference approximation of scalar conservation laws by addressing some considerations about stability, which clarify the behavior of the approximate solution compared to the original model and represent a necessary requirement to make sure that the numerical approximation converges to the exact solution u. We briefly address the **CFL condition** (from Courant-Friedrichs-Lewy). It requires that the speed at which the scheme propagates the initial state must not be smaller than the characteristic speed of the model, namely $|a|$ for the case $u_t + au_x = 0$. As shown in Fig. 2.23, for any interval (t^n, t^{n+1}) on the computational mesh, such condition is equivalent to

$$|a\tau| \leq h \quad \text{that is} \quad |a\lambda| \leq 1.$$

The importance of the CFL condition could be summarized as follows. We notice that we could affect the exact solution of the model in (x_i, t^{n+1}) by suitably modifying the initial state in a point $x_0 = x_i - at^{n+1}$. If this condition is not satisfied, the point x_0 falls far enough from x_i such that the numerical solution remains unchanged, namely, x_0 falls outside the domain of dependence of the numerical scheme relative to the node (x_i, t^{n+1}). As a result of that, we conclude that the numerical scheme can not converge to the exact solution for $\tau, h \to 0$ outside the range of space and time discretization steps, τ, h respectively, that satisfy the CFL condition.

In Appendix C we present a more rigorous, yet incomplete analysis of finite difference methods for scalar and linear initial value problems.[11]

2.8 Exercises

2.1. Consider the Green light problem described in equation (2.25), with the initial condition $\rho(x, 0) = g(x)$, where g is assigned in (2.32) and calculate the car density at the light for $t > 0$. Then, find the time that a car located at t_0 in the position $x_0 = -v_m t_0$ takes to reach the light.

2.2. Use the traffic dynamics introduced in equation (2.25) to describe the density $\rho = \rho(x, t)$ of cars on a straight highway supposing that the initial density is

$$\rho_0 = \begin{cases} a\,\rho_m & x < 0 \\ \rho_m/2 & x > 0. \end{cases}$$

Describe, with respect to the parameter $a \in [0, 1]$, the evolution of ρ as $t > 0$: find the characteristics, the shock curve and find a solution in the half plane (x, t), for $t > 0$. Give an interpretation of the result.

2.3. Study the problem (Burgers equation)

$$\begin{cases} u_t + uu_x = 0 & x \in \mathbb{R}, t > 0 \\ u(x, 0) = g(x) & x \in \mathbb{R} \end{cases}$$

when the initial data $g(x)$, respectively, is:

a) $\begin{cases} 0 & \text{if } x < 0 \\ 1 & \text{if } 0 < x < 1 \\ 0 & \text{if } x > 1 \end{cases}$ b) $\begin{cases} 1 & \text{if } x < 0 \\ 2 & \text{if } 0 < x < 1 \\ 0 & \text{if } x > 1 \end{cases}$ c) $\begin{cases} 1 & \text{if } x \leq 0 \\ 1 - x & \text{if } 0 < x < 1 \\ 0 & \text{if } x \geq 1. \end{cases}$

2.4. The conservation law

$$u_t + u^3 u_x = 0 \quad x \in \mathbb{R}, t > 0$$

---

[11] We refer the reader to *Quarteroni* [43] and *Le Veque* [40] for a detailed treatment of this matter.

is given. Find the solution using the characteristics technique (highlighting rarefaction or shock waves) associated to the initial condition

$$u(x,0) = g(x) = \begin{cases} 0 & \text{if } x < 0 \\ 1 & \text{if } 0 < x < 1 \\ 0 & \text{if } x > 1. \end{cases}$$

2.5. Draw the characteristics and describe the evolution as $t \to +\infty$ of the solution of the initial value problem

$$\begin{cases} u_t + uu_x = 0 & t > 0, x \in \mathbb{R} \\ u(x,0) = \begin{cases} \sin x & 0 < x < \pi \\ 0 & x \le 0 \text{ or } x \ge \pi. \end{cases} \end{cases}$$

2.6. Consider the following problem $(a > 0)$

$$\begin{cases} u_t + au_x = f(x,t) & 0 < x < R, t > 0 \\ u(0,t) = 0 & t > 0 \\ u(x,0) = 0 & 0 < x < R. \end{cases}$$

Prove the stability estimate

$$\int_0^R u^2(x,t)dx \le e^t \int_0^t \int_0^R f^2(x,s)dx\,ds, \qquad t > 0.$$

2.7 (Traffic in a tunnel). A rather realistic model for the car speed in a very long tunnel is the following:

$$v(\rho) = \begin{cases} v_m & 0 \le \rho \le \rho_c \\ \lambda \log\left(\frac{\rho_m}{\rho}\right) & \rho_c \le \rho \le \rho_m \end{cases}$$

where

$$\lambda = \frac{v_m}{\log(\rho_m/\rho_c)}.$$

Observe that v is continuous also at $\rho_c = \rho_m e^{-v_m/\lambda}$, which represents a *critical density*: if $\rho \le \rho_c$, the drivers are free to reach the speed limit. Typical values are: $\rho_c = 7\text{car}/\text{Km}$, $v_m = 90 \text{ KM/h}$, $\rho_m = 110 \text{ car}/\text{Km}$, $v_m/\lambda = 2.75$.

Assume that the entrance is placed at $x = 0$ and that the cars are waiting (with the maximum density) the tunnel is open to the traffic at time $t = 0$. Thus, the initial density is

$$\rho = \begin{cases} \rho_m & x < 0 \\ 0 & x > 0. \end{cases}$$

a) Determine density and car speed; draw their graphs as a function of time.
b) Determine and draw in the x, t plane the trajectory of a car initially at $x = x_0$, and compute the time it takes to enter the tunnel.

2.8. Consider the conservation law

$$u_t + u^5 u_x = 0$$

in $\mathbb{R} \times (0, +\infty)$. Find an explicit solution of the problem associated with the initial data

$$g_1(x) = \begin{cases} 0 & x \le 0 \\ 1 & x > 0 \end{cases} \qquad g_2(x) = \begin{cases} 1 & x \le 0 \\ 0 & x > 0. \end{cases}$$

Then, represent the solutions at time $t = 1$.

2.9. Show that, for every $\alpha > 1$, the function

$$u_\alpha(x,t) = \begin{cases} -1 & 2x < (1-\alpha)t \\ -\alpha & (1-\alpha)t < 2x < 0 \\ \alpha & 0 < 2x < (\alpha-1)t \\ 1 & (\alpha-1)t < 2x \end{cases}$$

is a generalized solution of the problem

$$\begin{cases} u_t + u u_x = 0 & t > 0, x \in \mathbb{R} \\ u(x,0) = \begin{cases} -1 & x < 0 \\ 1 & x > 0. \end{cases} \end{cases}$$

Is it also an entropy solution, at least for some α?

2.10. Using the Hopf-Cole transformation, solve the following problem for the viscous Burgers equation

$$\begin{cases} u_t + u u_x = \epsilon u_{xx} & t > 0, x \in \mathbb{R} \\ u(0, x) = \mathcal{H} & x \in \mathbb{R} \end{cases}$$

where $\mathcal{H}$ is the Heavyside function. Show that, as $t \to +\infty$, $u(x,t)$ converges to a traveling wave similar to (2.65).

2.8.1 Numerical approximation of a constant coefficient scalar conservation law

We apply the upwind scheme to the discretization of the following problem, where periodic boundary conditions are equivalent to consider a periodic initial state over the real line,

$$\begin{cases} u_t + u_x = 0 & \text{for } -1 < x < 15, \ t > 0 \\ u(-1, t) = u(15, t) & \text{for } t > 0 \\ u(x, 0) = \begin{cases} \sin(x) & 0 < x < \pi \\ 0 & \text{elsewhere.} \end{cases} \end{cases}$$

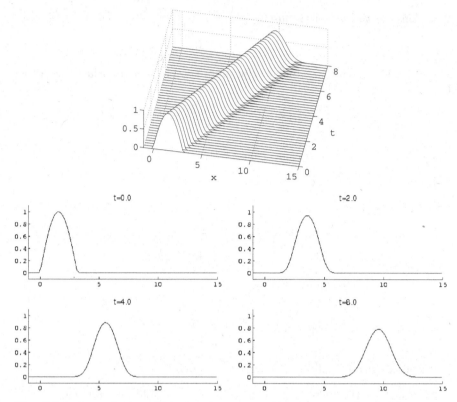

Fig. 2.24. A constant coefficient scalar conservation law approximated by means of the upwind scheme. The numerical solution is represented on the (x, t) plane at the top and at times $t = 0, 2, 4, 8$ on the bottom panels

Fig. 2.24 shows the initial state and the numerical solution at times $t = 0, 2, 4, 8$, obtained on a computational mesh characterized by the discretization steps $h = \tau = 0.2$.

These results show that the scheme correctly captures the propagation of the initial state from left to right, when $a > 0$. However, the peak of the numerical solution progressively decreases, in contrast with the fact that the initial state should simply propagate by translation. This is put into evidence by the plots in Fig. 2.24. We say that the upwind method features a **diffusive** behavior, which can be justified by means of a straightforward manipulation of the scheme. Let us assume that the exact solution is regular enough to perform a Taylor expansion of $u(x_i, t^n)$. It is easily proved that

$$\frac{1}{2} \frac{a}{h} \left(u(x_{i+1}, t^n) - u(x_{i-1}, t^n) \right) - \frac{1}{2} \frac{|a|}{h} \left(u(x_{i+1}, t^n) - 2u(x_i, t^n) + u(x_{i-1}, t^n) \right)$$

$$= a u_x(x_i, t^n) - \frac{|a|h}{2} u_{xx}(x_i, t^n) + \mathcal{O}(h^2).$$

Then, we observe that the upwind method can be reinterpreted as an approximation scheme for the model $u_t + au_x - \frac{1}{2}|a|\lambda u_{xx} = 0$. However, the upwind method turns out to be more accurate with respect to the new equation than to the original model. This shows that the upwind scheme naturally embeds into the approximation an artificial diffusion term with diffusivity proportional to $\frac{1}{2}|a|\lambda$. Although such effect is at the basis of the good stability properties of the scheme, it is not desirable for the approximation of steep gradients or discontinuous solutions, as it will be put into evidence by the forthcoming examples.

2.8.2 Numerical approximation of Burgers equation

Let us apply upwind method to the following problem governed by Burgers equation,

$$
\begin{cases}
u_t + uu_x = 0 \quad \text{for } -1 < x < 15, \ t > 0 \\[2mm]
u(-1,t) = u(15,t) \quad \text{for } t > 0 \\[2mm]
u(x,0) = \begin{cases} \sin(x) & 0 < x < \pi \\ 0 & \text{elsewhere} \end{cases}
\end{cases}
\tag{2.75}
$$

whose solution, explored in Exercise 2.5 by means of the method of characteristics, is given by a combination of a rarefaction fan with a compression, the latter giving rise to a shock wave. For this example of nonlinear scalar conservation law, computing numerical approximation is by far more interesting than for the linear case. In particular, focusing on the approximation of the shock wave, from Fig. 2.25 (lower left panel) we notice that the upwind scheme tends to smooth out the discontinuous exact solution. In order to obtain a more accurate approximation, advanced *high-resolution* methods should be applied (see *Le Veque* [40], Cap. 16). They consist on second order accurate schemes suitably modified to minimize the spurious oscillations that may appear when approximating very steep gradients. Fig. 2.25 (right) shows the improvement obtained using these methods instead of the upwind scheme for the Burgers equation.

2.8.3 Numerical approximation of traffic dynamics

Let us consider the model for traffic flow summarized in (2.24). By applying for simplicity unit coefficients $v_m = \rho_m = 1$, we get

$$
\begin{cases}
\rho_t + (1 - 2\rho)\rho_x = 0 & \text{for } -5 < x < 5, \ t > 0 \\
\rho(-3,t) = \rho(3,t) & \text{for } t > 0 \\
\rho(x,0) = 0.2\exp\left(-x^2\right) & \text{for } -5 \le x \le 5.
\end{cases}
\tag{2.76}
$$

The initial state of the system features a localized higher vehicle density because of an obstacle to flow. The numerical simulations reported in Fig. 2.26

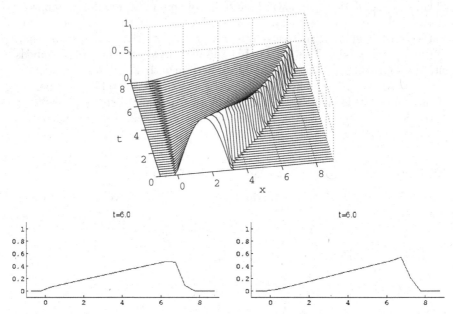

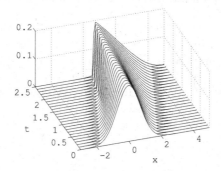

Fig. 2.25. High-resolution numerical approximation of Burgers problem (2.75) represented on the (x, t) plane. The bottom panels highlight the comparison between the upwind method (left) and the high-resolution scheme (right) in the approximation of steep solution gradients obtained at $t = 6$

Fig. 2.26. High-resolution numerical approximation of the problem (2.76) reported on the (x, t) plane

confirm that the traffic evolves towards a more dangerous situation where drivers observe a sudden significant change of vehicle density combined with a corresponding decrease in flow velocity. The converging characteristics give rise to a compression fan evolving into a shock wave after a finite time, precisely at $t = 2$. Conversely, once the obstacle has been overtaken, the flow accelerates to normal conditions.

3

Diffusion

3.1 The Diffusion Equation

3.1.1 Introduction

The one-dimensional **diffusion equation** is the *linear second order partial differential equation*

$$u_t - D u_{xx} = f$$

where $u = u(x,t)$, x is a real space variable, t a time variable and D a positive constant, called **diffusion coefficient**. In space dimension $n > 1$, that is when $\mathbf{x} \in \mathbb{R}^n$, the diffusion equation reads

$$u_t - D\Delta u = f \tag{3.1}$$

where Δ denotes the *Laplace operator*:

$$\Delta = \sum_{k=1}^{n} \frac{\partial^2}{\partial x_k^2}.$$

When $f \equiv 0$ the equation is said to be **homogeneous** and in this case the **superposition principle** holds: if u and v are solutions of (3.1) and a, b are real (or complex) numbers, $au + bv$ also is a solution of (3.1). More generally, if $u_k(\mathbf{x},t)$ is a family of solutions depending on the parameter k (integer or real) and $g = g(k)$ is a function rapidly vanishing at infinity, then

$$\sum_{k=1}^{\infty} u_k(\mathbf{x},t)\, g(k) \quad \text{and} \quad \int_{-\infty}^{+\infty} u_k(\mathbf{x},t)\, g(k)\, dk$$

are still solutions.

A common example of diffusion is given by *heat conduction* in a solid body. Conduction comes from molecular collision, transferring heat by kinetic energy, without macroscopic material movement. If the medium is homogeneous and isotropic with respect to the heat propagation, the evolution of the

Salsa S., Vegni F.M.G., Zaretti A., Zunino P.: *A Primer on PDEs. Models, Methods, Simulations.*
Unitext – La Matematica per il 3+2 65.
DOI 10.1007/978-88-470-2862-3_3, © Springer-Verlag Italia 2013

temperature is described by equation (3.1); f represents the intensity of an external distributed source. For this reason equation (3.1) is also known as **the heat equation.**

On the other hand equation (3.1) constitutes a much more general diffusion model, where by **diffusion** we mean, for instance, the *transport of a substance due to the molecular motion of the surrounding medium*. In this case, u could represent the concentration of a polluting material or of a solute in a liquid or a gas (dye in a liquid, smoke in the atmosphere) or even a probability density. We may say that the diffusion equation unifies at a macroscopic scale a variety of phenomena, that look quite different when observed at a microscopic scale.

Through equation (3.1) and some of its variants we will explore the deep connection between probabilistic and deterministic models, according (roughly) to the scheme

diffusion processes ↔ probability density ↔ differential equations.

The *star* in this field is *Brownian motion,* derived from the name of the botanist Brown, who observed in the middle of the 19th century, the apparently chaotic behavior of certain particles on a water surface, due to the molecular motion. This irregular motion is now modeled as a *stochastic process* under the terminology of *Wiener process or Brownian motion.* The operator

$$\frac{1}{2}\varDelta$$

is strictly related to Brownian motion[1] and indeed it captures and synthesizes the microscopic features of that process.

Under equilibrium conditions, that is when there is no time evolution, the solution u depends only on the space variable and satisfies the *stationary* version of the diffusion equation (letting $D = 1$)

$$-\varDelta u = f \tag{3.2}$$

($-u_{xx} = f$, in dimension $n = 1$). Equation (3.2) is known as the *Poisson equation*. When $f = 0$, it is called *Laplace's equation* and its solutions are so important in so many fields that they have deserved the special name of **harmonic functions.** This equation will be considered in the next chapter.

3.1.2 The conduction of heat

Heat is a form of energy which it is frequently convenient to consider as separated from other forms. For historical reasons, *calories* instead of Joules are used as units of measurement, each *calorie* corresponding to 4.182 Joules.

We want to derive a mathematical model for the heat conduction in a solid body. We assume that the body is homogeneous and isotropic, with constant

---

[1] In the theory of stochastic processes, $\frac{1}{2}\varDelta$ represents the *infinitesimal generator of the Brownian motion.*

mass density ρ, and that it can receive energy from an external source (for instance, from an electrical current or a chemical reaction or from external absorption/radiation). Denote by r the time rate per unit mass at which heat is supplied[2] by the external source.

Since heat is a form of energy, it is natural to use the law of conservation of energy, that we can formulate in the following way:

Let V be an arbitrary control volume inside the body. *The time rate of change of thermal energy in V equals the net flux of heat through the boundary ∂V of V, due to the conduction, plus the time rate at which heat is supplied by the external sources.*

If we denote by $e=e(\mathbf{x},t)$ the thermal energy per unit mass, the total quantity of thermal energy inside V is given by

$$\int_V e\rho \, d\mathbf{x}$$

so that its time rate of change is[3]

$$\frac{d}{dt}\int_V e\rho \, d\mathbf{x} = \int_V e_t\rho \, d\mathbf{x}.$$

Denote by $\mathbf{q}$ the *heat flux* vector[4], which specifies the heat flow direction and the magnitude of the rate of flow across a unit area. More precisely, if d is an area element contained in ∂V with *outer* unit normal $\boldsymbol{\nu}$, then $\mathbf{q}\cdot\boldsymbol{\nu}d\sigma$ is the energy flow rate through $d\sigma$ and therefore the *total inner heat flux* through ∂V is given by

$$-\int_{\partial V} \mathbf{q}\cdot\boldsymbol{\nu}\, d\sigma \underset{\text{(divergence theorem)}}{=} -\int_V \operatorname{div}\mathbf{q}\, d\mathbf{x}.$$

Finally, the contribution due to the external source is given by

$$\int_V r\rho \, d\mathbf{x}.$$

Thus, conservation of energy requires:

$$\int_V e_t\rho \, d\mathbf{x} = -\int_V \operatorname{div}\mathbf{q}\, d\mathbf{x} + \int_V r\rho \, d\mathbf{x}. \tag{3.3}$$

The arbitrariness of V allows us to convert the integral equation (3.3) into the pointwise relation

$$e_t\rho = -\operatorname{div}\mathbf{q} + r\rho \tag{3.4}$$

----

[2] Dimensions of r: $[r] = [cal] \times [time]^{-1} \times [mass]^{-1}$.
[3] Assuming that the time derivative can be carried inside the integral.
[4] $[\mathbf{q}] = [cal] \times [lenght]^{-2} \times [time]^{-1}$.

that constitutes a basic law of heat conduction. However, e and $\mathbf{q}$ are unknown and we need additional information through *constitutive relations* for these quantities. We assume the following:

• **Fourier law** of heat conduction. Under "normal" conditions, for many solid materials, the heat flux is a linear function of the temperature gradient, that is:

$$\mathbf{q} = -\kappa \nabla u \tag{3.5}$$

where u is the absolute temperature and $\kappa > 0$, the *thermal conductivity*[5], depends on the properties of the material. In general, κ may depend on u, $\mathbf{x}$ and t, but often varies so little in cases of interest that it is reasonable to neglect its variation. Here we consider κ *constant* so that

$$\text{div}\mathbf{q} = -\kappa \Delta u. \tag{3.6}$$

The minus sign in the law (3.5) reflects the tendency of heat to flow from hotter to cooler regions.

• The thermal energy is a linear function of the absolute temperature:

$$e = c_v u \tag{3.7}$$

where c_v denotes the *specific heat*[6] (at constant volume) of the material. In many cases of interest c_v can be considered constant. The relation (3.7) is reasonably true over not too wide ranges of temperature.

Using (3.6) and (3.7), equation (3.4) becomes

$$u_t = \frac{\kappa}{c_v \varrho} \Delta u + \frac{1}{c_v} r \tag{3.8}$$

which is the diffusion equation with $D = \kappa/(c_v \varrho)$ and $f = r/c_v$. As we will see, the coefficient D, called *thermal diffusivity*, encodes the thermal response time of the material.

3.1.3 Well posed problems ($n = 1$)

The governing equations in a mathematical model have to be supplemented by additional information in order to obtain a *well posed problem*, i.e. a problem that has exactly one solution, depending continuously on the data.

On physical grounds, it is not difficult to outline some typical well posed problems for the heat equation. Consider the evolution of the temperature u inside a cylindrical bar, whose lateral surface is *perfectly insulated* and whose length L is much larger than its cross-sectional area A. Although the bar is three dimensional, we may assume that heat moves only down the length of

---

[5] $[\kappa] = [cal] \times [deg]^{-1} \times [time]^{-1} \times [length]^{-1}$ (deg stays for degree, Celsius or Kelvin).

[6] $[c_v] = [cal] \times [deg]^{-1} \times [mass]^{-1}$.

the bar and that the heat transfer intensity is uniformly distributed in each section of the bar. Thus we may assume that $e = e(x, t)$, $r = r(x, t)$, with $0 \leq x \leq L$. Accordingly, the constitutive relations (3.5) and (3.7) read

$$e(x, t) = c_v u(x, t), \quad \mathbf{q} = -\kappa u_x \mathbf{i}.$$

By choosing $V = A \times [x, x + \Delta x]$ as the control volume in (3.3), the cross-sectional area A cancels out, and we obtain

$$\int_x^{x+\Delta x} c_v \rho u_t \, dx = \int_x^{x+\Delta x} \kappa u_{xx} \, dx + \int_x^{x+\Delta x} r\rho \, dx$$

that yields for u the one-dimensional heat equation

$$u_t - D u_{xx} = f.$$

We want to study the temperature evolution during an interval of time, say, from $t = 0$ until $t = T$. It is then reasonable to prescribe its initial distribution inside the bar: different initial configurations will correspond to different evolutions of the temperature along the bar. Thus we need to prescribe **the initial condition**

$$u(x, 0) = g(x)$$

where g models the initial temperature profile.

This is not enough to determine a unique evolution; it is necessary to know how the bar interacts with the surroundings. Indeed, starting with a given initial temperature distribution, we can change the evolution of u by controlling the temperature or the heat flux at the two ends of the bar[7]; for instance, we could keep the temperature at a certain fixed level or let it vary in a certain way, depending on time. This amounts to prescribing

$$u(0, t) = h_1(t), \quad u(L, t) = h_2(t) \tag{3.9}$$

at any time $t \in (0, T]$. The (3.9) are called **Dirichlet boundary conditions**.

We could also prescribe the heat flux at the end points. Since from Fourier law we have

inward heat flow at $x = 0 : -\kappa u_x(0, t)$

inward heat flow at $x = L : \kappa u_x(L, t)$

the heat flux is assigned through the **Neumann boundary conditions**

$$-u_x(0, t) = h_1(t), \quad u_x(L, t) = h_2(t)$$

at any time $t \in (0, T]$.

Another type of boundary condition is the **Robin** or **radiation condition**. Let the surroundings be kept at temperature U and assume that the

---

[7] Remember that the bar has perfect lateral thermal insulation.

inward heat flux from one end of the bar, say $x = L$, depends linearly on the difference $U - u$, that is[8]

$$\kappa u_x = \gamma(U - u) \qquad (\gamma > 0). \qquad (3.10)$$

Letting $\alpha = \gamma/\kappa > 0$ e $h = \gamma U/\kappa$, the Robin condition at $x = L$ reads

$$u_x + \alpha u = h.$$

Clearly, it is possible to assign **mixed conditions**: for instance, at one end a Dirichlet condition and at the other one a Neumann condition.

The problems associated with the above boundary conditions have a corresponding nomenclature. Summarizing, we can state the most common problems for the one-dimensional heat equation as follows: *given $f = f(x,t)$ (external source) and $g = g(x)$ (initial or Cauchy data), determine $u = u(x,t)$ such that:*

$$\begin{cases} u_t - D u_{xx} = f & 0 < x < L, 0 < t < T \\ u(x,0) = g(x) & 0 \le x \le L \\ + \text{ boundary conditions} & 0 < t \le T \end{cases}$$

where the boundary conditions may be:

- *Dirichlet:*
$$u(0,t) = h_1(t), \ u(L,t) = h_2(t).$$

- *Neumann:*
$$-u_x(0,t) = h_1(t), \ u_x(L,t) = h_2(t).$$

- *Robin or radiation:*
$$-u_x(0,t) + \alpha u(0,t) = h_1(t), \ u_x(L,t) + \alpha u(L,t) = h_2(t) \qquad (\alpha > 0),$$

or *mixed* conditions. Accordingly, we have the initial-Dirichlet problem, the initial-Neumann problem and so on. When $h_1 = h_2 = 0$, we say that the boundary conditions are **homogeneous**.

Remark 3.1. Observe that only a special part of the boundary of the rectangle

$$Q_T = (0,L) \times (0,T),$$

called the *parabolic boundary of Q_T*, carries the data (see Fig. 3.1). *No final condition (for $t = T, 0 < x < L$) is required.*

---

[8] Formula (3.10) is based on *Newton's law of cooling*: the heat loss from the surface of a body is a linear function of the temperature drop $U - u$ from the surroudings to the surface. It represents a good approximation to the radiative loss from a body when $|U - u|/u \ll 1$.

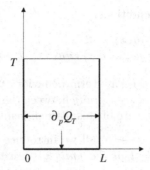

Fig. 3.1. The parabolic boundary of Q_T

In important applications, for instance in financial mathematics, x varies over unbounded intervals, typically $(0, \infty)$ or $\mathbb{R}$. In these cases one has to require that the solution does not grow too much at infinity. We will later consider the global Cauchy problem

$$\begin{cases} u_t - D u_{xx} = f & x \in \mathbb{R}, 0 < t < T \\ u(x, 0) = g(x) & x \in \mathbb{R} \\ + \text{ conditions as } x \to \pm\infty. \end{cases}$$

3.1.4 A solution by separation of variables

We will prove that under reasonable hypotheses the initial Dirichlet, Neumann or Robin problems are well posed. Sometimes this can be shown using elementary techniques like *the separation of variables method* that we describe below through a simple example of heat conduction.

As in the previous section, consider a bar (that we can consider one-dimensional) of length L, initially (at time $t = 0$) at constant temperature u_0. Thereafter, the end point $x = 0$ is kept at the same temperature while the other end $x = L$ is kept at a constant temperature $u_1 > u_0$. We want to know how the temperature evolves inside the bar.

Before making any computations, let us try to conjecture what could happen. Given that $u_1 > u_0$, heat starts flowing from the hotter end, raising the temperature inside the bar and causing a heat outflow into the cold boundary. On the other hand, the interior increase of temperature causes the hot inflow to decrease in time, while the outflow increases. We expect that sooner or later the two fluxes balance each other and that the temperature eventually reaches a steady state distribution. It would also be interesting to know how fast the steady state is reached.

We show that this is exactly the behavior predicted by our mathematical model, given by the heat equation

$$u_t - D u_{xx} = 0 \quad t > 0, 0 < x < L$$

with the initial-Dirichlet conditions

$$u(x,0) = g(x) \qquad\qquad 0 \le x \le L$$
$$u(0,t) = u_0,\ u(L,t) = u_1 \qquad t > 0.$$

Since we are interested in the long term behavior of our solution, we leave t unlimited. Notice the *jump discontinuity* between the initial and the boundary data at $x = L$; we will take care of this little difficulty later.

• *Dimensionless variables.* First of all we introduce dimensionless variables, that is variables *independent of the units of measurement.* To do that we rescale space, time and temperature with respect to quantities that are characteristic of our problem. For the space variable we can use the length L of the bar as rescaling factor, setting

$$y = \frac{x}{L}$$

which is clearly dimensionless, being a ratio of lengths. Notice that

$$0 \le y \le 1.$$

How can we rescale time? Observe that the dimensions of the diffusion coefficient D are

$$[length]^2 \times [time]^{-1}.$$

Thus the constant $\tau = L^2/D$ gives a characteristic time scale for our diffusion problem. Therefore we introduce the dimensionless time

$$s = \frac{t}{\tau}. \tag{3.11}$$

Finally, we rescale the temperature by setting

$$z(y,s) = \frac{u(Ly, \tau s) - u_0}{u_1 - u_0}.$$

For the dimensionless temperature z we have:

$$z(y,0) = \frac{u(Ly, 0) - u_0}{u_1 - u_0} = 0, \quad 0 \le y \le 1$$

$$z(0,s) = \frac{u(0, \tau s) - u_0}{u_1 - u_0} = 0, \quad z(1,s) = \frac{u(L, \tau s) - u_0}{u_1 - u_0} = 1.$$

Moreover

$$(u_1 - u_0)z_s = \frac{\partial t}{\partial s}u_t = \tau u_t = \frac{L^2}{D}u_t$$

$$(u_1 - u_0)z_{yy} = \left(\frac{\partial x}{\partial y}\right)^2 u_{xx} = L^2 u_{xx}.$$

Hence, since $u_t = Du_{xx}$,

$$(u_1 - u_0)(z_s - z_{yy}) = \frac{L^2}{D}u_t - L^2 u_{xx} = \frac{L^2}{D}Du_{xx} - L^2 u_{xx} = 0.$$

In conclusion, we find

$$z_s - z_{yy} = 0 \qquad (3.12)$$

with the initial condition

$$z(y, 0) = 0 \qquad (3.13)$$

and the boundary conditions

$$z(0, s) = 0, \quad z(1, s) = 1. \qquad (3.14)$$

We see that in the dimensionless formulation the parameters L and D have disappeared, emphasizing the mathematical essence of the problem. On the other hand, we will show later the relevance of the dimensionless variables in test modelling.

• *The steady state solution.* We start solving problem (3.12), (3.13), (3.14) by first determining the steady state solution z^{St}, that satisfies the equation $z_{yy} = 0$ and the boundary conditions (3.14). An elementary computation gives

$$z^{St}(y) = y.$$

In terms of the original variables the steady state solution is

$$u^{St}(x) = u_0 + (u_1 - u_0)\frac{x}{L}$$

corresponding to a uniform heat flux along the bar given by the Fourier law:

$$\text{heat flux} = -\kappa u_x = -\kappa \frac{(u_1 - u_0)}{L}.$$

• *The transient regime.* Knowing the steady state solution, it is convenient to introduce the function

$$U(y, s) = z^{St}(y, s) - z(y, s) = y - z(y, s).$$

Since we expect our solution to eventually reach the steady state, U represents a *transient regime* that should converge to zero as $s \to \infty$. Furthermore, the rate of convergence to zero of U gives information on how fast the temperature reaches its equilibrium distribution. U satisfies (3.12) with initial condition

$$U(y, 0) = y \qquad (3.15)$$

and *homogeneous* boundary conditions

$$U(0, s) = 0 \quad \text{and} \quad U(1, s) = 0. \qquad (3.16)$$

• *The method of separation of variables.* We are now in a position to find an explicit formula for U using the method of separation of variables. The main idea is to exploit the linear nature of the problem constructing the solution by superposition of simpler solutions of the form $w(s) v(y)$ in which the variables s and y appear in *separated form*.

Step 1. We look for non-trivial solutions of (3.12) of the form

$$U(y, s) = w(s) v(y)$$

with $v(0) = v(1) = 0$. By substitution in (3.12) we find

$$0 = U_s - U_{yy} = w'(s) v(y) - w(s) v''(y)$$

from which, separating the variables,

$$\frac{w'(s)}{w(s)} = \frac{v''(y)}{v(y)}. \tag{3.17}$$

Now, the left hand side in (3.17) is a function of s only, while the right hand side is a function of y only and the equality must hold for every $s > 0$ and every $y \in (0, L)$. This is possible only when both sides are equal to a common constant λ, say. Hence we have

$$v''(y) - \lambda v(y) = 0 \tag{3.18}$$

with

$$v(0) = v(1) = 0 \tag{3.19}$$

and

$$w'(s) - \lambda w(s) = 0. \tag{3.20}$$

Step 2. We first solve problem (3.18), (3.19). There are three different possibilities for the general solution of (3.18):

a) If $\lambda = 0$,

$$v(y) = A + By \qquad (A, B \text{ arbitrary constants})$$

and the conditions (3.19) imply $A = B = 0$.

b) If λ is a positive real number, say $\lambda = \mu^2 > 0$, then

$$v(y) = Ae^{-\mu y} + Be^{\mu y}$$

and again it is easy to check that the conditions (3.19) imply $A = B = 0$.

c) Finally, if $\lambda = -\mu^2 < 0$, then

$$v(y) = A \sin \mu y + B \cos \mu y.$$

From (3.19) we get

$$v(0) = B = 0$$
$$v(1) = A \sin \mu + B \cos \mu = 0$$

from which

$$A \text{ arbitrary}, \ B = 0, \ \mu_m = m\pi, \ m = 1, 2, \dots .$$

Thus, only in case c) we find non-trivial solutions

$$v_m(y) = A \sin m\pi y. \tag{3.21}$$

In this context, (3.18), (3.19) is called an *eigenvalue problem*; the special values λ_m are the *eigenvalues* and the solutions v_m are the corresponding *eigenfunctions*.
With $\lambda_m = -\mu_m^2 = -m^2\pi^2$, the general solution of (3.20) is

$$w_m(s) = Ce^{-m^2\pi^2 s} \qquad (C \text{ arbitrary constant}). \tag{3.22}$$

From (3.21) and (3.22) we obtain damped sinusoidal waves of the form

$$U_m(y, s) = A_m e^{-m^2\pi^2 s} \sin m\pi y.$$

Step 3. Although the solutions U_m satisfy the homogeneous Dirichlet conditions, they do not match, in general, the initial condition $U(y, 0) = y$. As we already mentioned, we try to construct the correct solution superposing the U_m by setting

$$U(y, s) = \sum_{m=1}^{\infty} A_m e^{-m^2\pi^2 s} \sin m\pi y. \tag{3.23}$$

Some questions arise:

Q1. The initial condition requires

$$U(y, 0) = \sum_{m=1}^{\infty} A_m \sin m\pi y = y \qquad \text{for } 0 \le y \le 1. \tag{3.24}$$

Is it possible to choose the coefficients A_m in order to satisfy (3.24)? In which sense does U attain the initial data? For instance, is it true that

$$U(z, s) \to y \quad \text{if} \quad (z, s) \to (y, 0)?$$

Q2. Any finite linear combination of the U_m is a solution of the heat equation; can we make sure that the same is true for U? The answer is positive if we could differentiate term by term the infinite sum and get

$$(\partial_s - \partial_{yy}^2)U\,(y,s) = \sum_{m=1}^{\infty}(\partial_s - \partial_{yy}^2)U_m\,(y,s) = 0. \tag{3.25}$$

What about the boundary conditions?

Q3. Even if we have a positive answer to questions 1 and 2, are we confident that U is the unique solution of our problem and therefore that it describes the correct evolution of the temperature?

Q1. Question 1 is rather general and concerns the *Fourier series expansion*[9] of a function, in particular of the initial data $f\,(y) = y$, in the interval $(0,1)$. Due to the homogeneous Dirichlet conditions it is convenient to expand $f\,(y) = y$ in a *sine Fourier series*, whose coefficients are given by the formulas

$$A_m = 2\int_0^1 y\sin m\pi y\,dy = -\frac{2}{m\pi}[y\cos m\pi y]_0^1 + \frac{2}{m\pi}\int_0^1 \cos m\pi y\,dy =$$

$$= -2\frac{\cos m\pi}{m\pi} = (-1)^{m+1}\frac{2}{m\pi}.$$

The sine Fourier expansion of $f\,(y) = y$ is therefore

$$y = \sum_{m=1}^{\infty}(-1)^{m+1}\frac{2}{m\pi}\sin m\pi y. \tag{3.26}$$

Where is the expansion (3.26) valid? It cannot be true at $y = 1$ since $\sin m\pi = 0$ for every m and we would obtain $1 = 0$. This clearly reflects the jump discontinuity of the data at $y = 1$.

The theory of Fourier series implies that (3.26) is true at every point $y \in [0,1)$ and that the series converges uniformly in every interval $[0,a]$, $a < 1$. Moreover, equality (3.26) holds **in the least square sense** (or $L^2\,(0,1)$ sense), that is

$$\int_0^1 [y - \sum_{m=1}^{N}(-1)^{m+1}\frac{2}{m\pi}\sin m\pi y]^2 dy \to 0 \qquad \text{as } N \to \infty.$$

From (3.23) and the expression of A_m, we obtain the *formal* solution

$$U\,(y,s) = \sum_{m=1}^{\infty}(-1)^{m+1}\frac{2}{m\pi}e^{-m^2\pi^2 s}\sin m\pi y \tag{3.27}$$

---

[9] Appendix A.

that attains the initial data in the least squares sense, i.e.[10].

$$\lim_{s \to 0^+} \int_0^1 [U(y, s) - y]^2 dy = 0. \qquad (3.28)$$

In fact, from Parseval's equality[11], we can write

$$\int_0^1 [U(y, s) - y]^2 \, dy = \frac{4}{\pi^2} \sum_{m=1}^{\infty} \frac{\left(e^{-m^2\pi^2 s} - 1\right)^2}{m^2}. \qquad (3.29)$$

Since for $s \geq 0$

$$\frac{\left(e^{-m^2\pi^2 s} - 1\right)^2}{m^2} \leq \frac{1}{m^2}$$

and the series $\sum 1/m^2$ converges, then the series (3.29) converges uniformly by Weierstrass criterion in $[0, \infty)$ and we can take the limit under the sum, obtaining (3.28).

Q2. The analytical expression of U is rather reassuring: it is a superposition of sinusoids of increasing frequency m and of strongly damped amplitude because of the negative exponential, at least when $s > 0$. Indeed, for $s > 0$, the rapid convergence to zero of each term and its derivatives in the series (3.27) allows us to differentiate term by term. Precisely, we have

$$\frac{\partial U_m}{\partial s} = (-1)^{m+2} 2m\pi e^{-m^2\pi^2 s} \sin m\pi y, \qquad \frac{\partial^2 U_m}{\partial y^2} = (-1)^{m+2} 2e^{-m^2\pi^2 s} \sin m\pi y$$

so that, if $s \geq s_0 > 0$,

$$\left| \frac{\partial U_m}{\partial s} \right| \leq 2m\pi e^{-m^2\pi^2 s_0}, \qquad \left| \frac{\partial^2 U_m}{\partial y^2} \right| \leq 2m\pi e^{-m^2\pi^2 s_0}.$$

Since the numerical series

$$\sum_{m=1}^{\infty} m e^{-m^2\pi^2 s_0}$$

is convergent, we conclude by the Weierstrass test that the series

$$\sum_{m=1}^{\infty} \frac{\partial U_m}{\partial s} \quad \text{and} \quad \sum_{m=1}^{\infty} \frac{\partial^2 U_m}{\partial y^2}$$

converge uniformly in $[0, 1] \times [s_0, \infty)$ so that (3.25) is true and therefore U is a solution of (3.12).

---

[10] It is also true that $U(z, s) \to y$ in the pointwise sense, when $y \neq 1$ and $(z, s) \to (y, 0)$. We omit the proof.
[11] Appendix A.

It remains to check the Dirichlet conditions: if $s_0 > 0$,

$$U(z, s) \to 0 \quad \text{as } (z, s) \to (0, s_0) \text{ or } (z, s) \to (L, s_0).$$

This is true because we can take the two limits under the sum, due to the uniform convergence of the series (3.27) in any region $[0, L] \times (b, +\infty)$ with $b > 0$. For the same reason, U has continuous derivatives of any order, up to the lateral boundary of the strip $[0, L] \times (b, +\infty)$.

Note, in particular, that U *immediately* forgets the initial discontinuity and becomes smooth at any positive time.

Q3. To show that U is indeed the unique solution, we use the so-called *energy method*, that we will develop later in greater generality. Suppose W is another solution of problem (3.12), (3.15), (3.16). Then, by linearity,

$$v = U - W$$

satisfies

$$v_s - v_{yy} = 0 \tag{3.30}$$

and has zero initial-boundary data. Multiplying (3.30) by v, integrating in y over the interval $[0, 1]$ and keeping $s > 0$, fixed, we get

$$\int_0^1 vv_s \, dy - \int_0^1 vv_{yy} \, dy = 0. \tag{3.31}$$

Observe that

$$\int_0^1 vv_s \, dy = \frac{1}{2} \int_0^1 \partial_s \left(v^2\right) dy = \frac{1}{2} \frac{d}{ds} \int_0^1 v^2 dy. \tag{3.32}$$

Moreover, integrating by parts we can write

$$\int_0^1 vv_{yy} \, dy = [v(1, s) v_y(1, s) - v(0, s) v_y(0, s)] - \int_0^1 (v_y)^2 \, dy \tag{3.33}$$

$$= -\int_0^1 (v_y)^2 \, dy$$

since $v(1, s) = v(0, s) = 0$. From (3.31), (3.32) and (3.33) we get

$$\frac{1}{2} \frac{d}{ds} \int_0^1 v^2 dy = -\int_0^1 (v_y)^2 \, dy \le 0 \tag{3.34}$$

and therefore, the *nonnegative* function

$$E(s) = \int_0^1 v^2(y, s) \, dy$$

is non-increasing. On the other hand, using (3.28) for v instead of U, we get

$$E(s) \to 0 \qquad \text{as } s \to 0$$

which forces $E(s) = 0$, for every $s > 0$. But $v^2(y,s)$ is nonnegative and continuous in $[0,1]$ if $s > 0$, so that it must be $v(y,s) = 0$ for every $s > 0$ or, equivalently, $U = W$.

• *Back to the original variables.* In terms of the original variables, our solution is expressed as

$$u(x,t) = u_0 + (u_1 - u_0)\frac{x}{L} - \sum_{m=1}^{\infty} (-1)^{m+1} \frac{2}{m\pi} e^{\frac{-m^2\pi^2 D}{L^2}t} \sin\frac{m\pi}{L}x.$$

This formula confirms our initial guess about the evolution of the temperature towards the steady state. Indeed, each term of the series converges to zero exponentially as $t \to +\infty$ and it is not difficult to show[12] that

$$u(x,t) \to u_0 + (u_1 - u_0)\frac{x}{L} \qquad \text{as } t \to +\infty.$$

Moreover, among the various terms of the series, the first one $(m = 1)$ decays much more slowly than the others and very soon it determines the main deviation of u from the equilibrium, *independently of the initial condition.* This leading term is the damped sinusoid

$$\frac{2}{\pi} e^{\frac{-\pi^2 D}{L^2}t} \sin\frac{\pi}{L}x.$$

In this mode there is a concentration of heat at $x = L/2$ where the temperature reaches its maximum amplitude $2\exp(-\pi^2 Dt/L^2)/\pi$. At time $t = L^2/D$ the amplitude decays to $2\exp(-\pi^2)/\pi \simeq 3.3 \times 10^{-5}$, about 0.005 per cent of its initial value. This simple calculation shows that to reach the steady state a time of order L^2/D is required, a fundamental fact in heat diffusion.

Not surprisingly, the scaling factor in (3.11) was exactly $\tau = L^2/D$. The dimensionless formulation is extremely useful in experimental modelling tests. To achieve reliable results, these models must reproduce the same characteristics at different scales. For instance, if our bar were an experimental model of a much bigger beam of length L_0 and diffusion coefficient D_0, to reproduce the same heat diffusion effects, we must choose material (D) and length (L) for our model bar such that

$$\frac{L^2}{D} = \frac{L_0^2}{D_0}.$$

Fig. 3.2 shows the solution of the dimensionless problem (3.12) associated to the initial conditions (3.15), (3.16) for $0 < t \leq 1$.

3.1.5 Problems in dimension $n > 1$

The formulation of the well posed problems in Section 3.1.3 can be easily generalized to any spatial dimension $n > 1$, in particular to $n = 2$ or $n = 3$.

---

[12] The Weierstrass test works here for $t \geq t_0 > 0$.

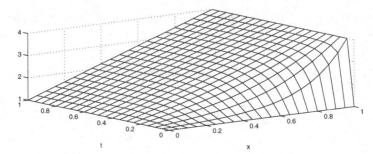

Fig. 3.2. The solution to the dimensionless problem (3.12), (3.13), (3.14)

Suppose we want to determine the evolution of the temperature in a heat conducting body that occupies a bounded domain[13] $\Omega \subset \mathbb{R}^n$, during an interval of time $[0,T]$. Under the hypotheses of Section 3.1.2, the temperature is a function $u = u(\mathbf{x},t)$ that satisfies the heat equation $u_t - D\Delta u = f$, in the *space-time cylinder*

$$Q_T = \Omega \times (0,T).$$

To select a unique solution we have to prescribe first of all the *initial distribution*

$$u(\mathbf{x},0) = g(\mathbf{x}) \quad \mathbf{x} \in \overline{\Omega},$$

where $\overline{\Omega} = \Omega \cup \partial\Omega$ denotes the *closure* of Ω.

The control of the interaction of the body with the surroundings is modeled through *suitable conditions* on $\partial\Omega$. The most common ones are:

Dirichlet condition. The temperature is kept at a prescribed level on $\partial\Omega$; this amounts to assigning

$$u(\boldsymbol{\sigma},t) = h(\boldsymbol{\sigma},t) \quad \boldsymbol{\sigma} \in \partial\Omega \text{ and } t \in (0,T].$$

Neumann condition. The heat flux through $\partial\Omega$ is assigned. To model this condition, we assume that the boundary $\partial\Omega$ is a smooth curve or surface, having a tangent line or plane at every point[14] with *outward* unit vector $\boldsymbol{\nu}$. From Fourier law we have

$$\mathbf{q} = \text{heat flux} = -\kappa\nabla u$$

so that the *inward heat flux* is

$$-\mathbf{q} \cdot \boldsymbol{\nu} = \kappa\nabla u \cdot \boldsymbol{\nu} = \kappa\partial_{\boldsymbol{\nu}} u.$$

---

[13] Recall that by *domain* we mean an *open connected set* in $\mathbb{R}^n$.

[14] We can also allow boundaries with corner points, like squares, cones, or edges, like cubes. It is enough that the set of points where the tangent plane does not exist has zero surface measure (zero length in two dimensions). Lipschitz domains have this property.

Thus the Neumann condition reads

$$\partial_\nu u\left(\boldsymbol{\sigma},t\right) = h\left(\boldsymbol{\sigma},t\right) \qquad \boldsymbol{\sigma} \in \partial\Omega \ \text{ and } \ t \in (0,T].$$

Radiation or Robin condition. The *inward* (say) heat flux through $\partial\Omega$ depends linearly on the difference[15] $U - u$:

$$-\mathbf{q} \cdot \boldsymbol{\nu} = \gamma\left(U - u\right) \qquad (\gamma > 0)$$

where U is the ambient temperature. From the Fourier law we obtain

$$\partial_\nu u + \alpha u = h \qquad \text{on } \partial\Omega \times (0,T]$$

with $\alpha = \gamma/\kappa > 0$, $h = \gamma U/\kappa$.

Mixed conditions. The boundary of Ω is decomposed into various parts where different boundary conditions are prescribed. For instance, a formulation of a mixed Dirichlet-Neumann problem is obtained by writing

$$\partial\Omega = \partial_D\Omega \cup \partial_N\Omega \quad \text{with} \quad \partial_D\Omega \cap \partial_N\Omega = \varnothing$$

with $\partial_D\Omega$ and $\partial_N\Omega$ "reasonable" subsets of $\partial\Omega$. Typically $\partial_N\Omega = \partial\Omega \cap A$, where A is open in $\mathbb{R}^n$. In this case we say that $\partial_N\Omega$ is a *relatively open* set in $\partial\Omega$. Then we assign

$$u = h_1 \text{ on } \partial_D\Omega \times (0,T]$$
$$\partial_\nu u = h_2 \text{ on } \partial_N\Omega \times (0,T].$$

Summarizing, we have the following typical problems: *given* $f = f\left(\mathbf{x},t\right)$ and $g = g\left(\mathbf{x}\right)$, *determine* $u = u\left(\mathbf{x},t\right)$ *such that:*

$$\begin{cases} u_t - D\Delta u = f & \text{in } Q_T \\ u\left(\mathbf{x},0\right) = g\left(\mathbf{x}\right) & \text{in } \overline{\Omega} \\ + \text{ boundary conditions on } \partial\Omega \times (0,T] \end{cases}$$

where the boundary conditions are:

• *Dirichlet:*
$$u = h.$$

• *Neumann:*
$$\partial_\nu u = h.$$

• *radiation or Robin:*
$$\partial_\nu u + \alpha u = h \qquad (\alpha > 0).$$

• *mixed:*
$$u = h_1 \text{ on } \partial_D\Omega, \qquad \partial_\nu u = h_2 \text{ on } \partial_N\Omega.$$

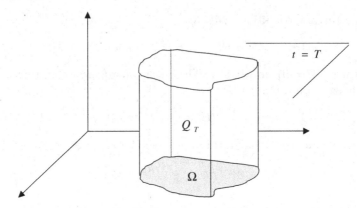

Fig. 3.3. The space-time cylinder Q_T

Also in dimension $n > 1$, the *global Cauchy problem* is important:

$$\begin{cases} u_t - D\Delta u = f & \mathbf{x} \in \mathbb{R}^n, 0 < t < T \\ u\,(\mathbf{x},0) = g\,(\mathbf{x}) & \mathbf{x} \in \mathbb{R}^n \\ + \text{ condition as } |\mathbf{x}| \to \infty. \end{cases}$$

Remark 3.2. We again emphasize that no final condition (for $t = T, \mathbf{x} \in \Omega$) is required. The data is assigned on the *parabolic boundary* $\partial_p Q_T$ of Q_T, given by the union of the bottom points $\bar{\Omega} \times \{t = 0\}$ and the side points $\partial\Omega \times (0,T]$:

$$\partial_p Q_T = \left(\bar{V} \times \{t = 0\}\right) \cup \left(\partial\Omega \times (0,T]\right).$$

3.2 Uniqueness

3.2.1 Integral method

Generalizing the energy method used in Section 3.1.4, it is easy to show that all the problems we have formulated in the previous section have at most one solution under reasonable conditions on the data. Suppose u and v are solutions of one of those problems, sharing the same boundary conditions, and let $w = u - v$; we want to show that $w \equiv 0$. For the time being we do not worry about the precise hypotheses on u e v; we assume they are sufficiently smooth in Q_T up to $\partial_p Q_T$ and observe that w satisfies the homogeneous equation

$$w_t - D\Delta w = 0 \tag{3.35}$$

in $Q_T = \Omega \times (0,T)$, with initial condition

$$w\,(\mathbf{x},0) = 0$$

---

[15] Linear Newton law of cooling.

in $\overline{\Omega}$, and one of the following conditions on $\partial\Omega \times (0,T]$:

$$w = 0 \qquad \text{(Dirichlet)} \tag{3.36}$$

or

$$\partial_\nu w = 0 \qquad \text{(Neumann)} \tag{3.37}$$

or

$$\partial_\nu w + \alpha w = 0 \qquad \alpha > 0, \qquad \text{(Robin)} \tag{3.38}$$

or

$$w = 0 \text{ on } \partial_D\Omega, \quad \partial_\nu w = 0 \text{ on } \partial_N\Omega \qquad \text{(mixed).} \tag{3.39}$$

Multiply equation (3.35) by w and integrate on Ω; we find

$$\int_\Omega ww_t \, d\mathbf{x} = D\int_\Omega w\Delta w \, d\mathbf{x}.$$

Now,

$$\int_\Omega ww_t \, d\mathbf{x} = \frac{1}{2}\frac{d}{dt}\int_\Omega w^2 d\mathbf{x} \tag{3.40}$$

and from Green's identity (1.12) with $u = v = w$,

$$\int_\Omega w\Delta w \, d\mathbf{x} = \int_{\partial\Omega} w\partial_\nu w \, d\sigma - \int_\Omega |\nabla w|^2 \, d\mathbf{x}. \tag{3.41}$$

Then, letting

$$E(t) = \int_\Omega w^2 d\mathbf{x},$$

(3.40) and (3.41) give

$$\frac{1}{2}E'(t) = D\int_{\partial\Omega} w\partial_\nu w \, d\sigma - D\int_\Omega |\nabla w|^2 \, d\mathbf{x}.$$

If Robin condition (3.38) holds,

$$\int_{\partial\Omega} w\partial_\nu w \, d\sigma = -\alpha\cdot\int_\Omega w^2 d\mathbf{x} \le 0.$$

If one of the (3.36), (3.37), (3.39) holds, then

$$\int_{\partial\Omega} w\partial_\nu w \, d\sigma = 0.$$

In any case it follows that

$$E'(t) \le 0$$

and therefore E is a nonincreasing function. Since

$$E(0) = \int_\Omega w^2(\mathbf{x},0) \, d\mathbf{x} = 0,$$

we must have $E(t) = 0$ for every $t \geq 0$ and this implies $w(\mathbf{x},t) \equiv 0$ in Ω for every $t > 0$. Thus $u = v$.

The above calculations are completely justified if Ω is a sufficiently smooth domain (C^1 or even Lipschitz domains) and, for instance, we require that u and v are continuous in $\overline{Q}_T = \overline{\Omega} \times [0, T]$, together with their first and second spatial derivatives and their first order time derivatives. We denote the set of these functions by the symbol (not too *appealing* ...)

$$C^{2,1}\left(\overline{Q}_T\right)$$

and synthesize everything in the following statement.

Theorem 3.1. *The initial-Dirichlet, Neumann, Robin and mixed problems have at most one solution belonging to $C^{2,1}\left(\overline{Q}_T\right)$.*

3.2.2 Maximum principles

The fact that heat flows from higher to lower temperature regions implies that a solution of the homogeneous heat equation attains its maximum and minimum values on $\partial_p Q_T$. This result is known as the *maximum principle*. Moreover the equation reflects the time irreversibility of the phenomena that it describes, in the sense that the future cannot have an influence on the past (*causality principle*). In other words, the value of a solution u at time t is independent of any change of the data after t.

The following simple theorem translates these principles and holds for functions in the class $C^{2,1}(Q_T) \cap C\left(\overline{Q}_T\right)$. These functions are continuous up to the boundary of Q_T, with derivatives continuous in the interior of Q_T.

Theorem 3.2. *Let $w \in C^{2,1}(Q_T) \cap C\left(\overline{Q}_T\right)$ such that*

$$w_t - D\Delta w = q \leq 0 \qquad in\ Q_T. \tag{3.42}$$

Then w attains its maximum on $\partial_p Q_T$:

$$\max_{\overline{Q}_T} w = \max_{\partial_p Q_T} w. \tag{3.43}$$

In particular, if w is negative on $\partial_p Q_T$, then is negative in all Q_T.

Proof. We recall that $\partial_p Q_T$ is the union of base and lateral boundary of Q_T (Remark 3.1). Let M' be the maximum of w on $\partial_p Q_T$ and assume that (3.43) is not true. Then there is a point $(\mathbf{x}_0, t_0)$, $\mathbf{x}_0 \in \Omega$, $0 < t_0 \leq T$, such that

$$w(\mathbf{x}_0, t_0) = \max_{\overline{Q}_T} w = M > M'.$$

We want to reach a contradiction. We break the proof into two steps.

Step 1. Let $\varepsilon > 0$ such that $T - \varepsilon > 0$. We prove that

$$\max_{\overline{Q}_{T-\varepsilon}} w \leq \max_{\partial_p Q_T} w + \varepsilon T. \tag{3.44}$$

Introduce the auxiliary function

$$v\left(\mathbf{x},t\right) = w\left(\mathbf{x},t\right) - \varepsilon t.$$

Then

$$u_t - D\Delta u = q - \varepsilon < 0. \tag{3.45}$$

Let us show that the maximum of u in $\overline{Q}_{T-\varepsilon}$ is attained at a point on $\partial_p Q_{T-\varepsilon}$. Suppose not. Then there exists a point $(\mathbf{x}_0, t_0)$, $\mathbf{x}_0 \in \Omega$, $0 < t_0 \leq T - \varepsilon$, at which u attains its maximum in $\overline{Q}_{T-\varepsilon}$.

Since $u_{x_j x_j}\left(\mathbf{x}_0, t_0\right) \leq 0$ for each $j = 1, ..., n$, we have

$$\Delta u\left(\mathbf{x}_0, t_0\right) \leq 0$$

and

$$u_t\left(\mathbf{x}_0, t_0\right) = 0 \qquad \text{if } t_0 < T - \varepsilon$$

while

$$u_t\left(\mathbf{x}_0, t_0\right) \geq 0 \qquad \text{if } t_0 = T - \varepsilon.$$

In both cases

$$w_t\left(\mathbf{x}_0, t_0\right) - \Delta w\left(\mathbf{x}_0, t_0\right) \geq 0,$$

contradicting (3.45). Thus

$$\max_{\overline{Q}_{T-\varepsilon}} u \leq \max_{\partial_p Q_{T-\varepsilon}} u \leq \max_{\partial_p Q_T} w$$

since $u \leq w$. On the other hand, $w \leq u + \varepsilon T$, so that

$$\max_{\overline{Q}_{T-\varepsilon}} w \leq \max_{\partial_p Q_{T-\varepsilon}} u + \varepsilon T \leq \max_{\partial_p Q_T} w + \varepsilon T \tag{3.46}$$

which is (3.44).

Step 2. Since w is continuous in $\overline{Q}_T$, we deduce that

$$\max_{\overline{Q}_{T-\varepsilon}} w \to \max_{\overline{Q}_T} w \qquad \text{as } \varepsilon \to 0.$$

By taking limits as $\varepsilon \to 0$ of both sides in (3.44) we get

$$\max_{\overline{Q}_T} w \leq \max_{\partial_p Q_T} w$$

which ends the proof. $\qquad\qquad\qquad\qquad\qquad\qquad\qquad\qquad\qquad\qquad\quad$ $\square$

As an immediate consequence of Theorem 3.2 we have that if

$$w_t - D\Delta w = 0 \qquad \text{in } Q_T$$

then w attains its maximum and its minimum on $\partial_p Q_T$. In particular

$$\min_{\partial_p Q_T} w \le w(\mathbf{x},t) \le \max_{\partial_p Q_T} w \qquad \text{for every } (\mathbf{x},t) \in Q_T.$$

Moreover:

Corollary 3.1 (Comparison and stability). *Let v and w satisfy*

$$v_t - D\Delta v = f_1 \qquad and \qquad w_t - D\Delta w = f_2.$$

Then:

1. *If $v \ge w$ on $\partial_p Q_T$ and $f_1 \ge f_2$ in Q_T then $v \ge w$ in all Q_T.*
2. *The following stability estimate holds*

$$\max_{\overline{Q}_T} |v - w| \le \max_{\partial_p Q_T} |v - w| + T \max_{\overline{Q}_T} |f_1 - f_2| . \tag{3.47}$$

In particular the initial-Dirichlet problem has at most one solution that, moreover, depends continuously on the data.

Remark 3.3. Corollary 3.1 gives uniqueness for the initial-Dirichlet problem under much less restrictive hypotheses than Theorem 3.1: indeed it does not require the continuity of any derivatives of the solution up to $\partial_p Q_T$.

Inequality (3.47) is a *uniform pointwise stability* estimate, extremely useful in several applications. In fact if $v = g_1$, $w = g_2$ on $\partial_p Q_T$ and

$$\max_{\partial_p Q_T} |g_1 - g_2| \le \varepsilon \quad \text{and} \quad \max_{\overline{Q}_T} |f_1 - f_2| \le \varepsilon,$$

we deduce

$$\max_{\overline{Q}_T} |v - w| \le \varepsilon (1 + T) .$$

Thus, in finite time, a small uniform distance between the data implies small uniform distance between the corresponding solutions.

Remark 3.4 (Strong maximum principle). Theorem 3.2 is a version of the so called weak maximum principle, weak because this result says nothing about the possibility that a solution achieves its maximum or minimum at an interior point as well. Actually a more precise result is known as *strong maximum principle* and states[16] that *if a solution of $u_t - D\Delta u = 0$ achieves its maximum M (minimum) at a point $(\mathbf{x}_1, t_1)$ with $x_1 \in V$, $0 < t_1 \le T$, then $u = M$ in $\bar{V} \times [0, t_1]$* (Fig. 3.4).

---

[16] We omit the rather long proof.

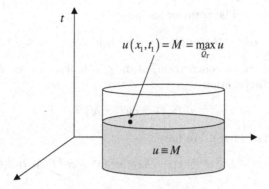

Fig. 3.4. The strong maximum principle

3.3 The Fundamental Solution

There are privileged solutions of the diffusion equation that can be used to construct many other solutions. In this section we are going to discover one of these special building blocks, the most important one.

3.3.1 Invariant transformations

The *homogeneous* diffusion equation has simple but important properties. Let $u = u(\mathbf{x}, t)$ be a solution of

$$u_t - D\Delta u = 0. \tag{3.48}$$

- *Time reversal.* The function

$$v(\mathbf{x},t) = u(\mathbf{x}, -t),$$

obtained by the change of variable $t \longmapsto -t$, is a solution of the **adjoint** or **backward** equation.

$$v_t + D\Delta v = 0.$$

Coherently, the (3.48) is sometimes called the **forward** equation. The non-invariance of (3.48) with respect to a change of sign in time is another aspect of time irreversibility.

- *Space and time translations invariance.* For $\mathbf{y},s$ fixed, the function

$$v(\mathbf{x},t) = u(\mathbf{x} - \mathbf{y}, t - s),$$

is still a solution of (3.48). Clearly, for $\mathbf{x}, t$ fixed the function $u(\mathbf{x} - \mathbf{y}, t - s)$ is a solution of the *backward* equation with respect to $\mathbf{y}$ and s.

• *Parabolic dilations* The transformation

$$\mathbf{x} \longmapsto a\mathbf{x}, \qquad t \longmapsto bt, \qquad u \longmapsto cu \qquad (a, b, c > 0)$$

represents a dilation (or contraction) of the graph of u. Let us check for which values of a, b, c the function

$$u^* (\mathbf{x},t) = cu\,(a\mathbf{x},bt)$$

is still a solution of (3.48). We have:

$$u_t^* (\mathbf{x},t) - D\Delta u^* (\mathbf{x},t) = cbu_t\,(a\mathbf{x},bt) - ca^2 D\Delta u\,(a\mathbf{x},bt)$$

and so u^* is a solution of (3.48) if

$$b = a^2. \tag{3.49}$$

The relation (3.49) suggests the name of *parabolic dilation* for the transformation

$$\mathbf{x} \longmapsto a\mathbf{x} \qquad t \longmapsto a^2 t \qquad (a, b > 0).$$

Under this transformation the expressions

$$\frac{|\mathbf{x}|^2}{Dt} \qquad \text{or} \qquad \frac{\mathbf{x}}{\sqrt{Dt}}$$

are left unchanged. Moreover, we already observed that they are a *dimensionless group*. Thus it is not surprising that these combinations of the independent variables occur frequently in the study of diffusion phenomena.

• *Dilations and conservation of mass* (or *energy*). Let $u = u\,(\mathbf{x}, t)$ be a solution of (3.48) in the half-space $\mathbb{R}^n \times (0, +\infty)$. Then we just checked that the function

$$u^* (\mathbf{x},t) = cu\,(a\mathbf{x},a^2 t) \qquad (a > 0)$$

is also a solution in the same set. Suppose u satisfies the condition

$$\int_{\mathbb{R}^n} u\,(\mathbf{x}, t)\,d\mathbf{x} = q \qquad \text{for every } t > 0. \tag{3.50}$$

If, for instance, u represents the concentration of a substance (density of mass), equation (3.50) states that the total mass is q at every time t. If u is a temperature, (3.50) says that the total internal energy is constant $(= q\rho c_v)$. We ask for which a, c the solution u^* still satisfies (3.50). We have

$$\int_{\mathbb{R}^n} u^* (\mathbf{x}, t)\,d\mathbf{x} = c \int_{\mathbb{R}^n} u\,(a\mathbf{x}, a^2 t)\,d\mathbf{x}.$$

Letting $\mathbf{y} = a\mathbf{x}$, so that $d\mathbf{y} = a^n d\mathbf{x}$, we find

$$\int_{\mathbb{R}^n} u^* (\mathbf{x}, t)\,d\mathbf{x} = ca^{-n} \int_{\mathbb{R}^n} u\,(\mathbf{y}, a^2 t)\,d\mathbf{y} = ca^{-n}$$

and for (3.50) to be satisfied we must have:

$$c = qa^n.$$

In conclusion, if $u = u(\mathbf{x},t)$ is a solution of (3.48) in the half-space $\mathbb{R}^n \times (0,+\infty)$ satisfying (3.50), the same is true for

$$u^*(\mathbf{x},t) = qa^n u(a\mathbf{x},a^2t). \tag{3.51}$$

3.3.2 Fundamental solution ($n = 1$)

We are now in position to construct our special solution, starting with dimension $n = 1$. To help intuition, think for instance of our solution as the concentration of a substance of total mass q and suppose we want to keep the total mass equal to q at any time.

We have seen that the combination of variables $x/\sqrt{Dt}$ is not only invariant with respect to parabolic dilations but also dimensionless. It is then natural to check if there are solutions of (3.48) involving such dimensionless group. Since $\sqrt{Dt}$ has the dimension of a length, the quantity $q/\sqrt{Dt}$ is a typical order of magnitude for the concentration, so that it makes sense to look for solutions of the form

$$u^*(x,t) = \frac{q}{\sqrt{Dt}} U\left(\frac{x}{\sqrt{Dt}}\right) \tag{3.52}$$

where U is a (dimensionless) function of a single variable.

Here is the main question: is it possible to determine $U = U(\xi)$ such that u^* is a solution of (3.48)? Solutions of the form (3.52) are called *similarity solutions*[17].

Moreover, since we are interpreting u^* as a concentration, we require $U \geq 0$ and the total mass condition yields

$$1 = \frac{1}{\sqrt{Dt}} \int_{\mathbb{R}} U\left(\frac{x}{\sqrt{Dt}}\right) dx \underset{\xi=x/\sqrt{Dt}}{=} \int_{\mathbb{R}} U(\xi) \, d\xi$$

so that we require that

$$\int_{\mathbb{R}} U(\xi) \, d\xi = 1. \tag{3.53}$$

Let us check if u^* is a solution to (3.48). We have

$$u_t^* = \frac{q}{\sqrt{D}}\left[-\frac{1}{2}t^{-\frac{3}{2}}U(\xi) - \frac{1}{2\sqrt{D}}xt^{-2}U'(\xi)\right] = -\frac{q}{2t\sqrt{Dt}}\left[U(\xi) + \xi U'(\xi)\right]$$

$$u_{xx}^* = \frac{q}{(Dt)^{3/2}}U''(\xi),$$

---

[17] A solution of a particular evolution problem is a *similarity* or *self-similar* solution if its spatial configuration (graph) remains similar to itself at all times during the evolution. In one space dimension, *self-similar* solutions have the general form

$$u(x,t) = a(t) F(x/b(t))$$

where, preferably, u/a and x/b are dimensionless quantity.

hence

$$u_t^* - Du_{xx}^* = -\frac{q}{t\sqrt{Dt}} \left\{ U''\left(\xi\right) + \frac{1}{2}\xi U'\left(\xi\right) + \frac{1}{2}U\left(\xi\right) \right\}.$$

We see that for u^* to be a solution of (3.48), U must be a solution in $\mathbb{R}$ of the ordinary differential equation

$$U''\left(\xi\right) + \frac{1}{2}\xi U'\left(\xi\right) + \frac{1}{2}U\left(\xi\right) = 0. \tag{3.54}$$

Since $U \geq 0$, (3.53) implies[18]:

$$U\left(-\infty\right) = U\left(+\infty\right) = 0.$$

On the other hand, (3.54) is invariant with respect to the change of variables

$$\xi \longmapsto -\xi$$

and therefore we look for *even solutions*: $U\left(-\xi\right) = U\left(\xi\right)$. Then we can restrict ourselves to $\xi \geq 0$, asking

$$U'\left(0\right) = 0 \text{ e } U\left(+\infty\right) = 0. \tag{3.55}$$

To solve (3.54) observe that it can be written in the form

$$\frac{d}{d\xi} \left\{ U'\left(\xi\right) + \frac{1}{2}\xi U\left(\xi\right) \right\} = 0$$

that yields

$$U'\left(\xi\right) + \frac{1}{2}\xi U\left(\xi\right) = C \qquad\qquad (C \in \mathbb{R}). \tag{3.56}$$

Letting $\xi = 0$ in (3.56) and recalling (3.55) we deduce that $C = 0$ and therefore

$$U'\left(\xi\right) + \frac{1}{2}\xi U\left(\xi\right) = 0. \tag{3.57}$$

The general integral of (3.57) is

$$U\left(\xi\right) = c_0 e^{-\frac{\xi^2}{4}} \qquad\qquad (c_0 \in \mathbb{R}).$$

This function is even, positive, integrable and vanishes at infinity. It only remains to choose c_0 in order to ensure (3.53). Since[19]

$$\int_{\mathbb{R}} e^{-\frac{\xi^2}{4}} d\xi \underset{\xi=2z}{=} 2\int_{\mathbb{R}} e^{-z^2} dz = 2\sqrt{\pi}$$

the choice is $c_0 = (4\pi)^{-1/2}$.

---

[18] Rigorously, the precise conditions are:

$$\liminf_{x \to \pm\infty} U\left(x\right) = 0.$$

[19] Recall that

$$\int_{\mathbb{R}} e^{-z^2} dz = \sqrt{\pi}.$$

Going back to the original variables, we have found the following solution of (3.48)

$$u^* (x,t) = \frac{q}{\sqrt{4\pi Dt}} e^{-\frac{x^2}{4Dt}}, \qquad x \in \mathbb{R}, t > 0$$

positive, even in x, and such that

$$\int_\mathbb{R} u^* (x,t)\, dx = q \qquad \text{for every } t > 0. \tag{3.58}$$

The choice $q = 1$ gives a family of *Gaussians*, parametrized with time, and it is natural to think of a *normal probability density*.

Definition 3.1. *The function*

$$\Gamma_D (x,t) = \frac{1}{\sqrt{4\pi Dt}} e^{-\frac{x^2}{4Dt}}, \qquad x \in \mathbb{R}, t > 0 \tag{3.59}$$

*is called the **fundamental solution** of equation (3.48).*

3.3.3 The Dirac distribution

It is worthwhile to examine the behavior of the fundamental solution. For every fixed $x \neq 0$,

$$\lim_{t \to 0^+} \Gamma_D (x,t) = \lim_{t \to 0^+} \frac{1}{\sqrt{4\pi Dt}} e^{-\frac{x^2}{4Dt}} = 0 \tag{3.60}$$

while

$$\lim_{t \to 0^+} \Gamma_D (0,t) = \lim_{t \to 0^+} \frac{1}{\sqrt{4\pi Dt}} = +\infty. \tag{3.61}$$

If we interpret Γ_D as a probability density, equations (3.60), (3.61) and (3.58) imply that when $t \to 0^+$ the fundamental solution tends to concentrate mass around the origin; eventually, the whole probability mass is concentrated at $x = 0$ (see Fig. 3.5).

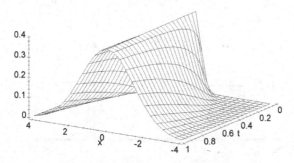

Fig. 3.5. The fundamental solution Γ_1 for $-4 < x < 4$, $0 < t < 1$

The limiting density distribution can be mathematically modeled by the so called *Dirac distribution* (or *measure*) *at the origin,* denoted by the symbol δ_0 or simply by δ. The Dirac distribution is not a function in the usual sense of Analysis; if it were, it should have the following properties:

- $\delta(0) = \infty$, $\delta(x) = 0$ for $x \neq 0$;
- $\int_{\mathbb{R}} \delta(x)\, dx = 1$,

clearly incompatible with any concept of classical function or integral. A rigorous definition of the Dirac measure requires the theory of *generalized functions* or *distributions of L. Schwartz,* that we will consider in Chapter 7. Here we restrict ourselves to some heuristic considerations.

Let

$$\mathcal{H}(x) = \begin{cases} 1 & \text{if } x \geq 0 \\ 0 & \text{if } x < 0, \end{cases}$$

be the characteristic function of the interval $[0, \infty)$, known as the Heaviside function. Observe that

$$\frac{\mathcal{H}(x+\varepsilon) - \mathcal{H}(x-\varepsilon)}{2\varepsilon} = \begin{cases} \frac{1}{2\varepsilon} & \text{if } -\varepsilon \leq x < \varepsilon \\ 0 & \text{otherwise.} \end{cases} \tag{3.62}$$

Denote by $I_\varepsilon(x)$ the quotient (3.62); the following properties hold:

i) For every $\varepsilon > 0$,

$$\int_{\mathbb{R}} I_\varepsilon(x)\, dx = \frac{1}{2\varepsilon} \times 2\varepsilon = 1.$$

We can interpret I_ε as a *unit impulse of extent* 2ε (Fig. 3.6).

ii)

$$\lim_{\varepsilon \downarrow 0} I_\varepsilon(x) = \begin{cases} 0 & \text{if } x \neq 0 \\ \infty & \text{if } x = 0. \end{cases}$$

iii) If $\varphi = \varphi(x)$ is a smooth function, vanishing outside a bounded interval, (a *test function*), we have

$$\int_{\mathbb{R}} I_\varepsilon(x)\, \varphi(x)\, dx = \frac{1}{2\varepsilon} \int_{-\varepsilon}^{\varepsilon} \varphi(x)\, dx \xrightarrow[\varepsilon \longrightarrow 0]{} \varphi(0).$$

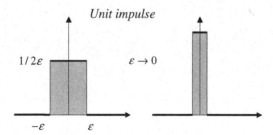

Unit impulse

$1/2\varepsilon$ $\varepsilon \to 0$

$-\varepsilon$ ε

Fig. 3.6. Approximation of the Dirac measure

Properties i) e ii) say that I_ε tends to a mathematical object that has precisely the formal features of the Dirac distribution at the origin. In particular iii) suggests how to identify this object, that is *through its action on test functions*.

Definition 3.2. *We call Dirac measure at the origin the generalized function, denoted by δ, that acts on a test function φ as follows*

$$\delta[\varphi] = \varphi(0). \tag{3.63}$$

Equation (3.63) is often written in the form $\langle \delta, \varphi \rangle = \varphi(0)$ or even

$$\int \delta(x)\,\varphi(x)\,dx = \varphi(0)$$

where the integral symbol is purely formal. Observe that property ii) shows that

$$\mathcal{H}' = \delta$$

whose meaning is given in the following computations, where an integration by parts is used and φ is a test function

$$\int_{\mathbb{R}} \varphi d\mathcal{H} = -\int_{\mathbb{R}} \mathcal{H}\varphi' = -\int_0^\infty \varphi' = \varphi(0), \tag{3.64}$$

since φ vanishes for large[20] x.

With the notion of Dirac measure at hand, we can say that Γ_D satisfies the initial conditions

$$\Gamma_D(x, 0) = \delta.$$

If the unit mass is concentrated at a point $y \neq 0$, we denote by δ_y or $\delta(x - y)$ the Dirac measure at y, defined through the formula

$$\int \delta(x - y)\,\varphi(x)\,dx = \varphi(y).$$

Then, by translation invariance, the fundamental solution $\Gamma_D(x - y, t)$ is a solution of the diffusion equation, that satisfies the initial condition

$$\Gamma_D(x - y, 0) = \delta(x - y).$$

Indeed it is the unique solution satisfying the total mass condition (3.58) with $q = 1$.

---

[20] The first integral in (3.64) is a Riemann-Stieltjes integral, that formally can be written as

$$\int \varphi(x)\,\mathcal{H}'(x)\,dx$$

and interpreted as *the action of the generalized function $\mathcal{H}'$ on the test function φ.*

As any solution u of (3.48) has several interpretations (concentration of a substance, probability density, temperature in a bar) so the fundamental solution can have several meanings.

We can think of it as a **unit source solution**: $\Gamma_D(x,t)$ gives the concentration at the point x at time t, generated by the diffusion of **a unit mass initially** $(t=0)$ **concentrated at the origin**. From another point of view, if we imagine a unit mass composed of a large number N of particles, $\Gamma_D(x,t)\,dx$ gives the probability that a single particle is placed between x and $x+dx$ at time t or equivalently, the percentage of particles inside the interval $(x,x+dx)$ at time t.

Initially Γ_D is zero outside the origin. As soon as $t>0$, Γ_D becomes positive everywhere: this amounts to saying that the unit mass diffuses instantaneously all over the $x-$axis and therefore with *infinite speed of propagation*. This could be a problem in using (3.48) as a realistic model, although (see Fig. 3.5) for $t>0$, small, Γ_D is practically zero outside an interval centered at the origin of length $4D$.

3.3.4 Pollution in a channel. Diffusion, drift and reaction

Let us go back to the simple model of pollution on the surface of a narrow channel, considered in Section 2.1. If c denotes the pollutant concentration the law of mass conservation leads to the equation

$$c_t = -q_x. \tag{3.65}$$

Let us examine the combination of *diffusion* and *drift*. If we adopt the *Fick law of diffusion*: we get for q the constitutive law

$$q(x,t) = vc(x,t) - Dc_x(x,t).$$

From (3.65) we infer that

$$c_t = Dc_{xx} - vc_x. \tag{3.66}$$

Since D and v are constant, it is easy to determine the evolution of a mass Q of pollutant, initially located at the origin (say). Its concentration is the solution of (3.66) with initial condition

$$c(x,0) = Q\delta(x)$$

where δ is the Dirac measure at the origin. To find an explicit formula, we can get rid of the drift term $-vc_x$ by setting

$$w(x,t) = c(x,t)\,e^{hx+kt}$$

with h,k to be chosen suitably. We have:

$$w_t = [c_t + kc]e^{hx+kt}$$
$$w_x = [c_x + hc]e^{hx+kt}, \qquad w_{xx} = [c_{xx} + 2hc_x + h^2c]e^{hx+kt}.$$

Using the equation $c_t = Du_{xx} - vc_x$, we can write

$$w_t - Dw_{xx} = e^{hx+kt}[c_t - Dc_{xx} - 2Dhc_x + (k - Dh^2)c] =$$
$$= e^{hx+kt}[(-v - 2Dh)c_x + (k - Dh^2)c].$$

Thus if we choose

$$h = -\frac{v}{2D} \quad \text{and} \quad k = \frac{v^2}{4D},$$

w is a solution of the diffusion equation $w_t - Dw_{xx} = 0$, with the initial condition

$$w(x,0) = c(x,0) e^{-\frac{v}{2D}x} = Q\delta(x) e^{-\frac{v}{2D}x}.$$

In Chapter 7 we show that $\delta(x) e^{-\frac{v}{2D}x} = \delta(x)$, so that $w(x,t) = Q\Gamma_D(x,t)$ and finally

$$c(x,t) = Qe^{\frac{v}{2D}\left(x-\frac{v}{2}t\right)}\Gamma_D(x,t). \tag{3.67}$$

The concentration c is thus given by the fundamental solution Γ_D, "carried" by the travelling wave $\exp\left\{\frac{v}{2D}\left(x - \frac{v}{2}t\right)\right\}$, in motion to the right with speed $v/2$.

In realistic situations, the pollutant undergoes some sort of decay, due for instance to biological decomposition. The resulting equation for the concentration becomes

$$c_t = Dc_{xx} - vc_x - \gamma c \tag{3.68}$$

where γ is a rate of decay[21].

It is useful to look separately at the effect of the three terms in the right hand side of (3.68).

- $c_t = Dc_{xx}$ models pure diffusion. The typical effects are spreading and smoothing, as shown by the typical behavior of the fundamental solution Γ_D.

- $c_t = bc_x$ is a transport equation, that we have considered in detail in Chapter 2. The solutions are travelling waves of the form $g(x + bt)$.

- $c_t = -\gamma c$ models linear reaction. The solutions are multiples of $e^{-\gamma t}$, exponentially decaying (increasing) if $\gamma > 0$ ($\gamma < 0$).

3.3.5 Fundamental solution ($n > 1$)

In space dimension greater than 1, we can more or less repeat the same arguments. We look for positive, radial, self-similar solutions u^* to (3.48), with total mass equal to q at every time, that is

$$\int_{\mathbb{R}^n} u^*(\mathbf{x},t)\,d\mathbf{x} = q \qquad \text{for every } t > 0. \tag{3.69}$$

---

[21] $[\gamma] = [time]^{-1}$.

Since $q/\left(Dt\right)^{n/2}$ is a concentration per unit volume, we set

$$u^{*}\cdot(\mathbf{x},t) = \frac{q}{(Dt)^{n/2}}U\left(\xi\right), \qquad \xi = |\mathbf{x}|/\sqrt{Dt}.$$

Arguing as in the one-dimensional case we have obtain solutions of the form

$$u^{*}\left(\mathbf{x},t\right) = \frac{q}{(4\pi Dt)^{n/2}}\exp\left(-\frac{|\mathbf{x}|^{2}}{4Dt}\right), \qquad (t>0).$$

Once more, the choice $q=1$ is special.

Definition 3.3. *The function*

$$\Gamma_{D}\left(\mathbf{x},t\right) = \frac{1}{(4\pi Dt)^{n/2}}\exp\left(-\frac{|\mathbf{x}|^{2}}{4Dt}\right) \qquad (t>0)$$

*is called the **fundamental solution** of the diffusion equation (3.48).*

It is also possible to define the n-dimensional Dirac measure at a point $\mathbf{y}$ through the formula[22]

$$\int \delta\left(\mathbf{x}-\mathbf{y}\right)\varphi\left(\mathbf{x}\right)dx = \varphi\left(\mathbf{y}\right) \tag{3.70}$$

that expresses the action on the *test function* φ, smooth in $\mathbb{R}^{n}$ and vanishing outside a *compact* set. For fixed $\mathbf{y}$, the fundamental solution $\Gamma_{D}\left(\mathbf{x}-\mathbf{y},t\right)$ is the unique solution of the global Cauchy problem

$$\begin{cases} u_{t} - D\Delta_{\mathbf{x}}u = 0 & \mathbf{x}\in\mathbb{R}^{n}, t>0 \\ u\left(\mathbf{x},0\right) = \delta\left(\mathbf{x}-\mathbf{y}\right) & \mathbf{x}\in\mathbb{R}^{n} \end{cases}$$

which satisfies (3.69) with $q=1$.

3.4 The Global Cauchy Problem $(n=1)$

3.4.1 The homogeneous case

In this section we consider the global Cauchy problem

$$\begin{cases} u_{t} - Du_{xx} = 0 & \text{in } \mathbb{R}\times(0,\infty) \\ u\left(x,0\right) = g\left(x\right) & \text{in } \mathbb{R} \end{cases} \tag{3.71}$$

where g, the *initial data,* is given. We will limit ourselves to the one-dimensional case; techniques, ideas and formulas can be extended without too much effort to the n-dimensional case.

---

[22] As in dimension $n=1$, in (3.70) the integral has a symbolic meaning only.

The problem (3.71) models the evolution of the temperature or of the concentration of a substance along a very long (infinite) bar or channel, respectively, given the initial $(t = 0)$ distribution.

By heuristic considerations, we can guess what could be a candidate solution. Consider a unit mass composed of a large number $M \gg 1$ of particles and interpret the solution u as their concentration (or percentage). Then, $u(x, t)\, dx$ gives the mass inside the interval $(x, x + dx)$ at time t.

We want to determine the concentration $u(x, y)$, due to the diffusion of a mass whose initial concentration is given by g.

Thus, the quantity $g(y)\, dy$ represents the mass concentrated in the interval $(y, y + dy)$ at time $t = 0$. As we have seen, $\Gamma(x - y, t)$ is a *unit source solution*, representing the concentration at x at time t, due to the diffusion of a unit mass, initially concentrated in the same interval. Accordingly,

$$\Gamma_D (x - y, t)\, g(y)\, dy$$

gives the concentration at x at time t, due to the diffusion of the mass $g(y)\, dy$.

Thanks to the linearity of the diffusion equation, we can use the *superposition principle* and compute the solution as the sum of all contributions. In this way, we get the formula

$$u(x, t) = \int_{\mathbb{R}} g(y)\, \Gamma_D (x - y, t)\, dy = \frac{1}{\sqrt{4\pi Dt}} \int_{\mathbb{R}} g(y)\, e^{-\frac{(x-y)^2}{4Dt}}\, dy. \qquad (3.72)$$

Clearly, one has to check rigorously that, under reasonable hypotheses on the initial data g, formula (3.72) really gives the unique solution of the Cauchy problem. This is not a negligible question. First of all, if g grows too much at infinity, more than an exponential of the type e^{ax^2}, $a > 0$, in spite of the rapid convergence to zero of the Gaussian, the integral in (3.72) could be divergent and formula (3.72) loses any meaning. Even more delicate is the question of the uniqueness of the solution, as we will see later.

3.4.2 Existence of a solution

The following theorem states that (3.72) is indeed a solution of the global Cauchy problem under rather general hypotheses on g, satisfied in most of the interesting applications[23].

Theorem 3.3. *Assume that g is a bounded function with a finite number of jump discontinuities in $\mathbb{R}$. Then:*

i) *the function (3.72) is well-defined, belongs to $C^\infty (\mathbb{R} \times (0, +\infty))$ and*

$$u_t - Du_{xx} = 0;$$

ii) *if x_0 is a point of continuity for g, then:*

$$u(y, t) \to g(x_0) \qquad if \ (y, t) \to (x_0, 0)\,, \ t > 0;$$

---

[23] We omit the long and technical proof.

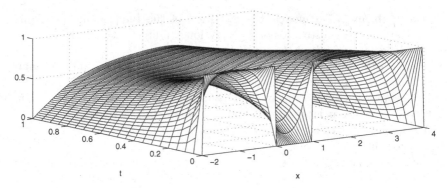

Fig. 3.7. Smoothing effect of the diffusion equation

iii)
$$|u\left(x,t\right)| \le \max_{\mathbb{R}} |g| \qquad \forall (x,t) \in \mathbb{R} \times \left(0,\infty\right).$$

Remark 3.5. The property *i)* shows a typical and important phenomenon connected with the diffusion equation: even if the initial data is discontinuous at some point, immediately after the solution is smooth. The diffusion is therefore a **smoothing process**. In Fig. 3.7, this phenomenon is shown for the initial data $g\left(x\right) = \chi_{(-2,0)}\left(x\right) + \chi_{(1,4)}\left(x\right)$, where $\chi_{(a,b)}$ denotes the characteristic function of the interval (a,b). By *ii)*, if the initial data g is continuous in all $\mathbb{R}$, then the solution is continuous up to $t = 0$, that is in $\mathbb{R} \times [0,T)$.

3.4.3 The non homogeneous case. Duhamel's method

Consider now the *non homogeneous* Cauchy problem

$$\begin{cases} v_t - Dv_{xx} = f\left(x,t\right) \text{ in } \mathbb{R} \times \left(0,T\right) \\ v\left(x,0\right) = 0 \qquad\qquad \text{ in } \mathbb{R} \end{cases} \qquad (3.73)$$

where f models a distributed source on the half-plane $t > 0$, capable to produce mass density at the time rate $f\left(x,t\right)$. Precisely, $f\left(x,t\right)dxdt$ is the mass produced[24] between x and $x + dx$, over the time interval $(t, t + dt)$. If initially no mass is present, we motivate the form of the solution at the point (x,t) using heuristic considerations. Let us compute the contribution dv to $v\left(x,t\right)$ of a mass $f\left(y,s\right)dyds$. It is like having a source term of the form

$$f^*\left(x,t\right) = f\left(x,t\right)\delta\left(x - y, t - s\right)$$

and therefore, we have

$$dv\left(x,t\right) = \Gamma_D\left(x - y, t - s\right)f\left(y,s\right)dyds. \qquad (3.74)$$

---

[24] Negative production ($f < 0$) means removal.

We obtain the solution $v(x,t)$ by superposition, summing all the contributions (3.74). We split it into the following two steps:

- we sum over y the contributions for fixed s, to get the total density at (x,t), due to the diffusion of mass produced at time s. The result is $w(x,t,s)\,ds$, where

$$w(x,t,s) = \int_{\mathbb{R}} \Gamma_D(x-y,t-s)\,f(y,s)\,dy; \tag{3.75}$$

- we sum the above contributions for s ranging from 0 to t:

$$v(x,t) = \int_0^t \int_{\mathbb{R}} \Gamma_D(x-y,t-s)\,f(y,s)\,dy\,ds.$$

The above construction is an example of application of the *Duhamel method*, that we state below:

Duhamel's method. *The procedure to solve problem (3.73) consists in the following two steps:*

1. *Construct a family of solutions of homogeneous Cauchy problems, with variable initial time $s > 0$, and initial data $f(x,s)$.*
2. *Integrate the above family with respect to s, over $(0,t)$.*

Indeed, let us examine the two steps.

1. Consider the homogeneous Cauchy problems

$$\begin{cases} w_t - Dw_{xx} = 0 & x \in \mathbb{R},\ t > s \\ w(x,s,s) = f(x,s) & x \in \mathbb{R} \end{cases} \tag{3.76}$$

where the initial time s plays the role of a parameter.

The function $\Gamma^{y,s}(x,t) = \Gamma_D(x-y,t-s)$ is the fundamental solution of the diffusion equation that satisfies for $t = s$, the initial condition

$$\Gamma^{y,s}(x,s) = \delta(x-y).$$

Hence, the solution of (3.76) is given by the function (3.75)

$$w(x,t,s) = \int_{\mathbb{R}} \Gamma_D(x-y,t-s)\,f(y,s)\,dy.$$

Thus, $w(x,t,s)$ is the required family.

2. Integrating w over $(0,t)$ with respect to s, we find

$$v(x,t) = \int_0^t w(x,t,s)\,ds = \int_0^t \int_{\mathbb{R}} \Gamma_D(x-y,t-s)\,f(y,s)\,dy\,ds. \tag{3.77}$$

Using (3.76) we have

$$v_t - Dv_{xx} = w(x,t,t) + \int_0^t [w_t(x,t,s) - Dw_{xx}(x,t,s)]\,ds = f(x,t).$$

Moreover, $v(x,0) = 0$ and therefore v is a solution to (3.73).

Everything works under rather mild hypotheses on f. More precisely:

Theorem 3.4. *If f and its derivatives f_t, f_x, f_{xx} are continuous and bounded in $\mathbb{R} \times [0,T)$, then (3.77) gives a solution v of problem (3.73) in $\mathbb{R} \times (0,T)$, continuous up to $t = 0$, with derivatives v_t, v_x, v_{xx} continuous in $\mathbb{R} \times (0,T)$.*

The formula for the general Cauchy problem

$$\begin{cases} u_t - Du_{xx} = f(x,t) & \text{in } \mathbb{R} \times (0,T) \\ u(x,0) = g(x) & \text{in } \mathbb{R} \end{cases} \tag{3.78}$$

is obtained by superposition of (3.72) and (3.77)

$$u(x,t) = \int_{\mathbb{R}} \Gamma_D(x-y,t) g(y)\, dy + \int_0^t \int_{\mathbb{R}} \Gamma(x-y,t-s) f(y,s)\, dy ds. \tag{3.79}$$

Under the hypotheses on f and g stated in Theorems 3.3 and 3.4, (3.79) is a solution of (3.78) in $\mathbb{R} \times (0,\infty)$ continuous with its derivatives u_t, u_x, u_{xx}.

The initial condition means that $u(x,t) \to g(x_0)$ as $(x,t) \to (x_0,0)$ at any point x_0 of continuity of g. In particular, if g is continuous in $\mathbb{R}$ then u is continuous in $\mathbb{R} \times [0,T)$.

Moreover, if f is as in Theorem 3.4 and

$$v(x,t) = \int_0^t \int_{\mathbb{R}} \Gamma_D(x-y,t-s) f(y,s)\, dy ds,$$

we easily get the estimate

$$t \inf_{\mathbb{R}} f \leq v(x,t) \leq t \sup_{\mathbb{R}} f, \tag{3.80}$$

for every $x \in \mathbb{R}$, $0 \leq t \leq T$. In fact:

$$v(x,t) \leq \sup_{\mathbb{R}} f \int_0^t \int_{\mathbb{R}} \Gamma_D(x-y,t-s)\, dy ds = t \sup_{\mathbb{R}} f$$

since

$$\int_{\mathbb{R}} \Gamma_D(x-y,t-s)\, dy = 1$$

for every x,t,s, $t > s$. In the same way it can be shown that $v(x,t) \geq t \inf_{\mathbb{R}} f$. As a consequence, we have:

Corollary 3.2 (Uniqueness). *Let g be continuous and bounded in $\mathbb{R}$ and let f be as in Theorem 3.4. Then the Cauchy problem (3.78) has a unique bounded solution u in $\mathbb{R} \times (0,T)$. This solution is given by (3.79) and moreover*

$$\inf_{\mathbb{R}} g + t \inf_{\mathbb{R}} f \leq u(x,t) \leq \sup_{\mathbb{R}} g + t \sup_{\mathbb{R}} f. \tag{3.81}$$

Proof. If u and v are solutions of the same Cauchy problem (3.78), then $w = u - v$ is a solution of (3.78) with $f = g = 0$ and satisfies the hypotheses of Corollary 3.2. It follows that $w(x, t) \equiv 0$.   □

• *Stability and comparison.* The inequality (3.81) is a stability estimate for the correspondence

$$data \longmapsto solution.$$

Indeed, let u_1 and u_2 be solutions of (3.78) with data g_1, f_1 and g_2, f_2, respectively. Under the hypotheses of Corollary 3.2, from (3.81) we can write

$$\sup_{\mathbb{R}\times[0,T]} |u_1 - u_2| \leq \sup_{\mathbb{R}} |g_1 - g_2| + T \sup_{\mathbb{R}\times[0,T]} |f_1 - f_2|.$$

Therefore if

$$\sup_{\mathbb{R}\times[0,T]} |f_1 - f_2| \leq \varepsilon, \quad \sup_{\mathbb{R}} |g_1 - g_2| \leq \varepsilon$$

also

$$\sup_{\mathbb{R}\times[0,T]} |u_1 - u_2| \leq \varepsilon(1 + T)$$

that means *uniform pointwise stability*.

This is not the only consequence of (3.81). We can use it to compare two solutions. For instance, from the left inequality we immediately deduce that if $f \geq 0$ and $g \geq 0$, also $u \geq 0$.

Similarly, if $f_1 \geq f_2$ and $g_1 \geq g_2$, then

$$u_1 \geq u_2.$$

• *Backward equations* arise in several applied contexts, from *control theory* and *dynamic programming* to *probability* and *finance*. An example is the celebrated *Black–Scholes equation*.

Due to the time irreversibility, to have a well posed problem for the backward equation in the time interval $[0, T]$ we must prescribe a *final condition*, that is for $t = T$, rather than an initial one. On the other hand, the change of variable $t \longmapsto T - t$ transforms the backward into the forward equation, so that, from the mathematical point of view, the two equations are equivalent. Except for this remark the theory we have developed so far remains valid.

3.5 An example of Nonlinear diffusion. The porous medium equation

All the mathematical models we have examined so far are *linear*. On the other hand, the nature of most real problems is nonlinear. For example, *nonlinear diffusion* has to be taken into account in filtration problems, *non linear drift* terms are quite important in fluid dynamics while *nonlinear reaction* terms occur frequently in population dynamics and kinetics chemistry.

The presence of a nonlinearity in a mathematical model gives rise to many interesting phenomena that cannot occur in the linear case; typical instances are finite speed of diffusion, finite time blow-up or existence of travelling wave solutions of certain special profiles, each one with its own characteristic velocity.

In this section we try to convey some intuition of what could happen in a typical and important example from filtration through a porous medium and population dynamics.

Consider a gas of density $\rho = \rho(\mathbf{x}, t)$ flowing through a porous medium. Denote by $\mathbf{v} = \mathbf{v}(\mathbf{x}, t)$ the velocity of the gas and by κ the *porosity* of the medium, representing the volume fraction filled with gas. Conservation of mass reads, in this case:

$$\kappa \rho_t + \operatorname{div}(\rho \mathbf{v}) = 0. \tag{3.82}$$

Besides (3.82), the flow is governed by the two following constitutive (empirical) laws.

• **Darcy's law:**

$$\mathbf{v} = -\frac{\mu}{\nu} \nabla p \tag{3.83}$$

where $p = p(\mathbf{x}, t)$ is the pressure, μ is the *permeability* of the medium and ν is the *viscosity* of the gas. We assume μ and ν are positive constants.

• **Equation of state:**

$$p = p_0 \rho^\alpha \qquad p_0 > 0, \alpha > 0. \tag{3.84}$$

From (3.83) and (3.84) we have, since $p^{1/\alpha} \nabla p = (1 + 1/\alpha)^{-1} \Delta(p^{1+1/\alpha})$,

$$\operatorname{div}(\rho \mathbf{v}) = -\frac{\mu}{(1 + 1/\alpha)\nu p_0^{1/\alpha}} \Delta(p^{1+1/\alpha}) = -\frac{(m-1)\mu p_0}{m\nu} \Delta(\rho^m)$$

where $m = 1 + \alpha > 1$. From (3.82) we obtain

$$\rho_t = \frac{(m-1)\mu p_0}{\kappa m \nu} \Delta(\rho^m).$$

Rescaling time $(t \mapsto \dfrac{(m-1)\mu p_0}{\kappa m \nu} t)$ we finally get the **porous medium equation**

$$\rho_t = \Delta(\rho^m). \tag{3.85}$$

Since

$$\Delta(\rho^m) = \operatorname{div}\left(m\rho^{m-1}\nabla\rho\right)$$

we see that the diffusion coefficient is $D(\rho) = m\rho^{m-1}$, showing that the diffusive effect increases with the density.

The porous medium equation can be written in terms of the pressure variable

$$u = p/p_0 = \rho^{m-1}.$$

It is not difficult to check that the equation for u is given by

$$u_t = u \Delta u + \frac{m}{m-1} |\nabla u|^2 \qquad (3.86)$$

showing once more the dependence on u of the diffusion coefficient.

One of the basic questions related to the equation (3.85) or (3.86) is to understand how an initial data ρ_0, confined in a small region Ω, evolves with time. The key object to examine is therefore the unknown boundary $\partial\Omega$ (the *free boundary*) of the gas, whose speed of expansion we expect to be proportional to $|\nabla p|$ (from (3.83)). This means that we expect a *finite speed of propagation*, in contrast with the classical case $m = 1$.

The porous media equation cannot be treated by elementary means, since at very low density the diffusion has a very low effect and the equation degenerates. However we can get some clue of what happens by examining a sort of fundamental solutions, the so called *Barenblatt solutions*, in spatial dimension 1.

The equation is

$$\rho_t = (\rho^m)_{xx}. \qquad (3.87)$$

We look for *nonnegative self-similar* solutions of the form

$$\rho(x,t) = t^{-\alpha} U\left(xt^{-\beta}\right) \equiv t^{-\alpha} U(\xi)$$

satisfying

$$\int_{-\infty}^{+\infty} \rho(x,t)\, dx = 1.$$

This condition requires

$$1 = \int_{-\infty}^{+\infty} t^{-\alpha} U\left(xt^{-\beta}\right) dx = t^{\beta - \alpha} \int_{-\infty}^{+\infty} U(\xi)\, d\xi$$

so that we must have $\alpha = \beta$ and $\int_{-\infty}^{+\infty} U(\xi)\, d\xi = 1$. Substituting into (3.87), we find

$$\alpha t^{-\alpha-1}(-U - \xi U') = t^{-m\alpha - 2\alpha}(U^m)''.$$

Thus, if we choose $\alpha = 1/(m+1)$, we get for U the differential equation

$$(m+1)(U^m)'' + \xi U' + U = 0$$

that can be written in the form

$$\frac{d}{d\xi}\left[(m+1)(U^m)' + \xi U\right] = 0.$$

Thus, we have

$$(m + 1) (U^m)' + \xi U = \text{constant}.$$

Choosing the constant equal to zero, we get

$$(m + 1) (U^m)' = (m + 1) m U^{m-1} U' = -\xi U$$

or

$$(m + 1) m U^{m-2} U' = -\xi.$$

This in turn is equivalent to

$$\frac{(m + 1) m}{m - 1} (U^{m-1})' = -\xi$$

whose solution is

$$U(\xi) = \left[A - B_m \xi^2 \right]^{1/(m-1)}$$

where A is an arbitrary constant and $B_m = (m - 1)/2m(m + 1)$. Clearly, to have a physical meaning, we must have $A > 0$ and $A - B_m \xi^2 \geq 0$.

In conclusion we have found solutions of the porous medium equation of the form

$$\rho(x, t) = \begin{cases} \dfrac{1}{t^\alpha} \left[A - B_m \dfrac{x^2}{t^{2\alpha}} \right]^{1/(m-1)} & \text{if } x^2 \leq A t^{2\alpha}/B_m \\ 0 & \text{if } x^2 > A t^{2\alpha}/B_m \end{cases} \qquad (\alpha = 1/(m+1))$$

known as *Barenblatt solutions*. The points

$$x = \pm \sqrt{A/B_m}\, t^\alpha \equiv \pm r(t)$$

represent the gas interface between the part filled by gas and the empty part. Its speed of propagation is therefore

$$\dot{r}(t) = \alpha \sqrt{A/B_m}\, t^{\alpha-1}.$$

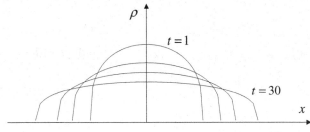

Fig. 3.8. The Barenblatt solution $\rho(x, t) = t^{-1/5} \left[1 - x^2 t^{-2/5} \right]_+^{1/3}$ for $t = 1, 4, 10, 30$

3.6 Numerical methods

3.6.1 Finite difference approximation of the heat equation

Let us consider the heat equation in $x \in (0,1)$, $t \in \mathbb{R}^+$ complemented with mixed type boundary conditions,

$$
\begin{cases}
u_t - u_{xx} = f & 0 < x < 1, \ t > 0 \\
u(0,t) = 0, \ u_x(1,t) = 0 & t > 0 \\
u(x,0) = u_0(x) & 0 \le x \le 1.
\end{cases} \tag{3.88}
$$

To build up a finite difference approximation scheme we proceed as in Chapter 2. After defining a uniform partition of the domain, characterized by spatial nodes x_i and time levels t^n,

$$
x_i = i\,h \text{ with } h = \frac{1}{N} \text{ ed } i, N \in \mathbb{N}, \quad t^n = n\,\tau \text{ with } n \in \mathbb{N},
$$

we introduce the following centered approximation for the second derivative of u on a generic node x_i,

$$
u_{xx}(x_i,t) = \frac{1}{h^2}\big(u(x_{i+1},t) - 2u(x_i,t) + u(x_{i-1},t)\big) + \mathcal{O}(h^2). \tag{3.89}
$$

For the discretization of the Neumann boundary condition on $x_N = 1$, it is convenient to use a one-sided approximation. More precisely, to be coherent with the accuracy of (3.89), we use the following second order approximation

$$
u_x(x_N,t) = \frac{1}{2h}\big(-3u(x_{N-2},t) + 2u(x_{N-1},t) - u(x_N,t)\big) + \mathcal{O}(h^2). \tag{3.90}
$$

Both (3.89) and (3.90) could be easily verified by means of Taylor expansions.

Now, let $u_i(t)$ be a discrete function in space approximating $u(x_i,t)$. By rewriting (3.89) and (3.90) in terms of $u_i(t)$ and replacing into (3.88), we obtain,

$$
\begin{cases}
\dot{u}_i(t) - \frac{1}{h^2}\big(u_{i-1}(t) - 2u_i(t) + u_{i-1}(t)\big) = f(x_i,t), & i = 1, N-1 \\
u_0(t) = 0, \\
\frac{1}{2h}\big(-3u_{N-2}(t) + 2u_{N-1}(t) - u_N(t)\big) = 0, \\
u_i(0) = u_0(x_i), & i = 1, N+1
\end{cases} \tag{3.91}
$$

which is a **system of ordinary differential equations** for $u_i(t)$, also known as **semi-discrete problem**, because it represents an intermediate step where only the space variable has been discretized. We notice that problem (3.91) can be equivalently formulated in matrix form. Precisely, let $\mathbf{U}(t) = \{u_i(t)\}_{i=1}^{N} \in \mathbb{R}^N$ be the vector of the nodal unknowns (from which we have subtracted the node x_0, because the value $u(t,x_0)$ is known), let $\mathbf{A}_h^M$ be the discrete counterpart of the operator $-\partial_{xx}$ (where the apex M reminds that we address a problem with mixed Dirichlet/Neumann boundary conditions) and $\mathbf{F}(t)$ be the discrete right hand side, defined respectively as follows

$$
\mathbf{A}_h^M = \frac{1}{h^2}
\begin{bmatrix}
2 & -1 & 0 & \cdots & & \cdots & 0 \\
-1 & 2 & -1 & 0 & \cdots & & \cdots & 0 \\
0 & -1 & 2 & -1 & 0 & \cdots & \cdots & 0 \\
\vdots & & \ddots & \ddots & \ddots & & & \vdots \\
\vdots & & & \ddots & \ddots & \ddots & & \vdots \\
0 & \cdots & & & \cdots & -1 & 2 & -1 & 0 \\
0 & \cdots & & & & \cdots & -1 & 2 & -1 \\
0 & \cdots & & & & \cdots & -\frac{3}{2} & 2 & -\frac{1}{2}
\end{bmatrix}
, \quad
\mathbf{F}(t) =
\begin{bmatrix}
f(x_1, t) \\
\vdots \\
\\
\\
\\
\vdots \\
f(x_{N-1}, t) \\
0
\end{bmatrix}
. \quad (3.92)
$$

Then, problem (3.91) can be easily rewritten as

$$
\dot{\mathbf{U}}(t) + \mathbf{A}_h^M \mathbf{U}(t) = \mathbf{F}(t), \quad \mathbf{U}(0) = \{u_i(0)\}_{i=1}^N. \quad (3.93)
$$

From (3.93), to end up with a fully discrete scheme we address Euler type discretization in time. It consists in combining the following first order approximation of the time derivative at a reference time $t^* \in [t^n, t^{n+1}]$,

$$
u_t(x_i, t^*) = \frac{1}{\tau}\left(u(x_i, t^{n+1}) - u(x_i, t^n)\right) + \mathcal{O}(\tau), \quad \dot{\mathbf{U}}(t^*) = \frac{1}{\tau}\left(\mathbf{U}(t^{n+1}) - \mathbf{U}(t^n)\right)
$$
$$(3.94)$$

with the aforementioned space discretization of $u_{xx}(x_i, t^*)$. If the reference time is selected as $t^* = t^n$, then the difference quotient (3.94) can be seen as a **forward** approximation of $u_t(x_i, t^*)$, while choosing $t^* = t^{n+1}$ leads to a

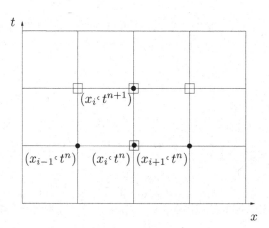

Fig. 3.9. Finite difference computational grid for the heat equation, where the nodes involved in the forward Euler (• marker) and backward Euler schemes (□ marker) have been put into evidence

backward approximation scheme. By this way, we obtain the **forward and backward Euler** schemes, respectively given by

$$\frac{1}{\tau}(\mathbf{U}_{n+1} - \mathbf{U}_n) + \mathbf{A}_h^M \begin{pmatrix} \mathbf{U}_n \\ \mathbf{U}_{n+1} \end{pmatrix} = \begin{pmatrix} \mathbf{F}(t^n) \\ \mathbf{F}(t^{n+1}) \end{pmatrix}. \qquad (3.95)$$

Then, considering for simplicity the homogeneous heat equation with $\mathbf{F}(t) = 0$, given the initial state $\mathbf{U}_0 = \mathbf{U}(0)$, the fully discrete scheme consists to find a sequence of vectors $\mathbf{U}_n$, by solving the following linear system of equations

$$\mathbf{U}_{n+1} = \mathbf{C}_h^\tau \mathbf{U}_n \quad \text{where} \quad \begin{cases} \mathbf{C}_h^\tau = I - \tau \mathbf{A}_h^M & \text{forward Euler (explicit)} \\ \mathbf{C}_h^\tau = (I + \tau \mathbf{A}_h^M)^{-1} & \text{backward Euler (implicit).} \end{cases}$$
$$(3.96)$$

We immediately notice that the forward scheme gives rise to an **explicit** set of equations, where updating the solution from t^n to t^{n+1} is obtained by multiplying U_n for a given matrix $\mathbf{C}_h^\tau$. Conversely, the backward Euler scheme is said to be **implicit**, because to determine $\mathbf{U}_{n+1}$ from $\mathbf{U}_n$ we need to solve a linear system of equations governed by the matrix $\mathbf{C}_h^\tau = I + \tau \mathbf{A}_h^M$. As we will see in the next section, this is a substantial difference that significantly affects the stability properties of the schemes.

3.6.2 Stability analysis for Euler methods

Let us consider for simplicity the homogeneous Cauchy-Dirichlet problem, namely the heat equation complemented with $u(0,t) = u(1,t) = 0$ and $f = 0$, whose discretization by means of finite differences leads to the matrix $\mathbf{A}_h \in \mathbb{R}^{(N-1)\times(N-1)}$ obtained removing the $N - th$ row and column from $\mathbf{A}_h^M \in \mathbb{R}^{N \times N}$.

The exact solution of the Cauchy-Dirichlet problem is such that

$$\lim_{t \to \infty} \max_{x \in (0,1)} |u(x,t)| = 0 \qquad (3.97)$$

and for Euler type schemes, ensuring stability consists in enforcing that the discrete solution $\mathbf{U}_n$ satisfies an equivalent property at the discrete level,

$$\lim_{n \to \infty} \|\mathbf{U}_n\|_\infty = 0, \qquad (3.98)$$

where $\|\cdot\|_\infty$ denotes the maximum norm, namely $\|\mathbf{U}\|_\infty = \max_{i=1,\dots,N-1} |u_i|$. We refer to Appendix C and in particular to (3.98) for a more rigorous definition of stability.

For both forward and backward Euler schemes we observe that (3.96) implies $\mathbf{U}_n = (\mathbf{C}_h^\tau)^n \mathbf{U}_0$. As a consequence, $(\mathbf{C}_h^\tau)^n \to 0$ as $n \to \infty$ is a sufficient condition to ensure that $\|\mathbf{U}_n\|_\infty \to 0$. Then, stability substantially depends on the spectrum of the iteration matrix $\mathbf{C}_h^\tau$, which is mainly determined by the one of $\mathbf{A}_h$. The following properties are useful to analyze the stability of Euler schemes.

Theorem 3.5. *The matrix* $\mathbf{A}_h \in \mathbb{R}^{N-1 \times N-1}$ *is symmetric and positive definite. The spectrum of* $\mathbf{A}_h$ *consists of the following* $N-1$ *real distinct eigenvalues,*

$$\lambda_i = \frac{4}{h^2} \sin^2\left(\frac{\pi}{2}ih\right), \quad i = 1, \ldots, N-1.$$

Corollary 3.3. *The eigenvalues of the forward Euler matrix are* $\mu_i = 1 - \tau\lambda_i$, *while the backward Euler ones are* $\eta_i = \left(1 + \tau\lambda_i\right)^{-1}$.

To conclude, we apply the following property which puts into evidence the relation between the condition $(\mathbf{C}_h^\tau)^n \to 0$ as $n \to \infty$ and the matrix spectrum.

Theorem 3.6. *For any* $\mathbf{U}_0 \in \mathbb{R}^{(N-1)}$ *we have* $\lim_{n\to\infty} \|(\mathbf{C}_h^\tau)^n \mathbf{U}_0\|_\infty = 0$ *if and only if* $\max_{i=1,\ldots,N-1} |\lambda_i| < 1$, *where* λ_i *are the eigenvalues of* $\mathbf{C}_h^\tau$.

Since λ_i are monotonically increasing with respect to the index i and the entire sequence is upper bounded by $4/h^2$, we observe that to ensure stability it is sufficient to satisfy $|\lambda_i| < 1$ for the largest eigenvalue. Then, owing to Corollary 3.3 and Theorem 3.6 the following conclusions hold true. [25]

Corollary 3.4. *Backward Euler scheme is unconditionally stable, namely it satisfies* (3.98) *without restrictions on* h *and* τ.

Corollary 3.5. *Forward Euler scheme is only conditionally stable, because it satisfies* (3.98) *provided that* $\sqrt{2\tau} < h$.

Furthermore, we notice that the restriction $2\tau < h^2$ implies that forward Euler matrix $\mathbf{C}_h^\tau = I - \tau\mathbf{A}_h$ is **positive**, namely all its elements are non negative, with at least one being strictly positive. Then, the expression $\mathbf{U}_n = (\mathbf{C}_h^\tau)^n \mathbf{U}_0$ (valid for the simplified case $\mathbf{F}(t) = 0$) allows us to conclude that the discrete solution $\mathbf{U}_n$ is a positive vector, provided that $\mathbf{U}_0$ is positive. This property can be seen as the discrete counterpart of the maximum principle for a homogeneous Cauchy-Dirichlet problem and it is often called discrete maximum principle. It is also valid for the backward Euler scheme, without any restriction on h and τ.

We conclude this section observing that the stability analysis is a fundamental step toward proving the convergence of the discrete solution to the exact one. The role of stability in the error analysis of finite difference schemes is discussed in Appendix C.

3.6.3 The solution of the heat equation as a probability density function

Let us consider the random walk of a particle moving on a lattice of equidistributed nodes in $x \in \mathbb{R}$, $t \in \mathbb{R}^+$ spaced by h and τ along the space and time

---

[25] For further details we refer to *Quarteroni, Sacco, Saleri* (2007).

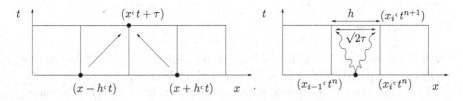

Fig. 3.10. Interpretation of the forward Euler stability condition $\sqrt{2\tau} < h$ as a restriction of random walks

axes, respectively. From the initial state $x = 0$ for $t = 0$, we assume that in each time interval the particle takes a single step on the left or on the right. Being (x,t) a generic node of the lattice, we aim to determine the probability $p(x, t + \tau)$ of finding the particle at $(x, t + \tau)$. To this purpose, we observe that the particle can reach $(x, t+\tau)$ only from $(x \pm h, t)$, as shown in Fig. 3.10 (left). Then, the probabilities of finding the particle in these nodes satisfy the following relation,

$$p(x, t + \tau) = \frac{1}{2}\left(p(x - h, t) + p(x + h, t)\right).$$

Assuming that $p(\cdot, \cdot)$ is a regular function of its arguments, we apply Taylor expansions

$$p(x, t + \tau) = p(x, t) + \tau p_t(x, t) + \mathcal{O}(\tau^2)$$
$$p(x \pm h, t) = p(x, t) \pm h p_x(x, t) + \frac{1}{2}h^2 p_{xx}(x, t) \pm \frac{1}{6}h^3 p_{xxx}(x, t) + \mathcal{O}(h^4)$$

to conclude that

$$p_t + \mathcal{O}(\tau) = \frac{h^2}{2\tau} p_{xx} + \frac{h^2}{\tau}\mathcal{O}(h^2).$$

Both left and right hand side of the previous expression remain bounded passing to the limit $\tau, h \to 0$, provided that the ratio $h^2/2\tau$ is kept constant (which consists in performing a parabolic dilation). Then, the constant $D = h^2/2\tau$ plays the role of diffusion coefficient and the probability density of finding the particle at (x, t) is governed by the heat equation

$$p_t = D p_{xx}, \quad (x, t) \in \mathbb{R} \times \mathbb{R}^+.$$

Furthermore, denoting with h^* the **mean free path** traveled by the particle in a time slab of length τ, we observe that $(h^*)^2 = 2\tau$ or $h^* = \sqrt{2\tau}$. Then, going back to the Euler schemes for the approximation of the heat equation, the stability condition for the forward Euler scheme corresponds to require that for each time slab the mean free path of this random walk must be smaller than the width of a computational cell, namely $h^* < h$, see also Fig. 3.10 (right).

3.7 Exercises

3.1. A homogeneous bar is located in the interval $0 \leq x \leq 1$; its section is negligible compared to its length. The lateral surface of the bar is adiabatic and its temperature u at the initial time $t = 0$ is $u(x, 0) = g(x)$. Furthermore, at time $t = 0$ the endpoints of the bar, $x = 0$ and $x = 1$ are settled and kept at constant temperatures, respectively, u_0 and u_1. The thermal diffusivity of the bar is $D\,m^2/s$. Write the mathematical model describing the evolution of u as $t > 0$. Then, calculate u for $g(x) = x$, $u_0 = 1$, $u_1 = 0$.

3.2. Consider the following Cauchy-Neumann problem:

$$\begin{cases} u_t - u_{xx} = 0 & 0 < x < L, t > 0 \\ u(x, 0) = x & 0 < x < L \\ u_x(0, t) = u_x(L, t) = 0 & t > 0. \end{cases}$$

Find the solution using the separation of variables and examine the asymptotic behavior of $u(x, t)$ as $t \to +\infty$.

3.3. Use the separation of variables to solve the following non homogeneous Cauchy-Neumann problem:

$$\begin{cases} u_t - u_{xx} = tx & 0 < x < \pi, t > 0 \\ u(x, 0) = 1 & 0 \leq x \leq \pi \\ u_x(0, t) = u_x(\pi, t) = 0 & t > 0. \end{cases}$$

3.4 (The evolution of a chemical solution). Consider a tube of length L and constant cross section A, where x is the symmetry-axis. The tube contains a saline solution of concentration c (dimensionally $[mass] \times [length]^{-3}$). Let A be small enough to assume that the concentration c depends only on x and t, in order that the diffusion of salt is one dimensional, in the direction x. Let also the velocity of the fluid be negligible.

From the left boundary of the pipe at $x = 0$ a solution of constant concentration C_0 ($[mass] \times [length]^{-3}$) enters with velocity R_0 ($[length]^3 \times [time]$), while at the other end $x = L$ the solution is removed at the same speed.

Using the *Fick's law*, show that c solves a diffusion Neumann-Robin problem. Thus, find the explicit solution and verify that for $t \to +\infty$, $c(x, t)$ tends to a steady state.

3.5 (Diffusion of concentrated source). Find the similarity solutions of the equation $u_t - u_{xx} = 0$ of the form $u(x, t) = U(x/\sqrt{t})$ and express the result in term of the *error function*

$$\operatorname{erf}(x) = \frac{2}{\sqrt{\pi}} \int_0^x e^{-z^2} dz.$$

Then, use the result to solve the following diffusion problem on the half straight line, with constant concentration in $x = 0$ as $t > 0$:

$$\begin{cases} u_t - u_{xx} = 0 & x > 0, t > 0 \\ u(0,t) = C, \ \lim_{x \to +\infty} u(x,t) = 0 & t > 0 \\ u(x,0) = 0 & x > 0. \end{cases}$$

3.6 (A problem in dimension $n = 3$). Assume that

$$B_R = \left\{ \mathbf{x} \in \mathbb{R}^3 : |\mathbf{x}| < 1 \right\}$$

is the volume occupied by a material that is homogeneous and has the constant temperature U at the initial time $t = 0$. When $t > 0$, the temperature on the boundary of the ball is brought and maintained at the value zero. Describe the evolution of temperature at the points of the ball, and make sure that the temperature at the center of the sphere tends to zero as $t \to +\infty$.

3.7. Prove that, if $w_t - D\Delta w = 0$ in Q_T and $w \in C\left(\overline{Q}_T\right)$, then

$$\min_{\partial_p Q_T} w \leq w(\mathbf{x},t) \leq \max_{\partial_p Q_T} w \qquad \text{for every } (\mathbf{x},t) \in Q_T.$$

3.8. Find an explicit formula for the solution of the global Cauchy problem

$$\begin{cases} u_t = D u_{xx} + b u_x + cu & x \in \mathbb{R}, t > 0 \\ u(x,0) = g(x) & x \in \mathbb{R} \end{cases}$$

where D, b, c are constant coefficients. Show that, if $c < 0$ and g is bounded, $u(x,t) \to 0$ as $t \to +\infty$.

3.9. Find an explicit formula for the solution of the Cauchy-Dirichlet problem

$$\begin{cases} u_t = u_{xx} & x > 0, t > 0 \\ u(x,0) = g(x) & x \geq 0 \\ u(0,t) = 0 & t > 0 \end{cases}$$

with g continuous and $g(0) = 0$.

3.10. Let $Q_T = \Omega \times (0,T)$, with Ω bounded domain in $\mathbb{R}^n$. Let $u \in C^{2,1}(Q_T) \cap C\left(\overline{Q}_T\right)$ satisfy the equation

$$u_t = D\Delta u + \mathbf{b}(\mathbf{x},t) \cdot \nabla u + c(\mathbf{x},t) u \qquad \text{in } Q_T$$

where $\mathbf{b}$ and c are continuous in $\overline{Q}_T$. Show that if $u \geq 0$ (resp. $u \leq 0$) on $\partial_p Q_T$ then $u \geq 0$ (resp. $u \leq 0$) in Q_T.

3.11. Solve the following initial-Dirichlet problem in $B_1 = \left\{ \mathbf{x} \in \mathbb{R}^3 : |\mathbf{x}| < 1 \right\}$:

$$\begin{cases} u_t = \Delta u & \mathbf{x} \in B_1, t > 0 \\ u(\mathbf{x},0) = 0 & \mathbf{x} \in B_1 \\ u(\boldsymbol{\sigma},t) = 1 & \boldsymbol{\sigma} \in \partial B_1, t > 0. \end{cases}$$

Compute $\lim_{t \to +\infty} u$.

3.12. Solve the following initial-Dirichlet problem

$$\begin{cases} u_t = \Delta u & \mathbf{x} \in K, t > 0 \\ u(\mathbf{x}, 0) = 0 & \mathbf{x} \in K \\ u(\sigma, t) = 1 & \sigma \in \partial K, t > 0 \end{cases}$$

where K is the rectangular box

$$K = \left\{ (x, y, z) \in \mathbb{R}^3 \colon 0 < x < a, 0 < y < b, 0 < z < c \right\}.$$

Compute $\lim_{t \to +\infty} u$.

3.13. Solve the following initial-Neumann problem in $B_1 = \left\{ \mathbf{x} \in \mathbb{R}^3 \colon |\mathbf{x}| < 1 \right\}$:

$$\begin{cases} u_t = \Delta u & \mathbf{x} \in B_1, t > 0 \\ u(\mathbf{x}, 0) = |\mathbf{x}| & \mathbf{x} \in B_1 \\ u_\nu(\sigma, t) = 1 & \sigma \in \partial B_1, t > 0. \end{cases}$$

3.14. Solve the following non homogeneous initial-Dirichlet problem in the unit sphere B_1 $(u = u(r, t), r = |\mathbf{x}|)$:

$$\begin{cases} u_t - (u_{rr} + \dfrac{2}{r} u_r) = q e^{-t} & 0 < r < 1, t > 0 \\ u(r, 0) = U & 0 \le r \le 1 \\ u(1, t) = 0 & t > 0. \end{cases}$$

3.7.1 Application of Euler methods to the discretization of the Cauchy-Dirichlet problem

We address the following problem,

$$\begin{cases} u_t - u_{xx} = 0 & -L < x < L, \tau < t < T \\ u(-L, t) = u(L, t) = 0 & \tau < t < T \\ u(x, \tau) = \Gamma_1(x, \tau) & -L \le x \le L. \end{cases}$$

Provided that the domain width L is large enough to make sure that the solution is almost insensitive to the boundary conditions, we assume that the fundamental solution of the one-dimensional heat equation, $\Gamma_1(x, t)$, satisfies the problem (see also Fig. 3.11). For the numerical approximation we apply the schemes (3.96), with the aim to verify the stability properties summarized in Corollary 3.4 and 3.5.

Let us start from forward Euler method defined on a computational mesh that satisfies the stability condition, for example $\tau = \frac{1}{4} h^2$. Fig. 3.11 (top-left panel) confirms the good behavior of the scheme in this case. The numerical solution obtained by progressively increasing τ until $\tau = \frac{3}{4} h^2$, reported in Fig. 3.11 (top-right panel), confirms that spurious oscillations appear above the threshold $\tau^* = \frac{1}{2} h^2$, as symptoms of the lack of stability. Conversely, the bottom panels of Fig. 3.11 show that the backward scheme is unconditionally stable.

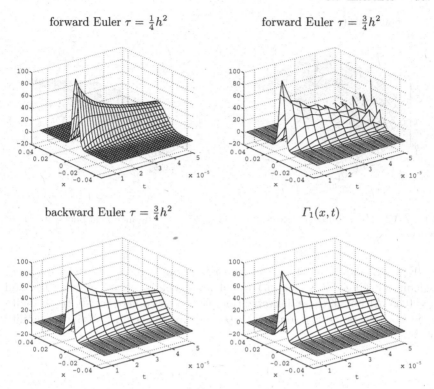

forward Euler $\tau = \frac{1}{4}h^2$

forward Euler $\tau = \frac{3}{4}h^2$

backward Euler $\tau = \frac{3}{4}h^2$

$\Gamma_1(x,t)$

Fig. 3.11. A comparison between forward and backward Euler methods

3.7.2 Application to the dynamics of chemicals

As in Exercise 3.4, we aim to study the evolution of a concentration field $c(x,t)$ such that

$$\begin{cases} c_t - c_{xx} = 0 & 0 < x < 1,\ t > 0 \\ c_x(0,t) = -1,\ c_x(1,t) + c(1,t) = 0 & t > 0 \\ c(x,0) = c_0(x) & 0 \le x \le 1. \end{cases} \qquad (3.99)$$

For any positive value of the initial concentration, problem (3.99) admits a steady state $c^*(x) = 2 - x$. Exploiting the change of variable $u(x,t) = c(x,t) - c^*(x)$, we rewrite (3.99) as a problem with a null steady state,

$$\begin{cases} u_t - u_{xx} = 0 & 0 < x < 1,\ t > 0 \\ u_x(0,t) = 0,\ u_x(1,t) + u(1,t) = 0 & t > 0 \\ u(x,0) = c_0(x) - c^*(x) & 0 \le x \le 1 \end{cases} \qquad (3.100)$$

where $c_0(x) = \frac{1}{2}x^2 - x + \frac{1}{2}$ for compatibility with the boundary conditions of (3.99).

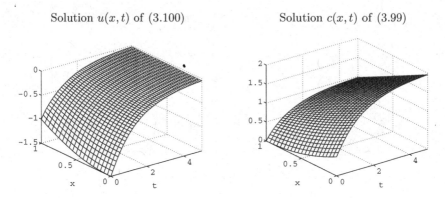

Solution $u(x,t)$ of (3.100)   Solution $c(x,t)$ of (3.99)

Fig. 3.12. The solution of Exercise 3.4 approximated by means of backward Euler scheme

Fig. 3.12 shows the numerical approximation of $u(x,t)$ by backward Euler scheme, obtained on a computational mesh characterized by $\tau = 0.1$ and $h = 0.05$. Although these discretization parameters do not satisfy the stability condition, the numerical solution is stable owing to the unconditional stability of the method.

4

The Laplace Equation

4.1 Introduction

The Laplace equation $\Delta u = 0$ occurs frequently in applied sciences, in particular in the study of the *steady state phenomena*. Its solutions are called *harmonic* functions. For instance, the equilibrium position of a perfectly elastic membrane is a harmonic function as it is the velocity potential of a homogeneous fluid. Also, the steady state temperature of a homogeneous and isotropic body is a harmonic function and in this case Laplace equation constitutes the stationary counterpart (time independent) of the diffusion equation.

Slightly more generally, Poisson's equation $\Delta u = f$ plays an important role in the theory of *conservative fields* (electrical, magnetic, gravitational, ...) where the vector field is derived from the gradient of a potential.

For example, let $\mathbf{E}$ be a force field due to a distribution of electric charges in a domain $\Omega \subset \mathbb{R}^3$. Then, in standard units, div $\mathbf{E} = 4\pi\rho$, where ρ represents the density of the charge distribution. When a *potential* u exists such that $\nabla u = -\mathbf{E}$, then $\Delta u = \text{div}\nabla u = -4\pi\rho$, which is Poisson's equation. If the electric field is created by charges located outside Ω, then $\rho = 0$ in Ω and u is harmonic therein. Analogously, the potential of a gravitational field due to a mass distribution is a harmonic function in a region free from mass.

In dimension two, the theories of harmonic and holomorphic functions are strictly connected[1]. Indeed, the real and the imaginary part of a holomorphic function are harmonic. For instance, since the functions

$$z^m = r^m \left(\cos m\theta + i \sin m\theta \right), \qquad m \in \mathbb{N},$$

---

[1] A complex function $f = f(z)$ is *holomorphic* in an open subset Ω of the complex plane if for every $z_0 \in \Omega$, the limit

$$\lim_{z \to z_0} \frac{f(z) - f(z_0)}{z - z_0} = f'(z_0)$$

exists and it is finite.

Salsa S., Vegni F.M.G., Zaretti A., Zunino P.: *A Primer on PDEs. Models, Methods, Simulations.*
Unitext – La Matematica per il 3+2 65.
DOI 10.1007/978-88-470-2862-3_4, © Springer-Verlag Italia 2013

$(r, \theta$ polar coordinates) are holomorphic in the whole plane $\mathbb{C}$, the functions

$$u(r, \theta) = r^m \cos m\theta \quad \text{and} \quad v(r, \theta) = r^m \sin m\theta \qquad m \in \mathbb{N},$$

are harmonic in $\mathbb{R}^2$ (called *elementary harmonics*). In Cartesian coordinates, they are harmonic polynomials; for $m = 1, 2, 3$ we find

$$x, \ y, \ xy, \ x^2 - y^2, \ x^3 - 3xy^2, \ 3x^2y - y^3.$$

Other examples are

$$u(x, y) = e^{\alpha x} \cos \alpha y, \quad v(x, y) = e^{\alpha x} \sin \alpha y \qquad (\alpha \in \mathbb{R}),$$

the real and imaginary parts of $f(z) = e^{i\alpha z}$, both harmonic in $\mathbb{R}^2$, and

$$u(r, \theta) = \log r, \quad v(r, \theta) = \theta,$$

the real and imaginary parts of $f(z) = \log_0 z = \log r + i\theta$, harmonic in $\mathbb{R}^2 \backslash (0, 0)$ and $\mathbb{R}^2 \backslash \{\theta = 0\}$, respectively.

In this chapter we present the formulation of the most important well posed problems and the classical properties of harmonic functions, focusing mainly on dimensions two and three. A central notion is the concept of *fundamental solution*, that we develop in conjunction with the very basic elements of the so called *potential theory*.

4.2 Well Posed Problems. Uniqueness

Consider the Poisson equation

$$\Delta u = f \quad \text{in } \Omega \tag{4.1}$$

where $\Omega \subset \mathbb{R}^n$ is a **bounded domain**. The well posed problems associated with equation (4.1) are the stationary counterparts of the corresponding problems for the diffusion equation. Clearly here there is no initial condition. On the boundary $\partial \Omega$ we may assign:

- *Dirichlet data*

$$u = g. \tag{4.2}$$

- *Neumann data*

$$\partial_\nu u = h \tag{4.3}$$

where ν is the outward normal unit vector to $\partial \Omega$.

- A *Robin (radiation) condition*

$$\partial_\nu u + \alpha u = h \qquad (\alpha > 0). \tag{4.4}$$

• A *mixed condition*; for instance,

$$u = g \qquad \text{on } \Gamma_D \qquad\qquad (4.5)$$
$$\partial_\nu u = h \qquad \text{on } \Gamma_N,$$

where $\Gamma_D \cup \Gamma_N = \partial\Omega$, $\Gamma_D \cap \Gamma_N = \varnothing$, and Γ_N is a relatively open subset of $\partial\Omega$.

When $g = h = 0$ we say that the above boundary conditions are *homogeneous*.

We give some interpretations. If u is the position of a perfectly flexible membrane and f is an external distributed load (vertical force per unit surface), then (4.1) models a steady state.

The Dirichlet condition corresponds to fixing the position of the membrane at its boundary. Robin condition describes an elastic attachment at the boundary while a homogeneous Neumann condition corresponds to a free vertical motion of the boundary.

If u is the steady state concentration of a substance, the Dirichlet condition prescribes the level of u at the boundary, while the Neumann condition assigns the flux of u through the boundary.

Using Green's identity (1.12) we can prove the following uniqueness result.

Theorem 4.1. *Let $\Omega \subset \mathbb{R}^n$ be a smooth, bounded domain. Then there exists at most one solution $u \in C^2(\Omega) \cap C^1(\overline{\Omega})$ of (4.1), satisfying on $\partial\Omega$ one of the conditions (4.2), (4.3), (4.4) or (4.5).*

In the case of the Neumann condition, that is when

$$\partial_\nu u = h \qquad \text{on } \partial\Omega,$$

two solutions differ by a constant.

Proof. Let u and v be solutions of the same problem, sharing the same boundary data, and let $w = u - v$. Then w is harmonic and satisfies homogeneous boundary conditions (one among (4.2)-(4.5)). Substituting $u = v = w$ into (1.12) we find

$$\int_\Omega |\nabla w|^2 \, d\mathbf{x} = \int_{\partial\Omega} w \partial_\nu w \, d\sigma.$$

If Dirichlet or mixed conditions hold, we have

$$\int_{\partial\Omega} w \partial_\nu w \, d\sigma = 0.$$

When a Robin condition holds

$$\int_{\partial\Omega} w \partial_\nu w \, d\sigma = - \int_{\partial\Omega} \alpha w^2 d\sigma \leq 0.$$

In any case we obtain that

$$\int_\Omega |\nabla w|^2 \, d\mathbf{x} \leq 0. \tag{4.6}$$

From (4.6) we infer $\nabla w = \mathbf{0}$ and therefore $w = u - v = $ constant. This concludes the proof in the case of Neumann condition. In the other cases, the constant must be zero (why?), hence $u = v$.   □

Remark 4.1. Consider the Neumann problem $\Delta u = f$ in Ω, $\partial_\nu u = h$ on $\partial\Omega$. Integrating the equation on Ω and using Gauss' formula we find

$$\int_\Omega f \, d\mathbf{x} = \int_{\partial\Omega} h \, d\sigma. \tag{4.7}$$

The relation (4.7) appears as a *compatibility* condition on the data f and h, that has *necessarily* to be satisfied in order for the Neumann problem to admit a solution. Thus, when having to solve a Neumann problem, the first thing to do is to check the validity of (4.7). If it does not hold, the problem does not have any solution. We will examine later the physical meaning of (4.7).

4.3 Harmonic Functions

4.3.1 Mean value properties

Guided by their discrete characterization, we want to establish some fundamental properties of harmonic functions. To be precise, we say that a function u is *harmonic* in a domain $\Omega \subseteq \mathbb{R}^n$ if $u \in C^2(\Omega)$ and $\Delta u = 0$ in Ω.

Since $d-$harmonic functions are defined through a mean value property, we expect that harmonic functions inherit a mean value property of the following kind: the value at the center of any ball $B \subset\subset \Omega$, i.e. compactly contained in Ω, equals the average of the values on the boundary ∂B. Actually, something more is true.

Theorem 4.2. *Let u be harmonic in $\Omega \subseteq \mathbb{R}^n$. Then, for any ball $B_R(\mathbf{x}) \subset\subset \Omega$ the following mean value formulas hold:*

$$u(\mathbf{x}) = \frac{n}{\omega_n R^n} \int_{B_R(\mathbf{x})} u(\mathbf{y}) \, d\mathbf{y} \tag{4.8}$$

$$u(\mathbf{x}) = \frac{1}{\omega_n R^{n-1}} \int_{\partial B_R(\mathbf{x})} u(\boldsymbol{\sigma}) \, d\sigma \tag{4.9}$$

where ω_n is the surface of ∂B_R.

Proof ($n = 2$). Let us start from the second formula. For $r < R$ define

$$g\left(r\right) = \frac{1}{2\pi r}\int_{\partial B_r(\mathbf{x})} u\left(\boldsymbol{\sigma}\right)d\sigma.$$

Perform the change of variables $\boldsymbol{\sigma} = \mathbf{x} + r\boldsymbol{\sigma}'$. Then $\boldsymbol{\sigma}' \in \partial B_1\left(\mathbf{0}\right)$, $d\sigma = rd\sigma'$ and

$$g\left(r\right) = \frac{1}{2\pi}\int_{\partial B_1(\mathbf{0})} u\left(\mathbf{x} + r\boldsymbol{\sigma}'\right)d\sigma'.$$

Let $v\left(\mathbf{y}\right) = u\left(\mathbf{x} + r\mathbf{y}\right)$ and observe that

$$\nabla v\left(\mathbf{y}\right) = r\nabla u\left(\mathbf{x} + r\mathbf{y}\right)$$
$$\Delta v\left(\mathbf{y}\right) = r^2 \Delta u\left(\mathbf{x} + r\mathbf{y}\right).$$

Then we have

$$g'\left(r\right) = \frac{1}{2\pi}\int_{\partial B_1(\mathbf{0})} \frac{d}{dr}u\left(\mathbf{x} + r\boldsymbol{\sigma}'\right)d\sigma' = \frac{1}{2\pi}\int_{\partial B_1(\mathbf{0})} \nabla u\left(\mathbf{x} + r\boldsymbol{\sigma}'\right)\cdot\boldsymbol{\sigma}'d\sigma'$$

$$= \frac{1}{2\pi r}\int_{\partial B_1(\mathbf{0})} \nabla v\left(\boldsymbol{\sigma}'\right)\cdot\boldsymbol{\sigma}'d\sigma' = \quad \text{(divergence theorem)}$$

$$= \frac{1}{2\pi r}\int_{B_1(\mathbf{0})} \Delta v\left(\mathbf{y}\right)d\mathbf{y} = \frac{r}{2\pi}\int_{B_1(\mathbf{0})} \Delta u\left(\mathbf{x} + r\mathbf{y}\right)d\mathbf{y} = 0.$$

Thus, g is constant and since $g\left(r\right) \to u\left(\mathbf{x}\right)$ for $r \to 0$, we get (4.9).

To obtain (4.8), let $R = r$ in (4.9), multiply by r and integrate both sides between 0 and R. We find

$$\frac{R^2}{2}u\left(\mathbf{x}\right) = \frac{1}{2\pi}\int_0^R dr\int_{\partial B_r(\mathbf{x})} u\left(\boldsymbol{\sigma}\right)d\sigma = \frac{1}{2\pi}\int_{B_R(\mathbf{x})} u\left(\mathbf{y}\right)d\mathbf{y}$$

from which (4.8) follows. $\qquad\qquad\qquad\qquad\qquad\qquad\qquad\qquad\qquad\square$

Even more significant is a converse of Theorem 4.2. We say that a **continuous** function u satisfies the *mean value property* in Ω, if (4.8) or (4.9) holds for any ball $B_R\left(\mathbf{x}\right) \subset\subset \Omega$. It turns out that if u is continuous and possesses the mean value property in a domain Ω, then u is harmonic in Ω. Thus we obtain a characterization of harmonic functions through a mean value property, as in the discrete case. As a by product, we deduce that every harmonic function in a domain Ω is continuously differentiable of any order in Ω, that is, it belongs to $C^\infty\left(\Omega\right)$. Notice that this is not a trivial fact since it involves derivatives not appearing in the expression of the Laplace operator. For instance, $u\left(x, y\right) = x + y\left|y\right|$ is a solution of $u_{xx} + u_{xy} = 0$ in all $\mathbb{R}^2$ but it is not twice differentiable with respect to y at $\left(0, 0\right)$.

Theorem 4.3. *Let* $u \in C\left(\Omega\right)$. *If* u *satisfies the mean value property, then* $u \in C^\infty\left(\Omega\right)$ *and it is harmonic in* Ω.

We postpone the proof to the end of Section 4.3.3.

4.3.2 Maximum principles

As in the discrete case, a function satisfying the mean value property in a domain[2] Ω cannot attain its maximum or minimum at an *interior point of* Ω, unless it is constant. In case Ω is bounded and u (non constant) is continuous up to the boundary of Ω, it follows that u attains both its maximum and minimum **only on** $\partial\Omega$. This result expresses a maximum principle that we state precisely in the following theorem.

Theorem 4.4. *Let $u \in C(\Omega)$, $\Omega \subseteq \mathbb{R}^n$. If u has the mean value property and attains its maximum or minimum at $\mathbf{p} \in \Omega$, then u is constant. In particular, if Ω is bounded and $u \in C(\overline{\Omega})$ is not constant, then, for every $\mathbf{x} \in \Omega$,*

$$u(\mathbf{x}) < \max_{\partial\Omega} u \quad and \quad u(\mathbf{x}) > \min_{\partial\Omega} u \quad (strong\ maximum\ principle).$$

Proof (n = 2). Let $\mathbf{p}$ be a minimum point[3] for u:

$$m = u(\mathbf{p}) \leq u(\mathbf{y}), \qquad \forall \mathbf{y} \in \Omega.$$

We want to show that $u \equiv m$ in Ω. Let $\mathbf{q}$ be another arbitrary point in Ω. Since Ω is connected, it is possible to find a finite sequence of circles $B(\mathbf{x}_j) \subset\subset \Omega$, $j = 0, \ldots, N$, such that (Fig. 4.1):

- $\mathbf{x}_j \in B(\mathbf{x}_{j-1})$, $j = 1, \ldots, N$;
- $x_0 = \mathbf{p}$, $x_N = \mathbf{q}$.

The mean value property gives

$$m = u(\mathbf{p}) = \frac{1}{|B(\mathbf{p})|} \int_{B(\mathbf{p})} u(\mathbf{y})\,d\mathbf{y}.$$

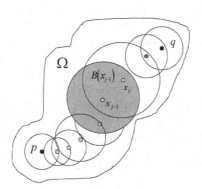

Fig. 4.1. A sequence of overlapping circles connecting the points **p** and **q**

---

[2] Recall that a *domain* is an open *connected* set.
[3] The argument for the maximum is the same.

Suppose there exists $\mathbf{z} \in B(\mathbf{p})$ such that $u(\mathbf{z}) > m$. Then, given a circle $B_r(\mathbf{z}) \subset B(\mathbf{p})$, we can write:

$$m = \frac{1}{|B(\mathbf{p})|} \int_{B(\mathbf{p})} u(\mathbf{y}) \, dy \tag{4.10}$$

$$= \frac{1}{|B(\mathbf{p})|} \left\{ \int_{B(\mathbf{p}) \backslash B_r(\mathbf{z})} u(\mathbf{y}) \, dy + \int_{B_r(\mathbf{z})} u(\mathbf{y}) \, dy \right\}.$$

Since $u(\mathbf{y}) \geq m$ for every $\mathbf{y}$ and, by the mean value again,

$$\int_{B_r(\mathbf{z})} u(\mathbf{y}) \, dy = u(\mathbf{z}) |B_r(\mathbf{z})| > m |B_r(\mathbf{z})|,$$

continuing from (4.10) we obtain

$$> \frac{1}{|B(\mathbf{p})|} \{ m |B(\mathbf{p}) \backslash B_r(\mathbf{z})| + m |B_r(\mathbf{z})| \} = m$$

and therefore the contradiction $m > m$.

Thus it must be that $u \equiv m$ in $B(\mathbf{p})$ and in particular $u(\mathbf{x}_1) = m$. We repeat now the same argument with $\mathbf{x}_1$ in place of $\mathbf{p}$ to show that $u \equiv m$ in $B(\mathbf{x}_1)$ and in particular $u(\mathbf{x}_2) = m$. Iterating the procedure we eventually deduce that $u(\mathbf{x}_N) = u(\mathbf{q}) = m$. Since $\mathbf{q}$ is an arbitrary point of Ω, we conclude that $u \equiv m$ in Ω. $\square$

An important consequence of the maximum principle is the following corollary.

Corollary 4.1. *Let $\Omega \subset \mathbb{R}^n$ be a bounded domain and $g \in C(\partial\Omega)$. The problem*

$$\begin{cases} \Delta u = 0 & \text{in } \Omega \\ u = g & \text{on } \partial\Omega \end{cases} \tag{4.11}$$

has at most a solution $u_g \in C^2(\Omega) \cap C(\overline{\Omega})$. Moreover, let u_{g_1} and u_{g_2} be the solutions corresponding to the data $g_1, g_2 \in C(\partial\Omega)$. Then:

(a) *(Comparison). If $g_1 \geq g_2$ on $\partial\Omega$ and $g_1 \neq g_2$, then*

$$u_{g_1} > u_{g_2} \qquad \text{in } \Omega. \tag{4.12}$$

(b) *(Stability).*

$$|u_{g_1}(\mathbf{x}) - u_{g_2}(\mathbf{x})| \leq \max_{\partial\Omega} |g_1 - g_2| \qquad \text{for every } \mathbf{x} \in \Omega. \tag{4.13}$$

Proof. We first show (a) and (b). Let $w = u_{g_1} - u_{g_2}$. Then w is harmonic and $w = g_1 - g_2 \geq 0$ on $\partial\Omega$. Since $g_1 \neq g_2$, w is not constant and from Theorem 4.4

$$w(\mathbf{x}) > \min_{\partial\Omega}(g_1 - g_2) \geq 0 \quad \text{for every } \mathbf{x} \in \Omega.$$

This is (4.12). To prove (b), apply Theorem 4.4 to w and $-w$ to find

$$\pm w\left(\mathbf{x}\right) \le \max_{\partial\Omega}\left|g_1 - g_2\right| \quad \text{for every } \mathbf{x} \in \Omega$$

which is equivalent to (4.13).

Now if $g_1 = g_2$, (4.13) implies $w = u_{g_1} - u_{g_2} \equiv 0$, so that the Dirichlet problem (4.11) has at most one solution. $\square$

Remark 4.2. Inequality (4.13) is a *stability estimate*. Indeed, suppose g is known within an absolute error less than ε, or, in other words, suppose g_1 is an approximation of g and $\max_{\partial\Omega}\left|g - g_1\right| < \varepsilon$; then (4.13) gives

$$\max_{\bar{\Omega}}\left|u_{g_1} - u_g\right| < \varepsilon$$

so that the approximate solution is known within the same absolute error.

4.3.3 The Dirichlet problem in a circle. Poisson's formula

To prove the existence of a solution to one of the boundary value problems we considered in Section 4.2 is not an elementary task. In Chapter 8, we solve this question in a general context, using the more advanced tools of Functional Analysis. However, in special cases, elementary methods, like separation of variables, work. We use it to compute the solution of the Dirichlet problem in a circle. Precisely, let $B_R = B_R\left(\mathbf{p}\right)$ the circle of radius R centered at $\mathbf{p} = \left(p_1, p_2\right)$ and $g \in C\left(\partial B_R\right)$. We want to prove the following theorem.

Theorem 4.5. *The unique solution* $u \in C^2\left(B_R\right) \cap C\left(\bar{B}_R\right)$ *of the problem*

$$\begin{cases} \Delta u = 0 \ \ in \ B_R \\ u = g \quad on \ \partial B_R \end{cases} \tag{4.14}$$

is given by ***Poisson's formula***

$$u\left(\mathbf{x}\right) = \frac{R^2 - \left|\mathbf{x} - \mathbf{p}\right|^2}{2\pi R} \int_{\partial B_R(\mathbf{p})} \frac{g\left(\boldsymbol{\sigma}\right)}{\left|\mathbf{x} - \boldsymbol{\sigma}\right|^2}d\sigma. \tag{4.15}$$

In particular, $u \in C^\infty\left(B_R\right)$.

Proof. For simplicity, we give the proof assuming that g is a smooth function. The symmetry of the domain suggests the use of polar coordinates

$$x_1 = p_1 + r\cos\theta \qquad x_2 = p_2 + r\sin\theta.$$

Accordingly, let

$$U\left(r, \theta\right) = u\left(p_1 + r\cos\theta, p_2 + r\sin\theta\right), \ G\left(\theta\right) = g\left(p_1 + R\cos\theta, p_2 + R\sin\theta\right).$$

The Laplace equation becomes[4]

$$U_{rr} + \frac{1}{r}U_r + \frac{1}{r^2}U_{\theta\theta} = 0, \qquad 0 < r < R, \, 0 \le \theta \le 2\pi, \qquad (4.16)$$

with the Dirichlet condition

$$U(R, \theta) = G(\theta), \qquad 0 \le \theta \le 2\pi.$$

Since we ask that u be continuous in $\overline{B}_R$, then U and G have to be continuous in $[0, R] \times [0, 2\pi]$ and $[0, 2\pi]$, respectively; moreover both have to be $2\pi-$periodic with respect to θ.

We use now the method of separation of variables, by looking first for solutions of the form

$$U(r, \theta) = v(r)w(\theta)$$

with v, w bounded and w $2\pi-$periodic. Substitution in (4.16) gives

$$v''(r)w(\theta) + \frac{1}{r}v'(r)w(\theta) + \frac{1}{r^2}v(r)w''(\theta) = 0$$

or, separating the variables,

$$-\frac{r^2v''(r) + rv'(r)}{v(r)} = \frac{w''(\theta)}{w(\theta)}.$$

This identity is possible only when the two quotients have a common constant value λ. Thus we are lead to the ordinary differential equation

$$r^2v''(r) + rv'(r) - \lambda v(r) = 0 \qquad (4.17)$$

and to the eigenvalue problem

$$\begin{cases} w''(\theta) - \lambda w(\theta) = 0 \\ w(0) = w(2\pi). \end{cases} \qquad (4.18)$$

We leave to the reader to check that problem (4.18) has only the zero solution for $\lambda \ge 0$. If $\lambda = -\mu^2$, $\mu > 0$, the differential equation in (4.18) has the general integral

$$w(\theta) = a\cos\mu\theta + b\sin\mu\theta \qquad (a, b \in \mathbb{R}).$$

The $2\pi-$periodicity forces $\mu = m$, a nonnegative integer.

The equation (4.17) is an Euler equation. The change of variables $s = \log r$ yields to the equation

$$v''(s) - m^2v(s) = 0$$

whose general solution is

$$v(r) = d_1 r^{-m} + d_2 r^m \qquad (d_1, d_2 \in \mathbb{R}).$$

Since v has to be bounded hence $d_1 = 0$.

---

[4] Appendix D.

We have found a countable number of 2π−periodic harmonic functions

$$U_m(r,\theta) = r^m \{a_m \cos m\theta + b_m \sin m\theta\} \qquad m = 0, 1, 2, \dots. \qquad (4.19)$$

We superpose now the (4.19) by writing

$$U(r,\theta) = a_0 + \sum_{m=1}^{\infty} r^m \{a_m \cos m\theta + b_m \sin m\theta\} \qquad (4.20)$$

with the coefficients a_m and b_m still to be chosen in order to satisfy the boundary condition

$$\lim_{(r,\theta)\to(R,\xi)} U(r,\theta) = G(\xi) \qquad \forall \xi \in [0, 2\pi]. \qquad (4.21)$$

Since G is smooth, it can be expanded in a uniformly convergent Fourier series

$$G(\xi) = \frac{\alpha_0}{2} + \sum_{m=1}^{\infty} \{\alpha_m \cos m\xi + \beta_m \sin m\xi\}$$

where

$$\alpha_m = \frac{1}{\pi} \int_0^{2\pi} G(\varphi) \cos m\varphi \, d\varphi, \qquad \beta_m = \frac{1}{\pi} \int_0^{2\pi} G(\varphi) \sin m\varphi \, d\varphi.$$

Then, the boundary condition (4.21) is satisfied if we choose

$$a_0 = \frac{\alpha_0}{2}, \qquad a_m = R^{-m} \alpha_m, \qquad b_m = R^{-m} \beta_m.$$

Substitution of these values of a_0, a_m, b_m into (4.20) gives, for $r \leq R$,

$$U(r,\theta) = \frac{\alpha_0}{2} + \frac{1}{\pi} \sum_{m=1}^{\infty} \left(\frac{r}{R}\right)^m \int_0^{2\pi} G(\varphi) \{\cos m\varphi \cos m\theta + \sin m\varphi \sin m\theta\} \, d\varphi$$

$$= \frac{1}{\pi} \int_0^{2\pi} G(\varphi) \left[\frac{1}{2} + \sum_{m=1}^{\infty} \left(\frac{r}{R}\right)^m \{\cos m\varphi \cos m\theta + \sin m\varphi \sin m\theta\} \right] d\varphi$$

$$= \frac{1}{\pi} \int_0^{2\pi} G(\varphi) \left[\frac{1}{2} + \sum_{m=1}^{\infty} \left(\frac{r}{R}\right)^m \cos m(\varphi - \theta) \right] d\varphi.$$

Note that in the second equality above, the exchange of sum and integration is possible because of the uniform convergence of the series. Moreover, for $r < R$, we can differentiate under the integral sign and then term by term as many times as we want (why?). Therefore, since for every $m \geq 1$ the functions

$$\left(\frac{r}{R}\right)^m \cos m(\varphi - \theta)$$

are smooth and harmonic, also $U \in C^\infty(B_R)$ and is harmonic for $r < R$.

To obtain a better formula, observe that

$$\sum_{m=1}^{\infty} \left(\frac{r}{R}\right)^m \cos m(\varphi - \theta) = \mathrm{Re}\left[\sum_{m=1}^{\infty}\left(e^{i(\varphi-\theta)}\frac{r}{R}\right)^m\right].$$

Since

$$\mathrm{Re}\sum_{m=1}^{\infty}\left(e^{i(\varphi-\theta)}\frac{r}{R}\right)^m = \mathrm{Re}\frac{1}{1-e^{i(\varphi-\theta)}\frac{r}{R}} - 1 = \frac{R^2 - rR\cos(\varphi-\theta)}{R^2 + r^2 - 2rR\cos(\varphi-\theta)} - 1$$

$$= \frac{rR\cos(\varphi-\theta) - r^2}{R^2 + r^2 - 2rR\cos(\varphi-\theta)}$$

we find

$$\frac{1}{2} + \sum_{m=1}^{\infty}\left(\frac{r}{R}\right)^m \cos m(\varphi-\theta) = \frac{1}{2}\frac{R^2 - r^2}{R^2 + r^2 - 2rR\cos(\varphi-\theta)}. \qquad (4.22)$$

Inserting (4.22) into the formula for U, we get **Poisson's formula** in polar coordinates:

$$U(r,\theta) = \frac{R^2 - r^2}{2\pi}\int_0^{2\pi}\frac{G(\varphi)}{R^2 + r^2 - 2Rr\cos(\theta-\varphi)}d\varphi. \qquad (4.23)$$

Going back to Cartesian coordinates[5] we obtain Poisson's formula (4.15). Corollary 3.1 assures that (4.23) is indeed the unique solution of the Dirichlet problem (4.14). $\qquad\square$

Remark 4.3 (Poisson's formula in dimension $n > 2$). Theorem 4.5 has an appropriate extension in any number of dimensions. When $B_R = B_R(\mathbf{p})$ is an n-dimensional ball, the solution of the Dirichlet problem (4.14) is given by

$$u(\mathbf{x}) = \frac{R^2 - |\mathbf{x}-\mathbf{p}|^2}{\omega_n R}\int_{\partial B_R(\mathbf{p})}\frac{g(\boldsymbol{\sigma})}{|\mathbf{x}-\boldsymbol{\sigma}|^n}d\sigma. \qquad (4.24)$$

Remark 4.4. The method of separation of variables can be used also in presence of a distributed source. Let us solve, for instance, the following nonhomogeneous Dirichlet problem.

$$\begin{cases} \Delta u = f & \text{in } B_R \\ u = 0 & \text{in } \partial B_R. \end{cases}$$

---

[5] With $\boldsymbol{\sigma} = R(\cos\varphi, \sin\varphi)$, $d\sigma = Rd\varphi$ and

$$\begin{aligned}|\mathbf{x}-\boldsymbol{\sigma}|^2 &= (r\cos\theta - R\cos\varphi)^2 + (r\sin\theta - R\sin\varphi)^2 \\ &= R^2 + r^2 - 2Rr(\cos\varphi\cos\theta + \sin\varphi\sin\theta) \\ &= R^2 + r^2 - 2Rr\cos(\theta-\varphi).\end{aligned}$$

Using again polar coordinates, we assume that $f = f(r, \theta)$ has a development in sine-Fouries series with respect to θ, in $[0, 2\pi]$:

$$f(r, \theta) = \sum_{m=1}^{\infty} f_m(r) \sin m\theta.$$

We write the candidate solution in the form

$$u(r, \theta) = \sum_{m=1}^{\infty} u_m(r) \sin m\theta$$

where the coefficients $u_m(r)$ are unknown. Substituting, we find:

$$\sum_{m=1}^{\infty} \left\{ u_m''(r) + \frac{1}{r} u'(r) - \frac{m^2}{r^2} u_m(r) \right\} \sin m\theta = \sum_{m=1}^{\infty} f_m(r) \sin m\theta$$

so that the mth coefficient u_m cand be found by solving the ordinary differential equation

$$u_m''(r) + \frac{1}{r} u'(r) - \frac{m^2}{r^2} u_m(r) = f_m(r) \qquad m \geq 1$$

with

$$u_m(R) = 0, \quad \text{and } u_m \text{ bounded in } [0, 1].$$

We are now in position to prove Theorem 4.3, the converse of the mean value property $(m.v.p.)$.

Proof (Theorem 4.3). First observe that if two functions satisfy the $m.v.p.$ in a domain Ω, their difference satisfies this property as well. Let $u \in C(\Omega)$ satisfying the $m.v.p.$ and consider a circle $B \subset\subset \Omega$. We want to show that u is harmonic and infinitely differentiable in Ω. Denote by v the solution of the Dirichlet problem

$$\begin{cases} \Delta v = 0 \text{ in } B \\ v = u \quad \text{on } \partial B. \end{cases}$$

From Theorem 4.5 we know that $v \in C^\infty(B) \cap C(\overline{B})$ and, being harmonic, it satisfies the $m.v.p.$ in B. Then, also $w = v - u$ satisfies the $m.v.p.$ in B and therefore it attains its maximum and minimum on ∂B. Since $w = 0$ on ∂B, we conclude that $u = v$ in B. Since B is arbitrary, $u \in C^\infty(\Omega)$ and harmonic in Ω. $\qquad \square$

4.4 Fundamental Solution and Newtonian Potential

4.4.1 The fundamental solution

In this section we shall derive formulas involving various types of *potentials*, constructed using a special function, called the *fundamental solution* of the Laplace operator.

As we did for the diffusion equation, let us look at the invariance properties characterizing the operator Δ: the invariances by *translations* and by *rotations*.

Let $u = u(\mathbf{x})$ be harmonic in $\mathbb{R}^n$. Invariance by translations means that the function $v(\mathbf{x}) = u(\mathbf{x} - \mathbf{y})$, for each fixed $\mathbf{y}$, is also harmonic, as it is immediate to check.

Invariance by rotations means that, given a rotation in $\mathbb{R}^n$, represented by an orthogonal matrix $\mathbf{M}$ (i.e. $\mathbf{M}^T = \mathbf{M}^{-1}$), also $v(\mathbf{x}) = u(\mathbf{Mx})$ is harmonic in $\mathbb{R}^n$. To check it, observe that, if we denote by $D^2 u$ the Hessian of u, we have

$$\Delta u = \mathrm{Tr} D^2 u = \text{ trace of the Hessian of } u.$$

Since

$$D^2 v(\mathbf{x}) = \mathbf{M}^T D^2 u(\mathbf{Mx}) \mathbf{M}$$

and $\mathbf{M}$ is orthogonal, we have

$$\Delta v(\mathbf{x}) = \mathrm{Tr}[\mathbf{M}^T D^2 u(\mathbf{Mx}) \mathbf{M}] = \mathrm{Tr} D^2 u(\mathbf{Mx}) = \Delta u(\mathbf{Mx}) = 0$$

and therefore v is harmonic.

Now, a typical rotation invariant quantity is *the distance function from a point*, for instance from the origin, that is $r = |\mathbf{x}|$. Thus, let us look for *radially symmetric* harmonic functions $u = u(r)$.

Consider first $n = 2$; using polar coordinates and recalling (4.16), we find

$$\frac{\partial^2 u}{\partial r^2} + \frac{1}{r}\frac{\partial u}{\partial r} = 0$$

so that

$$u(r) = C \log r + C_1.$$

In dimension $n = 3$, using spherical coordinates (r, ψ, θ), $r > 0$, $0 < \psi < \pi$, $0 < \theta < 2\pi$, the operator Δ has the following expression[6]:

$$\Delta = \underbrace{\frac{\partial^2}{\partial r^2} + \frac{2}{r}\frac{\partial}{\partial r}}_{\text{radial part}} + \frac{1}{r^2}\underbrace{\left\{ \frac{1}{(\sin\psi)^2}\frac{\partial^2}{\partial\theta^2} + \frac{\partial^2}{\partial\psi^2} + \cot\psi\frac{\partial}{\partial\psi} \right\}}_{\text{spherical part (Laplace-Beltrami operator)}}.$$

The Laplace equation for $u = u(r)$ becomes

$$\frac{\partial^2 u}{\partial r^2} + \frac{2}{r}\frac{\partial u}{\partial r} = 0$$

whose general integral is

$$u(r) = \frac{C}{r} + C_1 \qquad C, C_1 \text{ arbitrary constants.}$$

Choose $C_1 = 0$ and $C = \frac{1}{4\pi}$ if $n = 3$, $C = -\frac{1}{2\pi}$ if $n = 2$.

---

[6] Appendix D.

The function

$$\Phi(\mathbf{x}) = \begin{cases} -\frac{1}{2\pi}\log|\mathbf{x}| & n = 2 \\ \dfrac{1}{4\pi|\mathbf{x}|} & n = 3 \end{cases} \tag{4.25}$$

is called the **fundamental solution** for the Laplace operator Δ. The above choice of the constant C is made in order to have

$$\Delta\Phi(\mathbf{x}) = -\delta(\mathbf{x}) \qquad \text{in } \mathbb{R}^n$$

where $\delta(\mathbf{x})$ denotes *the Dirac measure at* $\mathbf{x} = \mathbf{0}$.

The physical meaning of Φ is remarkable: if $n = 3$, in standard units, $4\pi\Phi$ represents the electrostatic potential due to a unitary charge located at the origin and vanishing at infinity[7].

Clearly, if the origin is replaced by a point $\mathbf{y}$, the corresponding potential is $\Phi(\mathbf{x} - \mathbf{y})$ and

$$\Delta_{\mathbf{x}}\Phi(\mathbf{x} - \mathbf{y}) = -\delta(\mathbf{x} - \mathbf{y}).$$

By symmetry, we also have $\Delta_{\mathbf{y}}\Phi(\mathbf{x} - \mathbf{y}) = -\delta(\mathbf{x} - \mathbf{y})$.

4.4.2 The Newtonian potential

Suppose that $(4\pi)^{-1}f(\mathbf{x})$ is the density of a charge located inside a compact set in $\mathbb{R}^3$. Then $\Phi(\mathbf{x} - \mathbf{y})f(\mathbf{y})\,d\mathbf{y}$ represents the potential at $\mathbf{x}$ due to the charge $f(\mathbf{y})\,d\mathbf{y}$ inside a small region of volume $d\mathbf{y}$ around $\mathbf{y}$. The full potential is given by the sum of all the contributions; we get

$$\mathcal{N}_f(\mathbf{x}) = \int_{\mathbb{R}^3}\Phi(\mathbf{x} - \mathbf{y})f(\mathbf{y})\,d\mathbf{y} = \frac{1}{4\pi}\int_{\mathbb{R}^3}\frac{f(\mathbf{y})}{|\mathbf{x} - \mathbf{y}|}\,d\mathbf{y} \tag{4.26}$$

which is the *convolution between* f and Φ and it is called the **Newtonian potential** of f. Formally, we have

$$\Delta\mathcal{N}_f(\mathbf{x}) = \int_{\mathbb{R}^3}\Delta_{\mathbf{x}}\Phi(\mathbf{x} - \mathbf{y})f(\mathbf{y})\,d\mathbf{y} = -\int_{\mathbb{R}^3}\delta(\mathbf{x} - \mathbf{y})f(\mathbf{y})\,d\mathbf{y} = -f(\mathbf{x}). \tag{4.27}$$

Under suitable hypotheses on f, (4.27) is indeed true (see Theorem 4.6 below). Clearly, u is not the only solution of $\Delta v = -f$, since $u + c$, c constant, is a solution as well. However, from Liouville's Theorem, the Newtonian potential is the only solution vanishing at infinity[8]. All this is stated precisely in the

---

[7] In dimension 2,

$$2\pi\Phi(x_1, x_2) = -\log\sqrt{x_1^2 + x_2^2}$$

represents the potential due to a charge of density 1, distributed along the x_3 axis.

[8] Let $v \in C^2(\mathbb{R}^3)$ another solution to (4.28), vanishing at infinity. Then $u - v$ is a *bounded* harmonic function in $\mathbb{R}^3$ and therefore is constant. Since it vanishes at infinity it must be zero; thus $u = v$.

theorem below, where, for simplicity, we assume $f \in C^2\left(\mathbb{R}^3\right)$ with compact support[9]. We have:

Theorem 4.6. *Let $f \in C^2\left(\mathbb{R}^3\right)$ with **compact** support. Let u be the Newtonian potential of f, defined by (4.26). Then, u is the only solution in $\mathbb{R}^3$ of*

$$\Delta u = -f \qquad (4.28)$$

belonging to $C^2\left(\mathbb{R}^3\right)$ and vanishing at infinity.

Remark 4.5. An appropriate version of Theorem 4.6 holds in dimension $n = 2$, with the Newtonian potential replaced by the *logarithmic potential*

$$u\left(\mathbf{x}\right) = \int_{\mathbb{R}^2} \Phi\left(\mathbf{x} - \mathbf{y}\right) f\left(\mathbf{y}\right) d\mathbf{y} = -\frac{1}{2\pi} \int_{\mathbb{R}^2} \log\left|\mathbf{x} - \mathbf{y}\right| \ f\left(\mathbf{y}\right) d\mathbf{y}. \qquad (4.29)$$

The logarithmic potential does not vanish at infinity; its asymptotic behavior is

$$u\left(\mathbf{x}\right) = -\frac{M}{2\pi} \log\left|\mathbf{x}\right| + O\left(\frac{1}{\left|\mathbf{x}\right|}\right) \qquad \text{as } \left|\mathbf{x}\right| \to +\infty \qquad (4.30)$$

where

$$M = \int_{\mathbb{R}^2} f\left(\mathbf{y}\right) d\mathbf{y}.$$

Indeed, the logarithmic potential is the only solution of $\Delta u = -f$ in $\mathbb{R}^2$ satisfying (4.30).

4.5 The Green Function

4.5.1 An integral identity

Formula (4.26) gives a representation of the solution to Poisson's equation in all $\mathbb{R}^3$. In bounded domains, any representation formula has to take into account the boundary values, as indicated in the following theorem.

Theorem 4.7. *Let $\Omega \subset \mathbb{R}^n$ be a smooth, bounded domain and $u \in C^2\left(\overline{\Omega}\right)$. Then, for every $\mathbf{x} \in \Omega$,*

$$u\left(\mathbf{x}\right) = -\int_{\Omega} \Phi\left(\mathbf{x} - \mathbf{y}\right) \Delta u\left(\mathbf{y}\right) d\mathbf{y} +$$
$$+ \int_{\partial\Omega} \Phi\left(\mathbf{x} - \boldsymbol{\sigma}\right) \partial_{\nu_{\sigma}} u\left(\boldsymbol{\sigma}\right) d\sigma - \int_{\partial\Omega} u\left(\boldsymbol{\sigma}\right) \partial_{\nu_{\sigma}} \Phi\left(\mathbf{x} - \boldsymbol{\sigma}\right) d\sigma. \qquad (4.31)$$

The last two terms in the right hand side of (4.31) are called *single* and *double layer potentials*, respectively. We are going to examine these surface potentials later. The first one is the Newtonian potential of $-\Delta u$ in Ω.

---

[9] Recall that the *support* of a continuous function f is the *closure of the set where f is not zero*.

Proof. We give it for $n = 3$. Fix $\mathbf{x} \in \Omega$, and consider the fundamental solution

$$\Phi(\mathbf{x} - \mathbf{y}) = \frac{1}{4\pi r_{\mathbf{xy}}} \qquad r_{\mathbf{xy}} = |\mathbf{x} - \mathbf{y}|$$

as a function of $\mathbf{y}$: we write $\Phi(\mathbf{x} - \cdot)$.

We would like to apply *Green's identity* (1.14)

$$\int_{\Omega} (v\Delta u - u\Delta v)dx = \int_{\partial\Omega} (v\partial_\nu u - u\partial_\nu v)d\sigma \tag{4.32}$$

to u and $\Phi(\mathbf{x} - \cdot)$. However, $\Phi(\mathbf{x} - \cdot)$ has a singularity in $\mathbf{x}$, so that it cannot be inserted directly into (4.32). Let us isolate the singularity inside a ball $B_\varepsilon(\mathbf{x})$, with ε small. In the domain $\Omega_\varepsilon = \Omega\backslash\overline{B}_\varepsilon(\mathbf{x})$, $\Phi(\mathbf{x} - \cdot)$ is smooth and harmonic.

Thus, replacing Ω with Ω_ε, we can apply (4.32) to u and $\Phi(\mathbf{x} - \cdot)$. Since

$$\partial\Omega_\varepsilon = \partial\Omega \cup \partial B_\varepsilon(\mathbf{x}),$$

and $\Delta_\mathbf{y}\Phi(\mathbf{x} - \mathbf{y}) = 0$, we find

$$\int_{\Omega_\varepsilon} \frac{1}{r_{\mathbf{xy}}}\Delta u \, dy = \int_{\partial\Omega_\varepsilon} \left(\frac{1}{r_{\mathbf{x\sigma}}} \frac{\partial u}{\partial\nu_\sigma} - u\frac{\partial}{\partial\nu_\sigma}\frac{1}{r_{\mathbf{x\sigma}}} \right) d\sigma$$

$$= \int_{\partial\Omega} (\cdots) \, d\sigma + \int_{\partial B_\varepsilon(\mathbf{x})} \frac{1}{r_{\mathbf{x\sigma}}} \frac{\partial u}{\partial\nu_\sigma} d\sigma + \int_{\partial B_\varepsilon(\mathbf{x})} u\frac{\partial}{\partial\nu_\sigma}\frac{1}{r_{\mathbf{x\sigma}}} d\sigma. \tag{4.33}$$

We let now $\varepsilon \to 0$ in (4.33). We have

$$\int_{\Omega_\varepsilon} \frac{1}{r_{\mathbf{x\sigma}}}\Delta u \, dy \to \int_{\Omega} \frac{1}{r_{\mathbf{x\sigma}}}\Delta u \, dy \qquad \text{as } \varepsilon \to 0 \tag{4.34}$$

since $\Delta u \in C(\overline{\Omega})$ and $r_{\mathbf{x\sigma}}^{-1}$ is positive and integrable in Ω.

On $\partial B_\varepsilon(\mathbf{x})$, we have $r_{\mathbf{x\sigma}} = \varepsilon$ and $|\partial_\nu u| \le M$, since $|\nabla u|$ is bounded; then

$$\left| \int_{\partial B_\varepsilon(\mathbf{x})} \frac{1}{r_{\mathbf{x\sigma}}} \frac{\partial u}{\partial\nu_\sigma} d\sigma \right| \le 4\pi\varepsilon M \to 0 \qquad \text{as } \varepsilon \to 0. \tag{4.35}$$

The most delicate term is

$$\int_{\partial B_\varepsilon(\mathbf{x})} u\frac{\partial}{\partial\nu_\sigma}\frac{1}{r_{\mathbf{x\sigma}}} d\sigma.$$

On $\partial B_\varepsilon(\mathbf{x})$, the outward pointing (with respect to Ω_ε) unit normal at σ is $\nu_\sigma = \frac{\mathbf{x}-\sigma}{\varepsilon}$, so that

$$\frac{\partial}{\partial\nu_\sigma}\frac{1}{r_{\mathbf{x\sigma}}} = \nabla_\mathbf{y}\frac{1}{r_{\mathbf{x\sigma}}} \cdot \nu_\sigma = \frac{\mathbf{x}-\sigma}{\varepsilon^3}\frac{\mathbf{x}-\sigma}{\varepsilon} = \frac{1}{\varepsilon^2}.$$

As a consequence,

$$\int_{\partial B_\varepsilon(\mathbf{x})} u \, \frac{\partial}{\partial \boldsymbol{\nu}_\sigma} \frac{1}{r_{\mathbf{x}\sigma}} \, d\sigma = \frac{1}{\varepsilon^2} \int_{\partial B_\varepsilon(\mathbf{x})} u \, d\sigma \to u(\mathbf{x}) \qquad (4.36)$$

as $\varepsilon \to 0$, by the continuity of u.

Letting $\varepsilon \to 0$ in (4.33), from (4.34), (4.35), (4.36) we obtain (4.31). $\qquad\square$

4.5.2 The Green function for the Dirichlet problem

The function $\varPhi$ defined in (4.25) is the fundamental solution for the Laplace operator $\varDelta$ in all $\mathbb{R}^n$ ($n = 2, 3$). We can also define a fundamental solution for the Laplace operator in any open set and in particular in any *bounded* domain $\varOmega \subset \mathbb{R}^n$, representing the potential due to a unit charge placed at a point $\mathbf{x} \in \varOmega$ and equal to zero on $\partial\varOmega$.

This function, that we denote by $G(\mathbf{x}, \mathbf{y})$, is called the *Green function in* $\varOmega$, for the operator $\varDelta$; for fixed $\mathbf{x} \in \varOmega$, G satisfies

$$\varDelta_{\mathbf{y}} G(\mathbf{x}, \mathbf{y}) = -\delta_{\mathbf{x}} \qquad \text{in } \varOmega$$

and

$$G(\mathbf{x}, \boldsymbol{\sigma}) = 0, \qquad \boldsymbol{\sigma} \in \partial\varOmega.$$

More explicitly, the Green's function can be written in the form

$$G(\mathbf{x}, \mathbf{y}) = \varPhi(\mathbf{x} - \mathbf{y}) - \varphi(\mathbf{x}, \mathbf{y})$$

where φ, for fixed $\mathbf{x} \in \varOmega$, solves the Dirichlet problem

$$\begin{cases} \varDelta_{\mathbf{y}} \varphi = 0 & \text{in } \varOmega \\ \varphi(\mathbf{x}, \boldsymbol{\sigma}) = \varPhi(\mathbf{x} - \boldsymbol{\sigma}) & \text{on } \partial\varOmega. \end{cases} \qquad (4.37)$$

Two important properties of the Green's function are the following:

(a) *Positivity:* $G(\mathbf{x}, \mathbf{y}) > 0$ *for every* $\mathbf{x}, \mathbf{y} \in \varOmega$, *with* $G(\mathbf{x}, \mathbf{y}) \to +\infty$ *when* $\mathbf{x} - \mathbf{y} \to \mathbf{0}$;

(b) *Symmetry:* $G(\mathbf{x}, \mathbf{y}) = G(\mathbf{y}, \mathbf{x})$.

The existence of the Green's function for a particular domain depends on the solvability of the Dirichlet problem (4.37). From Theorem 4.7, we know that this is the case if $\varOmega$ is smooth and bounded, for instance.

Even if we know that the Green's function exists, explicit formulas are available only for special domains. Sometimes a technique known as works. In this method $\varphi(\mathbf{x}, \cdot)$ is considered as the potential due to an imaginary charge q placed at a suitable point $\mathbf{x}^*$, the *image of* $\mathbf{x}$, in the complement of $\varOmega$. The charge q and the point $\mathbf{x}^*$ have to be chosen so that $\varphi(\mathbf{x}, \cdot)$ on $\partial\varOmega$ is equal to the potential created by the unit charge in $\mathbf{x}$.

The simplest way to illustrate the method is to find the Green's function for the upper half-space, although this is an unbounded domain. Clearly, we require that G vanishes at infinity.

• *Green's function for the upper half space in* $\mathbb{R}^3$. Let $\mathbb{R}^3_+$ be the upper half space:
$$\mathbb{R}^3_+ = \{(x_1, x_2, x_3) : x_3 > 0\}.$$
Fix $\mathbf{x} = (x_1, x_2, x_3)$ and observe that if we choose $\mathbf{x}^* = (x_1, x_2, -x_3)$ then, on $y_3 = 0$ we have
$$|\mathbf{x}^* - \mathbf{y}| = |\mathbf{x} - \mathbf{y}|.$$
Thus, if $\mathbf{x} \in \mathbb{R}^3_+$, $\mathbf{x}^*$ belongs to the complement of $\mathbb{R}^3_+$, the function
$$\varphi(\mathbf{x}, \mathbf{y}) = \Phi(\mathbf{x}^* - \mathbf{y}) = \frac{1}{4\pi |\mathbf{x}^* - \mathbf{y}|}$$
is harmonic in $\mathbb{R}^3_+$ and $\varphi(\mathbf{x}, \mathbf{y}) = \Phi(\mathbf{x} - \mathbf{y})$ on the plane $y_3 = 0$. In conclusion,
$$G(\mathbf{x}, \mathbf{y}) = \frac{1}{4\pi |\mathbf{x} - \mathbf{y}|} - \frac{1}{4\pi |\mathbf{x}^* - \mathbf{y}|} \tag{4.38}$$
is the Green's function for the upper half space.

• *Green's function for sphere*. Let $\Omega = B_R = B_R(0) \subset \mathbb{R}^3$. To find the Green's function for B_R, set
$$\varphi(\mathbf{x}, \mathbf{y}) = \frac{q}{4\pi |\mathbf{x}^* - \mathbf{y}|},$$
$\mathbf{x}$ fixed in B_R, and try to determine $\mathbf{x}^*$, outside B_R, and q, so that
$$\frac{q}{4\pi |\mathbf{x}^* - \mathbf{y}|} = \frac{1}{4\pi |\mathbf{x} - \mathbf{y}|} \tag{4.39}$$
when $|\mathbf{y}| = R$. The (4.39) gives
$$|\mathbf{x}^* - \mathbf{y}|^2 = q^2 |\mathbf{x} - \mathbf{y}|^2 \tag{4.40}$$
or
$$|\mathbf{x}^*|^2 - 2\mathbf{x}^* \cdot \mathbf{y} + R^2 = q^2(|\mathbf{x}|^2 - 2\mathbf{x} \cdot \mathbf{y} + R^2).$$
Rearranging the terms we have
$$|\mathbf{x}^*|^2 + R^2 - q^2(R^2 + |\mathbf{x}|^2) = 2\mathbf{y} \cdot (\mathbf{x}^* - q^2\mathbf{x}). \tag{4.41}$$
Since the left hand side does not depend on $\mathbf{y}$, it must be that $\mathbf{x}^* = q^2\mathbf{x}$ and
$$q^4 |\mathbf{x}|^2 - q^2(R^2 + |\mathbf{x}|^2) + R^2 = 0$$
from which $q = R/|\mathbf{x}|$. This works for $\mathbf{x} \neq 0$ and gives
$$G(\mathbf{x}, \mathbf{y}) = \frac{1}{4\pi} \left[\frac{1}{|\mathbf{x} - \mathbf{y}|} - \frac{R}{|\mathbf{x}| |\mathbf{x}^* - \mathbf{y}|} \right], \qquad \mathbf{x}^* = \frac{R^2}{|\mathbf{x}|^2}\mathbf{x}, \; \mathbf{x} \neq 0. \tag{4.42}$$

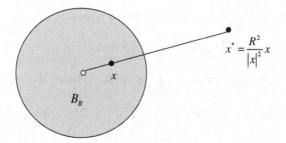

Fig. 4.2. The image $\mathbf{x}^*$ of $\mathbf{x}$ in the construction of the Green's function for the sphere

Since

$$|\mathbf{x}^* - \mathbf{y}| = |\mathbf{x}|^{-1}\left(R^4 - 2R^2\mathbf{x} \cdot \mathbf{y} + \mathbf{y}\,|\mathbf{x}|^2\right)^{1/2},$$

when $\mathbf{x} \to \mathbf{0}$ we have

$$\varphi\left(\mathbf{x}, \mathbf{y}\right) = \frac{1}{4\pi}\frac{R}{|\mathbf{x}|\,|\mathbf{x}^* - \mathbf{y}|} \to \frac{1}{4\pi R}$$

and therefore we can define

$$G\left(\mathbf{0}, \mathbf{y}\right) = \frac{1}{4\pi}\left[\frac{1}{|\mathbf{y}|} - \frac{1}{R}\right].$$

4.5.3 Green's representation formula

From Theorem 4.7 we know that every smooth function u can be written as the sum of a volume (Newtonian) potential with density $-\Delta u$, a single layer potential of *density* $\partial_\nu u$ and a double layer potential of *moment* u. Suppose u solves the Dirichlet problem

$$\begin{cases} \Delta u = f & \text{in } \Omega \\ u = g & \text{on } \partial\Omega. \end{cases} \tag{4.43}$$

Then (4.31) gives, for $\mathbf{x} \in \Omega$,

$$u\left(\mathbf{x}\right) = -\int_\Omega \Phi\left(\mathbf{x} - \mathbf{y}\right) f\left(\mathbf{y}\right) d\mathbf{y} +$$
$$+ \int_{\partial\Omega} \Phi\left(\mathbf{x} - \boldsymbol{\sigma}\right) \partial_{\nu_\sigma} u\left(\boldsymbol{\sigma}\right) d\sigma - \int_{\partial\Omega} g\left(\boldsymbol{\sigma}\right) \partial_{\nu_\sigma}\Phi\left(\mathbf{x} - \boldsymbol{\sigma}\right) d\sigma. \tag{4.44}$$

This representation formula for u is not satisfactory, since it involves the data f and g but also the normal derivative $\partial_{\nu_\sigma} u$, which is unknown. To get rid of $\partial_{\nu_\sigma} u$, let $G\left(\mathbf{x}, \mathbf{y}\right) = \Phi\left(\mathbf{x} - \mathbf{y}\right) - \varphi\left(\mathbf{x}, \mathbf{y}\right)$ be the Green's function in Ω. Since

$\varphi\left(\mathbf{x},\cdot\right)$ is harmonic in Ω, we can apply (4.32) to u and $\varphi\left(\mathbf{x},\cdot\right)$; we find

$$0 = \int_{\Omega}\varphi\left(\mathbf{x},\mathbf{y}\right)f\left(\mathbf{y}\right)d\mathbf{y} +$$
$$-\int_{\partial\Omega}\varphi\left(\mathbf{x},\boldsymbol{\sigma}\right)\partial_{\nu_\sigma}u\left(\boldsymbol{\sigma}\right)d\sigma + \int_{\partial\Omega}g\left(\boldsymbol{\sigma}\right)\partial_{\nu_\sigma}\varphi\left(\mathbf{x},\boldsymbol{\sigma}\right)d\sigma. \tag{4.45}$$

Adding (4.44), (4.45) and recalling that $\varphi\left(\mathbf{x},\boldsymbol{\sigma}\right) = \Phi\left(\mathbf{x}-\boldsymbol{\sigma}\right)$ on $\partial\Omega$, we obtain:

Theorem 4.8. *Let Ω be a smooth domain and u be a smooth solution of (4.43). Then:*

$$u\left(\mathbf{x}\right) = -\int_{\Omega}f\left(\mathbf{y}\right)G\left(\mathbf{x},\mathbf{y}\right)d\mathbf{y} - \int_{\partial\Omega}g\left(\boldsymbol{\sigma}\right)\partial_{\nu_\sigma}G\left(\mathbf{x},\boldsymbol{\sigma}\right)d\sigma. \tag{4.46}$$

Thus the solution of the Dirichlet problem (4.43) can be written as the sum of the two Green's potentials in the right hand side of (4.46) and it is known as soon as the Green's function in Ω is known. In particular, if u is harmonic, then

$$u\left(\mathbf{x}\right) = -\int_{\partial\Omega}g\left(\boldsymbol{\sigma}\right)\partial_{\nu_\sigma}G\left(\mathbf{x},\boldsymbol{\sigma}\right)d\sigma. \tag{4.47}$$

The function

$$P\left(\mathbf{x},\boldsymbol{\sigma}\right) = -\partial_{\nu_\sigma}G\left(\mathbf{x},\boldsymbol{\sigma}\right)$$

is called **Poisson's kernel**. Since $G\left(\cdot,\boldsymbol{\sigma}\right) > 0$ inside Ω and vanishes on Ω, P is *nonnegative* (actually positive).

On the other hand, the formula

$$u\left(\mathbf{x}\right) = -\int_{\Omega}f\left(\mathbf{y}\right)G\left(\mathbf{x},\mathbf{y}\right)d\mathbf{y}$$

gives the solution of the Poisson equation $\Delta u = f$ in Ω, vanishing on $\partial\Omega$. From the positivity of G we have that:

$$f \geq 0 \quad \text{in } \Omega \text{ implies } u \leq 0 \text{ in } \Omega,$$

which is another form of the maximum principle.

• *Poisson's kernel and Poisson's formula.* From (4.42) we can compute Poisson's kernel for the sphere $B_R\left(\mathbf{0}\right)$. We have, recalling that $\mathbf{x}^* = R^2\left|\mathbf{x}\right|^{-2}\mathbf{x}$, if $\mathbf{x} \neq \mathbf{0}$,

$$\nabla_{\mathbf{y}}\left[\frac{1}{\left|\mathbf{x}-\mathbf{y}\right|} - \frac{R}{\left|\mathbf{x}\right|\left|\mathbf{x}^*-\mathbf{y}\right|}\right] = \frac{\mathbf{x}-\mathbf{y}}{\left|\mathbf{x}-\mathbf{y}\right|^3} - \frac{R}{\left|\mathbf{x}\right|}\frac{\mathbf{x}^*-\mathbf{y}}{\left|\mathbf{x}^*-\mathbf{y}\right|^3}.$$

If $\boldsymbol{\sigma} \in \partial B_R\left(\mathbf{0}\right)$, from (4.40) we have $\left|\mathbf{x}^*-\boldsymbol{\sigma}\right| = R\left|\mathbf{x}\right|^{-1}\left|\mathbf{x}-\boldsymbol{\sigma}\right|$, therefore

$$\nabla_{\mathbf{y}}G\left(\mathbf{x},\boldsymbol{\sigma}\right) = \frac{1}{4\pi}\left[\frac{\mathbf{x}-\boldsymbol{\sigma}}{\left|\mathbf{x}-\boldsymbol{\sigma}\right|^3} - \frac{\left|\mathbf{x}\right|^2}{R^2}\frac{\mathbf{x}^*-\boldsymbol{\sigma}}{\left|\mathbf{x}-\boldsymbol{\sigma}\right|^3}\right] = \frac{-\boldsymbol{\sigma}}{4\pi\left|\mathbf{x}-\boldsymbol{\sigma}\right|^3}\left[1 - \frac{\left|\mathbf{x}\right|^2}{R^2}\right].$$

Since on $\partial B_R(\mathbf{0})$ the exterior unit normal is $\boldsymbol{\nu}_\sigma = \boldsymbol{\sigma}/R$, we have

$$P(\mathbf{x}, \boldsymbol{\sigma}) = -\partial_{\nu_\sigma} G(\mathbf{x}, \boldsymbol{\sigma}) = -\nabla_\mathbf{y} G(\mathbf{x}, \boldsymbol{\sigma}) \cdot \boldsymbol{\nu}_\sigma = \frac{R^2 - |\mathbf{x}|^2}{4\pi R} \frac{1}{|\mathbf{x} - \boldsymbol{\sigma}|^3}.$$

As a consequence, we obtain Poisson's formula

$$u(\mathbf{x}) = \frac{R^2 - |\mathbf{x}|^2}{4\pi R} \int_{\partial B_R(0)} \frac{g(\boldsymbol{\sigma})}{|\mathbf{x} - \boldsymbol{\sigma}|^3} d\sigma \qquad (4.48)$$

for the unique solution of the Dirichlet problem $\Delta u = 0$ in $B_R(\mathbf{0})$ and $u = g$ on $\partial B_R(\mathbf{0})$.

4.5.4 The Neumann function

We can find a representation formula for the solution of a Neumann problem as well. Let u be a smooth solution of the problem

$$\begin{cases} \Delta u = f & \text{in } \Omega \\ \partial_\nu u = h & \text{on } \partial\Omega \end{cases} \qquad (4.49)$$

where f and h have to satisfy the solvability condition

$$\int_{\partial\Omega} h(\boldsymbol{\sigma}) d\sigma = \int_\Omega f(\mathbf{y}) d\mathbf{y}, \qquad (4.50)$$

keeping in mind that u is uniquely determined up to an additive constant. From Theorem 4.7 we can write

$$u(\mathbf{x}) = -\int_\Omega \Phi(\mathbf{x} - \mathbf{y}) f(\mathbf{y}) d\mathbf{y} +$$
$$+ \int_{\partial\Omega} h(\boldsymbol{\sigma}) \Phi(\mathbf{x} - \boldsymbol{\sigma}) d\sigma - \int_{\partial\Omega} u(\boldsymbol{\sigma}) \partial_{\nu_\sigma} \Phi(\mathbf{x} - \boldsymbol{\sigma}) d\sigma \qquad (4.51)$$

and now we should get rid of the second integral, containing the unknown data u on $\partial\Omega$. Mimicking what we have done for the Dirichlet problem, we try to find an analog of the Green's function, that is a function $N = N(\mathbf{x}, \mathbf{y})$ given by

$$N(\mathbf{x}, \mathbf{y}) = \Phi(\mathbf{x} - \mathbf{y}) - \psi(\mathbf{x}, \mathbf{y})$$

where, for $\mathbf{x}$ fixed, ψ is a solution of

$$\begin{cases} \Delta_\mathbf{y} \psi = 0 & \text{in } \Omega \\ \partial_{\nu_\sigma} \psi(\mathbf{x}, \boldsymbol{\sigma}) = \partial_{\nu_\sigma} \Phi(\mathbf{x} - \boldsymbol{\sigma}) & \text{on } \partial\Omega, \end{cases}$$

in order to have $\partial_{\nu_\sigma} N(\mathbf{x}, \boldsymbol{\sigma}) = 0$ on $\partial\Omega$. But this Neumann problem has no solution because the compatibility condition

$$\int_{\partial\Omega} \partial_{\nu_\sigma} \Phi(\mathbf{x} - \boldsymbol{\sigma}) d\sigma = 0$$

is not satisfied. In fact, letting $u \equiv 1$ in (4.31), we get

$$\int_{\partial\Omega} \partial_{\nu_\sigma} \Phi\,(\mathbf{x} - \boldsymbol{\sigma})\,d\sigma = -1. \tag{4.52}$$

Thus, taking into account (4.52), we require ψ to satisfy

$$\begin{cases} \Delta_\mathbf{y} \psi = 0 & \text{in } \Omega \\ \partial_{\nu_\sigma} \psi\,(\mathbf{x}, \boldsymbol{\sigma}) = \partial_{\nu_\sigma} \Phi\,(\mathbf{x} - \boldsymbol{\sigma}) + \frac{1}{|\partial\Omega|} & \text{on } \partial\Omega. \end{cases} \tag{4.53}$$

In this way,

$$\int_{\partial\Omega} \left(\partial_{\nu_\sigma} \Phi\,(\mathbf{x} - \boldsymbol{\sigma}) + \frac{1}{|\partial\Omega|} \right) d\sigma = 0 \cdot$$

and (4.53) is solvable. Note that, with this choice of ψ, we have

$$\partial_{\nu_\sigma} N\,(\mathbf{x}, \boldsymbol{\sigma}) = -\frac{1}{|\partial\Omega|} \qquad \text{on } \partial\Omega. \tag{4.54}$$

Apply now (4.32) to u and $\psi\,(\mathbf{x}, \cdot)$; we find:

$$0 = -\int_{\partial\Omega} \psi\,(\mathbf{x}, \boldsymbol{\sigma})\,\partial_{\nu_\sigma} u\,(\boldsymbol{\sigma})\,d\sigma + \int_{\partial\Omega} h\,(\boldsymbol{\sigma})\,\partial_{\nu_\sigma} \psi\,(\boldsymbol{\sigma})\,d\sigma + \int_\Omega \psi\,(\mathbf{y})\,f\,(\mathbf{y})\,d\mathbf{y}. \tag{4.55}$$

Adding (4.55) to (4.51) and using (4.54) we obtain:

Theorem 4.9. *Let Ω be a smooth domain and u be a smooth solution of (4.49). Then:*

$$u\,(\mathbf{x}) - \frac{1}{|\partial\Omega|} \int_{\partial\Omega} u\,(\boldsymbol{\sigma})\,d\sigma = \int_{\partial\Omega} h\,(\boldsymbol{\sigma})\,N\,(\mathbf{x}, \boldsymbol{\sigma})\,d\sigma - \int_\Omega f\,(\mathbf{y})\,N\,(\mathbf{x}, \mathbf{y})\,d\mathbf{y}.$$

Thus, the solution of the Neumann problem (4.49) can also be written as the sum of two potentials, up to the additive constant $c = \frac{1}{|\partial\Omega|} \int_{\partial\Omega} u\,(\boldsymbol{\sigma})\,d\sigma$, the mean value of u.

The function N is called *Neumann function* (also Green's function for the Neumann problem) and it is defined up to an additive constant.

4.6 Numerical methods

4.6.1 The 5 point finite difference scheme for the Poisson problem

Let us consider a two-dimensional Poisson problem complemented with Dirichlet boundary conditions,

$$\begin{cases} -\Delta u = f & \text{in } \Omega = (0, L_x) \times (0, L_y) \\ u = g & \text{on } \partial\Omega. \end{cases} \tag{4.56}$$

The finite difference discretization is easily applicable to tensor product domains, such as $\Omega = (0, L_x) \times (0, L_y)$, because in this case the computational grid is obtained as a tensor product of the one-dimensional partition of $(0, L_x)$ and $(0, L_y)$. More precisely, given two integers N_x and N_y, we define

$$x_i = i \cdot h_x, \; h_x = L_x/(N_x + 1), \; i = 0, \ldots, N_x + 1$$
$$y_j = j \cdot h_y, \; h_y = L_y/(N_y + 1), \; j = 0, \ldots, N_y + 1$$

being (x_i, y_j) the computational nodes where we aim to approximate the solution of (4.56), namely $u_{ij} \simeq u(x_i, y_j)$.

Since in the Cartesian coordinate system the Laplace operator is simply given by $\Delta u = \partial_{xx} u + \partial_{yy} u$, a finite difference discretization can be achieved by exploiting the three-point approximation for second order derivatives, namely (3.89),

$$\partial_{xx} u(x_i, y_j) = \frac{1}{h_x^2} \left(u(x_{i+1}, y_j) - 2u(x_i, y_j) + u(x_{i-1}, y_j) \right) + \mathcal{O}(h_x^2)$$

$$\partial_{yy} u(x_i, y_j) = \frac{1}{h_y^2} \left(u(x_i, y_{j+1}) - 2u(x_i, y_j) + u(x_i, y_{j-1}) \right) + \mathcal{O}(h_y^2)$$

that can be combined to obtain the so called **5 point approximation** of the Laplace operator in terms of pointwise values of u,

$$\Delta u(x_i, y_j) = \frac{1}{h^2} \left(u(x_{i+1}, y_j) + u(x_i, y_{j+1}) - 4u(x_i, y_j) \right.$$
$$\left. + u(x_{i-1}, y_j) + u(x_i, y_{j-1}) \right) + \mathcal{O}(h^2) \qquad (4.57)$$

where for simplicity we have assumed $h_x = h_y = h$. Equation (4.57) also shows that the the 5 point scheme is second order accurate. Then, rewriting (4.57) for the discrete unknowns, the finite difference approximation of the Poisson problem (4.56) consists in finding u_{ij} such that

$$\begin{cases} \frac{1}{h^2} \left(u_{i+1,j} + u_{i,j+1} - 4u_{ij} + u_{i-1,j} + u_{i,j-1} \right) & i = 1, \ldots, N_x \\ \quad = f(x_i, y_j) & j = 1, \ldots, N_y \\ u_{i,0} = g(x_i, 0), \; u_{i,N_y+1} = g(x_i, L_x) & i = 0, \ldots, N_x + 1 \\ u_{0,j} = g(0, y_j), \; u_{N_x+1,j} = g(L_y, y_j) & j = 0, \ldots, N_y + 1. \end{cases} \qquad (4.58)$$

Now, the main difficulty consists to reformulate (4.58) in matrix form. In contrast to the one-dimensional case, where the ordering of the discrete unknowns is straightforwardly dictated by the orientation of the x axis, in multiple space dimensions the ordering of the unknowns is relevant. Denoting by $\mathbf{U} = \{U_k\}_{k=1}^N$ with $N = N_x \times N_y$ the collection of discrete degrees of freedom, the problem of ordering corresponds to define a mapping from the couple (i, j) to the integer k. In the classical cases of ordering by rows or by columns we obtain

ordering by rows $\qquad$ $\mathbf{U}_k = u_{ij}$ with $k = (j-1) \cdot N_x + i$

ordering by columns $\qquad$ $\mathbf{U}_k = u_{ij}$ with $k = (i-1) \cdot N_y + j$.

From now on we will adopt the row ordering, although switching i with j and N_x with N_y leads to the matrix formulation corresponding to column ordering. We aim to rewrite (4.58) as follows

$$\mathbf{A}_h \mathbf{U} = \mathbf{F}_h, \quad \mathbf{A}_h \in \mathbb{R}^{N \times N}, \ \mathbf{U} \in \mathbb{R}^N, \ \mathbf{F}_h \in \mathbb{R}^N \qquad (4.59)$$

where $\mathbf{A}_h$ and $\mathbf{F}_h$ have to be suitably defined and the subscript h reminds that their coefficients depend on $1/h_x^2$ and $1/h_y^2$. For the assembly of $\mathbf{A}_h$ we proceed by blocks. In particular, we define as $\mathbf{D}_h \in \mathbb{R}^{N_x \times N_x}$ the block that preforms the coupling of nodes on the same row, while $\mathbf{E}_h \in \mathbb{R}^{N_x \times N_x}$ is the block accounting for the remaining degrees of freedom in (4.58),

$$\mathbf{D}_h = \begin{bmatrix} \alpha & \beta & 0 & \cdots & & & \cdots & 0 \\ \beta & \alpha & \beta & 0 & \cdots & & \cdots & 0 \\ 0 & \beta & \alpha & \beta & 0 & \cdots & \cdots & 0 \\ \vdots & & \ddots & \ddots & \ddots & & & \vdots \\ \vdots & & & \ddots & \ddots & \ddots & & \vdots \\ 0 & \cdots & & \cdots & \beta & \alpha & \beta & 0 \\ 0 & \cdots & & & \cdots & \beta & \alpha & \beta \\ 0 & \cdots & & & \cdots & 0 & \beta & \alpha \end{bmatrix},$$

$$\mathbf{E}_h = \begin{bmatrix} \gamma & 0 & 0 & \cdots & & & \cdots & 0 \\ 0 & \gamma & 0 & 0 & \cdots & & \cdots & 0 \\ 0 & 0 & \gamma & 0 & 0 & \cdots & \cdots & 0 \\ \vdots & & \ddots & \ddots & \ddots & & & \vdots \\ \vdots & & & \ddots & \ddots & \ddots & & \vdots \\ 0 & \cdots & & \cdots & 0 & \gamma & 0 & 0 \\ 0 & \cdots & & & \cdots & 0 & \gamma & 0 \\ 0 & \cdots & & & \cdots & 0 & 0 & \gamma \end{bmatrix}$$

$$\alpha = \frac{2}{h_x^2} + \frac{2}{h_y^2}, \ \beta = -\frac{1}{h_x^2}, \ \gamma = -\frac{1}{h_y^2}.$$

Analogously to the block decomposition of $\mathbf{A}_h$, we proceed to the decomposition of $\mathbf{F}_h$ into sub-vectors $\mathbf{F}_{h,j} \in \mathbb{R}^{N_x}$ relative to the j–th row of the system,

$$\mathbf{F}_{h,1} = \begin{bmatrix} f(x_1,y_1) + \frac{g(0,y_1)}{h_x^2} + \frac{g(x_1,0)}{h_y^2} \\ f(x_2,y_1) + \frac{g(x_2,0)}{h_y^2} \\ \vdots \\ f(x_i,y_1) + \frac{g(x_i,0)}{h_y^2} \\ \vdots \\ f(x_{N_x-1},y_1) + \frac{g(x_{N_x-1},0)}{h_y^2} \\ f(x_{N_x},y_1) + \frac{g(L_x,y_1)}{h_x^2} + \frac{g(L_x,0)}{h_y^2} \end{bmatrix},$$

$$\mathbf{F}_{h,N_y} = \begin{bmatrix} f(x_1,y_{N_y}) + \frac{g(0,y_{N_y})}{h_x^2} + \frac{g(x_1,L_y)}{h_y^2} \\ f(x_2,y_{N_y}) + \frac{g(x_2,L_y)}{h_y^2} \\ \vdots \\ f(x_i,y_{N_y}) + \frac{g(x_i,L_y)}{h_y^2} \\ \vdots \\ f(x_{N_x-1},y_{N_y}) + \frac{g(x_{N_x-1},L_y)}{h_y^2} \\ f(x_{N_x},y_{N_y}) + \frac{g(L_x,y_{N_y})}{h_x^2} + \frac{g(L_x,L_y)}{h_y^2} \end{bmatrix},$$

$$\mathbf{F}_{h,j} = \begin{bmatrix} f(x_1,y_j) + \frac{g(0,y_j)}{h_x^2} \\ f(x_2,y_j) \\ \vdots \\ f(x_i,y_j) \\ \vdots \\ f(x_{N_x-1},y_j) \\ f(x_{N_x},y_j) + \frac{g(L_x,y_j)}{h_x^2} \end{bmatrix}, \quad j = 2,\ldots,N_y-1.$$

Casting $\mathbf{D}_h$, $\mathbf{E}_h$ and $\mathbf{F}_{h,j}$ into the global $\mathbf{A}_h$ and the right hand side $\mathbf{F}_h$, we obtain

$$\mathbf{A}_h = \begin{bmatrix} \mathbf{D}_h & \mathbf{E}_h & 0 & \cdots & & \cdots & 0 \\ \mathbf{E}_h & \mathbf{D}_h & \mathbf{E}_h & 0 & \cdots & & \cdots & 0 \\ 0 & \mathbf{E}_h & \mathbf{D}_h & \mathbf{E}_h & 0 & \cdots & \cdots & 0 \\ \vdots & & \ddots & \ddots & \ddots & & & \vdots \\ \vdots & & & \ddots & \ddots & \ddots & & \vdots \\ 0 & \cdots & & \cdots & \mathbf{E}_h & \mathbf{D}_h & \mathbf{E}_h & 0 \\ 0 & \cdots & & & \cdots & \mathbf{E}_h & \mathbf{D}_h & \mathbf{E}_h \\ 0 & \cdots & & & \cdots & 0 & \mathbf{E}_h & \mathbf{D}_h \end{bmatrix}, \quad \mathbf{F}_h = \begin{bmatrix} \mathbf{F}_{h,1} \\ \mathbf{F}_{h,2} \\ \mathbf{F}_{h,3} \\ \vdots \\ \vdots \\ \mathbf{F}_{h,N_y-2} \\ \mathbf{F}_{h,N_y-1} \\ \mathbf{F}_{h,N_y} \end{bmatrix}. \quad (4.60)$$

As observed in Chapter 3, problem (4.56) can be generalized to the unsteady case,

$$\begin{cases} u_t - \Delta u = f & \text{in } \Omega \times \mathbb{R}^+ \\ u = g & \text{on } \partial\Omega \times \mathbb{R}^+ \\ u(t = 0) = u_0 & \text{on } \Omega \times \{t = 0\} \end{cases}$$

that can be discretized by means of the 5 point scheme for the Laplace operator combined with one of the Euler schemes addressed for the one-dimensional heat equation. We remind that the application of the forward Euler time advancing method is restricted to the range of space and time discretization steps, h and τ respectively, that satisfy a stability condition. Theorem 3.6 shows that such condition is related to the spectrum of the forward Euler iteration matrix $\mathbf{C}_h^\tau = \mathbf{I} - \tau \mathbf{A}_h$. To facilitate the extension of corollaries 3.4 and 3.5 to the present case, we briefly summarize below the spectral properties of $\mathbf{A}_h$.

Theorem 4.10. *The matrix* $\mathbf{A}_h$ *in* (4.60) *is symmetric and positive definite. The spectrum of* $\mathbf{A}_h$ *consists of the following* $N_x \times N_y$ *distinct eigenvalues,*

$$\lambda_k = \frac{4}{h_x^2} \sin^2\left(\frac{\pi}{2}\frac{i h_x}{L_x}\right) + \frac{4}{h_y^2} \sin^2\left(\frac{\pi}{2}\frac{j h_y}{L_y}\right)$$

where $k = (j - 1) \cdot N_x + i$ *for* $i = 1, \ldots, N_x$ *and* $j = 1, \ldots, N_y$.

4.7 Exercises

4.1 (Separation of variables on the rectangle). Consider the rectangle $R = \{(x, y) \in \mathbb{R}^2 : 0 < x < L, 0 < y < H\}$. Solve the following problem:

$$\begin{cases} \Delta u = 0 & \text{in } R \\ u(0, y) = g_1(y), u(L, y) = g_2(y) & 0 < y < H \\ u(x, 0) = g_3(x), u(x, H) = g_4(x) & 0 < x < L \end{cases}$$

where $g_i \in C(\mathbb{R})$, $i = 1, \cdots, 4$ and $g_1(0) = g_3(0)$, $g_3(L) = g_2(0)$, $g_2(H) = g_4(L)$, $g_1(H) = g_4(0)$.

4.2 (Harmonic functions). Assume that u is harmonic in a domain $\Omega \subseteq \mathbb{R}^n$; show that the derivatives of u of any order are harmonic in Ω.

4.3. Consider the unit circle B_R centered in $(0, 0)$ in $\mathbb{R}^2$ and use the method of separation of variables to solve non homogeneous problem

$$\begin{cases} \Delta u = f & \text{in } B_R \\ u = 0 & \text{on } \partial B_R. \end{cases}$$

4.4. Solve the following non homogeneous problem with nonhomogeneus boundary conditions

$$\begin{cases} \Delta u\,(x,y) = y & \text{in } B_1 \subset \mathbb{R}^2 \\ u = 1 & \text{on } \partial B_1. \end{cases}$$

4.5 (Separation of the variables on the ring). Considering the domain $B_{1,R} = \{(r,\theta) \text{ such that } 1 < r < R\}$, find the solution of the Dirichlet problem

$$\begin{cases} \Delta u = 0 & \text{in } B_{1,R} \\ u(1,\theta) = g(\theta) & 0 \le \theta \le 2\pi \\ u(R,\theta) = h(\theta) & 0 \le \theta \le 2\pi \end{cases}$$

where $g, h \in C^1(\mathbb{R})$ and are 2π-periodic functions. Then, write the solution when $g(\theta) = \sin\theta$ and $h(\theta) = 1$.

4.6 (Schwarz's reflection principle). Denote

$$B_1^+ = \{(x,y) \in \mathbb{R}^2 : x^2 + y^2 < 1, \, y > 0\}$$

and assume $u \in C^2\left(B_1^+\right) \cap C(\overline{B_1^+})$ is harmonic in B_1^+ and such that $u\,(x,0) = 0$. Show that the function

$$U\,(x,y) = \begin{cases} u\,(x,y) & y \ge 0 \\ -u\,(x,-y) & y < 0 \end{cases}$$

obtained from u by odd reflection with respect to y is harmonic in B_1.

4.7 (Sub and superharmonic functions). A function $u \in C^2(\Omega)$, $\Omega \subseteq \mathbb{R}^n$ is *subharmonic* (resp. *superharmonic*) *in* Ω *if* $\Delta u \ge 0$ $(\Delta u \le 0)$ in Ω. For $n = 2$, prove the following statements.

a) If u is subharmonic[10], then, for every $B_R\,(\mathbf{x}) \subset\subset \Omega$, the following inequalities hold

$$u\,(\mathbf{x}) \le \frac{1}{2\pi R} \int_{\partial B_R(\mathbf{x})} u\,(\mathbf{y})\,d\mathbf{y} \quad \text{and} \quad u\,(\mathbf{x}) \le \frac{1}{\pi R^2} \int_{B_R(\mathbf{x})} u\,(\mathbf{y})\,d\mathbf{y}.$$

b) If u is subharmonic (superharmonic) and $u \in C\left(\overline{\Omega}\right)$, then the maximum (minimum) of u is attained at one unique point of $\partial\Omega$, unless u is constant.

c) If u is harmonic in Ω then u^2 is subharmonic.

d) Let u be subharmonic in Ω, and consider a function $F : \mathbb{R} \to \mathbb{R}$, with $F \in C^2\,(\mathbb{R})$. Which conditions on F guarantee that the composition function $F \circ u$ is subharmonic?

---

[10] If u were superharmonic, the direction of the inequalities has to be reverted, in fact, the statement can be applied to $-u$.

4.8 (Torsion problem). Assume $\Omega \subset \mathbb{R}^2$ is a bounded domain and let $v \in C^2(\Omega) \cap C^1(\overline{\Omega})$ be the solution to the problem

$$\begin{cases} v_{xx} + v_{yy} = -2 & \text{in } \Omega \\ v = 0 & \text{on } \partial\Omega. \end{cases} \tag{4.61}$$

Prove that $|\nabla v|^2$ attains its maximum on $\partial\Omega$.

4.9. Let $B_{1,2} = \{(r,\theta) \in \mathbb{R}^2 ; 1 < r < 2\}$. Examine the solvability of the Neumann problem

$$\begin{cases} \Delta u = -1 & \text{in } B_{1,2} \\ u = \cos\theta & \text{on } r = 1 \\ u = \lambda(\cos\theta)^2 & \text{on } r = 2 \end{cases} \qquad (\lambda \in \mathbb{R})$$

and write an explicit formula for the solution, when it exists.

4.10. Let u be harmonic in $\mathbb{R}^3$ such that

$$\int_{\mathbb{R}^3} |u(\mathbf{x})|^2 \, d\mathbf{x} < \infty.$$

Show that $u \equiv 0$.

4.11. Let u be harmonic in $\mathbb{R}^n$ and $\mathbf{M}$ an orthogonal matrix of order n. Using the mean value property, show that $v(\mathbf{x}) = u(\mathbf{Mx})$ is harmonic in $\mathbb{R}^n$.

4.12. Compute the Green function for the circle of radius R.

4.13. Let $\Omega \subset \mathbb{R}^n$ be a bounded smooth domain and G be the Green function in Ω. Prove that, for every $\mathbf{x}, \mathbf{y} \in \Omega, \mathbf{x} \neq \mathbf{y}$:

(a) $G(\mathbf{x}, \mathbf{y}) > 0$;
(b) $G(\mathbf{x}, \mathbf{y}) = G(\mathbf{y}, \mathbf{x})$.

4.14. Compute the Green function for the half plane $\mathbb{R}^2_+ = \{(x,y) ; y > 0\}$ and (formally) derive the Poisson formula

$$u(x,y) = \frac{y}{\pi} \int_{\mathbb{R}} \frac{u(x,0)}{(x-\xi)^2 + y^2} d\xi$$

for a bounded harmonic function in $\mathbb{R}^2_+$.

4.7.1 Approximation of an elastic membrane using the 5 point scheme

The Poisson equation is a good model to study the deformation of an elastic membrane under a given load. For this case Dirichlet, Neumann or Robin of

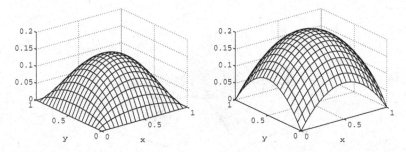

Fig. 4.3. Numerical approximation of Poisson problem with Dirichlet boundary conditions, $u = 0$ (left panel), and Neumann conditions, $\partial_\nu u = -1/2$ (right panel)

boundary conditions represent different ways to constrain the membrane at its boundaries.

In particular, let us consider a unit square membrane, $L_x = L_y = 1$, deformed by an upward constant load, $f = 2$. We approximate the Poisson problem with the 5 point scheme over a computational mesh characterized by $h_x = h_y = 0.05$, that is equivalent to set $N_x = N_y = 19$ in (4.58). We compare the numerical simulations obtained in the case of homogeneous Dirichlet boundary conditions, $u = 0$ on $\partial\Omega$, with the results obtained with Neumann conditions, $\partial_\nu u = -1/2$ on $\partial\Omega$. We notice that in the latter case, the external load and the boundary forces are in equilibrium, as prescribed by the compatibility condition (4.7). Finally, to remove the ambiguity with respect to rigid body motions, we introduce the additional constraint $u(0,0) = 0$ to the Neumann problem.

Results, reported in Fig. 4.3, match common sense intuition about membrane deformation. A more rigorous validation can be performed observing that the exact solution of the homogeneous Dirichlet problem is $u(x,y) = \frac{1}{2}x(1-x) + \frac{1}{2}y(1-y)$.

4.7.2 Numerical simulations for testing maximum principles

We address the following problem,

$$\begin{cases} -\Delta u = 0 & \text{in } \Omega = (0,1) \times (0,1) \\ \partial_\nu u = \sin(2\pi s) & \text{on } \partial\Omega \end{cases} \tag{4.62}$$

where s is the arc length along $\partial\Omega$ originating at $(0,0)$. Observing that the solution of (4.62) is an harmonic function, we expect that maxima and minima take place on the boundary.

Fig. 4.4 shows that the maximum principle is satisfied. Furthermore, we notice that peaks of the solution correspond to peaks of the normal derivatives. This property is related to the Hopf's principle, which states that the solution must satisfy $\partial_\nu u > 0$ on its minima and $\partial_\nu u < 0$ on its maxima.

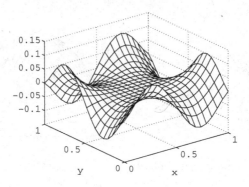

Fig. 4.4. Numerical approximation of (4.62) obtained with $h = 0.05$ and setting the additional constraint $u(0,0) = 0$

5

Reaction-diffusion models

In this chapter we shall focus on models in which reaction and diffusion are in competition. Of particular interest is the study of the asymptotic behavior of the solutions as time goes on and to explore the existence and the stability properties of limiting *steady states*.

In the first section we present some models of pure reaction, governed by ordinary differential equations. In this connections we briefly recall the notion of equilibrium point and the linearized stability criterion (more details can be found in Appendix B).

In Section 2 we examine some elementary examples of linear reaction introducing the concept of *critical dimension*. Section 3 is devoted to the method of super and subsolutions to control the time evolution of semilinear models. Our reference model is the celebrated Fisher-Kolmogoroff equation on population dynamics.

In the last two sections we describe the Turing instability phenomenon, arising in competition population dynamics in presence of different speeds of diffusion.

5.1 Reaction Models

5.1.1 The mass action law

Consider a system consisting of m chemical species $C_1, C_2, ..., C_m$, reciprocally reacting. The statistical mechanics law of mass action establish that these substances react according to the following scheme

$$\lambda_1 C_1 + \lambda_2 C_2 + ... + \lambda_m C_m \rightarrow \mu_1 C_1 + \mu_2 C_2 + ... + \mu_m C_m.$$

Moreover, the *rate of reaction* r, whose physical dimensions are $mol \times m^{-3} \times s^{-1}$, is given by

$$\dot{r} = k c_1^{\lambda_1} c_2^{\lambda_2} \cdots c_j^{\lambda_m} \tag{5.1}$$

where $c_j = [C_j]$ denotes the *concentration* of C_j.

Salsa S., Vegni F.M.G., Zaretti A., Zunino P.: *A Primer on PDEs. Models, Methods, Simulations.*
Unitext – La Matematica per il 3+2 65.
DOI 10.1007/978-88-470-2862-3_5, © Springer-Verlag Italia 2013

In (5.1) k is a dimensional constant and λ_j, μ_j are the so called *stoichiometric coefficients*. Conservation of mass implies that

$$\sum_{j=1}^{m} \lambda_j m_j = \sum_{j=1}^{m} \mu_j m_j$$

where m_j is the *molar mass* of the $j - th$ substance[1].

For a single chemical species, conservation of mass gives

$$\frac{dc_j}{dt} = \left(\mu_j - \lambda_j \right) \dot{r}. \tag{5.2}$$

Example 5.1. Hydrogen combustion is described by the reaction

$$2H_2 + O_2 \rightarrow 2H_2O$$

with reaction rate proportional to $h^2 o$ ($h = [h_2], o = [O_2]$).

Example 5.2 (An autocatalytic reaction). Autocatalysis is a process in which a substance is involved in its own production. For instance

$$A + B \rightarrow 2B.$$

Denoting with a and b the concentration of A and B, the reaction rate is $\dot{r} = kab$. Hence the higher the concentration of B is the faster the production of B is. From (5.2) we have

$$\frac{da}{dt} = -kab \qquad \frac{db}{dt} = kab.$$

Summing the two equations we get, letting $a_0 = a(0), b_0 = b(0)$:

$$\frac{d(a+b)}{dt} = 0$$

so that $a(t) + b(t) = a_0 + b_0$. From $a(t) = a_0 + b_0 - b(t)$ we find

$$\frac{db}{dt} = kb(a_0 + b_0 - b)$$

which can be solved by separation of variables. We write

$$\frac{1}{b(a_0 + b_0 - b)} db = kdt$$

and integrate to find

$$\int \frac{1}{b(a_0 + b_0 - b)} db = kt + c \qquad (c \in \mathbb{R}). \tag{5.3}$$

---

[1] Recall that, for Avogadro's law, a mole of any substance contains the same number of molecules $N = 6.022 \times 10^{23}$ (**Avogadro's number**).

Since

$$\frac{1}{b(a_0 + b_0 - b)} = \frac{1}{a_0 + b_0}\left(\frac{1}{b} + \frac{1}{a_0 + b_0 - b}\right)$$

we have

$$\int \frac{1}{b(a_0 + b_0 - b)}\,db = \frac{1}{a_0 + b_0}\log\frac{b}{a_0 + b_0 - b}.$$

Substituting into (5.3) we obtain

$$\log\frac{b}{a_0 + b_0 - b} = (a_0 + b_0)(kt + c).$$

Letting $t = 0$ we find

$$\log\frac{b_0}{a_0} = c(a_0 + b_0)$$

or

$$c = \frac{1}{a_0 + b_0}\log\frac{b_0}{a_0}.$$

Thus:

$$\log\frac{b}{a_0 + b_0 - b} = (a_0 + b_0)kt + \log\frac{b_0}{a_0}.$$

Taking exponential of both sides we get

$$\frac{b}{a_0 + b_0 - b} = \frac{b_0}{a_0}e^{k(a_0+b_0)t}$$

and finally

$$b(t) = \frac{b_0(a_0 + b_0)\,e^{k(a_0+b_0)t}}{a_0 + b_0 e^{k(a_0+b_0)t}} \qquad a(t) = \frac{a_0(a_0 + b_0)}{a_0 + b_0 e^{k(a_0+b_0)t}}.$$

Note $a(t) \to 0$ as $t \to \infty$ while $b(t) \to a_0 + b_0$.

Example 5.3 (Logistic reaction). Consider the double reaction

$$A + B \rightleftarrows 2B.$$

If the concentration of A is kept constant, then the equation for b reads

$$\frac{db}{dt} = kab - k^- b^2 = akb\left(1 - \frac{k^-}{ak}b\right)$$

where $k^- b^2$ is the rate of the inverse reaction.

This is known as the logistic model; b evolves towards the asymptotically stable steady state concentration given by $b = ka/k^-$.

Fig. 5.1 shows a typical behavior of a solution ($ka/k^- = 1$, $b_0 = 1/3$).

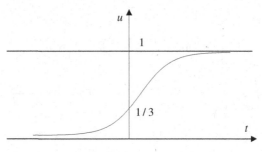

Fig. 5.1. Logistic solution for $ka/k^- = 1$ and $b_0 = 1/3$

5.1.2 Inhibition, activation

Biochemical reactions continuously occur in any living organism and the most part of them involves certain proteins, called *enzymes*, that act as *catalyzers*. The enzymes react selectively on composite substances, called *substrates*. For instance, blood *hemoglobin* is an enzyme and react with oxygen, a substrate. Enzymes are important agents for *activating or inhibiting* a reaction.

We present some models that describe some aspects, mostly kinetics, of complex biochemical reactions.

Indeed, in almost all biological processes, often the occurring reactions are not known with sufficient precision. What is known is the qualitative effects of the variation of a given reactant and this is what a model should try to reproduce, as a useful tool to make predictions. These models, when one takes into account only the chemical reactions are constituted by ordinary differential equations.

In presence of two reactant, one finds a system of the type

$$\frac{du}{dt} = f(u,v) \qquad \frac{dv}{dt} = g(u,v).$$

We say that u is an *activator (inhibitor)* of v if $g_u > 0$ (< 0) while v is an *inhibitor (activator)* of u if $f_v < 0$ (> 0).

The constant solutions $u(t) \equiv u_0$, $v(t) \equiv v_0$ are called **equilibria** (*steady states*) and can be found solving the algebraic system

$$f(u,v) = g(u,v) = 0.$$

The *linear stability* of a steady state (u_0, v_0) can be established by looking at the eigenvalues of the Jacobian matrix:

$$J(u_0, v_0) = \frac{\partial f(u_0, v_0)}{\partial g(u_0, v_0)} = \begin{pmatrix} f_u(u_0, v_0) & f_v(u_0, v_0) \\ g_u(u_0, v_0) & g_v(u_0, v_0) \end{pmatrix}.$$

If λ_1, λ_2 are the eigenvalues, we have

$$\mathrm{Tr}J = f_u + g_u = \lambda_1 + \lambda_2 \qquad \det J = |J| = f_u g_v - f_v g_u = \lambda_1 \lambda_2.$$

Then:

Proposition 5.1. *If $TrJ(u_0, v_0) < 0$ and $detJ(u_0, v_0) > 0$ then (u_0, v_0) is (locally) asymptotically stable.*
If $detJ(u_0, v_0) < 0$ or $TrJ(u_0, v_0) > 0$ then (u_0, v_0) is unstable.

These results are better described in Appendix B.

Example 5.4. Let us consider the mechanism governed by the following model.

$$\frac{du}{dt} = \frac{a}{b+v} - cu = f(u, v)$$

$$\frac{dv}{dt} = du - ev = g(u, v)$$

where a, b, c, d are positive constants.

The biological interpretation of the model is the following: u *activates* v through the term du ($g_u = d > 0$) and both u and v decrease linearly (terms $-cu$ and $-ev$). This behavior is called *first order kinetics decay*.

The term $a/(b+v)$ denotes a negative feedback of v on the production of u, since any increase of v slows the growth of u and in turn of itself, indirectly ($f_v = -a/(b+v)^2 < 0$). This is an example of *retroactive inhibition*.

The equilibria are the positive solutions of

$$\begin{cases} \dfrac{a}{b+v} - cu = 0 \\ du - ev = 0. \end{cases}$$

The unique solution is

$$u_0 = \frac{1}{2}\frac{-eb}{d} + \sqrt{\frac{e^2b^2}{d^2} + \frac{4ae}{cd}}, \quad v_0 = \frac{du_0}{e}.$$

The Jacobian matrix at (u_0, v_0) is given by

$$J(u_0, v_0) = \begin{pmatrix} -c & -\dfrac{a}{(v_0 + b)^2} \\ d & -e \end{pmatrix}.$$

Thus:

$$TrJ(u_0, v_0) = -c - e < 0 \qquad detJ(u_0, v_0) = ec + \frac{ad}{(v_0 + b)^2} > 0$$

so that (u_0, v_0) is asymptotically stable. It is also globally asymptotically stable since the vector field $(\dot{u}, \dot{v})$ points inward on the boundary of any rectangle in the first quadrant (check it).

Example 5.5. Consider the following model of *Schnakenberg*

$$\begin{cases} \dfrac{du}{dt} = \gamma(a - u + u^2 v) = f\,(u,v) \\ \dfrac{dv}{dt} = \gamma(b - u^2 v) = g\,(u,v) \end{cases}$$

where a, b and γ are positive constant. We have, for $u > 0, v > 0$,

$$f_v = \gamma u^2 > 0 \quad \text{and} \quad g_u = -2\gamma uv < 0.$$

Here there is a autocatalytic production of the activator u through the term $u^2 v$. The only equilibrium point is

$$u_0 = a + b, \quad v_0 = \sqrt{\dfrac{b}{a+b}}.$$

The Jacobian matrix at (u_0, v_0) is

$$J\,(u_0, v_0) = \begin{pmatrix} -\gamma + 2\gamma\sqrt{b\,(a+b)} & \gamma\,(a+b)^2 \\ -2\gamma\sqrt{b\,(a+b)} & -\gamma\,(a+b)^2 \end{pmatrix}.$$

Hence:

$$\mathrm{Tr}J\,(u_0, v_0) = -\gamma(1 - 2\sqrt{b\,(a+b)} + (a+b)^2) < 0$$
$$\det J\,(u_0, v_0) = \gamma^2\,(a+b)^2 > 0$$

and therefore (u_0, v_0) is asymptotically stable.

Example 5.6. We examine the dynamics described by the following model (*Thomas* 1975):

$$\dfrac{du}{dt} = a - u - \rho R\,(u, v) = f\,(u, v)$$
$$\dfrac{dv}{dt} = \alpha(b - v) - \rho R\,(u, v) = g\,(u, v)$$

where a, b, α, ρ are positive constants and

$$R\,(u, v) = \dfrac{uv}{1 + u + Ku^2}.$$

In this model, u represents the concentration of the *uric acid* (activator) which reacts in the presence of the substrate v (concentration of *oxygen*, inhibitor). Here u and v are supplied at constant rates a and αb and they linearly decay ($-u$ and $-\alpha v$). Both oxygen and uric acid take part in the reaction described by $R\,(u, v)$.

The kinetics of the reaction R is that of a *substrate inhibition*. For fixed values of v, as u goes to zero, $R\,(u, v) \sim uv$ and therefore it is linear with

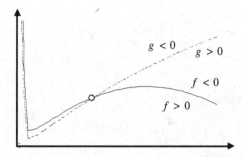

Fig. 5.2. Thomas Model. Isocline $f(u, v) = 0$ (solid line) and $g(u, v) = 0$ (discontinuous line) for $a = 150$, $b = 100$, $\alpha = 1.5$, $\rho = 13$, $K = 0.05$

respect to u (R is increasing with respect too u); on the other hand if $u \to +\infty$, $R(u, v) \sim v/Ku$ (R is decreasing with respect to u).

The two isoclines $f(u, v) = 0$ and $g(u, v) = 0$ intersect at the equilibrium point, which is an asymptotically stable focus (Fig. 5.2, see also Appendix B). We see that at the equilibrium point $f_u > 0$ and $g_v < 0$; indeed, $f(u, v_0)$ switches from negative values ($u < u_0$) to positive values ($u > u_0$), while $g(u_0, v)$ switches from positive values ($v < v_0$) to negative values ($v > v_0$).

5.2 Diffusion and linear reaction

5.2.1 Pure diffusion. Asymptotic behavior

Consider the 1-d diffusion equation

$$u_t = Du_{xx} \quad \text{in } (0, L) \tag{5.4}$$

with initial condition

$$u(x, 0) = g(x). \tag{5.5}$$

Here u may represent the concentration of a substance or the density of a population.

Let us compute the solution of (5.4), (5.5) with the following boundary conditions

1. **Dirichlet:** $u(0, t) = u(L, t) = 0$ (*hostile external environment*).
2. **Neumann:** $u_x(0, t) = u_x(0, t) = 0$ (*insulation, no flux condition*).

Using separation of variables, we look for solutions of the type $u(x, t) = U(x)V(t)$. Substituting into (5.4), we find

$$\frac{V'(t)}{DV(t)} = \frac{U''(x)}{U(x)} = \lambda = \text{constant}.$$

The equation for V is

$$V'(t) = \lambda D V(t)$$

so that

$$V(t) = e^{\lambda D t}.$$

U solves the eigenvalue problem

$$U''(x) - \lambda U(x) = 0 \qquad\qquad (5.6)$$

with the Dirichlet conditions

$$U(0) = U(L) = 0$$

or the Neumann conditions

$$U_x(0) = U_x(L) = 0.$$

We find:

$$
\begin{array}{ll}
U(x) = c_1 e^{-\sqrt{\lambda}x} + c_2 e^{\sqrt{\lambda}x} & \text{if } \lambda > 0 \\
U(x) = c_1 + c_2 x & \text{if } \lambda = 0 \;\; (c_1, c_2 \text{ arbitrary}) \\
U(x) = c_1 \sin\left(\sqrt{-\lambda}x\right) + c_2 \cos\left(\sqrt{-\lambda}x\right) & \text{if } \lambda < 0.
\end{array}
$$

Dirichlet conditions. If $\lambda \geq 0$, imposing $U(0) = U(L) = 0$ we find $U(x) = 0$. When $\lambda < 0$, we find

$$U(0) = c_2 = 0$$
$$U(L) = c_1 \sin\left(\sqrt{-\lambda}L\right) + c_2 \cos\left(\sqrt{-\lambda}L\right) = 0$$

from which

$$\sin\left(\sqrt{-\lambda}L\right) = 0.$$

We get the eigenvalues $\lambda_k = -k^2\pi^2/L^2$ and the **eigenfunctions**

$$U_k(x) = \sin\frac{k\pi x}{L}.$$

We now develop the initial data in sine Fourier series

$$g(x) = \sum_{k=1}^{\infty} b_k \sin\frac{k\pi x}{L} \qquad b_k = \frac{2}{L}\int_0^L g(x)\sin\frac{k\pi x}{L}dx.$$

Then, the solution u is given by

$$u(x,t) = \sum_{k=1}^{\infty} b_k \exp\left(-D\frac{k^2\pi^2}{L^2}t\right)\sin\frac{k\pi x}{L}.$$

From this formula we see that every term tends exponentially to zero as $t \to +\infty$. The asymptotic profile of u is given by the slowest decaying term, corresponding to $k = 1$:

$$u(x,t) \sim b_1 \exp\left(-D\frac{\pi^2}{L^2}t\right) \sin\frac{\pi x}{L} \qquad \text{as } t \to +\infty.$$

We conclude that *The null solution is asymptotically stable.*

Neumann conditions. If $\lambda \geq 0$, imposing the conditions $U_x(0) = U_x(L) = 0$ we find again $U(x) = 0$. If $\lambda < 0$, we find

$$U_x(0) = c_1\sqrt{-\lambda} = 0$$
$$U_x(L) = c_1\sqrt{-\lambda}\cos\left(\sqrt{-\lambda}L\right) - c_2\sqrt{-\lambda}\sin\left(\sqrt{-\lambda}L\right) = 0$$

whence

$$\sin\left(\sqrt{-\lambda}L\right) = 0.$$

Thus the eigenvalues are $\lambda_k = -k^2\pi^2/L^2$ with eigenfunctions

$$U_k(x) = \cos\frac{k\pi x}{L}.$$

Expanding the initial data in cosine Fourier series

$$g(x) = \frac{a_0}{2} + \sum_{k=1}^{\infty} a_k \sin\frac{k\pi x}{L} \qquad a_k = \frac{2}{L}\int_0^L g(x)\cos\frac{k\pi x}{L}dx,$$

we find the solution

$$u(x,t) = \frac{a_0}{2} + \sum_{k=1}^{\infty} a_k \exp\left(-D\frac{k^2\pi^2}{L^2}t\right)\cos\frac{k\pi x}{L}.$$

Now, each term in the above series goes to zero exponentially as $t \to +\infty$ so that $u(\cdot,t) \to a_0/2$. The asymptotic profile of u is given by

$$u(x,t) \sim \frac{a_0}{2} + a_1 \exp\left(-D\frac{\pi^2}{L^2}t\right)\cos\frac{\pi x}{L} \qquad \text{as } t \to +\infty.$$

Notice that

$$\frac{a_0}{2} = \frac{1}{L}\int_0^L g(x)\,dx$$

and therefore *the solution eventually converges to the mean value of the initial data.*

Conclusion. In both cases, the solutions converge to two constant *steady states*, in particular *spatially homogeneous.*

5.2.2 Asymptotic behavior in general domains

!By the method of separation of variables it is possible to compute explicitly
the solution of linear problems in one spatial dimension and in particularly
simple two or three-dimensional domains (squares, circular sectors, cubes,
spheres, etc.). Thus, in those cases, the asymptotic behavior of the solution
can be red directly from the resulting formula. This is not the case for general
geometries and one has to resort to more or less advanced methods. Sometimes, elementary *energy methods* can be useful, as shown in the following
theorem.

Theorem 5.1. *Let Ω be a bounded regular domain in $\mathbb{R}^n$ ($n = 1,2,3$) and u
be the solution of the problem*

$$
\begin{aligned}
u_t - D\Delta u &= 0 \quad \mathbf{x} \in \Omega,\, t > 0 \\
u(\mathbf{x},0) &= g(\mathbf{x}) \quad \mathbf{x} \in \Omega.
\end{aligned}
$$

(i)   *If (Dirichlet) $u = 0$ on $\partial\Omega$ then*

$$
u(\mathbf{x},t) \to 0 \quad as\ t \to \infty,
$$

   for every $\mathbf{x} \in \Omega$.

(ii)   *If (Neumann) $\partial_\nu u = 0$ on $\partial\Omega$ then*

$$
\int_\Omega u(\mathbf{x},t)\, d\mathbf{x} = \int_\Omega g(\mathbf{x})\, d\mathbf{x} \quad for\ every\ t \geq 0
$$

   and u converges to the mean value of the initial data:

$$
u(\mathbf{x},t) \to \frac{1}{|\Omega|} \int_\Omega g(\mathbf{x})\, d\mathbf{x} \quad as\ t \to \infty
$$

   for every $\mathbf{x} \in \Omega$.

(iii) *If (Robin) $\partial_\nu u + hu = 0$ on $\partial\Omega$ and $h > 0$, constant, then*

$$
u(\mathbf{x},t) \to 0 \quad as\ t \to \infty,
$$

   for every $\mathbf{x} \in \Omega$.

Proof. For simplicity we show it for $n = 1$, with $\Omega = (0, L)$. Multiply the
differential equation by u and integrate with respect to x over $(0, L)$:

$$
\int_0^L u u_t\, dx = D \int_0^L u u_{xx}\, dx. \tag{5.7}
$$

Note that:

$$
\int_0^L u u_t\, dx = \frac{1}{2} \int_0^L u_t^2\, dx = \frac{1}{2} \frac{d}{dt} \int_0^L u^2\, dx. \tag{5.8}
$$

Integrating by parts, we get

$$\int_0^L uu_{xx}dx = [u(L,t)u_x(L,t) - u(0,t)u_x(0,t)] - \int_0^L u_x^2 \, dx. \qquad (5.9)$$

(i) We now use the Dirichlet conditions. Since $u(L,t) = u(0,t) = 0$, we have

$$\int_0^L uu_{xx}dx = -\int_0^L u_x^2 \, dx \qquad (5.10)$$

and (5.7), (5.8), (5.9), (5.10) give

$$\frac{1}{2}\frac{d}{dt}\int_0^L u^2(x,t)\,dx = -D\int_0^L u_x^2(x,t)\,dx.$$

Set $E(t) = \int_0^L u^2(x,t)\,dx$. The last equation shows that

$$\dot{E}(t) = -2D\int_0^L u_x^2(x,t)\,dx < 0 \text{ for } t > 0 \qquad (5.11)$$

and therefore E decreases. Since $E(t) \geq 0$ we have

$$\lim_{t\to+\infty} E(t) \geq 0. \qquad (5.12)$$

From

$$\ddot{E}(t) = -4D\int_0^L u_x(x,t)u_{x,t}(x,t)\,dx =$$

$$= -4D[u_x(x,t)u_t(x,t)]_0^L + 4D\int_0^L u_{xx}(x,t)u_t(x,t)\,dx$$

$$= 4D^2\int_0^L u_t^2(x,t)\,dx > 0$$

since $u_t(0,t) = u_t(L,t) = 0$.

Using (5.12) we deduce that $\dot{E}(t) \to 0$ and (5.11) gives

$$\int_0^L (u_x(x,t))^2 dx \to 0 \qquad \text{as } t \to +\infty. \qquad (5.13)$$

Since $u(0,t) = 0$, we can write

$$u(x,t) = \int_0^x u_x(s,t)\,ds$$

and Schwarz inequality gives

$$|u(x,t)| \leq \int_0^L |u_x(s,t)|\,ds \leq \sqrt{L}\sqrt{\int_0^L u_x^2(x,t)\,dx}. \qquad (5.14)$$

From (5.13) we infer $u(x,t) \to 0$ as $t \to +\infty$, for every $x \in (0,L)$.

(ii) We now use the Neumann conditions $u_x(L,t) = u_x(0,t) = 0$ into (5.9). By arguing as in i) we deduce that (5.13) still holds.

On the other hand, notice that, integrating the equation $u_t = Du_{xx}$ with respect to x over $(0,L)$, we find

$$\frac{d}{dt}\int_0^L u(x,t)\,dx = \int_0^L u_t(x,t)\,dx = D\int_0^L u_{xx}(x,t)\,dx = 0$$

since

$$\int_0^L u_{xx}(x,t)\,dx = u_x(L,t) - u_x(0,t) = 0.$$

Then $\int_0^L u(x,t)\,dx$ is constant in time. Thus:

$$\frac{1}{L}\int_0^L u(x,t)\,dx = \frac{1}{L}\int_0^L g(x)\,dx \equiv K \quad \text{for every } t \geq 0.$$

We show now that $u(x,t) \to K$ as $t \to +\infty$. First note that by the Mean Value Theorem, for every t there exists $\tilde{x}(t) \in (0,L)$ such that

$$K = \frac{1}{L}\int_0^L u(x,t)\,dx = u(\tilde{x}(t),t).$$

Therefore we can write

$$u(x,t) - K = \int_{\tilde{x}(t)}^x u_x(s,t)\,ds.$$

By Schwarz inequality,

$$|u(x,t) - K| \leq \int_0^L |u_x(s,t)|\,ds \leq \sqrt{L}\sqrt{\int_0^L (u_x^2(x,t)\,dx}$$

and (5.13) gives $u(x,t) \to K$ for every $x \in (0,L)$.

(iii) Finally, let us use Robin'conditions $u_x(L,t) + hu(L,t) = u_x(0,t) + hu(0,t) = 0$ into (5.9). We have:

$$\int_0^L uu_{xx}\,dx = -h\left[u^2(L,t) + u^2(0,t)\right] - \int_0^L u_x^2\,dx. \tag{5.15}$$

Formulas (5.7), (5.8), (5.10) and (5.15) gives, since $h > 0$,

$$\frac{1}{2}\dot{E}(t) = -hD\left[u^2(L,t) + u^2(0,t)\right] - D\int_0^L u_x^2(x,t)\,dx < 0.$$

Once again we infer that $\dot{E}(t) \to 0$. Hence

$$hD\left[u^2(L,t) + u^2(0,t)\right] + \int_0^L u_x^2(x,t)\,dx \to 0 \quad \text{as } t \to +\infty. \tag{5.16}$$

Now, each term in (5.16) are non negative and therefore

$$u\left(L,t\right)\to 0,\, u\left(0,t\right)\to 0,\quad \int_0^L u_x^2\left(x,t\right)dx \to 0$$

as $t\to +\infty$, for every $x\in(0,L)$. Arguing as in (i) we get

$$\left|u\left(x,t\right)-u\left(0,t\right)\right| \le \int_0^L \left|u_x\left(s,t\right)\right|ds \le \sqrt{L}\sqrt{\int_0^L u_x^2\left(x,t\right)dx}.$$

We conclude that $u\left(x,t\right)\to 0$ as $t\to +\infty$, for every $x\in(0,L)$.

5.2.3 Linear reaction. Critical dimension

The simplest reaction mechanism is (Malthus) $\dot{u}=au$ with $a>0$. Let us examine the effect of the competition between diffusion and a Malthusian reaction over a population governed by the following system

$$\begin{cases} u_t - Du_{xx} = au & 0 < x < L, t > 0 \\ u\left(0,t\right) = u\left(L,t\right) = 0 & t > 0 \\ u\left(x,0\right) = g\left(x\right) & 0 < x < L. \end{cases}$$

Given the homogeneous Dirichlet conditions (hostile external environment), the population decays by diffusion while tends to increase by reaction. Thus the two effects compete. Let us examine which factors determine the overwhelming one.

First, since a is constant, setting

$$u\left(x,t\right) = e^{at}w\left(x,t\right)$$

we get

$$u_t = e^{at}(aw + w_t),\quad u_x = e^{at}w_x,\quad u_{xx} = e^{at}w_{xx}.$$

Thus the equation for w is:

$$e^{at}(aw + w_t) - De^{at}w_{xx} = ae^{at}w.$$

After simple algebra we get

$$w_t - Dw_{xx} = 0$$

with the same initial/boundary conditions

$$w\left(0,t\right) = w\left(L,t\right) = 0$$
$$w\left(x,0\right) = w\left(x\right).$$

We have already solved this problem in Section 5.2.1. The solution is

$$w\left(x,t\right) = \sum_{k=1}^{\infty} b_k \exp\left(-D\frac{k^2\pi^2}{L^2}t\right)\sin\frac{k\pi x}{L}.$$

Going back to u we find

$$u\left(x,t\right) = \sum_{k=1}^{\infty} b_k \exp\left\{(a - D\frac{k^2\pi^2}{L^2})t\right\} \sin\frac{k\pi x}{L}. \tag{5.17}$$

Formula (5.17) displays an important difference between pure diffusion ($a = 0$) and diffusion-reaction as far as asymptotic behavior is concerned. Assuming $b_1 \neq 0$, the evolution determining factor in (5.17) is the first exponential, corresponding to $k = 1$.

Now, if $a - D\frac{\pi^2}{L^2} < 0$, that is

$$\frac{aL^2}{D} < \pi^2 \tag{5.18}$$

then $\lim_{t \to +\infty} u\left(x,t\right) = 0$, since $a - D\frac{k^2\pi^2}{L^2} < a - D\frac{\pi^2}{L^2}$ for every $k > 1$.

On the opposite, if

$$a - D\frac{\pi^2}{L^2} > 0$$

then $\lim_{t \to +\infty} u\left(x,t\right) = \infty$, since the first term increases exponentially while the other terms are either of lower exponential order or negligible at infinity.

Remark 5.1. The coefficients a and D are intrinsic parameters, encoding the population features. When these parameters are fixed, the habitat size plays a major role.

Indeed, the value

$$L_0 = \pi\sqrt{\frac{D}{a}}$$

is a **critical value** for the population survival. If $L < L_0$ the habitat is too small to avoid extinction of the population; on the contrary, if $L > L_0$, one observe exponential growth.

L_0 is called a **bifurcation value**, since the **steady state solution** $u\left(x,t\right) = 0$ changes from **stability** ($L < L_0$) to **instability** ($L > L_0$).

What happens for $L = L_0$? In this case (5.17) becomes

$$u\left(x,t\right) = b_1 \sin\frac{\pi x}{L} + \sum_{k=2}^{\infty} b_k \exp\left\{(a - D\frac{k^2\pi^2}{L_0^2})t\right\} \sin\frac{k\pi x}{L}.$$

Note that

$$u_1\left(x,t\right) = b_1 \sin\frac{\pi x}{L}$$

is the eigenfunction corresponding to the first eigenvalue $k_1 = -\pi^2/L^2$ and, being time independent, is an **equilibrium solution**. Since $a - D\frac{k^2\pi^2}{L_0^2} < 0$ if $k > 1$, we infer that

$$u\left(x,t\right) \to b_1 \sin\frac{\pi x}{L} \quad \text{as } t \to +\infty.$$

Finally observe that, if for some $\overline{k} \geq 1$, $a - D\dfrac{\overline{k}\pi^2}{L^2} > 0$ then for every $1 \leq k < \overline{k}$ all numbers $a - D\dfrac{k\pi^2}{L_0^2}$ are positive and the corresponding terms contribute to the instability of the null solution: we say that *the modes of vibration*

$$b_k \exp\left\{ (a - D\frac{k^2\pi^2}{L_0^2})t \right\} \sin\frac{k\pi x}{L}$$

for $k = 1, 2, \ldots, \overline{k}$ *are activated.*

5.2.4 Linear reaction and diffusion in two dimensions

We examine an example of separation of variables to solve a reaction-diffusion problem in a rectangle

$$R = (0, p) \times (0, q).$$

Let us consider the following Neumann problem

$$\begin{cases} u_t = D\left(u_{xx} + u_{yy}\right) + \lambda u & \text{for } t > 0, (x, y) \in R \\ \nabla u \cdot \mathbf{n} = 0 & \text{on } \partial R \\ u\left(x, y, 0\right) = g\left(x, y\right) & (x, y) \in R \end{cases} \tag{5.19}$$

where D, λ, p, q are positive constants.

Look for solutions of the form

$$u\left(x, y, t\right) = U\left(t\right) V\left(x, y\right).$$

Substituting into the differential equation, we get

$$U'\left(t\right) V\left(x, y\right) = DU\left(t\right) \left(V_{xx}\left(x, y\right) + V_{yy}\left(x, y\right)\right) + \lambda U\left(t\right) V\left(x, y\right).$$

Dividing by $DU\left(t\right) V\left(x, y\right)$ we have

$$\frac{U'\left(t\right)}{DU\left(t\right)} - \frac{\lambda}{D} = \frac{V_{xx}\left(x, y\right) + V_{yy}\left(x, y\right)}{V\left(x, y\right)} = k \text{ (constant)}$$

so that U solves the equation

$$U'\left(t\right) = (\lambda + Dk)U\left(t\right)$$

that is

$$U\left(t\right) = ce^{(\lambda + Dk)t}.$$

Furthermore V is a solution of the eigenvalue problem

$$\begin{cases} V_{xx} + V_{yy} = kV & (x, y) \in R \\ \nabla V \cdot \mathbf{n} = 0 & \text{on } \partial R. \end{cases}$$

To solve this problem, we separate the spatial variables looking for solutions of the form

$$V(x,y) = W(x) Z(y).$$

Substituting into the differential equation, we get

$$\frac{W''(x)}{W(x)} + \frac{Z''(y)}{Z(y)} = k$$

which splits into the two Neumann problems

$$W''(x) = k_1 W(x), \qquad W'(0) = W'(p) = 0$$
$$Z''(y) = k_2 Z(y), \qquad Z'(0) = Z'(q) = 0$$

with

$$k_1 + k_2 = k.$$

Recalling the computations in the previous sections, we find

$$k_{1m} = -\frac{m^2 \pi^2}{p^2} \qquad W_m(x) = \cos \frac{m \pi x}{p} \qquad m = 0, 1, 2, ...$$

and

$$k_{2n} = -\frac{n^2 \pi^2}{q^2} \qquad Z_n(y) = \cos \frac{n \pi y}{q} \qquad n = 0, 1, 2, ...$$

Thus, we have the eigenvalues

$$k_{mn} = -\frac{\pi^2 m^2}{p^2} - \frac{n^2 \pi^2}{q^2}$$

with corresponding eigenfunctions

$$V_{mn}(x,y) = \cos \frac{m \pi x}{p} \cos \frac{n \pi y}{q}.$$

Setting

$$g_{mn} = \frac{4}{pq} \int_0^p \int_0^q g(x,y) \cos \frac{m \pi x}{p} \cos \frac{n \pi y}{q} dx dy$$

we can write the solution of problem (5.19)

$$u(x,y,t) = \sum_{m,n=0}^{\infty} g_{mn} e^{(Dk_{mn} + \lambda)t} \cos \frac{m \pi x}{p} \cos \frac{n \pi y}{q} dx dy.$$

Note that, since $\lambda > 0$, the first term in the above series ($m = n = 0$) is $g_{00} e^{\lambda t}$ which grows exponentially.

5.2.5 An Example in dimension $n = 3$

In the n-dimensional case, the *reaction-diffusion*-equation takes the form:

$$u_t = D\Delta u + cu.$$

In this section we examine a model of reaction-diffusion in a fissionable material. Although we deal with a greatly simplified model, some interesting implications can be drawn.

By shooting neutrons into an uranium nucleus it may happen that the nucleus breaks into two parts, releasing other neutrons already present in the nucleus and causing a chain reaction. Some macroscopic aspects of this phenomenon can be described by means of an elementary model.

Suppose a cylinder with height h and radius R is made of a fissionable material of constant density ρ, with total mass

$$M = \pi\varrho R^2 h.$$

At a macroscopic level, the free neutrons diffuse like a chemical in a porous medium, with a flux proportional and opposite to the density gradient. In other terms, if $N = N(x,y,z,t)$ is the *neutron density* and no fission occurs, the *flux of neutrons is equal to* $-k\nabla N$, where k is a positive constant depending on the material. The mass conservation then gives

$$N_t = k\Delta N.$$

When fission occurs at a constant rate $\gamma > 0$, we get the equation

$$N_t = D\Delta N + \gamma N, \tag{5.20}$$

where reaction and diffusion are competing: diffusion tends to slow down N, while, clearly, the reaction term tends to exponentially increase N. A crucial question is to examine the behavior of N in the long run (i.e. as $t \to +\infty$).

We look for *bounded* solutions satisfying a homogeneous Dirichlet condition on the boundary of the cylinder, with the idea that the density is higher at the center of the cylinder and very low near the boundary. Then it is reasonable to assume that N has a radial distribution with respect to the axis of the cylinder. More precisely, using the cylindrical coordinates (r,θ,z) with

$$x = r\cos\theta, \ y = r\sin\theta,$$

we can write $N = N(r,z,t)$ and the homogeneous Dirichlet condition on the boundary of the cylinder translates into

$$\begin{aligned} N(R,z,t) = 0 && 0 < z < h \\ N(r,0,t) = N(r,h,t) = 0 && 0 < r < R \end{aligned} \tag{5.21}$$

for every $t > 0$. Accordingly we prescribe an initial condition

$$N(r,z,0) = N_0(r,z) \tag{5.22}$$

such that

$$N(R, z) = 0 \text{ for } 0 < z < h, \text{ and } N(r, 0) = N(r, h) = 0. \tag{5.23}$$

To solve problem (5.20), (5.21), (5.22), let us first get rid of the reaction term by setting

$$N(r, z, t) = \mathcal{N}(r, z, t) e^{\gamma t}. \tag{5.24}$$

Then, writing the Laplace operator in cylindrical coordinates[2], $\mathcal{N}$ solves

$$\mathcal{N}_t = k \left[\mathcal{N}_{rr} + \frac{1}{r} \mathcal{N}_r + \mathcal{N}_{zz} \right] \tag{5.25}$$

with the same initial and boundary conditions of N. By maximum principle, we know that there exists only one solution, continuous up to the boundary of the cylinder. To find an explicit formula for the solution, we use the method of separation of variables, searching for bounded solutions of the form

$$\mathcal{N}(r, z, t) = u(r) v(z) w(t), \tag{5.26}$$

satisfying the homogeneous Dirichlet conditions $u(R) = 0$ and $v(0) = v(h) = 0$.

Substituting (5.26) into (5.25), we find

$$u(r) v(z) w'(t) = k[u''(r) v(z) w(t) + \frac{1}{r} u'(r) v(z) w(t) + u(r) v''(z) w(t)].$$

Dividing by $\mathcal{N}$ and rearranging the terms, we get,

$$\frac{w'(t)}{kw(t)} - \left[\frac{u''(r)}{u(r)} + \frac{1}{r} \frac{u'(r)}{u(r)} \right] = \frac{v''(z)}{v(z)}. \tag{5.27}$$

The two sides of (5.27) depend on different variables so that they must be equal to a common constant b. Then for v we have the eigenvalue problem

$$v''(z) - bv(z) = 0$$

$$v(0) = v(h) = 0.$$

The eigenvalues are $b_m \equiv -\nu_m^2 = -\frac{m^2 \pi^2}{h^2}$, $m \geq 1$ integer, with corresponding eigenfunctions

$$v(z) = c \sin \nu_m z.$$

The equation for w and u can be written in the form

$$\frac{w'(t)}{kw(t)} + \nu_m^2 = \frac{u''(r)}{u(r)} + \frac{1}{r} \frac{u'(r)}{u(r)} \tag{5.28}$$

---

[2] Appendix D.

where the variables r and t are again separated. This forces the two sides of (5.28) to be equal to a common constant μ. Therefore, for w we have the equation

$$w'(t) = k(\mu - \nu_m^2)w(t)$$

that gives

$$w(t) = c \exp\left[k\left(\mu - \nu_m^2\right)t\right] \qquad c \in \mathbb{R}. \tag{5.29}$$

Then the equation for u is

$$u''(r) + \frac{1}{r}u'(r) - \mu u(r) = 0 \tag{5.30}$$

with

$$u(R) = 0 \quad \text{and} \quad u \text{ bounded in } [0, R]. \tag{5.31}$$

The (5.30) is a *Bessel equation of order zero with parameter* $-\mu$; conditions (5.31) force[3] $\mu = -\lambda^2 < 0$. Then the only bounded solution of (5.30), (5.31) is $J_0(\lambda r)$, where

$$J_0(x) = \sum_{k=0}^{\infty} \frac{(-1)^k}{(k!)^2} \left(\frac{x}{2}\right)^{2k}$$

is the *Bessel function of first kind and order zero*. To match the boundary condition $u(R) = 0$ we require $J_0(\lambda R) = 0$. Now, J_0 has an infinite number of positive simple zeros[4] λ_n, $n \geq 1$

$$0 < \lambda_1 < \lambda_2 < \ldots < \lambda_n < \ldots$$

Thus, if $\lambda R = \lambda_n$, we find infinitely many solutions of (5.30), given by

$$u_n(r) = J_0\left(\frac{\lambda_n r}{R}\right).$$

Thus

$$\mu = \mu_n = -\frac{\lambda_n^2}{R^2}.$$

---

[3] In fact, write Bessel's equation (5.30) in the form

$$(ru')' - \mu ru = 0.$$

Multiplying by u and integrating over $(0, R)$, we have

$$\int_0^R (ru')' u \, dr = \mu \int_0^R u^2 \, dr. \tag{5.32}$$

Integrating by parts and using (5.31), we get

$$\int_0^R (ru')' u \, dr = \left[(ru')u\right]_0^R - \int_0^R (u')^2 \, dr = -\int_0^R (u')^2 \, dr < 0$$

and from (5.32) we get $\mu < 0$.

[4] The zeros of the Bessel functions are known with a considerable degree of accuracy. The first five zeros of J_0 are: $2.4048\ldots$, $5.5201\ldots$, $8.6537\ldots$, $11.7915\ldots$, $14.9309\ldots$.

To summarize, we have determined so far a countable number of solutions

$$\mathcal{N}_{mn}(r,z,t) = u_n(r)v_m(z)w_{m,n}(t) =$$

$$= J_0\left(\frac{\lambda_n r}{R}\right)\sin\nu_m z \, \exp\left[-k\left(\nu_m^2 + \frac{\lambda_n^2}{R^2}\right)t\right]$$

satisfying the homogeneous Dirichlet conditions. It remains to satisfy the initial condition. Due to the linearity of the problem, we look for a solution obtained by superposition of the $\mathcal{N}_{m,n}$, that is

$$\mathcal{N}(r,z,t) = \sum_{n,m=1}^{\infty} c_{mn}\mathcal{N}_{mn}(r,z,t).$$

Then, we choose the coefficients c_{mn} in order to have

$$\sum_{n,m=1}^{\infty} c_{mn}\mathcal{N}_{mn}(r,z,0) = \sum_{n,m=1}^{\infty} c_{mn}J_0\left(\frac{\lambda_n r}{R}\right)\sin\frac{m\pi}{h}z = N_0(r,z). \quad (5.33)$$

The second of (5.23) and (5.33) suggest an expansion of N_0 in sine Fourier series with respect to z. Let

$$c_m(r) = \frac{2}{h}\int_0^h N(r,z)\sin\frac{m\pi}{h}z, \qquad m \geq 1,$$

and

$$N_0(r,z) = \sum_{m=1}^{\infty} c_m(r)\sin\frac{m\pi}{h}z.$$

Then (5.33) shows that, for fixed $m \geq 1$, the c_{mn} are the coefficients of the expansion of $c_m(r)$ in the *Fourier-Bessel series*

$$\sum_{n=1}^{\infty} c_{mn}J_0\left(\frac{\lambda_n r}{R}\right) = c_m(r).$$

We are not really interested in the exact formula for the c_{mn}, however we will come back to this point in Remark 2.5 below.

In conclusion, recalling (5.24), the analytic expression of the solution of our original problem is the following

$$N(r,z,t) = \sum_{n,m=1}^{\infty} c_{mn}J_0\left(\frac{\lambda_n r}{R}\right)\exp\left\{\left(\gamma - k\nu_m^2 - k\frac{\lambda_n^2}{R^2}\right)t\right\}\sin\nu_m z.$$

$$(5.34)$$

Of course, (5.34) is only a formal solution, since we should check in which sense the boundary and initial condition are attained and that term by term differentiation can be performed. This can be done under reasonable smoothness properties of N_0 and we do not pursue the calculations here.

Rather, we notice that from (5.34) we can draw an interesting conclusion on the long range behavior of N. Consider for instance the value of N at the center of the cylinder, that is at the point $r = 0$ and $z = h/2$; we have, since $J_0(0) = 1$ and $\nu_m^2 = \frac{m^2\pi^2}{h^2}$,

$$N\left(0, \frac{h}{2}, t\right) = \sum_{n,m=1}^{\infty} c_{mn} \exp\left\{\left(\gamma - k\frac{m^2\pi^2}{h^2} - k\frac{\lambda_n^2}{R^2}\right)t\right\} \sin\frac{m\pi}{2}.$$

The exponential factor is maximized for $m = n = 1$, so the leading term in the sum is

$$c_{11} \exp\left\{\left(\gamma - k\frac{\pi^2}{h^2} - k\frac{\lambda_1^2}{R^2}\right)t\right\}.$$

If now

$$\gamma - k\left(\frac{\pi^2}{h^2} + \frac{\lambda_1^2}{R^2}\right) < 0,$$

each term in the series goes to zero as $t \to +\infty$ and the reaction dies out. On the opposite, if

$$\gamma - k\left(\frac{\pi^2}{h^2} + \frac{\lambda_1^2}{R^2}\right) > 0,$$

that is

$$\frac{\gamma}{k} > \frac{\pi^2}{h^2} + \frac{\lambda_1^2}{R^2}, \tag{5.35}$$

the leading term increases exponentially with time. To be true, (5.35) requires that the following relations be *both* satisfied:

$$h^2 > \frac{k\pi^2}{\gamma} \quad \text{and} \quad R^2 > \frac{k\lambda_1^2}{\gamma}. \tag{5.36}$$

The conditions (5.36) give a lower bound for the height and the radius of the cylinder. Thus, we deduce that *there exists a critical mass of material, below which the reaction cannot be sustained.*

Remark 5.2. A sufficiently smooth function f, for instance of class $C^1([0, R])$, can be expanded in a Fourier-Bessel series, where the Bessel functions $J_0\left(\frac{\lambda_n r}{R}\right)$, $n \geq 1$, play the same role of the trigonometric functions. More precisely, the functions $J_0(\lambda_n r)$ satisfy the following orthogonality relations:

$$\int_0^R x J_0(\lambda_m x) J_0(\lambda_n x) dx = \begin{cases} 0 & m \neq n \\ \frac{R^2}{2} c_n^2 & m = n \end{cases}$$

where

$$c_n = \sum_{k=0}^{\infty} \frac{(-1)^k}{k!\,(k+1)!} \left(\frac{\lambda_n}{2R}\right)^{2k+1}.$$

Then

$$f(x) = \sum_{n=0}^{\infty} f_n J_0(\lambda_n x) \tag{5.37}$$

with the coefficients f_n assigned by the formula

$$f_n = \frac{2}{R^2 c_n^2} \int_0^R x f(x) J_0(\lambda_n x)\, dx.$$

The series (5.37) converges in the following least square sense: if

$$S_N(x) = \sum_{n=0}^{N} f_n J_0(\lambda_n x)$$

then

$$\lim_{N \to +\infty} \int_0^R [f(x) - S_N(x)]^2 x\, dx = 0. \tag{5.38}$$

In Chapter 7, we will interpret (5.38) from the point of view of Hilbert space theory.

5.3 Diffusion and nonlinear reaction

5.3.1 Monotone methods

In this section we consider models of reaction-diffusion of the type

$$u_t - \Delta u = f(u) \tag{5.39}$$

in a space-time cylinder $D_T = \Omega \times (0, T)$, where Ω is a *regular* domain, complemented by Dirichlet or Neumann conditions on $S_T = \partial\Omega \times (0, T]$, that is

$$u = 0 \quad \text{or} \quad \frac{\partial u}{\partial \nu} = 0 \tag{5.40}$$

(ν exterior normal) and initial conditions

$$u(\mathbf{x}, 0) = g(\mathbf{x}). \tag{5.41}$$

We assume that

$$f \in C^1(\mathbb{R}), g \in C(\overline{\Omega}).$$

In the case of homogeneous Dirichlet we require that $g = 0$ on $\partial\Omega$ (*compatibility condition*).

Given the nonlinear reaction term, the well posedness of the problem in all D_T is not granted. For instance, even the o.d.e. Cauchy problem $\dot{u} = u^2$, $u(0) = 1$ has no solution in $(0, T)$ for $T > 1$. Thus it is useful to present some result in this direction. We need the notion of *super/subsolution*.

Definition 5.1. *A function* $\bar{u} \in C\left(\overline{D_T}\right) \cap C^{2,1}\left(D_T\right)$ *is a supersolution of problem* (5.39), (5.40), (5.41) *if it satisfies the following conditions:*

$$\bar{u}_t - \Delta \bar{u} \geq f(\bar{u}) \ \ in \ D_T$$

$$\bar{u} \geq 0 \ or \ \frac{\partial \bar{u}}{\partial \nu} \geq 0 \ on \ S_T \qquad (5.42)$$

$$\bar{u}(\mathbf{x}, 0) \geq g(\mathbf{x}) \ \ in \ \overline{\Omega}.$$

Analogously $\underline{u} \in C\left(\overline{D_T}\right) \cap C^{2,1}\left(D_T\right)$ *is a subsolution if it satisfies the opposite inequalities in* (5.42).

Clearly, u is a solution if and only if it is both a sub and a supersolution. Notice that sub/supersolution are bounded in $\overline{D_T}$. Thus,

Lemma 5.1. *Let* $\bar{u}$ *and* $\underline{u}$ *be super and subsolution of problem* (5.39), (5.40), (5.41), *respectively. Then*

$$\underline{u} \leq \bar{u} \qquad in \ D_T.$$

Proof. If $\bar{u}$ and $\underline{u}$ are super and subsolution, there always exists a finite interval $[a, b]$ such that

$$a \leq \underline{u}, \bar{u} \leq b \ \ in \ \overline{D_T}.$$

By the mean value theorem, we can write, for a suitable v between $\underline{u}$ and $\bar{u}$:

$$f(u_1) - f(u_2) = f'(v)(\underline{u} - \bar{u}).$$

Set $a(x, t) = f'(v(x, t))$. Then

$$|a| \leq M = \max_{s \in [a,b]} |f'(s)|$$

and $w = \bar{u} - \underline{u}$ is a solution of

$$w_t - \Delta w - aw \geq 0 \ in \ D_T$$

$$w \geq 0 \ or \ \frac{\partial w}{\partial \nu} \geq 0 \ \ on \ S_T$$

$$w(\mathbf{x}, 0) \geq 0 \qquad in \ \overline{\Omega}.$$

Set $z(x, t) = e^{2Mt} w(x, t)$. Then z is a solution of $z_t - \Delta z - a_0 z \geq 0$ where $a_0 = 2M - a > 0$ in $\overline{D_T}$.

We infer z cannot have a negative minimum at a point $(x_0, t_0) \in D_T$ with $0 < t_0 \leq T$. Indeed, if this is the case, we have[5] $u_t(x_0, t_0) \leq 0$ and $\Delta u(x_0, t_0) \geq 0$. From the differential inequality we get the contradiction

$$z_t(x_0, t_0) - \Delta z(x_0, t_0) \geq -a_0(x_0, t_0) z(x_0, t_0) > 0.$$

Thus, if the Dirichlet conditions hold, we infer that $z \geq 0$ in $\overline{D_T}$. If the Neumann conditions hold, the negative minimum can occur only on S_T. There one can prove that[6] $\frac{\partial w}{\partial \nu}(x_0, t_0) < 0$, contradicting the Neumann inequality. $\square$

---

[5] Assuming that $u_t(x_0, t_0)$ and $\Delta u(x_0, t_0)$ exist. Otherwise we argue as in Theorem 3.2.

[6] *Protter and Wienberger* [10].

In the next theorem, we build a monotone iteration procedure to construct a solution of problem (5.39), (5.40), (5.41).

Theorem 5.2. *Let $\bar{u}$ and $\underline{u}$ be super/subsolution of problem (5.39), (5.40), (5.41), respectively such that*

$$a \leq \underline{u}(\mathbf{x}, t) \leq g(\mathbf{x}) \leq \bar{u}(\mathbf{x}, t) \leq b \quad \text{in } D_T.$$

If Ω is a regular domain there exists a unique solution u such that

$$a \leq \underline{u}(\mathbf{x}, t) \leq u(\mathbf{x}, t) \leq \bar{u}(\mathbf{x}, t) \leq b \quad \text{in } D_T. \tag{5.43}$$

Proof. We give only a formal proof for the Dirichlet conditions. The case of Neumann conditions is similar.

We construct a recursive sequence of functions in the following way. Set $F(s) = f(s) + M$ where $M = \max_{s \in [a,b]} |f'(s)|$. Then

$$F'(s) = f'(s) + M \geq -M + M = 0$$

so that F is increasing in $[a, b]$. Write the differential equation in the form

$$u_t - \Delta u + Mu = f(u) + Mu \equiv F(u).$$

Define $u^{(0)} = \underline{u}$ and let $u^{(1)}$ be the solution of the linear problem

$$\begin{cases} u_t^{(1)} - \Delta u^{(1)} + Mu^{(1)} = F(u^{(0)}) & \text{in } D_T \\ u^{(1)} = 0 & \text{on } S_T \\ u^{(1)}(\mathbf{x}, 0) = g(\mathbf{x}) & \text{in } \overline{\Omega}. \end{cases}$$

Since $F(u^{(0)})$ is bounded and regular in D_T, there exists a unique solution $u^{(1)} \in C(\overline{D_T}) \cap C^{2,1}(D_T)$.

The function $w^{(1)} = u^{(1)} - u^{(0)}$ satisfies the inequality

$$w_t^{(1)} - \Delta w^{(1)} + Mw^{(1)} \geq F(u^{(0)}) - F(u^{(0)}) = 0 \text{ in } D_T$$

and moreover $u^{(1)} = 0$ on S_T and $u^{(1)}(\mathbf{x}, 0) = 0$ in $\overline{\Omega}$. Arguing as in Lemma 5.1, we infer that

$$a \leq u^{(0)} \leq u^{(1)} \leq \bar{u} \leq b \quad \text{in } \overline{D_T}.$$

Now let $u^{(2)}$ be the unique solution of the linear problem

$$\begin{cases} u_t^{(2)} - \Delta u^{(2)} + Mu^{(2)} = F(u^{(1)}) & \text{in } D_T \\ u^{(2)} = 0 & \text{on } S_T \\ u^{(2)}(\mathbf{x}, 0) = g(\mathbf{x}) & \text{in } \overline{\Omega}. \end{cases}$$

The function $w^{(2)} = u^{(2)} - u^{(2)}$ satisfies the inequality

$$w_t^{(2)} - \Delta w^{(2)} + Mw^{(2)} = F(u^{(1)}) - F(u^{(0)}) \quad \text{in } D_T.$$

and moreover $u^{(2)} = 0$ on S_T and $u^{(2)}(\mathbf{x}, 0) = 0$ in $\overline{\Omega}$. Since $a \le u^{(0)} \le u^{(1)} \le b$ and F is **increasing in** $[a, b]$ we have $F(u^{(1)}) - F(u^{(0)}) \ge 0$. Arguing once more as in Lemma 5.1 we infer that

$$a \le u^{(0)} \le u^{(1)} \le u^{(2)} \le \overline{u} \le b \quad \text{in } \overline{D_T}.$$

Iterating the above construction, we define a sequence $\left\{ u^{(k)} \right\}_{k \ge 1}$ such that

$$\begin{cases} u_t^{(k)} - \Delta u^{(k)} + Mw^{(k)} = F(u^{(k-1)}) & \text{in } D_T \\ u^{(k)} = 0 & \text{on } S_T \\ u^{(k)}(\mathbf{x}, 0) = g(\mathbf{x}) & \text{in } \overline{\Omega} \end{cases} \tag{5.44}$$

and

$$\underline{u} \le u^{(1)} \le u^{(2)} \le \cdots \le u^{(k)} \le u^{(k+1)} \le \cdots \le \overline{u}$$

in D_T. Being increasing and bounded, $\left\{ u^{(k)} \right\}$ converges to a function u satisfying (5.43). Taking the limit as $k \to +\infty$ in problem (5.44) one can show[7] that u solves (5.39), (5.40), (5.41).

The uniqueness follows again from Lemma 5.1: if u_1 and u_2 are solutions satisfying (5.43), since they are both super and sub solutions, we deduce $u_1 \ge u_2$ and $u_2 \ge u_1$. $\qquad\square$

Remark 5.3. Starting from $u^{(0)} = \overline{u}$ the sequence $\left\{ u^{(k)} \right\}$ is decreasing and converges to the solution u. Note that Theorem 5.2 reduces the existence and uniqueness of a solution to (5.39), (5.40), (5.41), to find bounded a super and a subsolution. In several interesting cases, it is possible to find super and subsolutions *time independent*. In this case Theorem 5.2 gives **global solutions**, that is defined for all $t > 0$.

5.3.2 The Fisher's equation

We apply the results in the last section to the Cauchy-Dirichlet problem

$$\begin{cases} w_\tau = Dw_{yy} + aw \left(1 - \dfrac{w}{N}\right) & 0 < y < L, \tau > 0 \\ w(y, 0) = g(y) & 0 < y < L \\ w(0, \tau) = w(L, \tau) = 0 & \tau > 0 \end{cases} \tag{5.45}$$

where $g \ge 0$. The differential equation in (5.45) (so called *Fisher's equation*) is a diffusion model with logistic reaction. In absence of diffusion, the resulting ordinary equation has two constant steady states $w = 0$, unstable and $w = 1$, asymptotically stable with basin of attraction $(0, +\infty)$.

The parameter a encodes the reaction speed (with dimension $(time)^{-1}$) while N represents the habitat carrying capacity, a threshold value for w.

---

[7] The justification of this fact is not elementary and out of the scope of this introductory text. See for instance [15].

To analyze the behavior of the solutions we introduce dimensionless quantities by suitably rescaling the variables y, τ and w. Since L is a typical length, set

$$x = \frac{y}{L}.$$

We may rescale time in two ways. For instance, recalling that the dimension of a is $(tima)^{-1}$ and that of D is $(length)^2 \times (tempo)^{-1}$, we may set

$$t = a\tau \quad \text{or} \quad t = \frac{D\tau}{L^2}.$$

Let us choose the second one. To rescale w we use N as a typical size and define

$$u(x,t) = \frac{1}{N} w\left(Lx, \frac{L^2 t}{D}\right).$$

We have

$$u_t = w_\tau \frac{d\tau}{dt} = w_\tau \frac{L^2}{ND}, \quad u_{xx} = \frac{1}{N} w_{yy} \frac{d^2 y}{dx^2} = w_{yy} \frac{L^2}{N}.$$

Substituting into (5.45) we get

$$u_t = u_{xx} + \lambda u (1 - u) \qquad 0 < x < 1$$

where

$$\lambda = \frac{aL^2}{D}$$

with initial condition

$$u(x,0) = g(Lx)/N \equiv G(x) \qquad 0 < x < 1$$

and

$$u(0,t) = u(1,t) = 0 \quad t > 0.$$

From the results of Section 3.1 we can easily prove that, if $0 \le g \le 1$, there exists for all $t > 0$ a unique solution u, such that $0 \le u \le 1$.

Indeed, it is enough to observe that $\underline{u} = 0$ and $u = 1$ are respectively sub and supersolution. What is mostly interesting is the asymptotic behavior of u as $t \to +\infty$. We shall see that λ plays the role of *bifurcation parameter*.

Notice that the same parameter appears in (5.18). We have seen that if $\lambda < \pi^2$, the solution of the *linearized equation*, obtained neglecting the nonlinear term $-\lambda u^2$, goes rapidly to zero. This occurs also to the Fisher-Kolmogoroff equation. In fact, heuristically, by adding $-\lambda u^2$ we enforce the decaying effect and therefore $u(x,t) \to 0$ for $t \to +\infty$.

What happens when $\lambda > \pi^2$? We shall see that, in this case, the solution u evolves towards a **steady state** $v = v(x)$, $0 \le v \le 1$, solution of the stationary problem

$$\begin{cases} v_{xx} + \lambda v (1 - v) = 0 & 0 < x < 1 \\ v(0) = v(1) = 0. \end{cases} \qquad (5.46)$$

5.3.3 Steady states, linearization and stability

To justify our conclusions on the Fisher's equations, we use a linearization technique that we describe for the following equation

$$u_t = \Delta u + f(u) \qquad \text{in } \Omega \times (0, +\infty) \tag{5.47}$$

where $f \in C^1(\mathbb{R})$, with initial condition

$$u(\mathbf{x}, 0) = g(\mathbf{x}) \qquad \text{in } \overline{\Omega}, \tag{5.48}$$

and *homogeneous* Dirichlet or Neumann conditions.

Let $v_s = v_s(\mathbf{x})$ be a *steady state* solution, that is a **bounded** solution of the stationary problem

$$\begin{cases} \Delta v + f(v) = 0 \quad \text{in } \Omega \\ v = 0 \text{ or } \dfrac{\partial v}{\partial \nu} = 0 \text{ on } \partial\Omega. \end{cases} \tag{5.49}$$

In particular, a steady state can be constant (spatially homogeneous). The **non constant** steady states correspond to heterogeneous spatial configurations. In the applications to biology these solutions (so called *patterns*) are particularly important.

We want to analyze the stability properties of v_s, that is to determine under which conditions $u(\mathbf{x}, t) \to v_s(\mathbf{x})$ as $t \to +\infty$, pointwise or uniformly in Ω.

Precisely, we say that v_s is (**neutrally**) **stable** if, for every $\varepsilon > 0$, there exists δ such that, if u is a *bounded* solution of (5.47), (5.48) with

$$|g(\mathbf{x}) - v_s(\mathbf{x})| < \delta, \ \forall \mathbf{x} \in \Omega,$$

then

$$|u(\mathbf{x}, t) - v_s(\mathbf{x})| < \varepsilon \qquad \forall \mathbf{x} \in \Omega, \forall t > 0.$$

If moreover

$$\lim_{t \to +\infty} \sup_{\mathbf{x} \in \Omega} |u(\mathbf{x}, t) - v_s(\mathbf{x})| = 0$$

then v_s is **asymptotically stable**. We say that v_s **unstable** is it is not **stable**.

Set

$$w(\mathbf{x}, t) = u(\mathbf{x}, t) - v_s(\mathbf{x}).$$

Then w solves

$$w_t = \Delta w + [f(u) - f(v_s)] \tag{5.50}$$

with

$$w(\mathbf{x}, 0) = g(\mathbf{x}) - v_s(\mathbf{x}) \qquad \text{in } \overline{\Omega}$$

and

$$w = 0 \text{ or } \frac{\partial w}{\partial \nu} = 0 \quad \text{on } S_T.$$

By the Mean Value Theorem, we can write

$$f\left(u\left(\mathbf{x},t\right)\right) - f\left(v\left(\mathbf{x}\right)\right) = f'\left(v_s\left(\mathbf{x}\right)\right)w\left(\mathbf{x},t\right) + R\left(\mathbf{x},t\right)$$

where $R\left(\mathbf{x},t\right)$ is of lower order as $w \to 0$.

We formally linearize the problem by neglecting the remainder R. Then (5.50) becomes

$$w_t = \Delta w + f'\left(v_s\left(\mathbf{x}\right)\right)w \tag{5.51}$$

which is **linear in** w. Note that $f'\left(v_s\left(\mathbf{x}\right)\right)$ is bounded in Ω.

If $\sup_{\mathbf{x} \in \Omega}\left|w\left(\mathbf{x},t\right)\right| \to 0$ as $t \to +\infty$ then we say that v is *linearly* asymptotically stable. Let us check under which conditions linear stability occurs. Using separation of variables we seek for solutions of the form $w\left(\mathbf{x},t\right) = U\left(\mathbf{x}\right)Z\left(t\right)$. Substituting into (5.51) we find

$$U\left(\mathbf{x}\right)Z'\left(t\right) = \Delta U\left(\mathbf{x}\right)Z\left(t\right) + f'\left(v_s\left(\mathbf{x}\right)\right)U\left(\mathbf{x}\right)Z\left(t\right)$$

and then

$$\frac{\Delta U\left(\mathbf{x}\right) + f'\left(v_s\left(\mathbf{x}\right)\right)U\left(\mathbf{x}\right)}{U\left(\mathbf{x}\right)} = \frac{Z'\left(t\right)}{Z\left(t\right)} = \mu$$

where μ is a constant. Thus, U is a solution of the *eigenvalue problem*

$$\Delta U\left(\mathbf{x}\right) + f'\left(v_s\left(\mathbf{x}\right)\right)U\left(\mathbf{x}\right) = \mu U\left(\mathbf{x}\right) \quad \text{in } \Omega \tag{5.52}$$

with boundary conditions

$$U = 0 \text{ or } \frac{\partial U}{\partial \nu} = 0 \quad \text{on } \partial\Omega. \tag{5.53}$$

If $f'\left(v_s\left(\mathbf{x}\right)\right)$ **is not identically zero**, the following theorem holds[8]:

Theorem 5.3. *Let $v_s = v_s\left(\mathbf{x}\right)$ be a solution of problem (5.49). Then:*

1. *There exists a sequence*

$$\cdots < \mu_{k+1} < \mu_k < \cdots < \mu_2 < \mu_1$$

    of real eigenvalues of problem (5.52), (5.53) such that $\mu_k \to -\infty$. In particular, there exists only a finite number of positive eigenvalues.

2. *All the eigenfunctions corresponding to the principal eigenvalue μ_1 are multiples of a single eigenfunction φ_1, positive in Ω in the case of Dirichlet conditions, positive in $\overline{\Omega}$ in the case of Neumann conditions.*

From this theorem we can get the information we want.

---

[8] See *Salsa* [15].

Theorem 5.4. *Let $v_s = v_s(\mathbf{x})$ be a solution of problem (5.49).*

(a) *If $\mu_1 < 0$ then v_s is asymptotically stable in the following sense: there exist positive numbers ρ and α such that if*

$$|g(\mathbf{x}) - v_s(\mathbf{x})| \leq \rho\varphi_1(\mathbf{x}) \qquad \forall \mathbf{x} \in \overline{\Omega} \tag{5.54}$$

then

$$|u(\mathbf{x},t) - v_s(\mathbf{x})| \leq \rho e^{-\alpha t}\varphi_1(\mathbf{x}) \qquad \forall \mathbf{x} \in \overline{\Omega}, \forall t > 0. \tag{5.55}$$

(b) *If $\mu_1 > 0$ then v_s is unstable unstable, and precisely, for every $\sigma \in (0,1)$, there exist positive numbers ρ and α such that, if*

$$g(\mathbf{x}) - v_s(\mathbf{x}) \geq \rho(1-\sigma)\varphi_1(\mathbf{x}) \qquad \forall \mathbf{x} \in \overline{\Omega} \tag{5.56}$$

then

$$u(\mathbf{x},t) - v(\mathbf{x}_s) \geq \rho(1 - \sigma e^{-\alpha t})\varphi_1(\mathbf{x}) \qquad \forall \mathbf{x} \in \overline{\Omega}, \forall t > 0. \tag{5.57}$$

Proof (a). Let $w(\mathbf{x},t) = v_s(\mathbf{x}) + \rho e^{-\alpha t}\varphi_1(\mathbf{x})$. We choose positive ρ and α to make w a supersolution to the problem (5.39), (5.40), (5.41).

We have, since $\Delta\varphi_1 + f'(v_s)\varphi_1 = \mu_1\varphi_1$

$$w_t - \Delta w = -\Delta v_s + (-\alpha\varphi_1 - \Delta\varphi_1)\rho e^{-\alpha t}$$
$$= f(v_s) + (-\alpha - \mu_1 + f'(v_s))\rho e^{-\alpha t}\varphi_1.$$

By the mean Value Theorem, for a suitable $\eta = \eta(\mathbf{x},t)$, $0 < \eta < \rho$, we can write

$$f(v_s) = f(w) - f'(v_s + \eta)\rho e^{-\alpha t}\varphi_1.$$

Since $f \in C^1(\mathbb{R})$ and $\mu_1 < 0$, if $-\alpha - \mu_1 > 0$ and ρ is sufficiently small, we have

$$-\alpha - \mu_1 + f'(v_s) > f'(v_s + \eta).$$

Recalling that $\varphi_1 > 0$,

$$w_t - \Delta w = f(w) + [-\alpha - \mu_1 + f'(v_s) - f'(v_s + \eta)]\rho e^{-\alpha t}\varphi_1 \geq f(w).$$

Observe that from (5.54), $v_s(\mathbf{x}) + \rho\varphi_1(\mathbf{x}) \geq u(\mathbf{x},0) = g(\mathbf{x})$ and $w = 0$ or $\partial_\nu w = 0$ on S_T. It follows that w is a supersolution of the problem (5.39), (5.40), (5.41) and therefore, Lemma 5.1 gives

$$u(\mathbf{x},t) \leq v_s(\mathbf{x}) + \rho e^{-\alpha t}\varphi_1(\mathbf{x}) \qquad t > 0, \mathbf{x} \in \overline{\Omega}.$$

Similarly for suitable positive ρ and α, $z(\mathbf{x},t) = v_s(\mathbf{x}) - \rho e^{-\alpha t}\varphi_1(\mathbf{x})$ is a subsolution of the same problem so that

$$u(\mathbf{x},t) \geq v_s(\mathbf{x}) - \rho e^{-\alpha t}\varphi_1(\mathbf{x}) \qquad t > 0, \mathbf{x} \in \overline{\Omega}.$$

This shows (5.55).

Proof (b). It is sufficient to check that $w(\mathbf{x},t) = v_s(\mathbf{x}) + \rho(1 - \sigma e^{-\alpha t})\varphi_1(\mathbf{x})$ is a subsolution for suitable ρ and α. We omit the details. $\square$

Remark 5.4. In the case of Neumann conditions, (5.55) expresses the asymptotic stability of v_s, since $\varphi_1 \geq c_0 > 0$ in $\overline{\Omega}$.

In the case of Dirichlet conditions, rigorously, the solution v_s attracts only solutions starting from an initial condition close to v_s, vanishing on the boundary, since in this case $\varphi_1 = 0$ on $\partial\Omega$. With a supplementary effort it is possible to show that this occurs also with general initial data, so that v_s is asymptotically stable.

5.3.4 Application to Fisher's equation (Dirichlet conditions)

We apply Theorem 5.4 to Fisher's equation with homogeneous Dirichlet conditions and initial data g, $0 \leq g \leq 1$. Consider first $v_s \equiv 0$. Since $f'(u) = (1 - 2u)$, we have $f'(0) = 1$ and the eigenvalue problem (5.52), (5.53) reduces to

$$U''(x) = (\mu - \lambda)U(x) \qquad 0 < x < 1$$

with

$$U(0) = U(1) = 0.$$

We have already solved this problem; we have

$$\mu - \lambda = -k^2\pi^2 \qquad k \geq 1$$

so that

$$\mu_1 = \lambda - \pi^2.$$

From Theorem 5.4 we infer that:

Proposition 5.2. *If $\lambda < \pi^2$ then $v_s = 0$ is asymptotically stable. If $\lambda > \pi^2$, $v_s = 0$ is unstable.*

We can also show that $v_s = 0$ is the unique nonnegative steady state if $\lambda < \pi^2$. In fact, multiplying the equation $v'' + \lambda f(v) = 0$ by $\sin \pi x$ and integrating over $(0,1)$ we find

$$\int_0^1 v'' \sin \pi x \, dx = -\int_0^1 \lambda v(1 - v) \sin \pi x \, dx.$$

An integration by parts gives, using the homogeneous Dirichlet conditions

$$\int_0^1 v'' \sin \pi x \, dx = -\int_0^1 \pi^2 v \sin \pi x \, dx$$

whence

$$(\lambda - \pi^2) \int_0^1 v \sin \pi x \, dx = \int_0^1 \lambda v^2 \sin \pi x \, dx. \qquad (5.58)$$

If $\lambda < \pi^2$, the (5.58) is possible only if $v \equiv 0$ in $(0, 1)$.

This fact implies that *every* solution with initial data g, bounded and nonnegative, converges to the zero solution as $t \to +\infty$.

Since in terms of the original parameters $\lambda = aL^2/D$, Proposition 5.2 shows that in presence of large diffusion coefficients or of too small habitat the population eventually goes towards extinction.

We now analyze the case $\lambda > \pi^2$. We shall show later on, that in this case there exists another steady state solution v_s Moreover, $0 \le v_s \le 1$ and v_s is positive in $(0,1)$. Assuming for the moment that such solution exists, we examine its stability using Theorem 5.4. Since v_s is not a constant, we cannot compute explicitly the eigenvalues of the stationary problem and we have to resort to an indirect method.

The following theorem holds.

Theorem 5.5. *Let $\lambda > \pi^2$ and v_s be a steady state solution, $0 \le v_s \le 1$, positive in $(0,1)$ of the problem*

$$\begin{cases} v'' + \lambda f(v) = 0 & 0 < x < 1 \\ v(0) = v(1) = 0. \end{cases} \tag{5.59}$$

Then v_s is asymptotically stable.

Proof. We must show that the first eigenvalue μ_1 of the problem (5.52), with $U(0) = U(1) = 0$ and $f(s) = s(1-s)$, is negative. Recall that the corresponding eigenfunction $\varphi_1(x)$ is positive in $(0,1)$.

Multiply the stationary Fisher's equation by φ_1 and integrate over $(0,1)$

$$\int_0^1 [v_s''(x) + \lambda f(v_s(x))]\varphi_1(x)\,dx = 0.$$

Multiply the equation

$$\varphi_1''(x) + \lambda f'(v_s(x))\varphi_1(x) = \mu_1\varphi_1(x)$$

by v and integrate over $(0,1)$:

$$\int_0^1 [\varphi_1''(x) + \lambda f'(v_s(x))\varphi_1(x)]v_s(x)\,dx = -\mu_1\int_0^1 \varphi_1(x)v_s(x)\,dx.$$

Subtracting the two equations we get.

$$\int_0^1 [v_s''(x)\varphi_1(x) - v_s(x)\varphi_1''(x)]\,dx + \lambda\int_0^1 [f(v_s(x))$$

$$-f'(v_s(x))v_s(x)]\varphi_1(x)\,dx = -\mu_1\int_0^1 \varphi_1(x)v_s(x)\,dx.$$

Integrating by parts the first terms we obtain

$$\int_0^1 [v_s''(x)\,\varphi_1(x) - v_s(x)\,\varphi_1''(x)]\,dx$$

$$= [v_s'(x)\,\varphi_1(x) - v_s(x)\,\varphi_1'(x)]_0^1 - \int_0^1 [v_s'(x)\,\varphi_1'(x) - v_s'(x)\,\varphi_1'(x)]\,dx$$

$$= 0.$$

Thus

$$\lambda \int_0^1 [f(v_s(x)) - f'(v_s(x))\,v_s(x)]\,\varphi_1(x)\,dx = -\mu_1 \int_0^1 \varphi_1(x)\,v_s(x)\,dx. \tag{5.60}$$

Since $\lambda > \pi^2$, $\varphi_1 > 0$, $v_s > 0$ in $(0,1)$ and moreover

$$f(v_s) - f'(v_s)\,v_s = v_s - v_s^2 - (1 - 2v_s)\,v_s = v_s^2 > 0 \quad \text{in } (0,1),$$

we deduce from (5.60) that $\mu_1 < 0$.

Theorem 5.4 implies that v_s is asymptotically stable.   □

We now prove the existence and uniqueness of the steady state $v_s = v_s(x)$. The following result holds[9].

Theorem 5.6. *There exists a unique solution v_s of problem (5.59), positive in $(0,1)$ and asymptotically stable (by Theorem 5.5).*

Proof. Set $v' = w$ and consider the system

$$\begin{cases} v' = w \\ w' = -\lambda v\,(1 - v) = -\lambda f(v). \end{cases} \tag{5.61}$$

Let $F(v) = \int_0^v f(s)\,ds = \frac{v^2}{2} - \frac{v^3}{3}$. The differential equations for the orbits in the phase plane v, w is

$$\frac{dw}{dv} = \frac{\lambda f(v)}{w}$$

or

$$w\,dw = -\lambda f(v)\,dv.$$

Integrating, we find the family of curves given by

$$w^2 = -2\lambda F(v) + c = -\lambda\left(v^2 - \frac{2}{3}v^3\right) + c \qquad c \in \mathbb{R}. \tag{5.62}$$

The orbits are symmetric with respect to the v axis; their configuration, depending on the parameter c, is described in Fig. 5.3.

---

[9] See *Murray* [24].

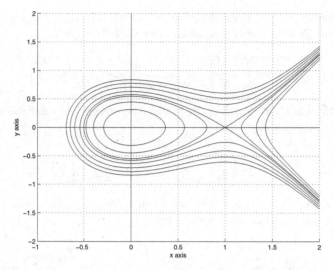

Fig. 5.3. Orbit configurations of equation (5.62) for $\lambda = 1$

A solution $v_s = v_s(x)$ vanishing at $x = 0$, $x = 1$ corresponds to a half of a periodic orbit and therefore it is always positive (right semiorbit) or negative (left semiorbit). Our task is to establish the range of λ for which such orbits exist and, in particular, to select the minimum λ in that range (*bifurcation value*).

Let us focus on the right semiorbits. By symmetry we must have $m = \max_{[0,1]} v_s = v_s(1/2)$ and $0 < m < 1$, while $w_s(1/2) = 0$. Moreover, $w_s > 0$ in $(0, 1/2)$ and $w_s < 0$ in $(1/2, 0)$. Thus, we can only consider $0 \le x \le 1/2$.

Note that, letting $x = 1/2$ into (5.62) we get

$$c = 2\lambda F\left(v_s\left(\frac{1}{2}\right)\right) = 2\lambda F(m) = \lambda\left(m^2 - \frac{2}{3}m^3\right)$$

hence we can write (5.62) in the form

$$w_s^2 = 2\lambda[F(m) - F(v_s)] = \frac{\lambda}{3}\left(3m^2 - 2m^3 - 3v_s^2 + 2v_s^3\right). \qquad (5.63)$$

From $v_s' = w_s$ and (5.63) we have

$$\frac{dv_s}{dx} = \sqrt{\frac{\lambda}{3}\left(3m^2 - 2m^3 - 3v_s^2 + 2v_s^3\right)}$$

whence

$$dx = \sqrt{\frac{3}{\lambda}}\frac{dv_s}{\sqrt{3m^2 - 2m^3 - 3v_s^2 + 2v_s^3}}.$$

When x goes from 0 to $1/2$), the solution v_s, when it exists, varies from 0 to m. Thus, integrating the last equation over $(0, 1/2)$ we find

$$\frac{1}{2} = \int_0^m \sqrt{\frac{3}{\lambda}} \frac{dv}{\sqrt{3m^2 - 2m^3 - 3v^2 + 2v^3}}. \tag{5.64}$$

This improper integral is finite[10] and therefore (5.64) defines a unique $\lambda = \lambda(m)$, for every m, $0 < m < 1$

$$\lambda(m) = 12 \left(\int_0^m \frac{dv}{\sqrt{3m^2 - 2m^3 - 3v^2 + 2v^3}} \right)^2.$$

The function $m \mapsto \lambda(m)$ is strictly increasing. To see it, let $v = my$. Then y varies from 0 to 1, while $dv = m\,dy$. Then

$$\int_0^m \frac{dv}{\sqrt{3m^2 - 2m^3 - 3v^2 + 2v^3}} = \int_0^1 \frac{dy}{\sqrt{3 - 2m - 3y^2 + 2my^3}}$$

$$\equiv \int_0^1 G(m, y)\, dy.$$

We compute

$$\frac{\partial G}{\partial m} = \frac{1 - y^3}{(3 - 2m - 3y^2 + 2my^3)^{3/2}} > 0 \quad \text{for every } y \in (0, 1)$$

so that $\lambda(m)$ is increasing Hence, the bifurcation value is

$$\lambda_{\min} = \lim_{m \to 0+} \lambda(m) = 4 \left(\int_0^1 \frac{dy}{\sqrt{1 - y^2}} \right)^2 = 4 \left([\arcsin y]_0^1 \right)^2 = \pi^2.$$

Moreover,

$$\lim_{m \to 1^-} \lambda(m) = +\infty.$$

We infer that, for every $\lambda > \pi^2$ there exists a unique $v = v_s(x)$, positive in $(0, 1)$, such that $\lambda(m) = \lambda$.   □

Remark 5.5. Given the uniqueness of the stationary solution v_s it is possible to prove that *the basin of attraction of v_s includes all the solutions of Fisher's equation with initial data $g > 0$ in $(0, 1)$.*

---

[10] We have
$$3m^2 - 2m^3 - 3v^2 + 2v^3 = (m - s)(3m + 3v - 2m^2 - 2mv - 2v^2).$$
If $v \to m$,
$$3m^2 - 2m^3 - 3v^2 + 2v^3 \sim 6(m - v)(m - m^2)$$
hence the integrand is of order $(m - v)^{-1/2}$ which is integrable near $v = m$.

5.3.5 Application to Fisher's equation (Neumann conditions)

In this section we analyze the asymptotic behavior of a solution of the equation $u_t = u_{xx} + f(u)$ with homogeneous Neumann conditions and initial condition g, $0 \le g \le 1$ in $(0, 1)$.

First observe that a nonconstant steady state (*pattern*) *cannot have constant sign*. This follows from the phase plane analysis in the previous section, since now the *patterns* corresponds to semiorbits connecting two point on the v axis, and all these orbits must cross the w axis.

Moreover, *a pattern cannot be asymptotically stable*. In fact, using Theorem 5.4 we can prove the following:

Theorem 5.7. *Let $v = v(x)$ be a non constant steady state, that is a solution of the problem*

$$v'' + \lambda f(v) = 0, \qquad 0 < x < 1,$$

with $v'(0) = v'(1) = 0$. Then v is unstable.

Proof. Let μ_1 and φ_1 the principal eigenvalue and the corresponding eigenfunction of the linearized Neumann problem

$$\varphi_1''(x) + \lambda f'(v(x)) \varphi_1(x) = \mu_1 \varphi_1(x) \tag{5.65}$$

and

$$\varphi_1'(0) = \varphi_1'(1) = 0.$$

We may choose φ_1 *positive in* $[0, L]$.

Since v changes his sign, we must argue differently from the proof of Theorem 5.7. Differentiating the stationary Fisher's equation, we get:

$$(v')'' + \lambda f'(v) v' = 0 \quad 0 < x < 1$$

which is an equation for v'. Multiply this equation by φ_1 and integrate over $(0, 1)$

$$\int_0^1 [(v')'' + \lambda f'(v) v'] \varphi_1 dx = 0. \tag{5.66}$$

Now multiply (5.65) by v' and integrate over $(0, 1)$:

$$\int_0^1 [\varphi_1'' + \lambda f'(v) \varphi_1] v' dx = \mu_1 \int_0^1 \varphi_1 v' dx. \tag{5.67}$$

Subtracting (5.66) from (5.67) we find

$$\int_0^1 [\varphi_1'' v' - (v')'' \varphi_1] dx = \mu_1 \int_0^1 \varphi_1 v' dx.$$

An integration by parts of the first integral gives, using the Neumann conditions

$$\int_0^1 [\varphi_1'' v' - (v')'' \varphi_1] dx = [\varphi_1' v' - v'' \varphi_1]_0^L = -v''(1) \varphi_1(1) + v''(0) \varphi_1(0).$$

We now distinguish 3 cases, recalling that $\varphi_1(1) > 0$ and $\varphi_1(0) > 0$.

a) v increasing in $[0,1]$. Then $v' \geq 0$, $v''(0) \geq 0$ and $v''(1) \leq 0$. Hence

$$-v''(1)\varphi_1(1) + v''(0)\varphi_1(0) \geq 0 \quad \text{while} \quad \int_0^1 \varphi_1 v' dx > 0.$$

Thus $\mu_1 > 0$.

b) v decreasing $[0,1]$. Then $v' \geq 0$, $v''(0) \leq 0$ and $v''(1) \leq 0$. Hence

$$-v''(1)\varphi_1(1) + v''(0)\varphi_1(0) \leq 0 \quad \text{and} \quad \int_0^1 \varphi_1 v' dx < 0$$

so that $\mu_1 > 0$ again.

c) v oscillating. Then there exists an interval $(a,b) \subset (0,1)$ such that $v'(a) = v'(b) = 0$, $v''(a) \geq 0$, $v''(b) \leq 0$ and $v' \geq 0$ in (a,b). If we integrate over (a,b) instead of $(0,1)$ and we use the same argument we infer again that $\mu_1 > 0$.

In any case we get $\mu_1 > 0$ so that v is unstable. □

Still we do not know which should be the asymptotic behavior of a solution? Here we have two constant steady states, $v_0 \equiv 0$ and $v_1 \equiv 1$.

Consider first $v_0 \equiv 0$. We have $f'(0) = 1$ and problem (5.52), (5.53) reduces to

$$U''(x) = (\mu - \lambda)U(x) \qquad 0 < x < 1$$

with

$$U'(0) = U'(1) = 0.$$

The eigenvalues are

$$\mu - \lambda = -k^2\pi^2 \qquad k \geq 0$$

whence

$$\mu_1 = \lambda > 0.$$

In **any case**, v_0 is **unstable**.

Let us try with v_1. We have $f'(1) = -1$ and problem (5.52), (5.53) becomes

$$U''(x) = (\mu + \lambda)U(x) \qquad 0 < x < 1$$

with

$$U'(0) = U'(1) = 0.$$

The eigenvalues are

$$\mu + \lambda = -k^2\pi^2 \qquad k \geq 0$$

whence

$$\mu_1 = -\lambda < 0.$$

We infer that $v_1 \equiv 1$ is **asymptotically stable**. Let us determine its basin of attraction. Let $g > 0$ in $[0,1]$. Choose $H \leq m = \min_{[0,1]} g$, $0 < H < 1$, and $K \geq M = \max_{[0,1]} g$. Let $U_1 = U_1(t)$, $U_2 = U_2(t)$ be the solutions of the logistic equation

$$\dot{U} = \lambda U(1-U)$$

with initial data $U_1(0) = H$ and $U_2(0) = K$, respectively.

Then U_1 is a subsolution and U_2 is a supersolution of problem (5.39), (5.40), (5.41) and both are bounded. Then, by Lemma 5.1 and Theorem 5.2, if u is the solution of Fisher's equation with initial data g, we have

$$U_1(t) \leq u(x,t) \leq U_2(t).$$

Since $\lim_{t \to \infty} U_1(t) = \lim_{t \to \infty} U_2(t) = 1$, we infer that

$$\sup_{x \in [0,1]} u(x,t) \to 1 \quad \text{as } t \to \infty.$$

Conclusion. *Any solution of Fisher's equation in $(0,1)$ with homogeneous Neumann conditions and positive initial data evolves uniformly in $[0,1]$ towards the constant steady state $v_s \equiv 1$.*

5.4 Turing instability

We have seen that solutions of Fisher's equation with Neumann boundary conditions and positive initial conditions evolves towards the constant steady state of the logistic o.d.e.

$$\frac{du}{dt} = \lambda u(1-u).$$

This fact an be generalized (with similar proof) to equations of the type

$$u_t = D \Delta u + f(u).$$

Indeed, if $u \equiv c$ is a an equilibrium solution of the o.d.e.

$$\frac{du}{dt} = f(u),$$

that is $f(c) = 0$, and $f'(c) < 0$, then $u \equiv c$ is asymptotically stable. In terms of population dynamics, this means that a *self organizing population* (zero boundary flux) evolves toward a spatially homogeneous steady state.

If instead, of a single equation we consider a system of two or more reaction-diffusion equations, the evolution can significantly change. Indeed a steady state which is asymptotically stable for the dynamics in absence of diffusion, can be turned into an unstable state in presence of diffusion.

This quite surprising effect was discovered by Alan Turing in 1952 (*The Chemical Basis of Morphogenesis*). He showed that, under certain conditions, specific chemical components may react and diffuse producing stable concentrations of chemical substances (*morphogenes*), spatially nonhomogeneous (*patterns*).

Here we consider a typical bidimensional example.

Assume that two chemical substances with concentrations $A = A(\xi,\eta,\tau)$ and $B = B(\xi,\eta,\tau)$ diffuse and react according to the system (τ denotes time)

$$\begin{cases} A_\tau = D_A \Delta A + F(A,B) \\ B_\tau = D_B \Delta A + G(A,B) \end{cases}$$

where F and G (in general nonlinear) model the chemical reactions involving A and B.

Turing intuition is the following. In absence of diffusion ($D_A = D_B = 0$) we obtain the o.d.e. system

$$\begin{cases} A_\tau = F(A,B) \\ B_\tau = G(A,B). \end{cases}$$

Assume that, A and B evolves as $t \to +\infty$ toward a constant state A_0, B_0, hence spatially homogeneous, and *asymptotically stable for the linearized system*

$$\begin{cases} a_\tau = F_A(A_0,B_0)(a-A_0) + F_B(A_0,B_0)(b-B_0) \\ b_\tau = G_A(A_0,B_0)(a-A_0) + G_A(A_0,B_0)(b-B_0). \end{cases}$$

Now, under certain conditions that we will describe later on, if $D_A \neq D_B$, for the original reaction-diffusion system the state A_0, B_0 becomes unstable and the solution evolves toward a steady state, spatially inhomogeneous, i. e. a *pattern*. This phenomenon is known as a *diffusion driven instability*

Since diffusion is usually a stabilizing factor, Turing's proof turned out to be rather surprising. Let us analyze the conditions under which the diffusion driven instability occurs.

As a typical model for the reaction terms F and G we choose the inhibitor/activator model of Schnakenberg:

$$F(A,B) = k_1 - k_2 A + k_3 A^2 B, \quad G(A,B) = k_4 - k_3 A^2 B$$

where the k_j are positive constants.

It is convenient to deal with dimensionless model. If we choose as our reference two-dimensional domain a rectangle and L is a typical length (for instance a side length), we can set:

$$A = \sqrt{\frac{k_2}{k_3}}u, \quad B = \sqrt{\frac{k_2}{k_3}}v, \quad x = \frac{\xi}{L}, \quad y = \frac{\eta}{L}, \quad t = \frac{D_A \tau}{L^2}$$

and

$$\gamma = \frac{L^2 k_2}{D_A}, \quad d = \frac{D_B}{D_A}, \quad a = \frac{k_1}{k_2}\sqrt{\frac{k_3}{k_2}}, b = \frac{k_4}{k_2}\sqrt{\frac{k_3}{k_2}}.$$

After simple computations, we obtain the dimensionless system

$$\begin{cases} u_t = u_{xx} + u_{yy} + \gamma(a - u + u^2 v) \\ v_t = d(v_{xx} + v_{yy}) + \gamma\left(b - u^2 v\right) \end{cases}$$

which is of the form

$$\begin{cases} u_t = \Delta u + \gamma f\left(u, v\right) \\ v_t = d\Delta v + \gamma g\left(u, v\right) \end{cases} \tag{5.68}$$

with

$$f_v > 0, \; g_u < 0. \tag{5.69}$$

Moreover, $u = u\left(x, y, t\right)$, $v = v\left(x, y, t\right)$ where $t > 0$ and x, y vary in a dimensionless rectangle

$$R = \{(x, y) : 0 < x < p, 0 < y < q\}$$

and the parameter γ has the following interpretations

a) γ is proportional to the habitat surface area;
b) γ encodes the intensity of the reaction terms;
c) increasing γ corresponds to decreasing the coefficient d.

To the system (5.68) we associate the initial condition

$$u\left(x, y, 0\right) = u_0\left(x, y\right), v\left(x, y, 0\right) = v_0\left(x, y\right) \qquad 0 < x < p, 0 < y < q$$

and homogeneous Neumann conditions on ∂R

$$u_x\left(0, y, t\right) = u_x\left(p, y, t\right) = u_y\left(x, 0, t\right) = u_y\left(x, q, t\right) = 0$$
$$v_x\left(0, y, t\right) = v_x\left(p, y, t\right) = v_y\left(x, 0, t\right) = v_y\left(x, q, t\right) = 0$$

which can be written in the compact form

$$\nabla u \cdot \mathbf{n} = \nabla v \cdot \mathbf{n} = 0 \; \text{ on } \partial R$$

where $\mathbf{n}$ denotes the exterior normal to ∂R.
    Setting

$$\mathbf{w} = \begin{pmatrix} u \\ v \end{pmatrix}, \; \mathbf{F}\left(\mathbf{w}\right) = \begin{pmatrix} f\left(u, v\right) \\ g\left(u, v\right) \end{pmatrix} \; \text{and} \; \mathbf{D} = \begin{pmatrix} 1 & 0 \\ 0 & d \end{pmatrix}$$

we can write our system in the form

$$\mathbf{w}_t = \mathbf{D}\Delta\mathbf{w} + \gamma\mathbf{F}\left(\mathbf{w}\right). \tag{5.70}$$

Consider the system

$$\mathbf{w}_t = \gamma\mathbf{F}\left(\mathbf{w}\right)$$

without diffusion and linearize it at the steady state $\mathbf{w}_0 = \begin{pmatrix} u_0 \\ v_0 \end{pmatrix}$. Setting $\mathbf{z} = \mathbf{w} - \mathbf{w}_0$ we get

$$\mathbf{z}_t = \gamma\mathbf{A}\mathbf{z} \tag{5.71}$$

with

$$\mathbf{A} = \begin{pmatrix} f_u & f_v \\ g_u & g_v \end{pmatrix}_{(u_0, v_0)}.$$

Assume now that $\mathbf{w}_0$ is asymptotically stable, or, equivalently, that $\mathbf{0}$ is **asymptotically stable for** (5.71). Then it must be

$$\mathrm{tr}\mathbf{A} = f_u + g_v < 0 \quad \text{and} \quad |\mathbf{A}| = f_u g_v - f_v g_u > 0 \tag{5.72}$$

where each derivative is computed at (u_0, v_0).

Let us linearize at $\mathbf{w}_0$ also the reaction-diffusion system (5.70); we find

$$\mathbf{z}_t = \mathbf{D}\Delta\mathbf{z} + \gamma\mathbf{A}\mathbf{z} \tag{5.73}$$

still with $\mathbf{z} = \mathbf{w} - \mathbf{w}_0$.

Our goal is now the following: **to determine under which conditions the solution $\mathbf{z} = \mathbf{0}$ is unstable for the system** (5.73).

Thus, we have to analyze the stability of $\mathbf{z} = \mathbf{0}$ for (5.73). We can use the separation of variables. To simplify simplify the computations, we introduce the eigenfunctions $\mathbf{W}_{mn}$ and the corresponding eigenvalues μ_{mn} for the Neumann problem

$$\Delta\mathbf{W}_{mn} + \mu_{mn}\mathbf{W}_{mn} = 0 \quad \text{in } R$$

and

$$\nabla\mathbf{W}_{mn} \cdot \mathbf{n} = \mathbf{0} \quad \text{on } \partial R.$$

From Section 5.2 we have

$$\mu_{mn} = \pi^2 \left(\frac{n^2}{p^2} + \frac{m^2}{q^2} \right), \qquad n, m = 0, 1, 2, \ldots$$

and

$$\mathbf{W}_{mn}(x, y) = \mathbf{c} \cos\frac{n\pi x}{p} \cos\frac{m\pi y}{q}, \qquad \mathbf{c} = \begin{pmatrix} c_1 \\ c_2 \end{pmatrix}.$$

Since we always find the an exponential t-dependence, let us look for solutions to (5.73) of the form

$$\mathbf{z}(x, y, t) = e^{\lambda t}\mathbf{W}_{mn}(x, y).$$

Substituting into (5.73) we find, after simplifying by $e^{\lambda t}$,

$$\lambda\mathbf{W}_{mn}(x, y) = \mathbf{D}\Delta\mathbf{W}_{mn} + \gamma\mathbf{A}\mathbf{W}_{mn}$$

or, since $\Delta\mathbf{W}_{mn} = -\mu_{mn}\mathbf{W}_{mn}$,

$$(\lambda\mathbf{I} - \gamma\mathbf{A} + \mathbf{D}\mu_{mn})\mathbf{W}_{mn} = \mathbf{0} \tag{5.74}$$

where $\mathbf{I}$ is the identity matrix.

Non trivial solutions of (5.74) exist if

$$|\lambda \mathbf{I} - \gamma \mathbf{A} + \mathbf{D}\mu_{mn}| = 0.$$

This condition gives an algebraic equation for λ as a function of μ_{mn} : $\lambda_{mn} = \lambda(\mu_{mn})$. Explicitly we find

$$\lambda^2 + \lambda \left[\mu_{mn}(1 + d) - \gamma \mathrm{tr}\mathbf{A}\right] + h(\mu_{mn}) = 0 \tag{5.75}$$

where

$$h(\mu_{mn}) = d\mu_{mn}^2 - \gamma \left(df_u + g_v\right)\mu_{mn} + \gamma^2 |\mathbf{A}|. \tag{5.76}$$

Even if μ_{mn} is a real number, λ_{mn} could be a complex one. Thus, the solution of the linearized problem is given by a formula of the following type

$$\mathbf{w}(x, y, t) = \sum_{m,n} \mathbf{c}_{n,m} e^{\lambda_{mn}t} \mathbf{W}_{mn}(x, y) \tag{5.77}$$

wherr the $\mathbf{c}_k$ are the cosine Fourier coefficients of the initial data.

To get instability we must have

$$\mathrm{Re}\,\lambda_{mn} > 0$$

for some couple m, n. This can happen in two cases: either $\mu_{mn}(1 + d) - \gamma \mathrm{tr}\mathbf{A} < 0$ or $h(\mu_{mn}) < 0$ for some mn.

Since $\mathrm{tr}\mathbf{A} < 0$ and $\mu_{mn}(1 + d) \geq 0$ the only possibility is the second one. From (5.76), since $|\mathbf{A}| > 0$, to have $h(\mu_{mn}) < 0$ it must be

$$df_u + g_v > 0.$$

Since $f_u + g_v = \mathrm{tr}\mathbf{A} < 0$, this implies $d \neq 1$ and that f_u and g_v have different signs

$$df_u + g_v > 0 \quad \Longrightarrow \quad d \neq 1 \text{ and } f_u g_v < 0. \tag{5.78}$$

Therefore, we have the following proposition.

Proposition 5.3. *For the existence of the driven diffusion instability it is necessary that $D_A \neq D_B$.*

The condition (5.78) is only necessary to guarantee that there exist couples m, n such that $h(\mu_{mn}) < 0$. Thus **the minimum of $h(\mu_{mn})$ must be negative.**

By examining the parabola

$$s \longmapsto h(s) = ds^2 - \gamma(df_u + g_v)s + \gamma^2 |A|$$

we see that its minimum is attained at

$$s_{\min} = \frac{\gamma(df_u + g_v)}{2d}$$

and is equal to

$$h_{\min} = h\left(s_{\min}\right) = \gamma^2 \left[|\mathbf{A}| - \frac{(df_u + g_v)^2}{4d}\right].$$

We infer that $h_{\min} < 0$ if

$$\frac{(df_u + g_v)^2}{4d} > |\mathbf{A}|.$$

Let us summarize **the necessary conditions for driven diffusion instability** we have found so far:

1. $f_u + g_v < 0$;
2. $df_u + g_v > 0$;
3. $f_u g_v - f_v g_u > 0$;
4. $(df_u + g_v)^2 - 4d(f_u g_v - f_v g_u) > 0$.

Keeping fixed the other parameters, the equation $(df_u + g_v)^2 = 4d|\mathbf{A}|$ defines the **critical value d_c for the ratio of the diffusion coefficients,** that is the solution > 1 of the equation $d_c^2 f_u^2 + 2\left(2f_v g_u - f_u g_v\right) d_c + g_v^2 = 0$.

Corresponding to the value d_c we have a **critical value s_c** given by

$$s_c = \frac{\gamma\left(d_c f_u + g_v\right)}{2d_c} = \gamma\left[\frac{|\mathbf{A}|}{d_c}\right]^{1/2}.$$

When $d > d_c$, $h\left(s\right)$ has two zeros s_1 and s_2 given by

$$s_1 = \frac{\gamma}{2d}\left[(df_u + g_v) - \left\{(df_u + g_v)^2 - 4d|\mathbf{A}|\right\}^{1/2}\right]$$

$$s_2 = \frac{\gamma}{2d}\left[(df_u + g_v) - \left\{(df_u + g_v)^2 - 4d|\mathbf{A}|\right\}^{1/2}\right].$$

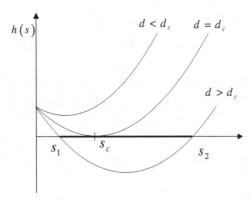

Fig. 5.4. Zeroes of the parabola $h(s)$, according to different values of the parameter d

The set of possible values of μ_{mn} corresponding to $\operatorname{Re}\lambda_{mn} > 0$ is located between s_1 and s_2. **It could be that this set is empty**, since the μ_{mn} assumes rather special values. In other terms, if there are values of μ_{mn} such that

$$s_1 \leq \mu_{mn} \leq s_2,$$

then these values correspond to exponentially growing terms in the series (5.77). This means that, for large t,

$$\mathbf{z}(x,y,t) \sim \sum_{s_1 \leq \mu_{mn} \leq s_2} \mathbf{c}_{mn} e^{\lambda_{mn}t} \mathbf{W}_{mn}(x,y).$$

Notice the dependence of the number of *activated modes* on the domain R. The larger are dimensions p or q the larger is the number of activated modes. As in the scalar case, the exponential growth is counterbalanced by the non-linear reaction terms and eventually spatially nonhomogeneous equilibrium configurations will appear[11].

5.5 Numerical methods

5.5.1 Numerical approximation of a nonlinear reaction-diffusion problem

We address the following problem, which is based on equation (5.39) in one space dimension, to be discretized by means of finite differences,

$$\begin{cases} u_t = u_{xx} + f(u) & 0 < x < 1,\ t > 0 \\ u(0,t) = u(1,t) = 0 \quad \text{or} \quad u_x(0,t) = u_x(1,t) = 0 & t > 0 \\ u(x,0) = g(x) & 0 \leq x \leq 1. \end{cases} \tag{5.79}$$

We observe that either Dirichlet or Neumann boundary conditions are addressed. Given a partition of $(0,1)$ in $N+1$ equally spaced nodes $x_i = ih$, $h = 1/N$, $i = 0,\ldots,N$, we denote by $\mathbf{A}_h^D \in \mathbb{R}^{(N-1)\times(N-1)}$ and $\mathbf{A}_h^N \in \mathbb{R}^{(N+1)\times(N+1)}$ the centred three point finite difference approximation of $-\partial_{xx}$ (see (3.92) Chapter 3) complemented with Dirichlet or Neumann boundary conditions respectively, as in (5.79). More precisely, given $\mathbf{U}^D(t) = \{u_i(t)\}_{i=1}^{N-1} \in \mathbb{R}^{(N-1)}$ or $\mathbf{U}^N(t) = \{u_i(t)\}_{i=0}^{N} \in \mathbb{R}^{(N+1)}$ the degrees of freedom for Dirichlet or Neumann problems, the **semi-discrete** counterpart of (5.79) reads as follows,

$$\dot{\mathbf{U}}^*(t) = f(\mathbf{U}^*) - \mathbf{A}_h^*\mathbf{U}^*, \quad \mathbf{U}^*(0) = \mathbf{g} = \{g(x_i)\} \tag{5.80}$$

where $* = D, N$ according to the choice of boundary conditions and $f(\mathbf{U}^*)$ denotes the component-wise application of function f to $\mathbf{U}^*$. Problem (5.80)

---

[11] The proof of this fact is not elementary and we refer to the original paper of Turing.

is an autonomous differential equation system to be discretized in time. The main difference with respect to the simple diffusion equation consists in the nonlinear term. The application of an explicit time stepping scheme, such as backward Euler method, allows to override the nonlinearity, by evaluating $f(\mathbf{U}^*)$ at the previous iterative step. As a result of this, we obtain,

$$\mathbf{U}^*_{n+1} = \mathbf{U}^*_n + \tau f(\mathbf{U}^*_n) - \tau \mathbf{A}^*_h \mathbf{U}^*_n. \tag{5.81}$$

In spite of its simplicity, the application of such scheme is severely limited by stability constraints and poor accuracy. An effective generalization which maintains the attractive feature of treating the nonlinear term explicitly is provided by the family of **explicit Runge-Kutta schemes**. As an example, we report here a third-order Runge-Kutta scheme.[12] Given $\mathbf{U}^*_n$ the updated solution $\mathbf{U}^*_{n+1}$ is provided through the following intermediate steps,

$$\begin{cases} \mathbf{U}^*_a = \mathbf{U}^*_n + \tau\big(f(\mathbf{U}^*_n) - \mathbf{A}^*_h \mathbf{U}^*_n\big) \\[2mm] \mathbf{U}^*_b = \big(\frac{3}{4}\mathbf{U}^*_n + \frac{1}{4}\mathbf{U}^*_a\big) + \frac{\tau}{4}\big(f(\mathbf{U}^*_a) - \mathbf{A}^*_h \mathbf{U}^*_a\big) \\[2mm] \mathbf{U}^*_c = \mathbf{U}^*_{n+1} = \big(\frac{1}{3}\mathbf{U}^*_n + \frac{2}{3}\mathbf{U}^*_b\big) + \frac{2\tau}{3}\big(f(\mathbf{U}^*_b) - \mathbf{A}^*_h \mathbf{U}^*_b\big). \end{cases} \tag{5.82}$$

We finally address the application of the aforementioned discretization methods to systems of equations, such as

$$\begin{cases} u_t = \Delta u + f(u, v) \\ v_t = d\Delta v + g(u, v) \end{cases} \tag{5.83}$$

(see (5.68) with $\gamma = 1$). Let $\mathbf{A}_h$ be the matrix corresponding to the five point discretization of Laplace operator in two space dimensions and Cartesian coordinates, complemented with suitable boundary conditions, $\mathbf{U}_n$ and $\mathbf{V}_n$ be the vectors of degrees of freedom for the discretization of $u(x, y, t)$ and $v(x, y, t)$ at time t^n, respectively. The backward Euler scheme applied to (5.83) reads as follows,

$$\begin{cases} \mathbf{U}_{n+1} = \mathbf{U}_n + \tau\big[f(\mathbf{U}_n, \mathbf{V}_n) - \mathbf{A}_h \mathbf{U}_n\big] \\ \mathbf{V}_{n+1} = \mathbf{V}_n + \tau\big[g(\mathbf{U}_n, \mathbf{V}_n) - \mathbf{A}_h \mathbf{V}_n\big]. \end{cases} \tag{5.84}$$

which can be extended to Runge-Kutta type time stepping as in (5.82).

5.6 Exercises

5.1 (An "invasion" problem). A population of density $P = P(x, y, t)$ and mass M is initially $(t = 0)$ located at a single point, for instance the origin $(0, 0)$. It increases with constant linear rate $a > 0$ and diffuses with constant D.

---

[12] See [43] for more details about these methods.

a) Write the problem describing the evolution of P and solve it.
b) Determine the evolution of the mass

$$M(t) = \int_{\mathbb{R}^2} P(x, y, t) \, dx dy.$$

c) Denoting B_R the ball centered at the origin $(0,0)$ and radius R. Find $R = R(t)$ in a way that

$$\int_{\mathbb{R}^2 \setminus B_{R(t)}} P(x, y, t) \, dx dy = M.$$

Considering the *metropolitan area* the region $B_{R(t)}$ and *rural area* the region $\mathbb{R}^2 \setminus B_{R(t)}$, find the speed of progression of the *metropolitan front*.

5.2. Consider the following initial-boundary value problem for the reaction diffusion equation

$$\begin{cases} u_t - u_{xx} - \alpha u = 0 & (0,1) \times (0, +\infty) \\ u(x,0) = x^2 - 1/3 & [0,1] \\ u_x(0,t) = 0, \ u_x(1,t) = 0 & (0, +\infty) \end{cases}$$

where $\alpha \in \mathbb{R}$.

a) After a change of variable, use the separation technique to find the solution u_α.
b) Calculate the limit of u_α as $t \to +\infty$ for every fixed $x \in [0,1]$.

5.6.1 Numerical simulation of Fisher's equations

We address the following Cauchy-Dirichlet problem

$$\begin{cases} u_t = u_{xx} + \lambda u(1-u) & 0 < x < 1, \ t > 0 \\ u(0,t) = u(1,t) = 0 & t > 0 \\ u(x,0) = c \sin(x) & 0 \le x \le 1 \end{cases} \tag{5.85}$$

$\lambda, c \in \mathbb{R}$ with the aim to apply numerical simulations based on scheme (5.82) to test the consequences of Theorem 5.5 and Theorem 5.7. Fig. 5.5 shows the results obtained for different combinations of λ and c. We observe that for $\lambda = \pi^2 - 1$, $c = 0.1$ the solution tends to a stable equilibrium state $u(x,t) = 0$ for $t \to \infty$, in agreement with Theorem 5.5. Conversely, when λ is larger than the critical threshold π^2, the only stable equilibrium solution is given by a positive function whose image belongs to $(0,1)$, which is an attractor for the evolution of the system starting from any initial state in $(0,1)$. This behaviour is confirmed by the numerical results obtained using $\lambda = \pi^2 + 1$, $c = 0.075$ and $c = 0.25$, with an initial state being either larger or smaller than the equilibrium state, as in Fig. 5.5.

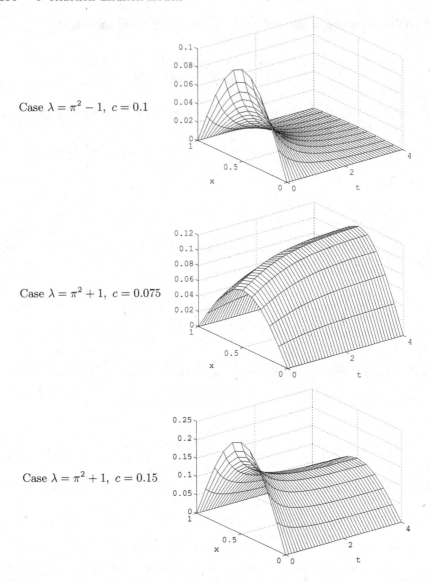

Case $\lambda = \pi^2 - 1,\ c = 0.1$

Case $\lambda = \pi^2 + 1,\ c = 0.075$

Case $\lambda = \pi^2 + 1,\ c = 0.15$

Fig. 5.5. Numerical approximation of problem (5.85) visualized on the (x,t) plane

Finally, we address the Cauchy-Neumann problem,

$$\begin{cases} u_t = u_{xx} + \lambda u(1-u) & 0 < x < 1,\ t > 0 \\ u_x(0,t) = u_x(1,t) = 0 & t > 0 \\ u(x,0) = c\exp(-10x^2) & 0 \le x \le 1 \end{cases} \qquad (5.86)$$

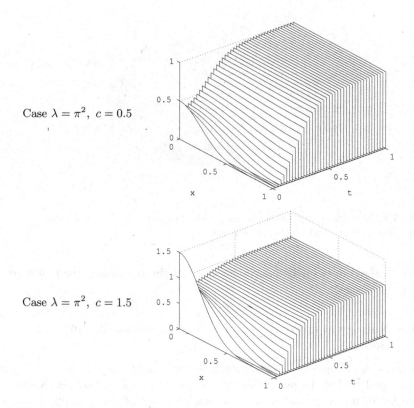

Case $\lambda = \pi^2$, $c = 0.5$

Case $\lambda = \pi^2$, $c = 1.5$

Fig. 5.6. Numerical approximation of (5.86) visualized on the (x,t) plane

whose solution evolves towards $u = 1$ for any possible value of λ and c. Indeed, Fig. 5.6 confirms the expected behaviour of the system for $\lambda = \pi^2$ and $c = 0.5$ or $c = 1.5$ for the initial state.

5.6.2 Numerical approximation of travelling wave solutions

We show that when reaction dominates, namely $\lambda \gg 1$, the Fisher-Kolmogoroff equation represents a propagation effect. Indeed, the equation has been proposed by R.A. Fisher in the paper entitled *The wave of advance of advantageous genes* appeared on Annals of Eugenics on 1937, with the aim to develop a mathematical model for the propagation of a dominating gene in a population. The model shows that, although the dominating gene is initially shared by a minority of individuals, it quickly propagates among others.

Fig. 5.7 shows the numerical simulations obtained with $\lambda = 50\pi^2 \simeq 500$ and the initial condition $g(x) = \exp(-50x^2)$, which describes the initial spatial distribution of individuals featuring a generic dominating gene. We see that the gene density propagates towards the entire domain starting from the initial

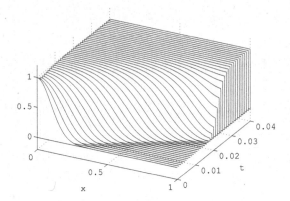

Fig. 5.7. Numerical approximation of problem (5.86) with $\lambda = 50\pi^2$ and $u(x,0) = \exp(-50x^2)$, visualized on the (x,t) plane

seed clustered in the neighbourhood of the origin, resembling the motion of a progressive wave.

5.6.3 Numerical approximation of Turing instability and pattern formation

We consider system (5.68) on a rectangular domain $\Omega = (0,1) \times (0,q)$ of aspect ratio equal to $1/q$. In particular, we choose $1/q = 5$ and we set Neumann type boundary conditions along the vertical sides of the domain, while we apply periodic conditions along the longitudinal boundaries. By this way, Ω is equivalent to a cylinder oriented along the x axis. In biology, these equations represent a model to study the formation of patterns on the mantle or the tail of various families of large animals, such as felines, bovine or reptiles.

We aim to find $u(x,y,t)$, $v(x,y,t)$ such that

$$
\begin{cases}
u_t = \Delta u + \gamma f(u,v) \quad v_t = d\Delta v + \gamma g(u,v) & (x,y,t) \in (0,1) \times (0,q) \times \mathbb{R}^+ \\
u_x(0,y,t) = u_x(1,y,t) = 0 & y \in (0,q), \ t \in \mathbb{R}^+ \\
v_x(0,y,t) = v_x(1,y,t) = 0 & y \in (0,q), t \in \mathbb{R}^+ \\
u(x,0,t) = u(x,q,t) & x \in (0,1), \ t \in \mathbb{R}^+ \\
v(x,0,t) = v(x,q,t) & x \in (0,1), t \in \mathbb{R}^+ \\
u(x,y,0) = u_0(x,y), \quad v(x,y,0) = v_0(x,y) & (x,y) \in (0,1) \times (0,q)
\end{cases}
\tag{5.87}
$$

where $f(u,v)$ and $g(u,v)$ correspond to the following kinetics, experimentally determined by D. Thomas on 1975, [13]

$$
f(u,v) = a - u - \frac{\rho u v}{1 + u + Ku^2}, \quad g(u,v) = \alpha(b-v) - \frac{\rho u v}{1 + u + Ku^2},
$$

---

[13] See Thomas D.: Artificial enzyme membranes, transport, memory, and oscillatory phenomena. In [26], pp. 115–150; or *Murray* [24].

with $a = 92$, $b = 64$, $K = 0.1$, $\alpha = 1.5$, $\rho = 18.5$. In what follows, we will focus on the behavior of solutions with respect to $\gamma \in \mathbb{R}^+$.

For the numerical approximation, we apply the Runge-Kutta method obtained merging (5.84) and (5.82), combined with the 5 point approximation of Laplace operator. We use numerical simulation to test that (5.87) allows for a non uniform steady solution, featuring small oscillations in the neighbourhood of $\bar{u}$, $\bar{v}$ such that $f(\bar{u}, \bar{v}) = g(\bar{u}, \bar{v}) = 0$. Turing instability is the primary reason of such non uniform equilibrium state, also called *pattern*. The numerical simulations also suggest that patterns are highly sensitive to the aspect ratio of the domain that is directly proportional to $\gamma \simeq L^2$ in the non-dimensional setting, where L is the characteristic length of the original domain $(0, L) \times (0, Lq)$. According to the principles of Turing instability, we expect that for smaller values of γ only oscillations along the longitudinal axis are activated, while increasing the parameter allows transversal patterns to appear.

Using the transformation $\xi = Lx$, $\eta = L(x)y$ from $(x, y) \in (0, 1) \times (0, q)$ into $(\xi, \eta) \in (0, L) \times (0, L(x)q)$, where $L(x)$ is an affine function of x, it is possible to study the pattern formation on a trapezoidal domain. In non-dimensional coordinates, it is equivalent to extend (5.87) with $\gamma = \gamma(x)$. As an example, the function $\gamma(x) = \gamma_{max}x + \gamma_{min}$, $x \in (0, 1)$, $\gamma_{max} > \gamma_{min} > 0$ takes into account of a domain that progressively dilates along with the longitudinal axis.

For better putting into evidence the formation of instabilities, we visualize the following indicator $\chi(u_\infty; \bar{u})$

$$\chi(u_\infty; \bar{u}) = \begin{cases} 0 \text{ if } u_\infty < \bar{u} \\ 1 \text{ if } u_\infty \geq \bar{u} \end{cases}.$$

where u_∞ is the steady state of (5.87).

Fig. 5.8 shows the approximation of $\chi(u_\infty; \bar{u})$ obtained by running the aforementioned scheme on a time interval large enough to reach equilibrium within a small tolerance. We observe that these results capture some typical features of animal mantle and in particular of their tail, since we are addressing the case of high aspect ratios. The results confirm that small animals, corresponding to $\gamma = 10, 100$ are more prone to develop stripes. Increasing the size, the stripe frequency increases, progressively transforming to spots.

Finally, in Fig. 5.9 we address the effect of tapering. We notice that the frequency of stripes increases when using increasing function $\gamma(x) = \gamma_{max}x + \gamma_{min}$. This corresponds to what is usually observed for felines such as the leopard or the jaguar. Although their mantle is characterized by spots, the increasing aspect ratio of the tail gives rise to stripes, which end up with a large black tip. Indeed, this behaviour is very similar to what happens in Fig. 5.9 nearby $x = 0$.

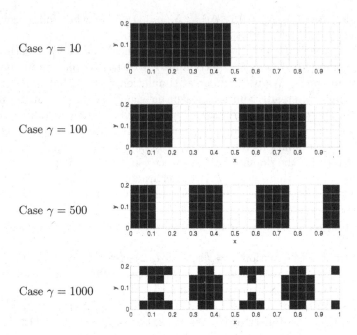

Case $\gamma = 10$

Case $\gamma = 100$

Case $\gamma = 500$

Case $\gamma = 1000$

Fig. 5.8. The characteristic function $\chi(u_\infty; \bar{u})$ quantified by numerical simulations of (5.87). Black pixels correspond to $\chi = 1$

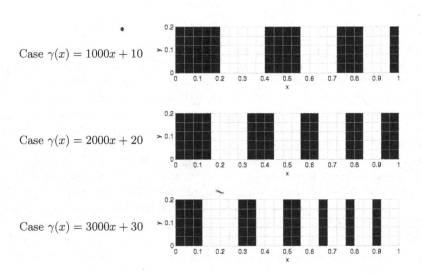

Case $\gamma(x) = 1000x + 10$

Case $\gamma(x) = 2000x + 20$

Case $\gamma(x) = 3000x + 30$

Fig. 5.9. Pattern formation on tapered domains, where the physical domain width proportionally increases with x. We notice that the frequency of oscillations follows the same trend

6

Waves and vibrations

6.1 General Concepts

6.1.1 Types of waves

Our daily experience deals with sound waves, electromagnetic waves (as radio or light waves), deep or surface water waves, elastic waves in solid materials. Oscillatory phenomena manifest themselves also in contexts and ways less macroscopic and known. This is the case, for instance, of rarefaction and shock waves in traffic dynamics or of electrochemical waves in human nervous system and in the regulation of the heart beat. In quantum physics, everything can be described in terms of wave functions, at a sufficiently small scale.

Although the above phenomena share many similarities, they show several differences as well. For example, progressive water waves propagate a disturbance, while standing waves do not. Sound waves need a supporting medium, while electromagnetic waves do not. Electrochemical waves interact with the supporting medium, in general modifying it, while water waves do not.

Thus, it seems too hard to give a general definition of *wave*, capable of covering all the above cases, so that we limit ourselves to introducing some terminology and general concepts, related to specific types of waves. We start with one-dimensional waves.

Progressive or **travelling** waves are disturbances described by a function of the following form:

$$u(x,t) = g(x - ct).$$

For $t = 0$, we have $u(x,0) = g(x)$, which is the "initial" profile of the perturbation. This profile propagates without change of shape with speed $|c|$, in the positive (negative) $x-$direction if $c > 0$ ($c < 0$). We have already met this kind of waves in Chapters 2 and 3.

Harmonic waves are particular progressive waves of the form

$$u(x,t) = A \exp\{i(kx - \omega t)\}, \qquad A, k, \omega \in \mathbb{R}. \tag{6.1}$$

Salsa S., Vegni F.M.G., Zaretti A., Zunino P.: *A Primer on PDEs. Models, Methods, Simulations*.
Unitext – La Matematica per il 3+2 65.
DOI 10.1007/978-88-470-2862-3_6, © Springer-Verlag Italia 2013

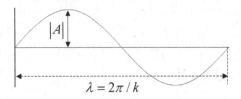

Fig. 6.1. Sinusoidal wave

Usually, only the *real part* (or the imaginary part) $A\cos(kx - \omega t)$ is of interest. Of course, the complex notation may often simplify the computations. Considering for simplicity ω and k positive in (6.1), waves are characterized by the following qualities:

- the wave *amplitude* $|A|$;

- the *wave number* k, which is the number of complete oscillations in the space interval $[0, 2\pi]$, and the *wavelength*

$$\lambda = \frac{2\pi}{k}$$

being the distance between successive maxima (*crest*) or minima (*troughs*) of the waveform;

- the *angular frequency* ω, and the *frequency*

$$f = \frac{\omega}{2\pi}$$

is the number of complete oscillations in one second (Hertz) at a fixed space position;

- the *wave or phase speed*

$$c_p = \frac{\omega}{k}$$

which is the crests (or troughs) speed.

Standing waves are of the form

$$u(x, t) = B\cos kx \cos \omega t.$$

In these disturbances, the basic sinusoidal wave, $\cos kx$, is modulated by the time dependent oscillation $B\cos \omega t$. A standing wave may be generated, for instance, by superposing two harmonic waves with the same amplitude, propagating in opposite directions:

$$A\cos(kx - \omega t) + A\cos(kx + \omega t) = 2A\cos kx \cos \omega t. \tag{6.2}$$

Consider now waves in dimension $n > 1$.

Plane waves. *Scalar* plane waves are of the form

$$u(\mathbf{x},t) = f(\mathbf{k} \cdot \mathbf{x} - \omega t).$$

The disturbance propagates in the direction of $\mathbf{k}$ with speed $c_p = \omega / |\mathbf{k}|$. The planes of equation

$$\theta(\mathbf{x},t) = \mathbf{k} \cdot \mathbf{x} - \omega t = \text{constant}$$

constitute the *wave-fronts*.

  Harmonic or monochromatic plane waves have the form

$$u(\mathbf{x},t) = A \exp\{i(\mathbf{k} \cdot \mathbf{x} - \omega t)\}.$$

Here $\mathbf{k}$ is the *wave number* vector and ω is the *angular frequency*. The vector $\mathbf{k}$ is orthogonal to the wave front and $|\mathbf{k}|/2\pi$ gives the number of waves per unit length. The scalar $\omega/2\pi$ still gives the number of complete oscillations in one second (Hertz) at a fixed space position.

Spherical waves are of the form

$$u(\mathbf{x},t) = v(r,t)$$

where $r = |\mathbf{x} - \mathbf{x}_0|$ and $\mathbf{x}_0 \in \mathbb{R}^n$ is a fixed point. In particular $u(\mathbf{x},t) = e^{i\omega t} v(r)$ represents a stationary spherical wave, while $u(\mathbf{x},t) = v(r - ct)$ is a progressive wave whose wavefronts are the spheres $r - ct = \text{constant}$, moving with speed $|c|$ (outgoing if $c > 0$, incoming if $c < 0$).

6.1.2 Group velocity and dispersion relation.

Many oscillatory phenomena can be modelled by linear equations whose solutions are superpositions of harmonic waves with angular frequency depending on the wave number:

$$\omega = \omega(k). \tag{6.3}$$

A typical example is the wave system produced by dropping a stone in a pond.

  If ω is linear, e.g. $\omega(k) = ck$, $c > 0$, the crests move with speed c, independent of the wave number. However, if $\omega(k)$ is not proportional to k, the crests move with speed $c_p = \omega(k)/k$, that *depends* on the wave number. In other words, the crests move at different speeds for different wavelengths. As a consequence, the various components in a wave packet given by the superposition of harmonic waves of different wavelengths will eventually separate or *disperse*. For this reason, (6.3) is called **dispersion relation**.

  In the theory of dispersive waves, the **group velocity**, given by

$$c_g = \omega'(k)$$

is a central notion, mainly for the following three reasons.

1. *It is the speed at which an isolated wave packet moves as a whole.*
2. *An observer that travels at the group velocity sees constantly waves of the same wavelength $2\pi/k$, after the transitory effects due to a localized initial*

perturbation (e.g. a stone thrown into a pond). In other words, c_g is the propagation speed of the wave numbers.

3. *Energy is transported at the group velocity by waves of wavelength $2\pi/k$.*

Let us comment on the first point. A wave packet may be obtained by the superposition of dispersive harmonic waves, for instance through a Fourier integral of the form

$$u(x,t) = \int_{-\infty}^{+\infty} a(k)\, e^{i[kx - \omega(k)t]}\, dk \tag{6.4}$$

where the real part only has a physical meaning. Consider a localized wave packet, with wave number $k \approx k_0$, almost constant, and with amplitude slowly varying with x. Then, the packet contains a large number of crests and the amplitudes $|a(k)|$ of the various Fourier components are negligible except that in a small neighborhood of k_0, $(k_0 - \delta, k_0 + \delta)$, say.

Fig. 6.2 shows the initial profile of a Gaussian packet,

$$\operatorname{Re} u(x, 0) = \frac{3}{\sqrt{2}} \exp\left\{-\frac{x^2}{32}\right\} \cos 14x,$$

slowly varying with x, with $k_0 = 14$, and its Fourier transform:

$$a(k) = 6 \exp\{-8(k - 14)^2\}.$$

As we can see, the amplitudes $|a(k)|$ of the various Fourier components are negligible except when k is near k_0.

Then we may write

$$\omega(k) \approx \omega(k_0) + \omega'(k_0)(k - k_0) = \omega(k_0) + c_g(k - k_0)$$

and

$$u(x,t) \approx e^{i\{k_0 x - \omega(k_0)t\}} \int_{k_0 - \delta}^{k_0 + \delta} a(k)\, e^{i(k - k_0)(x - c_g t)}\, dk. \tag{6.5}$$

Thus, u turns out to be well approximated by the product of two waves. The first one is a pure harmonic wave with relatively short wavelength $2\pi/k_0$ and

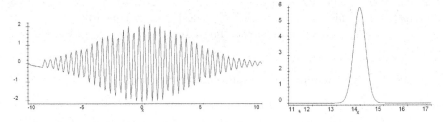

Fig. 6.2. Wave packet and its Fourier transform

phase speed $\omega(k_0)/k_0$. The second one depends on x, t through the combination $x - c_g t$, and is a superposition of waves of very small wavenumbers $k - k_0$, which correspond to very large wavelengths. We may interpret the second factor as a sort of envelope of the short waves of the packet, that is the packet as a whole, which therefore moves with the group speed.

In a wave packet like (6.5), the energy is proportional to[1]

$$\int_{k_0-\delta}^{k_0+\delta} |a(k)|^2 \, dk \simeq 2\delta |a(k_0)|^2$$

so that it moves at the same speed of k_0, that is c_g.

6.2 Transversal Waves in a String

6.2.1 The model

We derive a classical model for the small transversal vibration of a tightly stretched horizontal string (e.g. a string of a guitar). We assume the following hypotheses:

1. *Vibrations of the string have small amplitude.* This entails that the changes in the slope of the string from the horizontal equilibrium position are very small.
2. *Each point of the string undergoes vertical displacements only.* Horizontal displacements can be neglected, according to 1.
3. *The vertical displacement of a point depends on time and on its position on the string.* If we denote by u the vertical displacement of a point located at x when the string is at rest, then we have $u = u(x, t)$ and, according to 1, $|u_x(x, t)| \ll 1$.
4. *The string is perfectly flexible.* This means that it offers no resistance to bending. In particular, the stress at any point on the string can be modelled by a tangential[2] force **T** of magnitude τ, called *tension*. Fig. 6.3 shows how the forces due to the tension act at the end points of a small segment of the string.
5. *Friction is negligible.*

Under the above assumptions, the equation of motion of the string can be derived from *conservation of mass* and *Newton law*.

Let $\rho_0 = \rho_0(x)$ be the linear density of the string at rest and $\rho = \rho(x, t)$ be its density at time t. Consider an arbitrary part of the string between x and $x + \Delta x$ and denote by Δs the corresponding length element at time t. Then, conservation of mass yields

$$\rho_0(x) \Delta x = \rho(x, t) \Delta s. \tag{6.6}$$

---

[1] See *Segel* [25].

[2] Consequence of absence of distributed moments along the string.

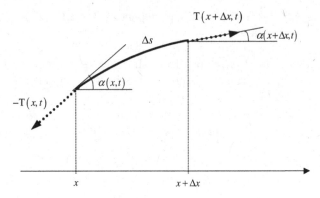

Fig. 6.3. Tension at the end points of a small segment of a string

To write Newton law of motion we have to determine the forces acting on our small piece of string. Since the motion is vertical, the horizontal forces have to balance. On the other hand they come from the tension only, so that if $\tau(x,t)$ denotes the magnitude of the tension at x at time t, we can write (Fig. 6.3):

$$\tau(x + \Delta x, t) \cos\alpha(x + \Delta x, t) - \tau(x,t) \cos\alpha(x,t) = 0.$$

Dividing by Δx and letting $\Delta x \to 0$, we obtain

$$\frac{\partial}{\partial x}\left[\tau(x,t) \cos\alpha(x,t)\right] = 0$$

from which

$$\tau(x,t) \cos\alpha(x,t) = \tau_0(t) \tag{6.7}$$

where $\tau_0(t)$ is *positive*[3].

The vertical forces are given by the vertical component of the tension and by body forces such as gravity and external loads.

Using (6.7), the scalar vertical component of the tension at x, at time t, is given by:

$$\tau_{vert}(x,t) = \tau(x,t)\sin\alpha(x,t) = \tau_0(t)\tan\alpha(x,t) = \tau_0(t)u_x(x,t).$$

Therefore, the (scalar) vertical component of the force acting on our small piece of string, due to the tension, is

$$\tau_{vert}(x + \Delta x, t) - \tau_{vert}(x,t) = \tau_0(t)\left[u_x(x + \Delta x, t) - u_x(x,t)\right].$$

Denote by $f(x,t)$ the magnitude of the (vertical) body forces per unit mass. Then, using (6.6), the the magnitude of the body forces acting on the string segment is given by:

$$\rho(x,t) f(x,t) \Delta s = \rho_0(x) f(x,t) \Delta x.$$

---

[3] It is the magnitude of a force.

Thus, using (6.6) again and observing that u_{tt} is the (scalar) vertical acceleration, Newton law gives:

$$\rho_0\left(x\right)\Delta x\, u_{tt} = \tau_0\left(t\right)\left[u_x\left(x + \Delta x\right) - u_x\left(x\right)\right] + \rho_0\left(x\right)f\left(x, t\right)\Delta x.$$

Dividing by Δx and letting $\Delta x \to 0$, we obtain the equation

$$u_{tt} - c^2\left(x, t\right)u_{xx} = f\left(x, t\right) \tag{6.8}$$

where $c^2\left(x, t\right) = \tau_0\left(t\right)/\rho_0\left(x\right)$.

If the string is homogeneous then ρ_0 is constant. If moreover it is **perfectly elastic**[4] then τ_0 is constant as well, since the horizontal tension is nearly the same as for the string at rest, in the horizontal position. We shall come back to equation (6.8) shortly.

6.2.2 Energy

Suppose that a *perfectly flexible and elastic* string has length L at rest, in the horizontal position. We may identify its initial position with the segment $[0, L]$ on the x axis. Since $u_t(x, t)$ is the vertical velocity of the point at x, the expression

$$E_{cin}\left(t\right) = \frac{1}{2}\int_0^L \rho_0 u_t^2\, dx \tag{6.9}$$

represents the total **kinetic energy during the vibrations**. The string stores **potential energy** too, due to the work of elastic forces. These forces stretch an element of string of length Δx at rest by[5]

$$\Delta s - \Delta x = \int_x^{x+\Delta x}\sqrt{1 + u_x^2}\, dx - \Delta x = \int_x^{x+\Delta x}\left(\sqrt{1 + u_x^2} - 1\right)dx \approx \frac{1}{2}u_x^2\Delta x$$

since $|u_x| \ll 1$. Thus, the work done by the elastic forces on that string element is

$$dW = \frac{1}{2}\tau_0 u_x^2\Delta x.$$

Summing all the contributions, the total **potential energy** is given by:

$$E_{pot}\left(t\right) = \frac{1}{2}\int_0^L \tau_0 u_x^2\, dx. \tag{6.10}$$

From (6.9) and (6.10) we find, for the total energy:

$$E\left(t\right) = \frac{1}{2}\int_0^L\left[\rho_0 u_t^2 + \tau_0 u_x^2\right]dx. \tag{6.11}$$

---

[4] For instance, guitar and violin strings are nearly homogeneous, perfectly flexible and elastic.

[5] Recall that, at first order, if $\varepsilon \ll 1$, $\sqrt{1 + \varepsilon} - 1 \simeq \varepsilon/2$.

Let us compute the variation of E. Taking the time derivative under the integral, we find (remember that $\rho_0 = \rho_0(x)$ and τ_0 is constant),

$$\dot{E}(t) = \int_0^L [\rho_0 u_t u_{tt} + \tau_0 u_x u_{xt}]\, dx.$$

By an integration by parts we get

$$\int_0^L \tau_0 u_x u_{xt}\, dx = \tau_0 [u_x(L,t) u_t(L,t) - u_x(0,t) u_t(0,t)] - \tau_0 \int_0^L u_t u_{xx} dx$$

whence

$$\dot{E}(t) = \int_0^L [\rho_0 u_{tt} - \tau_0 u_{xx}] u_t dx + \tau_0 [u_x(L,t) u_t(L,t) - u_x(0,t) u_t(0,t)].$$

Using (6.8), we find:

$$\dot{E}(t) = \int_0^L \rho_0 f u_t\, dx + \tau_0 [u_x(L,t) u_t(L,t) - u_x(0,t) u_t(0,t)]. \tag{6.12}$$

In particular, if $f = 0$ and u is constant at the end points 0 and L (therefore $u_t(L,t) = u_t(0,t) = 0$) we deduce $\dot{E}(t) = 0$. This implies

$$E(t) = E(0)$$

which expresses the *conservation of energy*.

6.3 The One-dimensional Wave Equation

6.3.1 Initial and boundary conditions

Equation (6.8) is called the *one-dimensional wave equation*. The coefficient c has the dimensions of a speed and in fact, we will shortly see that it represents the wave propagation speed along the string. When $f \equiv 0$, the equation is *homogeneous* and the *superposition principle holds:* if u_1 and u_2 are solutions of

$$u_{tt} - c^2 u_{xx} = 0 \tag{6.13}$$

and a, b are (real or complex) scalars, then $a u_1 + b u_2$ is a solution as well. More generally, if $u_k(\mathbf{x},t)$ is a family of solutions depending on the parameter k (integer or real) and $g = g(k)$ is a function rapidly vanishing at infinity, then

$$\sum_{k=1}^{\infty} u_k(\mathbf{x},t) g(k) \quad \text{and} \quad \int_{-\infty}^{+\infty} u_k(\mathbf{x},t) g(k)\, dk$$

are still solutions of (6.13).

Suppose we are considering the space-time region $0 < x < L, 0 < t < T$. In a well posed problem for the (one-dimensional) heat equation it is appropriate to assign the initial profile of the temperature, because of the presence of a first order time derivative, and a boundary condition at both ends $x = 0$ and $x = L$, because of the second order spatial derivative.

By analogy with the Cauchy problem for second order ordinary differential equations, the second order time derivative in (6.8) suggests that not only the initial profile of the string but the initial velocity has to be assigned as well.

Thus, our initial (or Cauchy) data are

$$u(x,0) = g(x), \qquad u_t(x,0) = h(x), \qquad x \in [0,L].$$

The boundary data are formally similar to those for the heat equation. Typically, we have the following.

• *Dirichlet data* describe the displacement of the end points of the string:

$$u(0,t) = a(t), \qquad u(L,t) = b(t), \qquad t > 0.$$

If $a(t) = b(t) \equiv 0$ (homogeneous data), both ends are fixed, with zero displacement.

• *Neumann data* describe the applied (scalar) vertical tension at the end points. As in the derivation of the wave equation, we may model this tension by $\tau_0 u_x$ so that the Neumann conditions take the form

$$\tau_0 u_x(0,t) = a(t), \qquad \tau_0 u_x(L,t) = b(t), \qquad t > 0.$$

In the special case of homogeneous data, $a(t) = b(t) \equiv 0$, both ends of the string are attached to a frictionless sleeve and are free to move vertically.

• *Robin data* describe a linear elastic attachment at the end points. One way to realize this type of boundary condition is to attach an end point to a linear spring[6] whose other end is fixed. This translates into assigning

$$\tau_0 u_x(0,t) = ku(0,t), \qquad \tau_0 u_x(L,t) = -ku(L,t), \qquad t > 0,$$

where k (positive) is the elastic constant of the spring.

In several concrete situations, *mixed conditions* have to be assigned. For instance, Robin data at $x = 0$ and Dirichlet data at $x = L$.

• *Global Cauchy problem*. We may think of a string of infinite length and assign only the initial data

$$u(x,0) = g(x), \qquad u_t(x,0) = h(x), \qquad x \in \mathbb{R}.$$

Although physically unrealistic, it turns out that the solution of the global Cauchy problem is of fundamental importance. We shall solve it in Section 6.4.

---

[6] Which obeys Hooke's law: the strain is a linear function of the stress.

A simple argument based of the energy formulas (6.12) and (6.11) shows that the above problems are well posed. Indeed, let u and v be solutions of (6.13). Then $w = u - v$ is a solution of the same problem with zero initial and boundary data. We want to show that $w \equiv 0$. Applying formula (6.12) to w, we find

$$\dot{E}(t) = \tau [w_x(L,t) w_t(L,t) - w_x(0,t) w_t(0,t)]$$

for all $t > 0$.

In the case of Dirichlet or Neumann lateral data, at both end points we have $w_t = 0$ or $w_x = 0$, respectively. Therefore $\dot{E}(t) = 0$ and E is constant. Since $E(0) = 0$ we infer

$$E(t) = E_{cin}(t) + E_{pot}(t) \equiv 0.$$

On the other hand, $E_{cin}(t) \geq 0$, $E_{pot}(t) \geq 0$, so that we deduce

$$E_{cin}(t) = 0, \ E_{pot}(t) = 0$$

which force $w_t = w_x = 0$. Thus w is constant and since $w(x,0) = 0$, we conclude that $w(x,t) = 0$ for every $t \geq 0$.

To find the solution one can often use the method of separation of variables as it is shown at the end of the chapter.

6.4 The d'Alembert Formula

6.4.1 The homogeneous equation

In this section we establish the celebrated formula of d'Alembert for the solution of the following global Cauchy problem:

$$\begin{cases} u_{tt} - c^2 u_{xx} = 0 & x \in \mathbb{R}, \ t > 0 \\ u(x,0) = g(x), \quad u_t(x,0) = h(x) & x \in \mathbb{R}. \end{cases} \tag{6.14}$$

To find the solution, we first factorize the wave equation in the following way:

$$(\partial_t - c\partial_x)(\partial_t + c\partial_x) u = 0. \tag{6.15}$$

Now, let

$$v = u_t + cu_x. \tag{6.16}$$

Then v solves the linear transport equation

$$v_t - cv_x = 0$$

whence

$$v(x,t) = \psi(x + ct)$$

where ψ is a differentiable arbitrary function. From (6.16) we have

$$u_t + cu_x = \psi\left(x + ct\right)$$

and formula (2.16) in sction 2.2.1 yields

$$u\left(x,t\right) = \int_0^t \psi\left(x - c\left(t - s\right) + cs\right)\,ds + \varphi\left(x - ct\right),$$

where φ is another arbitrary differentiable function.

Letting $x - ct + 2cs = y$, we find

$$u\left(x,t\right) = \frac{1}{2c}\int_{x-ct}^{x+ct}\psi\left(y\right)dy + \varphi\left(x - ct\right). \tag{6.17}$$

To determine ψ and φ we impose the initial conditions:

$$u\left(x,0\right) = \varphi\left(x\right) = g\left(x\right) \tag{6.18}$$

and

$$u_t\left(x,0\right) = \psi\left(x\right) - c\varphi'\left(x\right) = h\left(x\right)$$

whence

$$\psi\left(x\right) = h\left(x\right) + cg'\left(x\right). \tag{6.19}$$

Inserting (6.19) and (6.18) into (6.17) we get:

$$\begin{aligned}
u\left(x,t\right) &= \frac{1}{2c}\int_{x-ct}^{x+ct}\left[h\left(y\right) + cg'\left(y\right)\right]\,dy + g\left(x - ct\right)\\
&= \frac{1}{2c}\int_{x-ct}^{x+ct}h\left(y\right)dy + \frac{1}{2}\left[g\left(x + ct\right) - g\left(x - ct\right)\right] + g\left(x - ct\right)
\end{aligned}$$

and finally the **d'Alembert** formula

$$u\left(x,t\right) = \frac{1}{2}\left[g(x + ct) + g\left(x - ct\right)\right] + \frac{1}{2c}\int_{x-ct}^{x+ct}h\left(y\right)dy. \tag{6.20}$$

If $g \in C^2(\mathbb{R})$ and $h \in C^1\left(\mathbb{R}\right)$, formula (6.20) defines a C^2-solution in the half-plane $\mathbb{R}\times[0,+\infty)$. On the other hand, a C^2-solution u in $\mathbb{R}\times[0,+\infty)$ has to be given by (6.20), just because of the procedure we have used to solve the Cauchy problem. Thus the solution is *unique*. Observe however, that *no regularizing effect* takes place here: the solution u remains no more than C^2 for any $t > 0$. Thus, there is a striking difference with diffusion phenomena, governed by the heat equation.

Furthermore, let u_1 and u_2 be the solutions corresponding to the data g_1, h_1 and g_2, h_2, respectively. Then, the d'Alembert formula for $u_1 - u_2$ yields, for every $x \in \mathbb{R}$ and $t \in [0,T]$,

$$|u_1\left(x,t\right) - u_2\left(x,t\right)| \leq \|g_1 - g_2\|_\infty + T\|h_1 - h_2\|_\infty$$

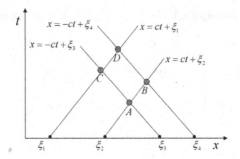

Fig. 6.4. Characteristic rectangle

where

$$\|g_1 - g_2\|_\infty = \sup_{x \in \mathbb{R}} |g_1(x) - g_2(x)|, \qquad \|h_1 - h_2\|_\infty = \sup_{x \in \mathbb{R}} |h_1(x) - h_2(x)|.$$

Therefore, we have stability in *pointwise uniform sense*, at least for finite time.

Rearranging the terms in (6.20), we may write u in the form[7]

$$u(x, t) = F(x + ct) + G(x - ct) \qquad (6.21)$$

which gives u as a *superposition of two progressive waves moving at constant speed c in the negative and positive x – direction*, respectively. Thus, these waves are not dispersive.

The two terms in (6.21) are respectively constant along the two families of straight lines γ^+ and γ^- given by

$$x + ct = \text{constant}, \qquad x - ct = \text{constant}.$$

These lines are called *characteristics*[8] and carry important information, as we will see in Section 6.5.

An interesting consequence of (6.21) comes from looking at Fig. 6.4. Consider the *characteristic parallelogram* with vertices at the point A, B, C, D. From (6.21) we have

$$F(A) = F(C), \quad G(A) = G(B)$$
$$F(D) = F(B), \quad G(D) = G(C).$$

---

[7] For instance:
$$F(x + ct) = \frac{1}{2}g(x + ct) + \frac{1}{2c} \int_0^{x+ct} h(y)\, dy$$

and

$$G(x - ct) = \frac{1}{2}g(x - ct) + \frac{1}{2c} \int_{x-ct}^0 h(y)\, dy.$$

[8] In fact they are the *characteristics* for the two first order factors in the factorization (6.15).

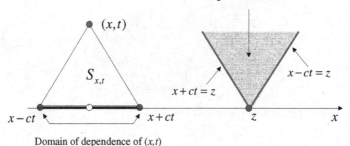

Fig. 6.5. Domain of dependence and range of influence

Summing these relations we get

$$[F(A) + G(A)] + [F(D) + G(D)] = [F(C) + G(C)] + [F(B) + G(B)]$$

which is equivalent to

$$u(A) + u(D) = u(C) + u(B). \tag{6.22}$$

Thus, knowing u at three points of a characteristic parallelogram, we can compute u at the fourth one.

From d'Alembert formula it follows that the value of u at the point (x, t) depends on the values of g at the points $x - ct$ e $x + ct$ and on the values of h over the whole interval $[x - ct, x + ct]$. This interval is called **domain of dependence of** (x, t) (Fig. 6.5).

From a different perspective, the values of g and h at a point z affect the value of u at the points (x, t) in the sector

$$z - ct \le x \le z + ct,$$

which is called **range of influence** of z (Fig. 6.5). This entails that a disturbance initially localized at z is not felt at a point x until time

$$t = \frac{|x - z|}{c}.$$

Remark 6.1. Physically realistic data include g *continuous* and h *bounded* only. On the other hand, observe that d'Alembert formula makes perfect sense even in these cases The question is in which sense the resulting function satisfies the wave equation, since, in principle, it is not even differentiable, only continuous. There are several ways to weaken the notion of solution to include this case; one of these is described in Chapter 8.

Fig. 6.6 shows the wave propagation along a chord of infinite length, plucked at the origin and originally at rest, modeled by the solution of the

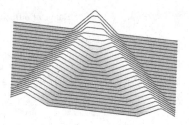

Fig. 6.6. Chord plucked at the origin ($c = 1$)

problem

$$\begin{cases} u_{tt} - u_{xx} = 0 & x \in \mathbb{R}, \, t > 0 \\ u(x,0) = g(x), \, u_t(x,0) = 0 & x \in \mathbb{R} \end{cases}$$

where g has a triangular profile. As we see, this generalized solution displays lines of discontinuities of the first derivatives, while outside these lines it is smooth. It can be shown that these lines are precisely the *characteristics*.

6.4.2 The nonhomogeneous equation. Duhamel's method

To solve the nonhomogeneous problem

$$\begin{cases} u_{tt} - c^2 u_{xx} = f(x,t) & x \in \mathbb{R}, \, t > 0 \\ u(x,0) = 0, \, u_t(x,0) = 0 & x \in \mathbb{R} \end{cases} \tag{6.23}$$

we use the Duhamel's method (see Section 3.4.3). For $s \geq 0$ fixed, let $w = w(x,t;s)$ be the solution of problem

$$\begin{cases} w_{tt} - c^2 w_{xx} = 0 & x \in \mathbb{R}, \, t \geq s \\ w(x,s;s) = 0, \, w_t(x,s;s) = f(x,s) & x \in \mathbb{R}. \end{cases} \tag{6.24}$$

Since the wave equation is invariant under (time) translations, from (6.20) we get

$$w(x,t;s) = \frac{1}{2c} \int_{x-c(t-s)}^{x+c(t-s)} f(y,s) \, dy.$$

Then, the solution of (6.23) is given by

$$u(x,t) = \int_0^t w(x,t;s) \, ds = \frac{1}{2c} \int_0^t ds \int_{x-c(t-s)}^{x+c(t-s)} f(y,s) \, dy.$$

In fact, $u(x,0) = 0$ and

$$u_t(x,t) = w(x,t;t) + \int_0^t w_t(x,t;s) \, ds = \int_0^t w_t(x,t;s) \, ds$$

since $w(x, t; t) = 0$. Thus $u_t(x, 0) = 0$. Moreover,

$$u_{tt}(x, t) = w_t(x, t; t) + \int_0^t w_{tt}(x, t; s)\ ds = f(x, t) + \int_0^t w_{tt}(x, t; s)\ ds$$

and

$$u_{xx}(x, t) = \int_0^t w_{xx}(x, t; s)\ ds.$$

Therefore, since $w_{tt} - c^2 w_{xx} = 0$,

$$u_{tt}(x, t) - c^2 u_{xx}(x, t) = f(x, t) + \int_0^t w_{tt}(x, t; s)\ ds - c^2 \int_0^t w_{xx}(x, t; s)\ ds$$
$$= f(x, t).$$

Everything works and gives the *unique* solution in $C^2(\mathbb{R} \times [0, +\infty))$, under rather natural hypotheses on f: we require f and f_x be continuous in $\mathbb{R} \times [0, +\infty)$.

Finally note that the value of u at the point (x, t) depends on the values of the forcing term f *in all the triangular sector* $S_{x,t}$ in Fig. 6.5.

6.4.3 Dissipation and dispersion

Dissipation and dispersion effects are quite important in wave propagation phenomena. Let us go back to our model for the vibrating string, assuming that its weight is negligible and that there are no external loads.

• *External damping.* External factors of dissipation like friction due to the medium may be included into the model through some empirical constitutive law. We may assume, for instance, a *linear law* of friction expressing a force proportional to the speed of vibration. Then, a force given by $-k\rho_0 u_t \Delta x \mathbf{j}$, where $k > 0$ is a damping constant, acts on the segment of string between x and $x + \Delta x$. The final equation takes the form

$$\rho_0 u_{tt} - \tau_0 u_{xx} + k\rho_0 u_t = 0. \tag{6.25}$$

For a string with fixed end points, the same calculations in Section 6.2.2 yield

$$\dot{E}(t) = -\int_0^L k\rho_0 u_t^2\ dx = -kE_{cin}(t) \leq 0 \tag{6.26}$$

which shows a rate of energy dissipation proportional to the kinetic energy.

For equation (6.25), the usual initial-boundary value problems are still well posed under reasonable assumptions on the data. In particular, the uniqueness of the solution follows from (6.26), since $E(0) = 0$ implies $E(t) = 0$ for all $t > 0$.

• *Internal damping.* The derivation of the wave equation in Section 6.2.1 leads to

$$\rho_0 u_{tt} = (\tau_{vert})_x$$

where τ_{vert} is the (scalar) vertical component of the tension. The hypothesis of vibrations of small amplitude corresponds to taking

$$\tau_{vert} \simeq \tau_0 u_x, \qquad (6.27)$$

where τ_0 is the (scalar) horizontal component of the tension. In other words, we assume that the vertical forces due to the tension at two end points of a string element are proportional to the relative displacement of these points. On the other hand, the string vibrations convert kinetic energy into heat, because of the friction among the particles. The amount of heat increases with the speed of vibration while, at the same time, the vertical tension decreases. Thus, the vertical tension depends not only on the relative displacements u_x, but also on how fast these displacements change with time[9]. Hence, we modify (6.27) by inserting a term proportional to u_{xt}

$$\tau_{vert} = \tau u_x + \gamma u_{xt} \qquad (6.28)$$

where γ is a *positive* constant. The positivity of γ follows from the fact that energy dissipation lowers the vertical tension, so that the slope u_x decreases if $u_x > 0$ and increases if $u_x < 0$. Using the law (6.28) we derive the third order equation

$$\rho_0 u_{tt} - \tau u_{xx} - \gamma u_{xxt} = 0. \qquad (6.29)$$

In spite of the presence of the term u_{xxt}, the usual initial-boundary value problems are again well posed under reasonable assumptions on the data. In particular, uniqueness of the solution follows once again from dissipation of energy, since, in this case[10],

$$\dot{E}(t) = -\int_0^L \gamma \rho_0 u_{xt}^2 \leq 0.$$

• *Dispersion.* When the string is under the action of a vertical elastic restoring force proportional to u, the equation of motion becomes

$$u_{tt} - c^2 u_{xx} + \lambda u = 0 \qquad (\lambda > 0) \qquad (6.30)$$

known as the *linearized Klein-Gordon equation*. To emphasize the effect of the zero order term λu, let us seek for *harmonic waves solutions* of the form

$$u(x,t) = Ae^{i(kx-\omega t)}.$$

Inserting u into (6.30) we find the *dispersion relation*

$$\omega^2 - c^2 k^2 = \lambda \qquad \Longrightarrow \qquad \omega(k) = \pm\sqrt{c^2 k^2 + \lambda}.$$

---

[9] In the movie *The Legend of 1900* there is a spectacular demo of this phenomenon.
[10] Check it.

Thus, this waves are dispersive with phase and group velocities given respectively by

$$c_p\left(k\right) = \frac{\sqrt{c^2 k^2 + \lambda}}{|k|}, \quad c_g = \frac{d\omega}{dk} = \frac{c^2\left|k\right|}{\sqrt{c^2 k^2 + \lambda}}.$$

Observe that $c_g < c_p$.

A wave packet solution can be obtained by an integration over all possible wave numbers k:

$$u\left(x, t\right) = \int_{-\infty}^{+\infty} A\left(k\right) e^{i[kx - \omega(k)t]} dk \qquad (6.31)$$

where $A\left(k\right)$ is the Fourier transform of the initial condition:

$$A\left(k\right) = \int_{-\infty}^{+\infty} u\left(x, 0\right) e^{-ikx} dx.$$

This entails that, even if the initial condition is *localized* inside a small interval, *all* the wavelength contribute to the value of u. These dispersive waves do not dissipate energy. For example, if the ends of the string are fixed, the total mechanical energy is given by

$$E\left(t\right) = \frac{\rho_0}{2} \int_0^L \left(u_t^2 + c^2 u_x^2 + \lambda u^2\right) dx$$

and one may check that $\dot{E}\left(t\right) = 0$, $t > 0$.

6.5 Second Order Linear Equations

6.5.1 Classification

To derive formula (6.21) we may use the characteristics in the following way. We change variables setting

$$\xi = x + ct, \qquad \eta = x - ct \qquad (6.32)$$

or

$$x = \frac{\xi + \eta}{2}, \qquad t = \frac{\xi - \eta}{2c}$$

and define

$$U\left(\xi, \eta\right) = u\left(\frac{\xi + \eta}{2}, \frac{\xi - \eta}{2c}\right).$$

Then

$$U_\xi = \frac{1}{2} u_x + \frac{1}{2c} u_t$$

and since $u_{tt} = c^2 u_{xx}$

$$U_{\xi\eta} = \frac{1}{4}u_{xx} - \frac{1}{4c}u_{xt} + \frac{1}{4c}u_{xt} - \frac{1}{4c^2}u_{tt} = 0.$$

The equation

$$U_{\xi\eta} = 0 \qquad\qquad (6.33)$$

is called the *canonical* form of the wave equation; its solution is immediate

$$U(\xi, \eta) = F(\xi) + G(\eta)$$

and going back to the original variables (6.21) follows.

Consider now a general equation of the form

$$au_{tt} + 2bu_{xt} + cu_{xx} + du_t + eu_x + hu = f \qquad\qquad (6.34)$$

with x, t varying, in general, in a domain Ω. We assume that the coefficients a, b, c, d, e, h, f are smooth functions[11] in Ω. The sum of second order terms

$$a(x, t) u_{tt} + 2b(x, t) u_{xt} + c(x, t) u_{xx} \qquad\qquad (6.35)$$

is called **principal part** of equation (6.34) and determines the *type* of equation according to the following classification. Consider the algebraic equation

$$H(p, q) = ap^2 + 2bpq + cq^2 = 1 \qquad (a > 0) \qquad\qquad (6.36)$$

in the plane p, q. If $b^2 - ac < 0$, (6.36) defines a hyperbola, if $b^2 - ac = 0$ a parabola and if $b^2 - ac < 0$ an ellipse. Accordingly, equation (6.34) is called

a) **hyperbolic** when $b^2 - ac < 0$;
b) **parabolic** when $b^2 - ac = 0$;
c) **elliptic** when $b^2 - ac > 0$.

Note that the quadratic form $H(p, q)$ is, in the three cases, *indefinite, non-negative, positive*, respectively. In this form, the above classification extends to equations in any number of variables, as we shall see later on.

It may happen that a single equation is of different type in different subdomains. For instance, the *Tricomi* equation $xu_{tt} - u_{xx} = 0$ is hyperbolic in the half plane $x > 0$, parabolic on $x = 0$ and elliptic in the half plane $x < 0$.

Basically all the equations in two variables we have met so far are particular cases of (6.34). Specifically,

- the *wave* equation

$$u_{tt} - c^2 u_{xx} = 0$$

is *hyperbolic:* $a(x, t) = 1$, $c(x, t) = -c^2$, and the other coefficients are zero;

---

[11] E.g. C^2 functions.

- the *diffusion* equation

$$u_t - D u_{xx} = 0$$

  is *parabolic*: $c(x,t) = -D$, $d(x,t) = 1$, and the other coefficients are zero;

- *Laplace* equation (using y instead of t)

$$u_{xx} + u_{yy} = 0$$

  is *elliptic*: $a(x,y) = 1$, $c(x,y) = 1$, and the other coefficients are zero.

May we reduce to a canonical form, similar to (6.33), the diffusion and the Laplace equation? Let us briefly examine why the change of variables (6.32) works for the wave equation. Decompose the wave operator as follows

$$\partial_{tt} - c^2 \partial_{xx} = (\partial_t + c\partial_x)(\partial_t - c\partial_x). \tag{6.37}$$

If we introduce the vectors $\mathbf{v} = (c,1)$ and $\mathbf{w} = (-c,1)$, then (6.37) can be written in the form

$$\partial_{tt} - c^2 \partial_{xx} = \partial_\mathbf{v} \partial_\mathbf{w}.$$

On the other hand, the characteristics

$$x + ct = 0, \quad x - ct = 0$$

of the two first order equations

$$\phi_t - c\phi_x = 0 \quad \text{and} \quad \psi_t + c\psi_t = 0,$$

corresponding to the two factors in (6.37), are straight lines in the direction of $\mathbf{w}$ and $\mathbf{v}$, respectively. The change of variables

$$\xi = \phi(x,t) = x + ct \qquad \eta = \psi(x,t) = x - ct$$

maps these straight lines into $\xi = 0$ and $\eta = 0$ and

$$\partial_\xi = \frac{1}{2c}(\partial_t + c\partial_x) = \frac{1}{2c}\partial_\mathbf{v}, \qquad \partial_\eta = \frac{1}{2c}(\partial_t - c\partial_x) = \frac{1}{2c}\partial_\mathbf{w}.$$

Thus, the wave operator is converted into a multiple of its canonical form

$$\partial_{tt} - c^2\partial_{xx} = \partial_\mathbf{v}\partial_\mathbf{w} = 4c^2\partial_{\xi\eta}.$$

Once the characteristics are known, the change of variables (6.32) reduces the wave equation to the form (6.33).

Proceeding in the same way, for the diffusion operator we would have

$$\partial_{xx} = \partial_x\partial_x.$$

Therefore we find only one family of characteristics, given by

$$t = \text{constant}.$$

Thus, no change of variables is necessary and the diffusion equation is already in its canonical form.

For the Laplace operator we find

$$\partial_{xx} + \partial_{yy} = (\partial_y + i\partial_x)(\partial_y - i\partial_x)$$

and there are two families of *complex* characteristics given by

$$\phi(x,y) = x + iy = \text{constant}, \qquad \psi(x,y) = x - iy = \text{constant}.$$

The change of variables

$$z = x + iy, \qquad \overline{z} = x - iy$$

leads to the equation

$$\partial_{z\overline{z}} U = 0$$

whose general solution is

$$U(z,\overline{z}) = F(z) + G(\overline{z}).$$

This formula may be considered as a characterization of the harmonic function in the complex plane.

It should be clear, however, that the characteristics for the diffusion and the Laplace equations do not play the same relevant role as they do for the wave equation.

6.5.2 Characteristics and canonical form

Let us go back to the equation in general form (6.34). Can we reduce to a canonical form its principal part? There are at least two substantial reasons to answer the question.

The first one is tied to the type of well posed problems associated with (6.34): which kind of data have to be assigned and where, in order to find a unique and stable solution? It turns out that hyperbolic, parabolic and elliptic equations share their well posed problems with their main prototypes: the wave, diffusion and Laplace equations, respectively. Also the choice of numerical methods depends very much on the type of problem to be solved.

The second reason comes from the different features the three types of equation exhibit. Hyperbolic equations model oscillatory phenomena with *finite speed of propagation of the disturbances*, while for parabolic equation, "information" travels with infinite speed. Finally, elliptic equations model stationary situations, with no evolution in time.

To obtain the canonical form of the principal part we try to apply the ideas at the end of the previous subsection. First of all, note that, if $a = c = 0$, the principal part is already in the form (6.33), so that we assume $a > 0$ (say).

Now we decompose the differential operator in (6.35) into the product of two first order factors, as follows[12]

$$a\partial_{tt} + 2b\partial_{xt} + c\partial_{xx} = a\left(\partial_t - \Lambda^+\partial_x\right)\left(\partial_t - \Lambda^-\partial_x\right) \tag{6.38}$$

where

$$\Lambda^\pm = \frac{-b \pm \sqrt{b^2 - ac}}{a}.$$

Case 1. $b^2 - ac > 0$, the equation is **hyperbolic**. The two factors in (6.38) represent derivatives along the direction fields

$$\mathbf{v}(x,t) = \left(-\Lambda^+(x,t), 1\right) \quad \text{and} \quad \mathbf{w}(x,t) = \left(-\Lambda^-(x,t), 1\right)$$

respectively, so that we may write

$$a\partial_{tt} + 2b\partial_{xt} + c\partial_{xx} = a\partial_\mathbf{v}\partial_\mathbf{w}.$$

The vector fields $\mathbf{v}$ and $\mathbf{w}$ are tangent at any point to the characteristics

$$\phi(x,t) = k_1 \quad \text{and} \quad \psi(x,t) = k_2 \tag{6.39}$$

of the following *quasilinear first-order* equations

$$\phi_t - \Lambda^+\phi_x = 0 \quad \text{and} \quad \psi_t - \Lambda^-\psi_x = 0. \tag{6.40}$$

Note that we may write the two equations (6.40) in the compact form

$$a v_t^2 + 2b v_x v_t + c v_x^2 = 0. \tag{6.41}$$

By analogy with the case of the wave equation, we expect that the change of variables

$$\xi = \phi(x,t), \quad \eta = \psi(x,t) \tag{6.42}$$

should straighten the characteristics, at least locally, converting $\partial_\mathbf{v}\partial_\mathbf{w}$ into a multiple of $\partial_{\xi\eta}$.

First of all, however, we have to make sure that the transformation (6.42) is *non-degenerate*, at least locally, or, in other words, that the Jacobian of the transformation does not vanish

$$\phi_t\psi_x - \phi_x\psi_t \neq 0. \tag{6.43}$$

On the other hand, this follows from the fact that the vectors $\nabla\phi$ and $\nabla\psi$ are orthogonal to $\mathbf{v}$ and $\mathbf{w}$, respectively, and that $\mathbf{v}, \mathbf{w}$ are nowhere colinear (since $b^2 - ac > 0$).

---

[12] Remember that

$$ax^2 + 2bxy + cy^2 = a(x - x_1)(x - x_2)$$

where

$$x_{1,2} = \left[-b \pm \sqrt{b^2 - ac}\right]/a.$$

Thus, at least locally, the inverse transformation

$$x = \Phi(\xi, \eta), \qquad t = \Psi(\xi, \eta)$$

exists. Let

$$U(\xi, \eta) = u(\Phi(\xi, \eta), \Psi(\xi, \eta)).$$

Then

$$u_x = U_\xi \phi_x + U_\eta \psi_x, \qquad u_t = U_\xi \phi_t + U_\eta \psi_t$$

and moreover

$$u_{tt} = \phi_t^2 U_{\xi\xi} + 2\phi_t \psi_t U_{\xi\eta} + \psi_t^2 U_{\eta\eta} + \phi_{tt} U_\xi + \psi_{tt} U_\eta$$

$$u_{xx} = \phi_x^2 U_{\xi\xi} + 2\phi_x \psi_x U_{\xi\eta} + \psi_x^2 U_{\eta\eta} + \phi_{xx} U_\xi + \psi_{xx} U_\eta$$

$$u_{xt} = \phi_t \phi_x U_{\xi\xi} + (\phi_x \psi_t + \phi_t \psi_x) U_{\xi\eta} + \psi_t \psi_x U_{\eta\eta} + \phi_{xt} U_\xi + \psi_{xt} U_\eta.$$

Then

$$a u_{tt} + 2b u_{xy} + c u_{xx} = A U_{\xi\xi} + 2B U_{\xi\eta} + C U_{\eta\eta} + D U_\xi + E U_\eta$$

where[13]

$$A = a\phi_t^2 + 2b\phi_t \phi_x + c\phi_x^2, \qquad C = a\psi_t^2 + 2b\psi_t \psi_x + c\psi_x^2$$
$$B = a\phi_t \psi_t + b(\phi_x \psi_t + \phi_t \psi_x) + c\phi_x \psi_x$$
$$D = a\phi_{tt} + 2b\phi_{xt} + c\phi_{xx}, \qquad E = a\psi_{tt} + 2b\psi_{xt} + c\psi_{xx}.$$

Now, $A = C = 0$, since ϕ and ψ both satisfy (6.41), so that

$$a u_{tt} + 2b u_{xt} + c u_{xx} = 2B U_{\xi\eta} + D U_\xi + E U_\eta.$$

We claim that $B \neq 0$; indeed, recalling that $\Lambda^+ \Lambda^- = c/a$, $\Lambda^+ + \Lambda^+ = -2b/a$
and

$$\phi_t = \Lambda^+ \phi_x, \qquad \psi_t = \Lambda^- \psi_x,$$

after elementary computations we find

$$B = \frac{2}{a} \left(ac - b^2 \right) \phi_x \psi_x.$$

From (6.43) we deduce that $B \neq 0$. Thus, (6.34) assumes the form

$$U_{\xi\eta} = \mathcal{F}(\xi, \eta, U, U_\xi, U_\eta)$$

which is its *canonical form*.

---

[13] It is understood that all the functions are evaluated at $x = \Phi(\xi, \eta)$ and $t = \Psi(\xi, \eta)$.

The curves (6.39) are called *characteristics* for (6.34) and are the solution curves of the ordinary differential equations

$$\frac{dx}{dt} = -\Lambda^+, \quad \frac{dx}{dt} = -\Lambda^-,$$ (6.44)

respectively. Note that the two equations (6.44) can be put into the compact form

$$a\left(\frac{dx}{dt}\right)^2 - 2b\frac{dx}{dt} + c = 0.$$ (6.45)

Example 6.1. Consider the equation

$$xu_{tt} - \left(1 + x^2\right)u_{xt} = 0.$$ (6.46)

Since $b^2 - ac = \left(1 + x^2\right)/4 > 0$, (6.46) is hyperbolic. Equation (6.45) is

$$x\left(\frac{dx}{dt}\right)^2 + \left(1 + x^2\right)\frac{dx}{dt} = 0$$

which yields, for $x \neq 0$,

$$\frac{dx}{dt} = -\frac{1 + x^2}{x} \quad \text{and} \quad \frac{dx}{dt} = 0.$$

Thus, the characteristics curves are:

$$\phi(x,t) = e^{2t}\left(1 + x^2\right) = k_1 \quad \text{and} \quad \psi(x,t) = x = k_2.$$

We set

$$\xi = e^{2t}\left(1 + x^2\right) \quad \text{and} \quad \eta = x.$$

After routine calculations, we find $D = E = 0$ so that the canonical form is

$$U_{\xi\eta} = 0.$$

The general solution of (6.46) is therefore

$$u(x,t) = F\left(e^{2t}\left(1 + x^2\right)\right) + G(x)$$

with F and G arbitrary C^2 functions.

Case 2. $b^2 - ac \equiv 0$, the equation is **parabolic**. There exists **only one** family of characteristics, given by $\phi(x,t) = k$, where ϕ is a solution of the first order equation

$$a\phi_t + b\phi_x = 0,$$

since $\Lambda^+ = \Lambda^- = -b/a$. If ϕ is known, choose any smooth function ψ such that $\nabla\phi$ and $\nabla\psi$ are linearly independent and

$$a\psi_t^2 + 2b\psi_t\psi_x + c\psi_x^2 = C \neq 0.$$

Set
$$\xi = \phi(x,t), \qquad \eta = \psi(x,t)$$

and
$$U(\xi,\eta) = u(\Phi(\xi,\eta), \Psi(\xi,\eta)).$$

For the derivatives of U we can use the computations done in **case 1**. However, observe that, since $b^2 - ac = 0$ and $a\phi_t + b\phi_x = 0$, we have

$$B = a\phi_t\psi_t + b(\phi_t\psi_x + \phi_x\psi_t) + c\phi_x\psi_x = \psi_t(a\phi_t + b\phi_x) + \psi_x(b\phi_t + c\phi_x)$$

$$= b\psi_x\left(\phi_t + \frac{c}{b}\phi_x\right) = b\psi_x\left(\phi_t + \frac{b}{a}\phi_x\right) = \frac{b}{a}\psi_x(a\phi_t + b\phi_x) = 0.$$

Thus, the equation for U becomes

$$CU_{\eta\eta} = \mathcal{F}(\xi,\eta,U,U_\xi,U_\eta)$$

which is the *canonical form*.

Example 6.2. The equation

$$u_{tt} - 6u_{xt} + 9u_{xx} = u$$

is parabolic. The family of characteristics is

$$\phi(x,t) = 3t + x = k.$$

Choose $\psi(x,t) = x$ and set

$$\xi = 3t + x, \qquad \eta = x.$$

Since $\nabla\phi = (3,1)$ and $\nabla\psi = (1,0)$, the gradients are independent and we set

$$U(\xi,\eta) = u\left(\frac{\xi - \eta}{3}, \eta\right).$$

We have, $D = E = 0$, so that the equation for U is

$$U_{\eta\eta} - U = 0$$

whose general solution is

$$U(\xi,\eta) = F(\xi)e^{-\eta} + G(\xi)e^{\eta}$$

with F and G arbitrary C^2 functions. Finally, we find

$$u(x,t) = F(3t + x)e^{-x} + G(3t + x)e^{x}.$$

Case 3. $b^2 - ac < 0$, the equation is **elliptic**. In this case there are no real characteristics. If the coefficients a, b, c are analytic functions[14] we can proceed as in case 1, with two families of complex characteristics. This yields the canonical form

$$U_{zw} = \mathcal{G}\left(z, w, U, U_z, U_w\right) \qquad z, w \in \mathbb{C}.$$

Letting

$$z = \xi + i\eta, \; w = \xi - i\eta$$

and $\tilde{U}\left(\xi, \eta\right) = U\left(\xi + i\eta, \xi - i\eta\right)$ we can eliminate the complex variables arriving at the real canonical form

$$\tilde{U}_{\xi\xi} + \tilde{U}_{\eta\eta} = \tilde{\mathcal{G}}\left(\xi, \eta, \tilde{U}, \tilde{U}_\xi, \tilde{U}_\eta\right).$$

6.6 The Multi-dimensional Wave Equation $(n > 1)$

6.6.1 Special solutions

The wave equation

$$u_{tt} - c^2 \Delta u = f, \tag{6.47}$$

constitutes a basic model for describing a remarkable number of oscillatory phenomena in dimension $n > 1$. Here $u = u\left(\mathbf{x}, t\right)$, $\mathbf{x} \in \mathbb{R}^n$ and, as in the one-dimensional case, c is the *speed of propagation*. If $f \equiv 0$, the equation is said *homogeneous* and the *superposition principle holds*. Let us examine some relevant solutions of (6.47).

• *Plane waves.* If $\mathbf{k} \in \mathbb{R}^n$ and $\omega^2 = c^2 |\mathbf{k}|^2$, the function

$$u\left(\mathbf{x}, t\right) = w\left(\mathbf{x} \cdot \mathbf{k} - \omega t\right)$$

is a solution of the homogeneous (6.47). Indeed,

$$u_{tt}\left(\mathbf{x}, t\right) - c^2 \Delta u\left(\mathbf{x}, t\right) = \omega^2 w''\left(\mathbf{x} \cdot \mathbf{n} - \omega t\right) - c^2 |\mathbf{k}|^2 w''\left(\mathbf{x} \cdot \mathbf{n} - \omega t\right) = 0.$$

We have already seen in Section 6.1.1 that the planes

$$\mathbf{x} \cdot \mathbf{k} - \omega t = \text{constant}$$

constitute the wave fronts, moving at speed $c_p = \omega / |\mathbf{k}|$ in the $\mathbf{k}$ direction. The scalar $\lambda = 2\pi / |\mathbf{k}|$ is the wavelength. If $w\left(z\right) = Ae^{iz}$, the wave is said *monochromatic* or *harmonic*.

• *Cylindrical waves* $(n = 3)$ are of the form

$$u\left(\mathbf{x}, t\right) = w\left(r, t\right)$$

---

[14] I.e. they can be locally expanded in Taylor series.

where $\mathbf{x} = (x_1, x_2, x_3)$, $r = \sqrt{x_1^2 + x_2^2}$. In particular, solutions like $u(\mathbf{x},t) = e^{i\omega t} w(r)$ represent stationary cylindrical waves, that can be found solving the homogeneous equation (6.47) using the separation of variables, in axially symmetric domains.

If the axis of symmetry is the x_3 axis, it is appropriate to use the cylindrical coordinates $x_1 = r \cos\theta$, $x_2 = r \sin\theta$, x_3. Then, the wave equation becomes[15]

$$u_{tt} - c^2 \left(u_{rr} + \frac{1}{r} u_r + \frac{1}{r^2} u_{\theta\theta} + u_{x_3 x_3} \right) = 0.$$

Looking for standing waves of the form $u(r, t) = e^{i\lambda ct} w(r)$, $\lambda \geq 0$, we find, after dividing by $c^2 e^{i\lambda ct}$,

$$w''(r) + \frac{1}{r} w' + \lambda^2 w = 0.$$

This is a Bessel equation of zero order. We know that the only solutions bounded at $r = 0$ are

$$w(r) = a J(\lambda r), \qquad a \in \mathbb{R}$$

where, we recall,

$$J_0(x) = \sum_{k=0}^{\infty} \frac{(-1)^k}{(k!)^2} \left(\frac{x}{2} \right)^{2k}$$

is the Bessel function of first kind of zero order. In this way we obtain waves of the form

$$u(r, t) = a J_0(\lambda r) e^{i\lambda ct}.$$

• *Spherical waves* $(n = 3)$ are of the form

$$u(\mathbf{x},t) = w(r, t)$$

where $\mathbf{x} = (x_1, x_2, x_3)$, $r = |\mathbf{x}| = \sqrt{x_1^2 + x_2^2 + x_3^2}$. In particular $u(\mathbf{x},t) = e^{i\omega t} w(r)$ represent standing spherical waves and can be determined by solving the homogeneous equation (6.47) using separation of variables in spherically symmetric domains. In this case, spherical coordinates

$$x_1 = r \cos\theta \sin\psi, \quad x_2 = r \sin\theta \sin\psi, \quad x_3 \cos\psi,$$

are appropriate and the wave equation becomes[16]

$$\frac{1}{c^2} u_{tt} - u_{rr} - \frac{2}{r} u_r - \frac{1}{r^2} \left\{ \frac{1}{(\sin\psi)^2} u_{\theta\theta} + u_{\psi\psi} + \frac{\cos\psi}{\sin\psi} u_\psi \right\} = 0. \qquad (6.48)$$

---

[15] Appendix D.
[16] Appendix D.

Let us look for solution of the form $u(r,t) = e^{i\lambda ct} w(r)$, $\lambda \geq 0$. We find, after simplifying out $c^2 e^{i\lambda ct}$,

$$w''(r) + \frac{2}{r} w' + \lambda^2 w = 0$$

which can be written[17]

$$(rw)'' + \lambda^2 rw = 0.$$

Thus, $v = rw$ is solution of

$$v'' + \lambda^2 v = 0$$

which gives $v(r) = a\cos(\lambda r) + b\sin(\lambda r)$ and hence the attenuated spherical waves

$$w(r,t) = ae^{i\lambda ct} \frac{\cos(\lambda r)}{r}, \qquad w(r,t) = be^{i\lambda ct} \frac{\sin(\lambda r)}{r}. \qquad (6.49)$$

Let us now determine the general form of a spherical wave in $\mathbb{R}^3$. Inserting $u(\mathbf{x},t) = w(r,t)$ into (6.48) we obtain

$$w_{tt} - c^2 \left\{ w_{rr}(r) + \frac{2}{r} w_r \right\} = 0$$

which can be written in the form

$$(rw)_{tt} - c^2 (rw)_{rr} = 0. \qquad (6.50)$$

Then, formula (6.21) gives

$$w(r,t) = \frac{F(r+ct)}{r} + \frac{G(r-ct)}{r} \equiv w_i(r,t) + w_o(r,t) \qquad (6.51)$$

which represents the superposition of two attenuated progressive spherical waves. The wave fronts of u_o are the spheres $r - ct = k$, expanding as time goes on. Hence, w_o represents an *outgoing wave*. On the contrary, the wave w_i is *incoming*, since its wave fronts are the contracting spheres $r + ct = k$.

6.6.2 Well posed problems. Uniqueness

The well posed problems in dimension one, are still well posed in any number of dimensions. Let

$$Q_T = \Omega \times (0, T)$$

a *space-time cylinder*, where Ω is a bounded C^1–domain[18] in $\mathbb{R}^n$. A solution $u(\mathbf{x},t)$ is uniquely determined by assigning initial data and appropriate boundary conditions on the boundary $\partial\Omega$ of Ω.

---

[17] Thanks to the miraculous presence of the factor 2 in the coefficient of w'!
[18] As usual we can afford corner points (e.g. a triangle or a cone) and also some edges (e.g. a cube or a hemisphere).

More specifically, we may pose the following problems: *Determine* $u = u(\mathbf{x}, t)$ *such that:*

$$\begin{cases} u_{tt} - c^2 \Delta u = f & \text{in } Q_T \\ u(\mathbf{x}, 0) = g(\mathbf{x}), \, u_t(\mathbf{x}, 0) = h(\mathbf{x}) & \text{in } \Omega \\ + \text{ boundary conditions} & \text{on } \partial\Omega \times [0, T) \end{cases} \quad (6.52)$$

where the boundary conditions are:

(a) $u = h$ (Dirichlet);

(b) $\partial_\nu u = h$ (Neumann);

(c) $\partial_\nu u + \alpha u = h$ ($\alpha \geq 0$, Robin);

(d) $u = h_1$ on $\partial_D\Omega$ and $\partial_\nu u = h_2$ on $\partial_N\Omega$ (mixed problem) with $\partial_N\Omega$ a relatively open subset of $\partial\Omega$ and $\partial_D\Omega = \partial\Omega \backslash \partial_N\Omega$.

The *global Cauchy problem*

$$\begin{cases} u_{tt} - c^2 \Delta u = f & \mathbf{x} \in \mathbb{R}^n, t > 0 \\ u(\mathbf{x}, 0) = g(\mathbf{x}), \, u_t(\mathbf{x}, 0) = h(\mathbf{x}) & \mathbf{x} \in \mathbb{R}^n \end{cases} \quad (6.53)$$

is quite important also in dimension $n > 1$. We will examine it with some details later on. Particularly relevant are the different features that the solutions exhibit for $n = 2$ and $n = 3$.

Under rather natural hypotheses on the data, problem (6.52) has at most one solution. To see it, we may use once again the conservation of energy, which is proportional to

$$E(t) = \frac{1}{2} \int_\Omega \left\{ u_t^2 + c^2 |\nabla u|^2 \right\} d\mathbf{x}.$$

The growth rate is

$$\dot{E}(t) = \int_\Omega \left\{ u_t u_{tt} + c^2 \nabla u_t \cdot \nabla u \right\} d\mathbf{x}.$$

Integrating by parts, we have

$$\int_\Omega c^2 \nabla u_t \cdot \nabla u \, d\mathbf{x} = c^2 \int_{\partial\Omega} u_\nu u_t \, d\sigma - \int_\Omega c^2 u_t \Delta u \, d\mathbf{x}$$

whence, since $u_{tt} - c^2 \Delta u = f$,

$$\dot{E}(t) = \int_\Omega \left\{ u_{tt} - c^2 \Delta u \right\} u_t \, d\mathbf{x} + c^2 \int_{\partial\Omega} u_\nu u_t \, d\sigma$$

$$= \int_\Omega f u_t \, d\mathbf{x} + c^2 \int_{\partial\Omega} u_\nu u_t \, d\sigma.$$

Now it is easy to prove the following result, where we use the symbol $C^{h,k}(D)$ to denote the set of functions h times continuously differentiable with respect to space and k times with respect to time in D.

Theorem 6.1. *Problem* (6.52), *coupled with one of the boundary conditions* (a)–(d) *above, has at most one solution in* $C^{2,2}(Q_T) \cap C^{1,1}(\overline{Q}_T)$.

Proof. Let u_1 and u_2 be solutions of the same problem, sharing the same data. Their difference $w = u_1 - u_2$ is a solution of the homogeneous equation, with zero data. We show that $w(\mathbf{x},t) \equiv 0$.

In the case of Dirichlet, Neumann and mixed conditions, since either $w_\nu = 0$ or $w_t = 0$ on $\partial \Omega \times [0,T)$, we have $\dot{E}(t) = 0$. Thus, since $E(0) = 0$, we infer

$$E(t) = \frac{1}{2} \int_\Omega \left\{ w_t^2 + c^2 |\nabla w|^2 \right\} d\mathbf{x} = 0, \qquad \forall t > 0.$$

Therefore, for each $t > 0$, both w_t and $|\nabla w(\mathbf{x},t)|$ vanish so that $w(\mathbf{x},t)$ is constant. Then $w(\mathbf{x},t) \equiv 0$, since $w(\mathbf{x}, 0) = 0$.

For Robin problem

$$\dot{E}(t) = -c^2 \int_{\partial \Omega} \alpha w w_t \, d\sigma = -\frac{c^2}{2} \frac{d}{dt} \int_{\partial \Omega} \alpha w^2 \, d\sigma$$

that is

$$\frac{d}{dt} \left\{ E(t) + \frac{c^2}{2} \int_{\partial \Omega} \alpha w^2 \, d\sigma \right\} = 0.$$

Hence,

$$E(t) + \frac{c^2}{2} \int_{\partial \Omega} \alpha w^2 d\sigma = \text{constant}$$

and, being zero initially, it is zero for all $t > 0$. Since $\alpha \geq 0$, we again conclude that $w \equiv 0$. $\qquad \square$

6.6.3 Small vibrations of an elastic membrane.

In Section 6.2.1 we derived a model for the small transversal vibrations of a string. Similarly, we may derive the governing equation of the small transversal vibrations of a highly stretched membrane (think e.g. of a drum), at rest in the horizontal position. We briefly sketch the derivation leaving it to the reader to fill in the details. Assume the following hypotheses.

1. *The vibrations of the membrane are small and vertical.* This means that the changes from the plane horizontal shape are very small and horizontal displacements are negligible.
2. *The vertical displacement of a point of the membrane depends on time and on its position at rest.* Thus, if u denotes the vertical displacement of a point located at rest at (x, y), we have $u = u(x, y, t)$.
3. *The membrane is perfectly flexible and elastic.* There is no resistance to bending. In particular, the stress in the membrane can be modelled by

a tangential force $\mathbf{T}$ of magnitude τ, called *tension*[19]. Perfect elasticity means that τ is a constant.

4. *Friction is negligible.*

Under the above assumptions, the equation of motion of the membrane can be derived from *conservation of mass* and *Newton's law*.

Let $\rho_0 = \rho_0(x, y)$ be the surface mass density of the membrane at rest and consider a small "rectangular" piece of membrane, with vertices at the points A, B, C, D of coordinates (x, y), $(x + \Delta x, y)$, $(x, y + \Delta y)$ and $(x + \Delta x, y + \Delta y)$, respectively. Denote by ΔS the corresponding area at time t. Then, conservation of mass yields

$$\rho_0(x, y)\, \Delta x \Delta y = \rho(x, y, t)\, \Delta S. \tag{6.54}$$

To write Newton's law of motion we have to determine the forces acting on our small piece of membrane. Since the motion is vertical, the horizontal forces have to balance.

The vertical forces are given by body forces (e.g. gravity and external loads) and the vertical component of the tension.

Denote by $f(x, y, t)\,\mathbf{k}$ the resultant of the body forces per unit mass. Then, using (6.54), the body forces acting on the membrane element are given by

$$\rho(x, y, t)\, f(x, y, t)\, \Delta S\ \mathbf{k} = \rho_0(x, y)\, f(x, y, t)\, \Delta x \Delta y\ \mathbf{k}.$$

Along the edges AB and CD, the tension is perpendicular to the x−axis and almost parallel to the y−axis. Its (scalar) vertical components are respectively given by

$$\tau_{vert}(x, y, t) \simeq \tau u_y(x, y, t)\, \Delta x, \qquad \tau_{vert}(x, y + \Delta y, t) \simeq \tau u_y(x, y + \Delta y, t)\, \Delta x.$$

Similarly, along the edge AC, the tension is perpendicular to the y−axis and almost parallel to the x−axis. Its (scalar) vertical components are respectively given by

$$\tau_{vert}(x, y, t) \simeq \tau u_x(x, y, t)\, \Delta y, \qquad \tau_{vert}(x + \Delta x, y, t) \simeq \tau u_x(x + \Delta x, y, t)\, \Delta y.$$

Thus, using (6.54) again and observing that u_{tt} is the (scalar) vertical accel-

---

[19] The tension $\mathbf{T}$ has the following meaning. Consider a small region on the membrane, delimited by a closed curve γ. The material on one side of γ exerts on the material on the other side a *force per unit length* $\mathbf{T}$ *(pulling)* along γ. A constitutive law for $\mathbf{T}$ is

$$\mathbf{T}(x, y, t) = \tau(x, y, t)\, \mathbf{N}(x, y, t) \qquad (x, y) \in \gamma \ .$$

where $\mathbf{N}$ is the outward unit normal vector to γ, tangent to the membrane. Again, the tangentiality of the tension force is due to the absence of distributed moments over the membrane.

eration, Newton's law gives:

$$\rho_0\left(x,y\right)\Delta x\Delta y\, u_{tt} =$$
$$= \tau[u_y\left(x, y+\Delta y, t\right) - u_y\left(x, y, t\right)]\Delta x + \tau[u_x\left(x+\Delta x, y, t\right) - u_x\left(x, y, t\right)]\Delta y +$$
$$+ \rho_0\left(x, y\right) f\left(x, y, t\right)\Delta x\Delta y.$$

Dividing for $\Delta x\Delta y$ and letting $\Delta x, \Delta y \to 0$, we obtain the equation

$$u_{tt} - c^2(u_{yy} + u_{xx}) = f\left(x, y, t\right) \tag{6.55}$$

where $c^2\left(x, y, t\right) = \tau/\rho_0\left(x, y\right)$.

Example 6.3 (Square membrane). Consider a membrane occupying at rest a square of side a, pinned at the boundary. We want to study its vibration when the membrane is initially horizontal, with speed $h = h\left(x, y\right)$. If there is no external load and the weight of the membrane is negligible, the vibrations are governed by the following initial-boundary value problem:

$$\begin{cases} u_{tt} - c^2\Delta u = 0 & 0 < x < a,\, 0 < y < a,\, t > 0 \\ u\left(x, y, 0\right) = 0,\, u_t\left(x, y, 0\right) = h\left(x, y\right) & 0 < x < a,\, 0 < y < a \\ u\left(0, y, t\right) = u\left(a, y, t\right) = 0 & 0 \le y \le a,\, t \ge 0 \\ u\left(x, 0, t\right) = u\left(x, a, t\right) = 0 & 0 \le x \le a,\, t \ge 0. \end{cases}$$

The square shape of the membrane and the homogeneous boundary conditions suggest the use of separation of variables. Let us look for solution of the form

$$u\left(x, y, t\right) = v\left(x, y\right) q\left(t\right)$$

with $v = 0$ at the boundary. Substituting into the wave equation we find

$$q''\left(t\right) v\left(x, y\right) - c^2 q\left(t\right)\Delta v\left(x, y\right) = 0$$

and, separating the variables,

$$\frac{q''\left(t\right)}{c^2 q\left(t\right)} = \frac{\Delta v\left(x, y\right)}{v\left(x, y\right)} = -\lambda^2$$

whence[20] the equation

$$q''\left(t\right) + c^2\lambda^2 q\left(t\right) = 0 \tag{6.56}$$

and the *eigenvalue problem*

$$\Delta v + \lambda^2 v = 0 \tag{6.57}$$

$$v\left(0, y\right) = v\left(a, y\right) = v\left(x, 0\right) = v\left(x, a\right) = 0, \qquad 0 \le x, y \le a.$$

---

[20] The two ratios must be equal to the same constant. The choice of $-\lambda^2$ is by our former experience.

We first solve the eigenvalue problem, using once more separation of variables and setting $v(x, y) = X(x)Y(y)$, with the conditions

$$X(0) = X(a) = 0, \qquad Y(0) = Y(a) = 0.$$

Substituting into (6.57), we obtain

$$\frac{Y''(y)}{Y(y)} + \lambda^2 = -\frac{X''(x)}{X(x)} = \mu^2$$

where μ is a new constant.

Letting $\nu^2 = \lambda^2 - \mu^2$, we have to solve the following two one-dimensional eigenvalue problems, in $0 < x < a$ and $0 < y < a$, respectively:

$$\begin{cases} X''(x) + \mu^2 X(x) = 0 \\ X(0) = X(a) = 0 \end{cases} \qquad \begin{cases} Y''(y) + \nu^2 Y(y) = 0 \\ Y(0) = Y(a) = 0. \end{cases}$$

The solutions are:

$$X(x) = A_m \sin \mu_m x, \qquad \mu_m = \frac{m\pi}{a}$$

$$Y(y) = B_n \sin \nu_n y, \qquad \nu_n = \frac{n\pi}{a}$$

where $m, n = 1, 2, \ldots$. Since $\lambda^2 = \nu^2 + \mu^2$, we have

$$\lambda_{mn}^2 = \frac{\pi^2}{a^2}\left(m^2 + n^2\right), \qquad m, n = 1, 2, \ldots \tag{6.58}$$

corresponding to the eigenfunctions

$$v_{mn}(x, y) = C_{mn} \sin \mu_m x \sin \nu_n y.$$

For $\lambda = \lambda_{mn}$, the general integral of (6.56) is

$$q_{mn}(t) = \bar{a}_{mn} \cos c\lambda_{mn}t + \bar{b}_m \sin c\lambda_{mn}t.$$

Thus we have found infinitely many special solutions to the wave equations, of the form,

$$u_{mn} = (a_{mn} \cos c\lambda_{mn}t + b_{mn} \sin c\lambda_{mn}t) \sin \mu_m x \sin \nu_n y$$

which, moreover, vanish on the boundary.

Every u_{mn} is a standing wave and corresponds to a particular mode of vibration of the membrane. The *fundamental frequency* is $f_{11} = c\sqrt{2}/2a$, while the other frequencies are $f_{mn} = c\sqrt{m^2 + n^2}/2a$, which are **not** integer multiple of the fundamental one (as they do for the vibrating string).

Going back to our problem, to find a solution which satisfies the initial conditions, we superpose the modes u_{mn} defining

$$u\left(x,y,t\right) = \sum_{m,n=1}^{\infty} \left(a_{mn} \cos c\lambda_{mn}t + b_{mn} \sin c\lambda_{mn}t\right) \sin \mu_m x \sin \nu_n y.$$

Since $u\left(x,y,0\right) = 0$ we choose $a_{mn} = 0$ for every $m,n \geq 1$. From $u_t\left(x,y,0\right) = h\left(x,y\right)$ we find the condition

$$\sum_{m,n=1}^{\infty} cb_{mn}\lambda_{mn} \sin \mu_m x \sin \nu_n x = h\left(x,y\right). \tag{6.59}$$

Therefore, we assume that h can be expanded in a double Fourier sine series as follows

$$h\left(x,y\right) = \sum_{m,n=1}^{\infty} h_{mn} \sin \mu_m x \sin \nu_n y,$$

where the coefficients h_{mn} are given by

$$h_{mn} = \frac{4}{a^2} \int_Q h\left(x,y\right) \sin \frac{m\pi}{a} x \sin \frac{n\pi}{a} y \, dxdy.$$

Then, if we choose $b_{mm} = h_{mm}/c\lambda_{mn}$, (6.59) is satisfied. Thus, we have constructed the *formal* solution

$$u\left(x,y,t\right) = \sum_{m,n=1}^{\infty} \frac{h_{mn}}{c\lambda_{mn}} \sin c\lambda_{mn}t \sin \mu_m x \sin \nu_n y. \tag{6.60}$$

If the coefficients $h_{mm}/c\lambda_{mn}$ vanish fast enough as $m,n \to +\infty$, it can be shown that (6.60) gives the unique solution[21].

6.6.4 Small amplitude sound waves

Sound waves are small disturbances in the density and pressure of a compressible gas. In an isotropic gas, their propagation can be described in terms of a single scalar quantity. Moreover, due to the small amplitudes involved, it is possible to *linearize* the equations of motion, within a reasonable range of validity. Three are the relevant equations: two of them are *conservation of mass* and *balance of linear momentum*, the other one is a *constitutive* relation between density and pressure.

Conservation of mass expresses the relation between the gas density $\rho = \rho\left(\mathbf{x},t\right)$ and its velocity $\mathbf{v} = \mathbf{v}\left(\mathbf{x},t\right)$

$$\rho_t + \mathrm{div}\left(\rho\mathbf{v}\right) = 0. \tag{6.61}$$

---

[21] We leave to the reader to find appropriate smoothness hypotheses on h, in order to assure that (6.60) is the unique solution.

The balance of linear momentum describes how the volume of gas occupying a region V reacts to the pressure exerted by the rest of the gas. Assuming that the viscosity of the gas is negligible, this force is given by the *normal pressure* $-p\boldsymbol{\nu}$ on the boundary of V ($\boldsymbol{\nu}$ is the exterior normal to ∂V).

Thus, if there are no significant external forces, the linear momentum equation is

$$\frac{D\mathbf{v}}{Dt} \equiv \mathbf{v}_t + (\mathbf{v}{\cdot}\nabla)\,\mathbf{v} = -\frac{1}{\rho}\nabla p. \tag{6.62}$$

The last equation is an empirical relation between p and ρ. Since the pressure fluctuations are very rapid, the compressions/expansions of the gas are *adiabatic, without any loss of heat*.

In these conditions, if $\gamma = c_p/c_v$ is the ratio of the specific heats of the gas ($\gamma \approx 1.4$ in air) then p/ρ^γ is constant, so that we may write

$$p = f(\rho) = C\rho^\gamma \tag{6.63}$$

with C constant.

The system of equations (6.61), (6.62), (6.63) is quite complicated and it would be extremely difficult to solve it in its general form. Here, the fact that sound waves are only small perturbation of normal atmospheric conditions allows a major simplification. Consider a static atmosphere, where ρ_0 and p_0 are constant density and pressure, with zero velocity field. We may write

$$\rho = (1+s)\,\rho_0 \approx \rho_0$$

where s is a small dimensionless quantity, called *condensation* and representing the fractional variation of the density from equilibrium. Then, from (6.63), we have

$$p - p_0 \approx f'(\rho_0)\,(\rho - \rho_0) = s\rho_0 f'(\rho_0) \tag{6.64}$$

and

$$\nabla p \approx \rho_0 f'(\rho_0)\,\nabla s.$$

Now, if $\mathbf{v}$ is also small, we may keep only first order terms in s and $\mathbf{v}$. Thus, we may neglect the convective acceleration $(\mathbf{v}{\cdot}\nabla)\,\mathbf{v}$ and approximate (6.62) and (6.61) with the linear equations

$$\mathbf{v}_t = -c_0^2\nabla s \tag{6.65}$$

and

$$s_t + \operatorname{div}\mathbf{v} = 0 \tag{6.66}$$

where we have set $c_0^2 = f'(\rho_0) = C\gamma\rho_0^{\gamma-1}$.

Let us pause for a moment to examine which implications the above linearization has. Suppose that V and S are average values of $|\mathbf{v}|$ and s, respectively. Moreover, let L and T typical order of magnitude for space and time in the wave propagation, such as wavelength and period. Rescale $\mathbf{v}$, s, $\mathbf{x}$ and t as follows

$$\boldsymbol{\xi} = \frac{\mathbf{x}}{L}, \quad \tau = \frac{t}{T}, \quad \mathbf{U}(\boldsymbol{\xi}, \tau) = \frac{\mathbf{v}(L\boldsymbol{\xi}, T\tau)}{V}, \quad \sigma(\boldsymbol{\xi}, \tau) = \frac{s(L\boldsymbol{\xi}, T\tau)}{S}. \tag{6.67}$$

Substituting (6.67) into (6.65) and (6.66) we obtain

$$\frac{V}{T}\mathbf{U}_\tau + \frac{c_0^2 S}{L}\nabla\sigma = \mathbf{0} \quad \text{and} \quad \frac{S}{T}\sigma_\tau + \frac{V}{L}\text{div}\mathbf{U} = 0.$$

In this equations the coefficients must be of the same order of magnitude, therefore

$$\frac{V}{T} \approx \frac{c_0^2 S}{L} \quad \text{and} \quad \frac{S}{T} \approx \frac{V}{L}$$

which implies

$$\frac{L}{T} \approx c_0.$$

As we see, c_0 is a typical propagation speed, namely it is **the sound speed**. Now, the convective acceleration is negligible with respect to (say) $\mathbf{v}_t$, if

$$\frac{V^2}{L}\mathbf{U}\cdot\nabla\mathbf{U} \ll \frac{V}{T}\mathbf{U}_\tau$$

or $V \ll c_0$.

Thus if the gas speed is much smaller than the sound speed, our linearization makes sense. The ratio $M = V/c_0$ is called **Mach number**.

We want to derive from (6.65) and (6.66) the following theorem in which we assume that both s and $\mathbf{v}$ are smooth functions.

Theorem 6.2. a) *The condensation s is a solution of the wave equation*

$$s_{tt} - c_0^2 \Delta s = 0 \tag{6.68}$$

where $c_0 = \sqrt{f'(\rho_0)} = \sqrt{\gamma p_0/\rho_0}$ is the speed of sound.

b) *If $\mathbf{v}(\mathbf{x}, 0) = \mathbf{0}$, there exists an acoustic potential ϕ such that $\mathbf{v} = \nabla\phi$. Moreover ϕ satisfies (6.68) as well.*

Proof (a). Taking the divergence on both sides of (6.65) and the t-derivative on both sides of (6.66) we get, respectively:

$$\text{div } \mathbf{v}_t = -c_0^2 \Delta s$$

and

$$s_{tt} = -(\text{div } \mathbf{v})_t.$$

Since $(\text{div } \mathbf{v})_t = \text{div } \mathbf{v}_t$, equation (6.68) follows.

Proof (b). From (6.65) we have

$$\mathbf{v}_t = -c_0^2 \nabla s.$$

Let

$$\phi(\mathbf{x},t) = -c_0^2 \int_0^t s(\mathbf{x},z)\, dz.$$

Then

$$\phi_t = -c_0^2 s$$

and we may write (6.65) in the form

$$\frac{\partial}{\partial t}\left[\mathbf{v} - \nabla \phi\right] = \mathbf{0}.$$

Hence, since $\phi(\mathbf{x},0) = 0$, $\mathbf{v}(\mathbf{x},0) = \mathbf{0}$, we infer

$$\mathbf{v}(\mathbf{x},t) - \nabla\phi(\mathbf{x},t) = \mathbf{v}(\mathbf{x},0) - \nabla\phi(\mathbf{x},0) = \mathbf{0}.$$

Thus $\mathbf{v} = \nabla\phi$. Finally, from (6.66),

$$\phi_{tt} = -c_0^2 s_t = c_0^2 \operatorname{div}\mathbf{v} = c_0^2 \Delta\phi$$

which is (6.68).   □

Once the potential ϕ is known, the velocity field $\mathbf{v}$, the condensation s and the pressure fluctuation $p - p_0$ can be computed from the following formulas

$$\mathbf{v} = \nabla\phi, \qquad s = -\frac{1}{c_0^2}\phi_t, \qquad p - p_0 = -\rho_0\phi_t.$$

Consider, for instance, a plane wave represented by the following potential

$$\phi(\mathbf{x},t) = w(\mathbf{x}\cdot\mathbf{k} - \omega t).$$

We know that if $c_0^2 |\mathbf{k}|^2 = \omega^2$, ϕ is a solution of (6.68). In this case, we have

$$\mathbf{v} = w'\mathbf{k}, \qquad s = \frac{\omega}{c_0^2}w', \qquad p - p_0 = \rho_0\omega w'.$$

Example 6.4 (Motion of a gas in a tube). Consider a straight cylindrical tube with axis along the x_1−axis, filled with gas in the region $x_1 > 0$. A flat piston, whose face moves according to $x_1 = h(t)$, sets the gas into motion. We assume that $|h(t)| \ll 1$ and $|h'(t)| \ll c_0$. Under these conditions, the motion of the piston generates sound waves of small amplitude and the acoustic potential ϕ is a solution of the homogeneous wave equation. To compute ϕ we need boundary conditions. The continuity of the normal velocity of the gas at the contact surface with the piston gives

$$\phi_{x_1}(h(t), x_2, x_3, t) = h'(t).$$

Since $h(t) \sim 0$, we may approximate this condition by

$$\phi_{x_1}(0, x_2, x_3, t) = h'(t). \tag{6.69}$$

At the tube walls the normal velocity of the gas is zero, so that, if $\boldsymbol{\nu}$ denotes the outward unit normal vector at the tube wall, we have

$$\nabla\phi \cdot \boldsymbol{\nu} = 0. \tag{6.70}$$

Finally since the waves are generated by the piston movement, we may look for *outgoing plane waves*[22] solution of the form

$$\phi(\mathbf{x}, t) = w(\mathbf{x} \cdot \mathbf{n} - ct)$$

where $\mathbf{n}$ is a unit vector. From (6.70) we have

$$\nabla\phi \cdot \boldsymbol{\nu} = w'(\mathbf{x} \cdot \mathbf{n} - ct)\, \mathbf{n} \cdot \boldsymbol{\nu} = 0$$

whence $\mathbf{n} \cdot \boldsymbol{\nu} = 0$ for every $\boldsymbol{\nu}$ orthogonal to the wall tube. Thus, we infer $\mathbf{n} = (1, 0, 0)$ and, as a consequence,

$$\phi(\mathbf{x}, t) = w(x_1 - ct).$$

From (6.69) we get

$$w'(-ct) = h'(t)$$

so that (assuming $h(0) = 0$),

$$w(s) = -ch\left(-\frac{s}{c}\right).$$

Hence, the acoustic potential is given by

$$\phi(\mathbf{x}, t) = -ch\left(t - \frac{x_1}{c}\right)$$

which represents a *progressive wave* propagating along the tube. In this case

$$\mathbf{v} = c\mathbf{i}, \quad s = \frac{1}{c}h'\left(t - \frac{x_1}{c}\right), \quad p = c\rho_0 h'\left(t - \frac{x_1}{c}\right) + p_0.$$

6.7 The Cauchy Problem

6.7.1 Fundamental solution ($n = 3$) and strong Huygens' principle

In this section we consider the global Cauchy problem for the three-dimensional homogeneous wave equation:

$$\begin{cases} u_{tt} - c^2 \Delta u = 0 & \mathbf{x} \in \mathbb{R}^3, t > 0 \\ u(\mathbf{x}, 0) = g(\mathbf{x}), \quad u_t(\mathbf{x}, 0) = h(\mathbf{x}) & \mathbf{x} \in \mathbb{R}^3. \end{cases} \tag{6.71}$$

---

[22] We do not expect *incoming* waves, which should be generated by sources placed far from the piston.

Our purpose here is to show that the solution u exists and to find an explicit formula for it, in terms of the data g and h. Our derivation is rather heuristic so that, for the time being, we do not worry too much about the correct hypotheses on h and g, which we assume as smooth as we need to carry out the calculations.

First we need a lemma that reduces the problem to the case $g = 0$ (and which actually holds in any dimension). Denote by w_h the solution of the problem

$$
\begin{cases}
w_{tt} - c^2 \Delta w = 0 & \mathbf{x} \in \mathbb{R}^3, t > 0 \\
w(\mathbf{x},0) = 0, \quad w_t(\mathbf{x},0) = h(\mathbf{x}) & \mathbf{x} \in \mathbb{R}^3.
\end{cases}
\tag{6.72}
$$

Lemma 6.1. *If w_g has continuous third-order partials, then $v = \partial_t w_g$ solves the problem*

$$
\begin{cases}
w_{tt} - c^2 \Delta w = 0 & \mathbf{x} \in \mathbb{R}^3, t > 0 \\
w(\mathbf{x},0) = g(\mathbf{x}), \quad w_t(\mathbf{x},0) = 0 & \mathbf{x} \in \mathbb{R}^3.
\end{cases}
\tag{6.73}
$$

Therefore the solution of (6.71) is given by

$$
u = \partial_t w_g + w_h.
\tag{6.74}
$$

Proof. Let $v = \partial_t w_g$. Differentiating the wave equation with respect to t we have
$$
0 = \partial_t(\partial_{tt} w_g - c^2 \Delta w_g) = (\partial_{tt} - c^2 \Delta)\partial_t w_g = v_{tt} - c^2 \Delta v.
$$

Moreover,
$$
v(\mathbf{x},0) = \partial_t w_g(\mathbf{x},0) = g(\mathbf{x}), \qquad v_t(\mathbf{x},0) = \partial_{tt} w_g(\mathbf{x},0) = c^2 \Delta w_g(\mathbf{x},0) = 0.
$$

Thus, v is a solution of (6.73) and $u = v + w_h$ is the solution of (6.71).   □

The lemma shows that, once the solution of (6.72) is determined, the solution of the complete problem (6.71) is given by (6.74).

Therefore, we focus on the solution of (6.72), first with a special h, given by the three-dimensional Dirac measure at $\mathbf{y}$, $\delta(\mathbf{x} - \mathbf{y})$. For example, in the case of sound waves, this initial data models a sudden change of the air density, concentrated at a point $\mathbf{y}$. If w represents the density variation with respect to a static atmosphere, then w solves the problem

$$
\begin{cases}
w_{tt} - c^2 \Delta w = 0 & \mathbf{x} \in \mathbb{R}^3, t > 0 \\
w(\mathbf{x},0) = 0, \quad w_t(\mathbf{x},0) = \delta(\mathbf{x} - \mathbf{y}) & \dot{\mathbf{x}} \in \mathbb{R}^3.
\end{cases}
\tag{6.75}
$$

The solution of (6.75), which we denote by $K(\mathbf{x},\mathbf{y},t)$, is called **fundamental solution** of the three-dimensional wave equation. To solve (6.75) we use ... *the heat equation* (!), approximating the Dirac measure with the *fundamental* solution of the three-dimensional diffusion equation. Indeed, from Section 3.3.5, (choosing $t = \varepsilon$, $D = 1$, $n = 3$) we know that

$$\Gamma\left(\mathbf{x} - \mathbf{y},\varepsilon\right) = \frac{1}{(4\pi\varepsilon)^{3/2}}\exp\left\{-\frac{|\mathbf{x} - \mathbf{y}|^2}{4\varepsilon}\right\} \rightarrow \delta\left(\mathbf{x} - \mathbf{y}\right)$$

as $\varepsilon \to 0$. Denote by w_ε the solution of (6.75) with $\delta\left(\mathbf{x} - \mathbf{y}\right)$ replaced by $\Gamma\left(\mathbf{x} - \mathbf{y},\varepsilon\right)$. Since $\Gamma\left(\mathbf{x} - \mathbf{y},\varepsilon\right)$ is radially symmetric with pole at $\mathbf{y}$, we expect that w_ε shares the same type of symmetry and is a spherical wave of the form $w_\varepsilon = w_\varepsilon\left(r,t\right)$, $r = |\mathbf{x} - \mathbf{y}|$. Thus, from (6.51) we may write

$$w_\varepsilon\left(r,t\right) = \frac{F\left(r + ct\right)}{r} + \frac{G\left(r - ct\right)}{r}. \tag{6.76}$$

The initial conditions require

$$F\left(r\right) + G\left(r\right) = 0 \qquad \text{and} \qquad c(F'\left(r\right) - G'\left(r\right)) = r\Gamma\left(r,\varepsilon\right)$$

or

$$F = -G \qquad \text{and} \qquad G'\left(r\right) = -r\Gamma\left(r,\varepsilon\right)/2c.$$

Integrating the second relation yields

$$G\left(r\right) = -\frac{1}{2c(4\pi\varepsilon)^{3/2}}\int_0^r s\exp\left\{-\frac{s^2}{4\varepsilon}\right\}ds = \frac{1}{4\pi c}\frac{1}{\sqrt{4\pi\varepsilon}}\left(\exp\left\{-\frac{r^2}{4\varepsilon}\right\} - 1\right)$$

and finally

$$w_\varepsilon\left(r,t\right) = \frac{1}{4\pi cr}\left\{\frac{1}{\sqrt{4\pi\varepsilon}}\exp\left\{-\frac{(r - ct)^2}{4\varepsilon}\right\} - \frac{1}{\sqrt{4\pi\varepsilon}}\exp\left\{-\frac{(r + ct)^2}{4\varepsilon}\right\}\right\}.$$

Now observe that the function

$$\tilde{\Gamma}\left(r, \varepsilon\right) = \frac{1}{\sqrt{4\pi\varepsilon}}\exp\left\{-\frac{r^2}{4\varepsilon}\right\}$$

is the fundamental solution of the one-dimensional diffusion equation with $x = r$ and $t = \varepsilon$. Letting $\varepsilon \to 0$ we find[23]

$$w_\varepsilon\left(r,t\right) \rightarrow \frac{1}{4\pi cr}\left\{\delta(r - ct) - \delta(r + ct)\right\}.$$

Since $r + ct > 0$ for every $t > 0$, we deduce that $\delta(r + ct) = 0$ and therefore we conclude that

$$K\left(\mathbf{x},\mathbf{y},t\right) = \frac{\delta(r - ct)}{4\pi cr} \qquad r = |\mathbf{x} - \mathbf{y}|. \tag{6.77}$$

---

[23] Here δ is one dimensional.

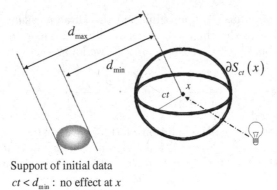

Support of initial data
$ct < d_{min}$: no effect at x

Fig. 6.7. Strong Huygen's principle

Thus, the fundamental solution is an *outgoing travelling wave*, initially concentrated at $\mathbf{y}$ and thereafter on

$$\partial B_{ct}\left(\mathbf{y}\right) = \left\{\mathbf{x} : \left|\mathbf{x} - \mathbf{y}\right| = ct\right\}.$$

The union of the surfaces $\partial B_{ct}\left(\mathbf{y}\right)$ is called the **support** of K and coincides with *the forward space-time cone, with vertex at* $\left(\mathbf{y}, 0\right)$ *and opening* $\theta = \tan^{-1} c$, given by

$$C_{\mathbf{y},0}^* = \left\{\left(\mathbf{x}, t\right) : \left|\mathbf{x} - \mathbf{y}\right| = ct,\, t > 0\right\}.$$

In the terminology of Section 6.4, $C_{\mathbf{y},0}^*$ constitutes the **range of influence of the point y**.

The fact that the range of influence of the point $\mathbf{y}$ is only the *boundary* of the forward cone and *not the full* cone has important consequences on the nature of the disturbances governed by the three-dimensional wave equation. The most striking phenomenon is that a perturbation generated at time $t = 0$ by a point source placed at $\mathbf{y}$ is felt at the point $\mathbf{x}_0$ **only at time** $t_0 = \left|\mathbf{x}_0 - \mathbf{y}\right| / c$ (Fig. 6.7). This is known as *strong Huygens' principle* and explains why *sharp signals* are propagated from a point source.

We will shortly see that this is not the case in two dimensions.

6.7.2 The Kirchhoff formula

Using the fundamental solution, we may derive a formula for the solution of (6.72) with a general h. Since

$$h\left(\mathbf{x}\right) = \int_{\mathbb{R}^3} \delta\left(\mathbf{x} - \mathbf{y}\right) h\left(\mathbf{y}\right) d\mathbf{y},$$

we may see h as a superposition of impulses $\delta\left(\mathbf{x} - \mathbf{y}\right) h\left(\mathbf{y}\right)$ localized at $\mathbf{y}$, of strength $h\left(\mathbf{y}\right)$. Accordingly, the solution of (6.72) is given by the superposition of the corresponding solutions $K(\mathbf{x}, \mathbf{y}, t) \, h(\mathbf{y})$, that is

$$
w_h(\mathbf{x}, t) = \int_{\mathbb{R}^3} K(\mathbf{x}, \mathbf{y}, t) \, h(\mathbf{y}) \, d\mathbf{y} = \int_{\mathbb{R}^3} \frac{\delta(|\mathbf{x} - \mathbf{y}| - ct)}{4\pi c \, |\mathbf{x} - \mathbf{y}|} \, h(\mathbf{y}) \, d\mathbf{y} =
$$

$$
= \int_0^\infty \frac{\delta(r - ct)}{4\pi c r} \, dr \int_{\partial B_r(\mathbf{x})} h(\boldsymbol{\sigma}) \, d\sigma = \frac{1}{4\pi c^2 t} \int_{\partial B_{ct}(\mathbf{x})} h(\boldsymbol{\sigma}) \, d\sigma
$$

where we have used the formula

$$
\int_0^\infty \delta(r - ct) \, f(r) \, dr = f(ct).
$$

Lemma 6.1 and the above intuitive argument lead to the following theorem:

Theorem 6.3 (Kirchhoff's formula). *Let $g \in C^3(\mathbb{R}^3)$ and $h \in C^2(\mathbb{R}^3)$. Then,*

$$
u(\mathbf{x}, t) = \frac{\partial}{\partial t} \left[\frac{1}{4\pi c^2 t} \int_{\partial B_{ct}(\mathbf{x})} g(\boldsymbol{\sigma}) \, d\sigma \right] + \frac{1}{4\pi c^2 t} \int_{\partial B_{ct}(\mathbf{x})} h(\boldsymbol{\sigma}) \, d\sigma \qquad (6.78)
$$

is the unique solution $u \in C^2(\mathbb{R}^3 \times [0, +\infty))$ of problem (6.71).

According to (6.78), $u(\mathbf{x}, t)$ depends upon the data g and h only on the surface

$$
\partial B_{ct}(\mathbf{x}) = \{|\boldsymbol{\sigma} - \mathbf{x}_0| = ct_0\}
$$

which therefore coincides with **the domain of dependence for** *(x,t).* This surface is the intersection of the space-time cone

$$
C_{\mathbf{x}_0, t_0} = \{|\boldsymbol{\sigma} - \mathbf{x}_0| = c(t_0 - t)\})
$$

with the hyperplane $t = 0$. This cone has vertex at $(\mathbf{x}_0, t_0)$ and extends backward in time (*backward characteristic cone*).

Assume now that the support of g and h is the compact set D. Then $u(\mathbf{x}, t)$ is different from zero only for $t_{min} < t < t_{max}$ where t_{min} and t_{max} are the *first* and the *last* time t such that $D \cap \partial B_{ct}(\mathbf{x}) \neq \emptyset$. In other words, a disturbance, initially localized inside D, starts affecting the point $\mathbf{x}$ at time t_{min} and ceases to affect it after time t_{max}. This is another way to express the *strong Huygens' principle*.

Fix t and consider the union of all the spheres $\partial B_{ct}(\boldsymbol{\xi})$ as $\boldsymbol{\xi}$ varies on ∂D. The envelope of these surfaces constitutes the *wave front* and bounds the support of u, which spreads at speed c-.

6.7.3 Cauchy problem in dimension 2

The solution of the Cauchy problem in two dimensions can be obtained from Kirchhoff formula, using the so called *Hadamard's method of descent*. Consider first the problem

$$\begin{cases} w_{tt} - c^2 \Delta w = 0 & \mathbf{x} \in \mathbb{R}^2, \ t > 0 \\ w(\mathbf{x}, 0) = 0, \quad w_t(\mathbf{x}, 0) = h(\mathbf{x}) & \mathbf{x} \in \mathbb{R}^2. \end{cases} \tag{6.79}$$

The key idea is to "immerse" the two-dimensional problem (6.79) in a three-dimensional setting. More precisely, write points in $\mathbb{R}^3$ as $(\mathbf{x}, x_3)$ and set $h(\mathbf{x}, x_3) = h(\mathbf{x})$. The solution U of the three-dimensional problem is given by Kirchhoff formula:

$$U(\mathbf{x}, x_3, t) = \frac{1}{4\pi c^2 t} \int_{\partial B_{ct}(\mathbf{x}, x_3)} h \, d\sigma. \tag{6.80}$$

We claim that, since h does not depend on x_3, U is independent of x_3 as well, and therefore the solution of (6.79) is given by (6.80) with, say, $x_3 = 0$.

To prove the claim, note that the spherical surface $\partial B_{ct}(\mathbf{x}, x_3)$ is a union of the two hemispheres whose equation are

$$y_3 = F_{\pm}(y_1, y_2) = x_3 \pm \sqrt{c^2 t^2 - r^2},$$

where $r^2 = (y_1 - x_1)^2 + (y_2 - x_2)^2$. On both hemispheres we have

$$d\sigma = \sqrt{1 + |\nabla F_{\pm}|^2} \, dy_1 dy_2$$

$$= \sqrt{1 + \frac{r^2}{c^2 t^2 - r^2}} \, dy_1 dy_2 = \frac{ct}{\sqrt{c^2 t^2 - r^2}} \, dy_1 dy_2$$

so that we may write $(d\mathbf{y} = dy_1 dy_2)$

$$U(\mathbf{x}, x_3, t) = \frac{1}{2\pi c} \int_{B_{ct}(\mathbf{x})} \frac{h(\mathbf{y})}{\sqrt{c^2 t^2 - |\mathbf{x} - \mathbf{y}|^2}} d\mathbf{y}$$

and U is independent of x_3 as claimed. From the above calculations and recalling Lemma 6.1 we deduce the following theorem.

Theorem 6.4 (Poisson's formula). *Let* $g \in C^3(\mathbb{R}^2)$ *and* $h \in C^2(\mathbb{R}^2)$. *Then,*

$$u(\mathbf{x}, t) = \frac{1}{2\pi c} \left\{ \frac{\partial}{\partial t} \int_{B_{ct}(\mathbf{x})} \frac{g(\mathbf{y}) \, d\mathbf{y}}{\sqrt{c^2 t^2 - |\mathbf{x} - \mathbf{y}|^2}} + \int_{B_{ct}(\mathbf{x})} \frac{h(\mathbf{y}) \, d\mathbf{y}}{\sqrt{c^2 t^2 - |\mathbf{x} - \mathbf{y}|^2}} \right\}$$

is the unique solution $u \in C^2(\mathbb{R}^2 \times [0, +\infty))$ *of the problem*

$$\begin{cases} u_{tt} - c^2 \Delta u = 0 & \mathbf{x} \in \mathbb{R}^2, t > 0 \\ u(\mathbf{x}, 0) = g(\mathbf{x}), \quad u_t(\mathbf{x}, 0) = h(\mathbf{x}) & \mathbf{x} \in \mathbb{R}^2. \end{cases}$$

Poisson's formula displays an important difference with respect to its three-dimensional analogue, Kirkhhoff's formula. In fact *the domain dependence* for the point $(\mathbf{x},t)$ is given by the **full circle** $B_{ct}(\mathbf{x}) = \{\mathbf{y}: |\mathbf{x}-\mathbf{y}| < ct\}$. This entails that a disturbance, initially localized at $\boldsymbol{\xi}$, starts affecting the point $\mathbf{x}$ at time $t_{\min} = |\mathbf{x}-\boldsymbol{\xi}|/c$. However, this effect does not vanish for $t > t_{\min}$, since $\boldsymbol{\xi}$ still belongs to the circle $B_{ct}(\mathbf{x})$ after $t_{\min}$.

It is the phenomenon one may observe by placing a cork on still water and dropping a stone not too far away. The cork remains undisturbed until it is reached by the wave front but its oscillations persist thereafter.

Thus, sharp signals do not exist in dimension two and *the strong Huygens principle does not hold.*

Remark 6.2. An examination of Poisson's formula reveals that the fundamental solution for the two-dimensional wave equation is given by

$$K(\mathbf{x},\mathbf{y},t) = \frac{1}{2\pi c}\frac{\mathcal{H}(ct-r)}{\sqrt{c^2t^2-r^2}}$$

where $r^2 = |\mathbf{x}-\mathbf{y}|$ and $\mathcal{H}$ is the Heaviside function. For $\mathbf{y}$ fixed, its support is the **full** forward space-time cone, with vertex at $(\mathbf{y},0)$ and opening $\theta = \tan^{-1} c$, given by

$$C_{\mathbf{y},0}^* = \{(\mathbf{x},t): |\mathbf{x}-\mathbf{y}| \le ct, t > 0\}.$$

6.7.4 Non homogeneous equation. Retarded potentials

The solution of the non-homogeneous Cauchy problem can be obtained via Duhamel's method. We give the details for $n = 3$ only. By linearity it is enough to derive a formula for the solution of the problem with zero initial data

$$\begin{cases} u_{tt} - c^2\Delta u = f(\mathbf{x},t) & \mathbf{x} \in \mathbb{R}^3, t > 0 \\ u(\mathbf{x},0) = 0, \quad u_t(\mathbf{x},0) = 0 & \mathbf{x} \in \mathbb{R}^3. \end{cases} \tag{6.81}$$

Assume that $f \in C^2(\mathbb{R}^3 \times [0,+\infty))$. For $s \ge 0$ fixed, let $w = w(\mathbf{x},t;s)$ be the solution of the problem

$$\begin{cases} w_{tt} - c^2\Delta w = 0 & \mathbf{x} \in \mathbb{R}^3, t \ge s \\ w(\mathbf{x},s;s) = 0, \quad w_t(\mathbf{x},s;s) = f(\mathbf{x},s) & \mathbf{x} \in \mathbb{R}^3. \end{cases}$$

Since the wave equation is invariant under time translations, w is given by Kirkhhoff's formula with t replaced by $t - s$

$$w(\mathbf{x},t;s) = \frac{1}{4\pi c^2(t-s)}\int_{\partial B_{c(t-s)}(\mathbf{x})} f(\boldsymbol{\sigma},s)\,d\sigma.$$

Then,

$$u(\mathbf{x},t) = \int_0^t w(\mathbf{x},t;s)\,ds = \frac{1}{4\pi c^2}\int_0^t \frac{ds}{(t-s)}\int_{\partial B_{c(t-s)}(\mathbf{x})} f(\boldsymbol{\sigma},s)\,d\sigma \quad (6.82)$$

is the unique solution $u \in C^2\left(\mathbb{R}^3 \times [0,+\infty)\right)$ of $(6.81)^{24}$.

Formula (6.82) shows that $u(\mathbf{x},t)$ depends on the values of f in the full **backward** cone

$$\mathbf{C}_{\mathbf{x},t} = \{(\mathbf{z},s): |\mathbf{z}-\mathbf{x}| \le c(t-s),\, 0 \le s \le t\}.$$

Note that (6.82) may be written in the form

$$u(\mathbf{x},t) = \frac{1}{4\pi}\int_{B_{ct}(\mathbf{x})} \frac{1}{|\mathbf{x}-\mathbf{y}|} f\left(\mathbf{y},t - \frac{|\mathbf{x}-\mathbf{y}|}{c}\right) d\mathbf{y} \quad (6.83)$$

which is a so called *retarded potential*. Indeed, $u(\mathbf{x},t)$ depends on the values of the source f at the earlier times

$$t' = t - \frac{|\mathbf{x}-\mathbf{y}|}{c}.$$

6.8 Numerical methods

6.8.1 Numerical approximation of the one-dimensional wave equation

We consider the homogeneous Cauchy-Dirichlet wave equation problem on the unit interval,

$$\begin{cases} u_{tt} - c^2 u_{xx} = 0 & 0 < x < 1,\ t > 0 \\ u(0,t) = u(1,t) = 0 & t > 0 \\ u(x,0) = g(x), \quad u_t(x,0) = h(x) & 0 \le x \le 1. \end{cases} \quad (6.84)$$

D'Alembert formula shows that the wave equation is related to a system of conservation laws. More precisely, the solution of the wave equation can be obtained as the superposition of propagating waves in opposite directions. The same conclusion can be obtained from a different point of view that may be helpful for the numerical approximation. In particular, we consider the change of variables $w_1 = u_x$ and $w_2 = u_t$, that combined with the identity of mixed derivatives leads to reformulate the wave equation as follows,

$$\begin{cases} u_{xt} = u_{tx} \\ u_{tt} - c^2 u_{xx} = 0 \end{cases} \Rightarrow \begin{cases} \partial_t w_1 - \partial_x w_2 = 0 \\ \partial_t w_2 - c^2 \partial_x w_1 = 0 \end{cases}$$

---

[24] Check it, mimicking the proof in dimension one (Section 6.4.2).

that is, $\mathbf{w}(x,t) = [w_1(x,t), w_2(x,t)] : (0,1) \times \mathbb{R}^+ \to \mathbb{R}^2$ satisfies a system of linear conservation laws,

$$\mathbf{w}_t - \mathbf{Z}\mathbf{w}_x = 0 \quad \text{where} \quad \mathbf{Z} = \begin{bmatrix} 0 & -1 \\ -c^2 & 0 \end{bmatrix}. \tag{6.85}$$

Since matrix $\mathbf{Z}$ features real eigenvalues $\mu_{1,2} = \pm c$, diagonalization allows to reformulate (6.85) as a system of independent scalar equations, whose solutions propagate along the real line with opposite velocities $\pm c$. These considerations suggest that the discretization of the wave equation may be addressed by means of the methods previously used for the approximation of scalar conservation laws, provided that (6.85) could be complemented by initial and boundary conditions compatible with the ones used for the original problem. While the initial conditions for w_1 and w_2 are immediately translated from the initial state of $u(t,x)$,

$$w_2(x,0) = u_t(x,0) = h(x), \quad w_1(x,0) = u_x(x,0) = g'(x)$$

the derivation of suitable boundary conditions represents the main limitation of this approach, because the characteristic variables w_1 and w_2 only depend on the space and time derivatives of the primal unknown u. As a result of that, for Cauchy-Dirichlet problems it is more convenient to address the discretization of the wave equation in the form (6.84).

We apply a finite difference discretization method defined on the nodes $(x_i, t^n) = (h \cdot i, \tau \cdot n)$ for $i = 0, \ldots, N$, $n \in \mathbb{N}$, uniformly distributed in the domain $(0,1) \times \mathbb{R}^+$ with space and time discretization steps given by h and τ, respectively. Using the usual centered difference quotients for both space and time second derivatives,

$$u_{tt}(x_i, t^n) = \frac{1}{\tau^2}\left(u(x_i, t^{n+1}) - 2u(x_i, t^n) + u(x_i, t^{n-1})\right) + \mathcal{O}(\tau^2)$$

$$u_{xx}(x_i, t^n) = \frac{1}{h^2}\left(u(x_{i+1}, t^n) - 2u(x_i, t^n) + u(x_{i-1}, t^n)\right) + \mathcal{O}(h^2)$$

and replacing them into the wave equation we obtain the so called **leapfrog scheme**, where $\lambda = \tau/h$,

$$\begin{cases} u_i^{n+1} - 2u_i^n + u_i^{n-1} = (c\lambda)^2\left(u_{i+1}^n - 2u_i^n + u_{i-1}^n\right) & i = 1, \ldots, N-1,\ n > 1 \\ u_0^{n+1} = u_N^{n+1} = 0 & n > 1 \\ u_i^0 = g(x_i),\ u_i^1 = u_i^0 + \tau h(x_i) & i = 0, \ldots, N. \end{cases}$$

$$\tag{6.86}$$

Leapfrog scheme is an explicit method, because u_i^{n+1} depends on the solution at the previous iterative steps solely, and not on solution values at t^{n+1} in the neighboring nodes, and it is locally second order accurate. Furthermore this scheme is prone to a straightforward implementation of Dirichlet boundary conditions. Conversely, the approximation of initial conditions is more delicate. In equation (6.86) we have applied a first order one step scheme to compute the initial values u_i^0 and u_i^1 on the basis of the data $g(x)$, $h(x)$. Given u_i^0 and u_i^1 in all nodes, the leapfrog scheme allows us to compute the numerical solution at any time level t^n. However, this solution strategy is slightly suboptimal, because the first order schemes used to initialize the method pollute its global second order accuracy. Furthermore, as for all explicit schemes, the correct application of the leapfrog method requires to satisfy some stability condition that restricts the range of space and time steps to be used. In particular, the scheme must satisfy a **CFL condition** such that $|c\lambda| \leq 1$. To overcome the restrictions due to conditional stability, one may switch to an implicit scheme.

The **Newmark scheme** is a popular implicit discretization method for the wave equation,

$$\begin{cases} u_i^{n+1} - 2u_i^n + u_i^{n-1} = \frac{(c\lambda)^2}{4}\left(z_i^{n+1} + 2z_i^n + z_i^{n-1}\right) \\ z_i^n = u_{i+1}^n - 2u_i^n + u_{i-1}^n \end{cases} \qquad (6.87)$$

complemented by the same initial and boundary conditions already used in (6.86). Since the term $z_i^{n+1} = u_{i+1}^{n+1} - 2u_i^{n+1} + u_{i-1}^{n+1}$ couples the unknowns on different nodes at the new time step t^{n+1}, using the Newmark scheme requires to solve a linear system of equations at each time step, in order to simultaneously determine all u_i^{n+1}, $i = 1,\dots,N-1$. The resulting additional computational cost is balanced by the **unconditional stability** of the scheme. Finally, the following property

$$z(x_i, t^n) = \frac{1}{4}\left(z(x_i, t^{n-1}) + 2z(x_i, t^n) + z(x_i, t^{n+1})\right) + \mathcal{O}(\tau^2)$$

shows that the Newmark method is second order accurate. Owing to its good stability properties, the forthcoming numerical tests on the wave equation are performed using the Newmark method.

6.9 Exercises

6.1 (The violin string modeling). We analyse the Cauchy-Dirichlet problem

$$\begin{cases} u_{tt} - c^2 u_{xx} = 0 & 0 < x < L, t > 0 \\ u(0,t) = u(L,t) = 0 & t \geq 0 \\ u(x,0) = g(x), u_t(x,0) = 0 & 0 \leq x \leq L \end{cases}$$

with $c^2 = \tau/\rho_0$ costant, describing the vibrating violin string. Write the formula for the solution using the separation of the variables technique.

6.2. Find the characteristics of the Tricomi equation $u_{tt} - tu_{xx} = 0$.

6.3. Classify the equation

$$u_{xx} - yu_{yy} - \frac{1}{2}u_y = 0$$

and find its characteristics. Rewrite it in canonical form and find its general solution.

6.4. Find the characteristics of the equation

$$t^2 u_{tt} + 2tu_{xt} + u_{xx} - u_x = 0.$$

Write it in canonical form and find its general solution.

6.5. Consider the problem

$$\begin{cases} u_{tt} - u_{xx} = 0 & (0,1) \times (0,+\infty) \\ u(x,0) = 1 - x & [0,1] \\ u_t(x,0) = 0 & [0,1] \\ u_x(0,t) = u_x(1,t) = 0 & (0,+\infty). \end{cases}$$

Find the solution and prove that it satisfies the following condition

$$\int_0^1 (|u_t(x,t)|^2 + |u_x(x,t)|^2)dx = 1.$$

6.6. Consider the problem

$$\begin{cases} u_{tt} - u_{xx} = 0 & (0,1) \times (0,+\infty) \\ u(x,0) = 0 & [0,1] \\ u_t(x,0) = x & [0,1] \\ u(0,t) = u(1,t) = 0 & (0,+\infty). \end{cases}$$

Find the solution. Is it continuous? Calculate the energy associated to the solution.

6.7 (Forced vibrations). Solve the problem

$$\begin{cases} u_{tt} - u_{xx} = g(t)\sin x & (0,\pi) \times (0,+\infty) \\ u(x,0) = u_t(x,0) = 0 & [0,\pi] \\ u(0,t) = u(\pi,t) = 0 & (0,+\infty). \end{cases}$$

6.8 (Circular membrane). A perfectly flexible and elastic membrane at rest has a shape of the circle $B_1 = \{(x,y) : x^2 + y^2 \le 1\}$. If the boundary is fixed and there are no external loads, the vibrations of the membrane are governed by the following system

$$\begin{cases} u_{tt} - c^2(u_{rr} + r^{-1}u_r + r^{-2}u_{\theta\theta}) = 0 & 0 < r < 1, 0 \le \theta \le 2\pi, t > 0 \\ u(r,\theta,0) = g(r,\theta), u_t(r,\theta,0) = h(r,\theta) & 0 < r < 1, 0 \le \theta \le 2\pi \\ u(1,\theta,t) = 0 & 0 \le \theta \le 2\pi, t > 0. \end{cases}$$

In the case $h = 0$ and $g = g(r)$, use the method of separation of the variables to find the solution.

236 6 Waves and vibrations

6.9. Check that the formula (6.22) may be written in the following form

$$u(x+c\xi-c\eta,t+\xi+\eta)-u(x+c\xi,t+\xi)-u(x-c\eta,t+\eta)+u(x,t)=0. \quad (6.88)$$

Show that if u is a C^2 function and satisfies (6.88), then $u_{tt}-c^2u_{xx}=0$.

6.10 (Membrane with dissipation). Consider the problem

$$\begin{cases} u_{tt}(\mathbf{x},t)+ku_t(\mathbf{x},t)-c^2\Delta u(\mathbf{x},t)=0 & \mathbf{x}\in\mathbb{R}^2, t>0 \\ u(\mathbf{x},t)=0, u_t(\mathbf{x},t)=g(\mathbf{x}) & \mathbf{x}\in\mathbb{R}^2. \end{cases}$$

a) Find $\alpha\in\mathbb{R}$ such that the function

$$v(\mathbf{x},t)=e^{\alpha t}u(\mathbf{x},t)$$

solves an equation without first order terms (but containing a zero degree term) in $\mathbb{R}^2\times(0,+\infty)$.

b) Find $\beta\in\mathbb{R}$ such that the function

$$w(x_1,x_2,x_3,t)=w(\mathbf{x},x_3,t)=e^{\beta x_3}v(\mathbf{x},t)$$

solves an equation with second order terms only in $\mathbb{R}^3\times(0,+\infty)$.

6.9.1 Numerical simulation of a vibrating string

We apply Newmark scheme for studying the behavior of a vibrating string in different conditions. We model the string with equations (6.84).

First, we use d'Alembert formula to build up reference exact solutions and compare them with the numerical approximations. We start from the case where the initial deformed configuration is given and the string is steady. Using (6.20) we obtain that the following $u(x,t)$ is a solution of the problem on the real line,

$$g(x)=\sin(2\pi x),\ h(x)=0 \quad \Rightarrow \quad u(x,t)=\frac{1}{2}\big(\sin(2\pi(x+t))+\sin(2\pi(x-t))\big).$$

We observe that the displacement of the string vanishes at the points $x=i/2$ with $i\in\mathbb{N}$. For this reason, $u(x,t)=\big(\sin(2\pi(x+t))+\sin(2\pi(x-t))\big)/2$ is also a solution of the Cauchy-Dirichlet problem based on (6.84) with initial conditions $g(x)=\sin(2\pi x)$ and $h(x)=0$. The problem is then discretized using Newmark scheme with $h=\tau=0.05$ and the results reported in Fig. 6.8 (left) show the computed displacement of the string. Although the discretization step is not particularly fine, we observe an excellent agreement with the exact solution. Similar conclusions hold true when we study the vibrations that originate from an undeformed configuration with a given initial velocity. According to d'Alembert formula the solution of this problem satisfies

$$g(x)=0,\ h(x)=\sin(2\pi x)\Rightarrow u(x,t)=\frac{1}{4\pi}\big(\cos(2\pi(x-t))-\cos(2\pi(x+t))\big),$$

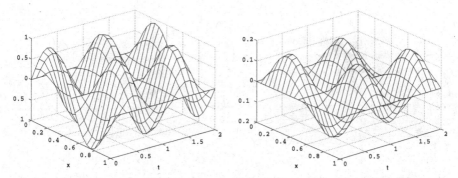

Fig. 6.8. Numerical approximation of (6.84) with initial conditions $g(x) = \sin(2\pi x)$, $h(x) = 0$ (left) and $g(x) = \sin(2\pi x)$, $h(x) = 0$ (right)

with the following roots $x = i/2$ for $i \in \mathbb{N}$. The corresponding numerical approximation on the unit interval is depicted in Fig. 6.8 (right).

Let us now consider a string plucked at a specific point, namely $x = 0$, which is modeled by a continuous but non differentiable initial state given by

$$g(x) = \max[-4|1 - 2x| + 1, 0], \quad h(x) = 0.$$

The formal application of d'Alembert formula suggests that the displacement of the string may be represented as $u(x,t) = \big(g(x+t) + g(x-t)\big)/2$. This is indeed a **generalized solution** of the wave equation. Using a finer computational mesh than for the previous tests, namely using $h = \tau = 0.02$, we approximate this solution with the Newmark scheme. We observe from Fig. 6.9 that the weak singularities (discontinuity of the solution derivatives) propagating along the characteristic lines are smoothed out. Furthermore, tiny wiggles appear in the plateau between the two diverging peaks.

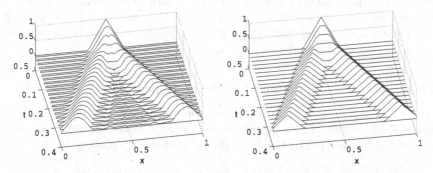

Fig. 6.9. The generalized solution of (6.84) with the initial state $g(x) = \max[-4|1 - 2x| + 1; 0]$ and $h(x) = 0$ (right) and its numerical approximation by the Newmark scheme (left)

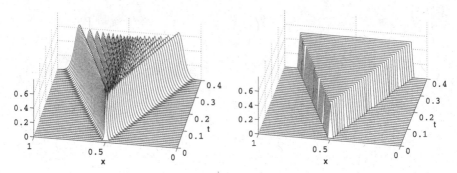

Fig. 6.10. Fundamental (generalized) solution of the wave equation (right) and its numerical approximation by the newmark scheme (left)

Finally, let us consider the case where the string is excited at a single point, as described by the following initial conditions,

$$g(x) = 0, \quad h(x) = \delta(x - 1/2)$$

where $\delta(x)$ denotes an impulse function (a Dirac delta function) located at the origin such that

$$\mathcal{H}(x) = \int_{-\infty}^{x} \delta(y)dy$$

being $\mathcal{H}(x)$ the Heaviside function. Then, the generalized solution of the wave equation originating from this initial state is again identified by means of the formal application of d'Alembert formula, and it is often called the **fundamental solution** of the wave equation in one space dimension,

$$u(x,t) = \frac{1}{2}\big(\mathcal{H}(x+t-1/2) - \mathcal{H}(x-t-1/2)\big).$$

The application of Neumark scheme with $h = \tau = 0.01$ to this case shows the intrinsic limitations of finite difference schemes to approximate solutions with strong singularities (jump discontinuity). Indeed, Fig. 6.10 shows that the spurious oscillations that appear in this case are remarkably larger than the ones observed for weak discontinuities (see Fig. 6.9).

6.9.2 Numerical simulation of a vibrating membrane

Let us consider an elastic membrane on the unit square $\Omega = (0,1) \times (0,1)$ whose displacement is constrained on its boudary. Small vibrations of that system can be modeled by the multi-dimensional wave equation,

$$\begin{cases} u_{tt} = \Delta u & \text{in } \Omega \times \mathbb{R}^+ \\ u = 0 & \text{on } \partial\Omega \times \mathbb{R}^+ \\ u(x,y,0) = g(x,y), \; u_t(x,y,0) = h(x,y) & \text{on } \Omega \end{cases}$$

$$m = 1, \ n = 1$$

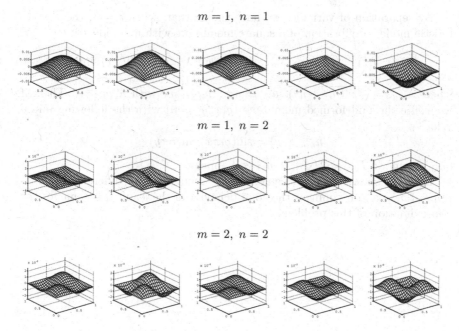

$$m = 1, \ n = 2$$

$$m = 2, \ n = 2$$

Fig. 6.11. Numerical simulation of a vibrating membrane with the leapfrog scheme for different combinations of n, m. Plots correspond to times $t_1 = T_{mn}/8$ to $t_7 = 7T_{mn}/8$ from left to right, where the plots of t_4 and t_5 have been omitted

where for simplicity we consider unitary coefficients, $c = 1$. For the space discretization, we apply the five point scheme addressed in Chapter 4. More precisely, given a uniform computational mesh of Ω with spacing equal to h on each axis, the discretization of $-\Delta u(t)$, complemented with homogeneous Dirichlet boundary conditions, is $(1/h^2)\mathbf{A}\mathbf{U}(t)$ where $\mathbf{A} \in \mathbb{R}^{N \times N}$ is a symmetric band matrix whose coefficients do not depend on h and $\mathbf{U}(t) \in \mathbb{R}^N$ is the vector that gathers the nodal approximations of $u(x, y, t)$, given a suitable ordering of the nodes. For time stepping, to avoid solving a linear system governed by $\mathbf{A}$ at each time step, we apply the leapfrog scheme, because it is explicit. We carefully select the uniform time discretization step, τ, in order to satisfy the CFL stability condition, namely $|c\lambda| \leq 1$ where $\lambda = \tau/h$. Then, the fully discrete scheme consists in computing a sequence of vectors $\mathbf{U}_n$ with $n > 1$ such that,

$$\begin{cases} \mathbf{U}_{n+1} = 2\mathbf{U}_n - \mathbf{U}_{n-1} - \lambda^2 \mathbf{A}\mathbf{U}_n \\ \mathbf{U}_0 = \mathbf{G} \quad \mathbf{U}_1 = \mathbf{G} + \tau\mathbf{H} \end{cases}$$

where $\mathbf{G}$ and $\mathbf{H}$ are the vectors gathering the nodal values of the initial displacement and velocity $g(x, y)$ and $h(x, y)$, respectively.

By separation of variables we have shown that, when $c = 1$, the characteristic modes of vibration of a square membrane with unit side are

$$u_{mn}(x, y, t) = \left(a_{mn} \cos(\lambda_{mn}t) + b_{mn} \sin(\lambda_{mn}t)\right) \sin(\mu_m x) \sin(\nu_n y)$$

where $\lambda_{mn} = \sqrt{\pi^2(m^2 + n^2)}$, $\mu_m = m\pi$, $\nu_n = n\pi$ with $n, m \in \mathbb{N}$. Then, if we excite the undeformed membrane $(g(x, y) = 0)$ with the following initial velocity

$$h_{mn}(x, y) = \sin(m\pi x) \sin(n\pi y)$$

only the mode $T_{mn} = 2/\sqrt{n^2 + m^2}$ will appear.

Fig. 6.11 shows the displacement of the membrane for different combinations of n, m and it confirms the good behavior of the leapfrog scheme for the approximation of this problem.

Part II
Functional Analysis Techniques for Differential Problems

7

Elements of Functional Analysis

The main purpose in the previous chapters has been to introduce part of the basic and classical theory of some important equations of mathematical physics. The emphasis on phenomenological aspects and the connection with a probabilistic point of view should have conveyed to the reader some intuition and feeling about the interpretation and the limits of those models.

However, these purposes are somehow in competition with one of the most important role of modern mathematics, which is to reach a unifying vision of large classes of problems under a common structure, capable not only of increasing theoretical understanding, but also of providing the necessary flexibility to guide the numerical methods which will be used to compute approximate solutions.

This conceptual jump requires a change of perspective, based on the introduction of abstract methods, historically originating from the vain attempts to solve basic problems (e.g. in electrostatics) at the end of the 19th century. It turns out that the new level of knowledge opens the door to the solution of complex problems in modern technology.

These abstract methods, in which analytical and geometrical aspects fuse, are the core of the branch of Mathematics, called Functional Analysis.

In this chapter, after a brief introduction to the Lebesgue integral and some of the functional spaces associated to it, we develop the tools of Functional Analysis, essential for a correct variational formulation of a wide variety of boundary value problems; in particular, we introduce the basic elements of the Hilbert spaces theory. The results we present constitute the theoretical basis for numerical methods such as *finite elements* or more generally, *Galerkin's methods*, and this makes the theory even more attractive and important.

Salsa S., Vegni F.M.G., Zaretti A., Zunino P.: *A Primer on PDEs. Models, Methods, Simulations.*
Unitext – La Matematica per il 3+2 65.
DOI 10.1007/978-88-470-2862-3_7, © Springer-Verlag Italia 2013

7.1 Lebesgue Measure and Integral

7.1.1 A counting problem

Two persons, that we denote by $\mathcal{R}$ ed $\mathcal{L}$, must compute the total value of M coins, ranging from 1 to 50 cents. $\mathcal{R}$ decides to group the coins arbitrarily in piles of, say, 10 coins each, then to compute the value of each pile and finally to sum all these values. $\mathcal{L}$, instead, decides to partition the coins according to their value, forming piles of 1-cent coins, of 5-cents coins and so on. Then he computes the value of each pile and finally sums all their values.

In more analytical terms, let

$$V : M \to \mathbb{N}$$

a *value function* that associates to each element of M (i.e. each coin) its value. $\mathcal{R}$ partitions the **domain** of V in disjoint subsets, sums the values of V in such subsets and then sums everything. $\mathcal{L}$ considers each point p in the **image** $\mathbb{N}$ of V (the value of a single coin), considers the inverse image $V^{-1}(p)$ (the pile of coins with the same value p), computes the corresponding value and finally sums over every p.

These two ways of counting correspond to the strategy behind the definitions of the integrals of Riemann and Lebesgue, respectively. Since V is defined on a discrete set and is integer valued, in both cases there is no problem in summing its values and the choice is determined by an efficiency criterion. Usually, the method of $\mathcal{L}$ is more efficient.

In the case of a real (or complex) function f, the "sums of its values" corresponds to an integration of f. While the construction of $\mathcal{R}$ remains rather elementary, the one of $\mathcal{L}$ turns out to be more complex and requires new tools. Let us examine the particular case of a *bounded* and *positive* function, defined on an interval $[a, b] \subset \mathbb{R}$. Thus, let

$$f : [a, b] \to [\inf f, \sup f].$$

To construct the Riemann integral, we partition $[a, b]$ in subintervals $I_1, \ldots, I_N$ (the piles of $\mathcal{R}$), then we choose in each interval I_k a point ξ_k and we compute $f(\xi_k)\, l(I_k)$, where $l(I_k)$ is the length of I_k (i.e. the value of the $k - th$ pile). Now we sum the values $f(\xi_k)\, l(I_k)$ and set

$$(\mathcal{R}) \int_a^b f = \lim_{\delta \to 0} \sum_{k=1}^N f(\xi_k)\, l(I_k),$$

where $\delta = \max \{l(I_1), \ldots, l(I_N)\}$. If the limit is finite and moreover is independent of the choice of the points ξ_k, then this limit defines the Riemann integral of f in $[a, b]$.

Now, let us examine the Lebesgue strategy. This time we partition the interval $[\inf f, \sup f]$ in subintervals $[y_{k-1}, y_k]$ (the values of each coin for $\mathcal{L}$)

with

$$\inf f = y_0 < y_1 < \ldots < y_{N-1} < y_N = \sup f.$$

Then we consider the inverse images $E_k = f^{-1}\left([y_{k-1}, y_k]\right)$ (the piles of homogeneous coins) and we would like to compute their ... *length*. However, in general E_k is *not* an interval or a union of intervals and, in principle, it could be a very irregular set so that it is not clear what is the length of E_k.

Thus, the need arises to associate with every E_k a *measure,* which replaces the length when E_k is an irregular set. This leads to the introduction of the *Lebesgue measure* of (practically every) set E, denoted by $|E|$.

Once we know how to measure E_k (the number of coins in the $k-th$ pile), we choose an arbitrary point $\overline{\alpha}_k \in [y_{k-1}, y_k]$ and we compute $\overline{\alpha}_k |E_k|$ (the value of the $k-th$ pile). Then, we sum all the values $\overline{\alpha}_k |E_k|$ and set

$$(L) \int_a^b f = \lim_{\rho \to 0} \sum_{k=1}^{N} \overline{\alpha}_k |E_k|$$

where ρ is the maximum among the lengths of the intervals $[y_{k-1}, y_k]$. It can be seen that under our hypotheses, the limit exists, is finite and is independent of the choice of $\overline{\alpha}_k$. Thus, we may always choose $\overline{\alpha}_k = y_{k-1}$. This remark leads to the definition of the Lebesgue integral in Section 7.1.3: the number $\sum_{k=1}^{N} y_{k-1} |E_k|$ is nothing else that the integral of a *simple function,* which approximates f from below and whose range is the finite set $y_0 < \ldots < y_{N-1}$. The integral of f is the supremum of these numbers.

The resulting theory has several advantages with respect to that of Riemann. For instance, the class of integrable functions is much wider and there is no need to distinguish among bounded or unbounded functions and integration domains.

Especially important are the convergence theorems presented in Section 7.1.4, which allow the possibility of interchanging the operation of limit and integration, under rather mild conditions.

7.1.2 Measures and measurable functions

A measure in a set Ω is a *set function,* defined on a particular class of subsets of Ω called *measurable set* which "behaves well" with respect to union, intersection and complementation. Precisely:

Definition 7.1. *A collection $\mathcal{F}$ of subsets of Ω is called $\sigma-$algebra if:*

(i) $\Omega \in \mathcal{F}$;

(ii) $A \in \mathcal{F}$ *implies* $\Omega \backslash A \in \mathcal{F}$;

(iii) *if* $\{A_k\}_{k \in \mathbb{N}} \subset \mathcal{F}$ *then also* $\cup A_k$ *and* $\cap A_k$ *belong to* $\mathcal{F}$.

Example 7.1. If $\Omega = \mathbb{R}^n$, the smallest $\sigma-$algebra containing all the open subsets of $\mathbb{R}^n$ is called the *Borel $\sigma-$algebra* and its elements are the *Borel* *sets.* Typical Borel sets are obtained by countable unions and/or intersections of open sets.

Definition 7.2. *Given a* $\sigma-$*algebra* $\mathcal{F}$ *in a set* Ω, *a measure on* $\mathcal{F}$ *is a function*

$$\mu : \mathcal{F} \to \mathbb{R}$$

such that:

(i) $\mu(A) \geq 0$ *for every* $A \in \mathcal{F}$;

(ii) *if* $A_1, A_2, \ldots$ *are pairwise disjoint sets in* $\mathcal{F}$, *then*

$$\mu\left(\cup_{k\geq1}A_k\right) = \sum_{k\geq1}\mu\left(A_k\right) \qquad (\sigma - additivity).$$

The elements of $\mathcal{F}$ *are called measurable sets.*

The Lebesgue measure in $\mathbb{R}^n$ is defined on a $\sigma-algebra$ $\mathcal{M}$ containing the Borel $\sigma-$algebra, through the following theorem.

Theorem 7.1. *There exists in* $\mathbb{R}^n$ *a* $\sigma-$*algebra* $\mathcal{M}$ *and a measure*

$$|\cdot|_n : \mathcal{M} \to [0, +\infty]$$

with the following properties:

1. *Each open and closed set belongs to* $\mathcal{M}$.
2. *If* $A \in \mathcal{M}$ *and* A *has measure zero, every subset of* A *belongs to* $\mathcal{M}$ *and has measure zero.*
3. *If*
$$A = \{\mathbf{x} \in \mathbb{R}^n : a_j < x_j < b_j; j = 1, \ldots, n\}$$
then $|A| = \prod_{j=1}^{n}(b_j - a_j)$.

The elements of $\mathcal{M}$ are called *Lebesgue measurable sets* and $|\cdot|_n$ (or simply $|\cdot|$ if no confusion arises) is called the *n-dimensional Lebesgue measure.* Unless explicitly said, from now on, *measurable* means *Lebesgue measurable* and the measure is the Lebesgue measure.

Not every subset of $\mathbb{R}^n$ is measurable. However, the nonmeasurable ones are quite ... pathological[1]!

Remark 7.1. The sets of measure zero are quite important. Here are some examples of sets with zero measure: all countable subsets, e.g. the set $\mathbb{Q}$ of rational numbers; straight lines or smooth curves in $\mathbb{R}^2$; straight lines, hyperplanes, smooth curves and surfaces in $\mathbb{R}^3$.

Notice that a straight line segment has measure zero in $\mathbb{R}^2$ but, of course not in $\mathbb{R}$.

---

[1] See e.g. *Rudin* [34].

We say that a *property holds almost everywhere in* $A \in \mathcal{M}$ (in short, a.e. in A) *if it holds at every point of A except that in a subset of measure zero.*

For instance, the sequence $f_n(x) = \exp\left(-n \sin^2 x\right)$ converges to zero a.e. in $\mathbb{R}$.

The Lebesgue integral is defined for *measurable* functions, characterized by the fact that the inverse image of every closed set is measurable.

Definition 7.3. *Let $A \subseteq \mathbb{R}^n$ be measurable, and $f : A \to \mathbb{R}$. We say that f is measurable if $f^{-1}(C) \in \mathcal{F}$ for any closed set $C \subseteq \mathbb{R}$.*

If f is continuous, is measurable. The sum and the product of a finite number of measurable functions are measurable. The pointwise limit of a sequence of measurable functions is measurable.

If $f : A \to \mathbb{R}$, is measurable, we define its *essential supremum* or *least upper bound* by the formula

$$\operatorname{ess\,sup} f = \inf\left\{K : f \le K \quad \text{a.e. in } A\right\}.$$

Note that, if $f = \chi_{\mathbb{Q}}$, the characteristic function of the rational numbers, we have $\sup f = 1$, but $\operatorname{ess\,sup} f = 0$, since $|\mathbb{Q}| = 0$.

Every measurable function may be approximated by **simple functions**. A function $s : A \subseteq \mathbb{R}^n \to \mathbb{R}$ is said to be **simple** if its range is constituted by a *finite number* of values $s_1, \ldots, s_N$, attained on measurable sets $A_1, \ldots, A_N$, contained in A. Introducing the characteristic functions χ_{A_j}, we may write

$$s = \sum_{j=1}^{N} s_j \chi_{A_j}.$$

We have the following theorem.

Theorem 7.2. *Let $f : A \to \mathbb{R}$, be measurable. There exists a sequence $\{s_k\}$ of simple functions converging pointwise to f in A. Moreover, if $f \ge 0$, we may choose $\{s_k\}$ increasing.*

7.1.3 The Lebesgue integral

We define the Lebesgue integral of a measurable function on a measurable set A. For a simple function $s = \sum_{j=1}^{N} s_j \chi_{A_j}$ we set:

$$\int_A s = \sum_{j=1}^{N} s_j |A_j|$$

with the convention that, if $s_j = 0$ and $|A_j| = +\infty$, then $s_j |A_j| = 0$.

If $f \geq 0$ is measurable, we define

$$\int_A f = \sup \int_A s$$

where the supremum is computed over the set of all simple functions s such that $s \leq f$ in A.

In general, if f is measurable, we write $f = f^+ - f^-$, where $f^+ = \max\{f, 0\}$ and $f^- = \max\{-f, 0\}$ are the positive and negative parts of f, respectively. Then we set:

$$\int_A f = \int_A f^+ - \int_A f^-$$

under the condition that at least one of the two integrals in the right hand side is finite.

If both these integrals are finite, the function f is said to be **integrable** or **summable** in A. From the definition, it follows immediately that a measurable functions f is *integrable if and only if* $|f|$ *is integrable.*[2]

All the functions Riemann integrable in a set A are Lebesgue integrable as well. An interesting example of non integrable function in $(0, +\infty)$ is given by $h(x) = \sin x / x$. In fact[3]

$$\int_0^{+\infty} \frac{|\sin x|}{x} dx = +\infty.$$

On the contrary, it may be proved that

$$\lim_{N \to +\infty} \int_0^N \frac{\sin x}{x} dx = \frac{\pi}{2}$$

and therefore the improper Riemann integral of h is finite.

7.1.4 Some fundamental theorems

The following two theorems are among the most important and useful in the theory of Lebesgue integration.

---

[2] The set of the functions which are Lebesgue integrable in A is denoted with $L^1(A)$; we will be more precise in Section 7.2.1. The set of all the functions which are Lebesgue integrable in every interval (a, b) is denoted by $L^1_{loc}(\mathbb{R})$.

[3] We may write

$$\int_0^{+\infty} \frac{|\sin x|}{x} dx = \sum_{k=1}^\infty \int_{(k-1)\pi}^{k\pi} \frac{|\sin x|}{x} dx \geq \sum_{k=1}^\infty \frac{1}{k\pi} \int_{(k-1)\pi}^{k\pi} |\sin x| \, dx = \sum_{k=1}^\infty \frac{2}{k\pi} = +\infty.$$

Theorem 7.3 (Dominated Convergence Theorem). *Let* $\{f_k\}$ *be a sequence of summable functions in* A *such that* $f_k \to f$ *a.e. in* A. *If there exists* $g \geq 0$, *summable in* A *and such that* $|f_k| \leq g$ *a.e. in* A, *then* f *is summable and*

$$\lim_{k \to \infty} \int_A f_k = \int_A f.$$

Theorem 7.4 (Monotone Convergence Theorem). *Let* $\{f_k\}$ *be a sequence of nonnegative, measurable functions in* A *such that*

$$f_1 \leq f_2 \leq \ldots \leq f_k \leq f_{k+1} \leq \ldots .$$

Then

$$\lim_{k \to \infty} \int_A f_k = \int_A \lim_{k \to \infty} f_k.$$

Example 7.2. A typical situations we often encounter in this book is the following. Let $f \in L^1(A)$ and, for $\varepsilon > 0$, set $A_\varepsilon = \{|f| > \varepsilon\}$. Then, we have

$$\int_{A_\varepsilon} f \to \int_A f \quad \text{as } \varepsilon \to 0.$$

This follows from Theorem 7.4 since, for every sequence $\varepsilon_j \to 0$, we have $|f| \chi_{A_{\varepsilon_j}} \leq |f|$ and therefore

$$\int_{A_{\varepsilon_j}} f = \int_A f \chi_{A_{\varepsilon_j}} \to \int_A f \quad \text{as } \varepsilon_j \to 0.$$

An important fact is that any summable function may be approximated by continuous function with compact support.

Theorem 7.5. *Let* $f \in L^1(A)$. *Then, for every* $\delta > 0$, *there exists a continuous function* $g \in C_0(A)$ *such that*

$$\int_A |f - g| < \delta.$$

The fundamental theorem of calculus extends to the Lebesgue integral in the following form:

Theorem 7.6 (Differentiation). *Let* $f \in L^1_{loc}(\mathbb{R})$. *Then*

$$\frac{d}{dx} \int_a^x f(t)\, dt = f(x) \qquad \text{a.e. in } x \in \mathbb{R}.$$

7.2 Hilbert Spaces

In the next sections we develop some elements of the theory of Hilbert spaces. This is the natural setting for the formulation and the resolution (both theoretical and numerical) of boundary value problems.

7.2.1 Normed spaces

We briefly recall the notion of *normed space*. Let X be a linear space over the scalar field $\mathbb{R}$ or $\mathbb{C}$. A *norm* in X, is a real function

$$\|\cdot\| : X \to \mathbb{R} \tag{7.1}$$

such that, for each scalar λ and every $x, y \in X$, the following properties hold:

1. $\|x\| \geq 0$; $\|x\| = 0$ if and only if $x = 0$ (*positivity*);
2. $\|\lambda x\| = |\lambda| \, \|x\|$ (*homogeneity*);
3. $\|x + y\| \leq \|x\| + \|y\|$ (*triangular inequality*).

A norm is introduced to measure the size (or the "length") of each vector $x \in X$, so that properties 1, 2, 3 should appear as natural requirements.

A *normed space* is a linear space X endowed with a norm $\|\cdot\|$. With a norm is associated the *distance* between two vectors given by

$$d(x, y) = \|x - y\|$$

which makes X a *metric space* and allows to define a notion of convergence in a very simple way.

We say that a sequence $\{x_n\} \subset X$ *converges* to x in X, and we write $x_m \to x$ in X, if

$$d(x_m, x) = \|x_m - x\| \to 0 \qquad \text{as } m \to \infty.$$

An important distinction is between convergent sequences and *Cauchy* sequences. A sequence $\{x_m\} \subset X$ is a *Cauchy* sequence if

$$d(x_m, x_n) = \|x_m - x_n\| \to 0 \qquad \text{as } m, n \to \infty.$$

If $x_m \to x$ in X, from the triangular inequality, we may write

$$\|x_m - x_n\| \leq \|x_m - x\| + \|x_n - x\| \to 0 \qquad \text{as } m, n \to \infty$$

and therefore we deduce

$$\{x_m\} \text{ is convergent } \textbf{implies} \; \{x_m\} \text{ is a Cauchy sequence.} \tag{7.2}$$

The converse in not true, in general. Take $X = \mathbb{Q}$, with the usual norm given by $|x|$. The sequence of rational numbers

$$x_m = \left(1 + \frac{1}{m}\right)^m$$

is a Cauchy sequence but it is *not* convergent in $\mathbb{Q}$, since its limit is the irrational number e.

A normed space in which every Cauchy sequence converges is called **complete** and deserves a special name.

Definition 7.4. *A complete, normed linear space is called* **Banach space**.

The notion of convergence (or of limit) can be extended to functions from a normed space into another, always reducing it to the convergence of distances, that are real functions.

Let X, Y linear spaces, endowed with the norms $\|\cdot\|_X$ and $\|\cdot\|_Y$, respectively, and let $F : X \to Y$. We say that F is continuous at $x \in X$ if

$$\|F(y) - F(x)\|_Y \to 0 \qquad \text{when } \|y - x\|_X \to 0$$

or, equivalently, if, for every sequence $\{x_m\} \subset X$,

$$\|x_m - x\|_X \to 0 \quad \text{implies} \quad \|F(x_m) - F(x)\|_Y \to 0 \quad \text{as } m \to \infty.$$

F is *continuous in* X if it is *continuous at every* $x \in X$. In particular:

Proposition 7.1. *Every norm in a linear space* X *is continuous in* X.

Proof. Let $\|\cdot\|$ be a norm in X. From the triangular inequality, we may write

$$\|y\| \leq \|y - x\| + \|x\| \quad \text{and} \quad \|x\| \leq \|y - x\| + \|y\|$$

whence

$$|\|y\| - \|x\|| \leq \|y - x\|.$$

Thus, if $\|y - x\| \to 0$ then $|\|y\| - \|x\|| \to 0$, which is the continuity of the norm. $\qquad\qquad\square$

Definition 7.5. *Two norms* $\|\cdot\|_1$ *and* $\|\cdot\|_2$ *defined in* X *are* **equivalent** *if there exist two positive numbers* c_1 *and* c_2 *such that*

$$c_1 \|x\|_2 \leq \|x\|_1 \leq c_2 \|x\|_2 \quad \text{for every } x \in X.$$

A sequence $\{x_n\} \subset X$ is fundamental with respect to $\|\cdot\|_1$ if and only if is fundamental with respect to $\|\cdot\|_2$. In particular, X is complete with respect to $\|\cdot\|_1$ if and only if it is complete with respect to $\|\cdot\|_2$.

Some examples are in order.

Spaces of continuous functions. Let $X = C(A)$ be the set of (real or complex) continuous functions on A, where A is a compact subset of $\mathbb{R}^n$, endowed with the norm (called *maximum norm*)

$$\|f\|_{C(A)} = \max_A |f|.$$

A sequence $\{f_m\}$ converges to f in $C(A)$ if

$$\max_A |f_m - f| \to 0 \quad \text{as } m \to \infty,$$

that is, if f_m *converges uniformly to* f in A. Since a uniform limit of continuous functions is continuous, $C(A)$ is a Banach space.

Note that other norms may be introduced in $C\left(A\right)$, for instance the *least squares*[4] or $L^2\left(A\right)$ *norm*

$$\|f\|_{L^2(A)} = \left(\int_A |f|^2\right)^{1/2}.$$

Equipped with this norm $C\left(A\right)$ is *not complete*. Let, for example $A = [-1,1] \subset \mathbb{R}$. The sequence

$$f_m(t) = \begin{cases} 0 & t \le 0 \\ mt & 0 < t \le \frac{1}{m} \\ 1 & t > \frac{1}{m} \end{cases} \quad (m \ge 1),$$

contained in $C\left([-1,1]\right)$, is a Cauchy sequence with respect to the L^2 norm. In fact (letting $m > k$),

$$\|f_m - f_k\|_{L^2(A)}^2 = \int_{-1}^{1} |f_m(t) - f_k(t)|^2 \, dt = (m-k)^2 \int_0^{1/m} t^2 \, dt$$

$$+ \int_0^{1/k} (1 - kt)^2 \, dt$$

$$= \frac{(m-k)^2}{3m^3} + \frac{1}{3k} < \frac{1}{3}\left(\frac{1}{m} + \frac{1}{k}\right) \to 0 \qquad \text{as } m, k \to \infty.$$

However, f_m converges in $L^2\left(-1,1\right)$ −norm (and pointwise) to the Heaviside function

$$\mathcal{H}(t) = \begin{cases} 1 & t \ge 0 \\ 0 & t < 0, \end{cases}$$

which is discontinuous at $t = 0$ and therefore does not belong to $C\left([-1,1]\right)$.

More generally, let $X = C^k\left(A\right)$, $k \ge 0$ integer, the set of functions continuously differentiable in A up to order k, included.

To denote a derivative of order m, it is convenient to introduce an $n - uple$ of nonnegative integers, $\alpha = (\alpha_1, \dots, \alpha_n)$, called *multi-index*, of *length*

$$|\alpha| = \alpha_1 + \dots + \alpha_n = m,$$

and set

$$D^\alpha = \frac{\partial^{\alpha_1}}{\partial x_1^{\alpha_1}} \cdots \frac{\partial^{\alpha_n}}{\partial x_n^{\alpha_n}}.$$

We endow $C^k\left(A\right)$ with the norm (*maximum norm of order k*)

$$\|f\|_{C^k(A)} = \|f\|_{C(A)} + \sum_{|\alpha|=1}^{k} \|D^\alpha f\|_{C(A)}.$$

---

[4] See also Section 7.2.2 and Example 7.4.

If $\{f_n\}$ is a Cauchy sequence in $C^k(A)$, all the sequences $\{D^\alpha f_n\}$ with $0 \le |\alpha| \le k$ are Cauchy sequences in $C(A)$. From the theorems on term by term differentiation of sequences, it follows that the resulting space is a Banach space.

Remark 7.2. With the introduction of *function spaces* we are actually making a step towards abstraction, regarding a function from a different perspective. In Calculus we see it as a point map while here we have to consider it as *a single element* (or a point or a vector) *of a vector space*.

Summable and bounded functions. Let Ω be an *open set* in $\mathbb{R}^n$ and $p \ge 1$ a real number. Let $X = L^p(\Omega)$ be the set of functions f such that $|f|^p$ is Lebesgue integrable in Ω. Identifying two functions f and g when they are *equal a.e.* in Ω, $L^p(\Omega)$ becomes a Banach space[5] when equipped with the norm (*integral norm of order p*)

$$\|f\|_{L^p(\Omega)} = \left(\int_\Omega |f|^p \right)^{1/p}.$$

The identification of two functions equal a.e. amounts to saying that an element of $L^p(\Omega)$ is not a single function but, actually, an equivalence class of functions, different from one another only on subsets of measure zero. At first glance, this fact could be annoying, but after all, the situation is perfectly analogous to considering a *rational number* as an equivalent class of fractions ($2/3$, $4/6$, $8/12$ ... represent the *same* number). For practical purposes one may always refer to the more convenient representative of the class.

Let $X = L^\infty(\Omega)$ the set of *essentially bounded* functions in Ω. Recall that $f : \Omega \to \mathbb{R}$ (or $\mathbb{C}$) is *essentially bounded* if there exists M such that

$$|f(x)| \le M \qquad \text{a.e. in } \Omega. \tag{7.3}$$

The infimum of all numbers M with the property (7.3) is called *essential supremum of f*, and denoted by

$$\|f\|_{L^\infty(\Omega)} = \operatorname*{ess\,sup}_\Omega |f|.$$

If we identify two functions when they are equal a.e., $\|f\|_{L^\infty(\Omega)}$ is a norm in $L^\infty(\Omega)$, and $L^\infty(\Omega)$ becomes a Banach space.

• *Hölder inequality.* The following inequality is of fundamental importance

$$\left| \int_\Omega fg \right| \le \|f\|_{L^p(\Omega)} \|g\|_{L^q(\Omega)}, \tag{7.4}$$

where $q = p/(p-1)$ is the *conjugate exponent of p*, allowing also the case $p = 1$, $q = \infty$.

---

[5] See e.g. *Yoshida* [37].

Note that, if Ω has *finite measure* and $1 \leq p_1 < p_2 \leq \infty$, from (7.4) we have, choosing $g \equiv 1$, $p = p_2/p_1$ and $q = p_2/(p_2 - p_1)$:

$$\left| \int_\Omega |f|^{p_1} \right| \leq |\Omega|^{1/q} \|f\|_{L^{p_2}(\Omega)}^{p_1}$$

and therefore $L^{p_2}(\Omega) \subset L^{p_1}(\Omega)$. If the measure of Ω is *infinite*, this inclusion is not true, in general; for instance, $f \equiv 1$ belongs to $L^\infty(\mathbb{R})$ but is not in $L^p(\mathbb{R})$ for $1 \leq p < \infty$.

7.2.2 Inner product and Hilbert Spaces

Let X be a linear space over $\mathbb{R}$. An *inner or scalar product* in X is a function

$$(\cdot, \cdot) : X \times X \to \mathbb{R}$$

with the following three properties. For every $x, y, z \in X$ and scalars $\lambda, \mu \in \mathbb{R}$:

1. $(x,x) \geq 0$ and $(x,x) = 0$ if and only if $x = 0$ (*positivity*);
2. $(x,y) = (y,x)$ (*symmetry*);
3. $(\mu x + \lambda y, z) = \mu(x,z) + \lambda(y,z)$ (*bilinearity*).

A linear space endowed with an inner product is called an *inner product space*. Property 3 shows that the inner product is linear with respect to its first argument. From 2, the same is true for the second argument as well. Then, we say that $(\cdot, \cdot)$ constitutes a *symmetric bilinear* form in X. When different inner product spaces are involved it may be necessary the use of notations like $(\cdot, \cdot)_X$, to avoid confusion.

Remark 7.3. If the scalar field is $\mathbb{C}$, then

$$(\cdot, \cdot) : X \times X \to \mathbb{C}$$

and property 2 has to be replaced by 2_{bis}: $(x,y) = \overline{(y,x)}$ where the bar denotes complex conjugation. As a consequence, we have

$$(z, \mu x + \lambda y) = \overline{\mu}(z,x) + \overline{\lambda}(z,y)$$

and we say that $(\cdot, \cdot)$ is *antilinear* with respect to its second argument or that it is a *sesquilinear form* in X.

An inner product *induces* a norm, given by

$$\|x\| = \sqrt{(x,x)}. \tag{7.5}$$

In fact, properties 1 and 2 in the definition of norm are immediate, while the triangular inequality is a consequence of the following quite important theorem.

Theorem 7.7. Let $x, y \in X$. Then:

(1) **Schwarz's inequality:**

$$|(x, y)| \leq \|x\| \, \|y\| . \tag{7.6}$$

Moreover equality holds in (7.6) if and only if x and y are linearly dependent.

(2) **Parallelogram law:**

$$\|x + y\|^2 + \|x - y\|^2 = 2 \|x\|^2 + 2 \|y\|^2 .$$

The parallelogram law generalizes an elementary result in euclidean plane geometry: *in a parallelogram, the sum of the squares of the sides length equals the sum of the squares of the diagonals length.* The Schwarz inequality implies that the inner product is continuous; in fact, writing

$$(w, z) - (x, y) = (w - x, z) + (x, z - y)$$

we have

$$|(w, z) - (x, y)| \leq \|w - x\| \, \|z\| + \|x\| \, \|z - y\|$$

so that, if $w \to x$ and $z \to y$, then $(w, z) \to (x, y)$.

Proof (1). We mimic the finite-dimensional proof. Let $t \in \mathbb{R}$ and $x, \, y \in X$. Using the properties of the inner product and (7.5), we may write:

$$0 \leq (tx + y, tx + y) = t^2 \|x\|^2 + 2t \, (x, y) + \|y\|^2 \equiv P(t) .$$

Thus, the second degree polynomial $P(t)$ is always nonnegative, whence

$$(x, y)^2 - \|x\|^2 \|y\|^2 \leq 0$$

which is the Schwarz inequality. Equality is possible only if $tx + y = 0$, i.e. if x and y are linearly dependent.

Proof (2). Just observe that

$$\|x \pm y\|^2 = (x \pm y, y \pm y) = \|x\|^2 \pm 2 \, (x, y) + \|y\|^2 . \tag{7.7}$$

$\square$

Definition 7.6. Let H be an inner product space. We say that H is a **Hilbert space** if it is complete with respect to the norm (7.5), induced by the inner product.

Two Hilbert spaces H_1 and H_2 are *isomorphic* if there exists a linear map $L : H_1 \to H_2$ which preserves the inner product, i.e.:

$$(x, y)_{H_1} = (Lx, Ly)_{H_2} \qquad \forall x, y \in H_1.$$

In particular

$$\|x\|_{H_1} = \|Lx\|_{H_2} .$$

Example 7.3. $\mathbb{R}^n$ is a Hilbert space with respect to the usual inner product

$$(\mathbf{x}, \mathbf{y})_{\mathbb{R}^n} = \mathbf{x} \cdot \mathbf{y} = \sum_{j=1}^{n} x_j y_j, \qquad \mathbf{x} = (x_1, \ldots, x_n), \, \mathbf{y} = (y_1, \ldots, y_n).$$

The induced norm is

$$\|\mathbf{x}\| = \sqrt{\mathbf{x} \cdot \mathbf{x}} = \sum_{j=1}^{n} x_j^2.$$

More generally, if $\mathbf{A} = (a_{ij})_{i,j=1,\ldots,n}$ is a square matrix of order n, *symmetric* and *positive*,

$$(\mathbf{x}, \mathbf{y})_{\mathbf{A}} = \mathbf{x} \cdot \mathbf{A}\mathbf{y} = \mathbf{A}\mathbf{x} \cdot \mathbf{y} = \sum_{i,j=1}^{n} a_{ij} x_i y_j \qquad (7.8)$$

defines another scalar product in $\mathbb{R}^n$. Actually, *every* inner product in $\mathbb{R}^n$ may be written in the form (7.8), with a suitable matrix $\mathbf{A}$.

$\mathbb{C}^n$ is a Hilbert space with respect to the inner product

$$\mathbf{x} \cdot \mathbf{y} = \sum_{j=1}^{n} x_j \bar{y}_j \qquad \mathbf{x} = (x_1, \ldots, x_n), \, \mathbf{y} = (y_1, \ldots, y_n).$$

It is easy to show that every real (resp. complex) linear space of dimension n is isomorphic to $\mathbb{R}^n$ (resp. $\mathbb{C}^n$).

Example 7.4. $L^2(\Omega)$ is a Hilbert space (perhaps the most important one) with respect to the inner product

$$(u, v)_{L^2(\Omega)} = \int_{\Omega} uv.$$

Remark 7.4. If Ω is fixed, we will simply use the notations $(u, v)_0$ instead of $(u, v)_{L^2(\Omega)}$ and $\|u\|_0$ instead of $\|u\|_{L^2(\Omega)}$.

Example 7.5. Let $l_{\mathbb{C}}^2$ be the set of complex sequences $\mathbf{x} = \{x_m\}$ such that

$$\sum_{m=1}^{\infty} |x_m|^2 < \infty.$$

For $\mathbf{x} = \{x_m\}$ and $\mathbf{y} = \{y_m\}$, define

$$(\mathbf{x}, \mathbf{y})_{l_{\mathbb{C}}^2} = \sum_{m=1}^{\infty} x_m \bar{y}_m.$$

Then $(\mathbf{x}, \mathbf{y})_{l_{\mathbb{C}}^2}$ is an inner product which makes $l_{\mathbb{C}}^2$ a Hilbert space over $\mathbb{C}$. This space constitutes the discrete analogue of $L^2(0, 2\pi)$. Indeed, each $u \in L^2(0, 2\pi)$ has an expansion in Fourier series (Appendix A)

$$u(x) = \sum_{m \in \mathbb{Z}} \hat{u}_m e^{imx},$$

where

$$\widehat{u}_m = \frac{1}{2\pi} \int_0^{2\pi} u(x) e^{-imx} dx.$$

Note that $\overline{\widehat{u}}_m = \widehat{u}_{-m}$, since u is a real function. From Parseval's identity, we have

$$(u,v)_0 = \int_0^{2\pi} uv = 2\pi \sum_{m \in \mathbb{Z}} \widehat{u}_m \widehat{v}_{-m}$$

and

$$\|u\|_0^2 = \int_0^{2\pi} u^2 = 2\pi \sum_{m \in \mathbb{Z}} |\widehat{u}_m|^2.$$

Example 7.6. A *Sobolev space*. It is possible to use the frequency space introduced in the previous example to define the derivative of a function in $L^2(0, 2\pi)$ in a weak or generalized sense. Let $u \in C^1(\mathbb{R})$, 2π−periodic. The Fourier coefficients of u' are given by

$$\widehat{u'}_m = im\widehat{u}_m$$

and we may write

$$\|u'\|_0^2 = \int_0^{2\pi} (u')^2 = 2\pi \sum_{m \in \mathbb{Z}} m^2 |\widehat{u}_m|^2. \tag{7.9}$$

Thus, both sequences $\{\widehat{u}_m\}$ and $\{m\widehat{u}_m\}$ belong to $l_\mathbb{C}^2$. But the right hand side in (7.9) does not involve u' directly, so that it makes perfect sense to define

$$H^1_{per}(0, 2\pi) = \left\{ u \in L^2(0, 2\pi) : \{\widehat{u}_m\}, \{m\widehat{u}_m\} \in l_\mathbb{C}^2 \right\}$$

and introduce the inner product

$$(u,v)_{1,2} = (2\pi) \sum_{m \in \mathbb{Z}} \left[1 + m^2\right] \widehat{u}_m \widehat{v}_{-m}$$

which makes $H^1_{per}(0, 2\pi)$ into a Hilbert space. Since

$$\{m\widehat{u}_m\} \in l_\mathbb{C}^2,$$

with each $u \in H^1_{per}(0, 2\pi)$ is associated the function $v \in L^2(0, 2\pi)$ given by

$$v(x) = \sum_{m \in \mathbb{Z}} im\widehat{u}_m e^{imx}.$$

We see that v may be considered as *a generalized derivative of u* and $H^1_{per}(0, 2\pi)$ as the space of functions in $L^2(0, 2\pi)$, together with their first derivatives. Let $u \in H^1_{per}(0, 2\pi)$ and

$$u(x) = \sum_{m \in \mathbb{Z}} \widehat{u}_m e^{imx}.$$

Since

$$\left| \widehat{u}_m e^{imx} \right| = \frac{1}{m} m \left| \widehat{u}_m \right| \leq \frac{1}{2} \left(\frac{1}{m^2} + m^2 \left| \widehat{u}_m \right|^2 \right)$$

the Weierstrass test entails that the Fourier series of u converges uniformly in $\mathbb{R}$. Thus u has a continuous, 2π-periodic extension to all $\mathbb{R}$. Finally observe that, if we use the symbol u' also for the generalized derivative of u, the inner product in $H^1_{per}(0,1)$ can be written in the form

$$(u,v)_{1,2} = \int_0^1 (u'v' + uv).$$

7.2.3 Projections

Hilbert spaces are the ideal setting to solve problems in infinitely many dimensions. They unify through the inner product and the induced norm, both an analytical and a geometric structure. As we shall shortly see, we may coherently introduce the concepts of orthogonality, projection and basis, prove a infinite-dimensional Pythagoras' Theorem and introduce other operations, extremely useful from both a theoretical and practical point of view.

As in finite-dimensional linear spaces, two elements x, y belonging to an inner product space are called **orthogonal or normal** if $(x,y) = 0$, and we write $x \perp y$.

Now, if we consider a subspace V of $\mathbb{R}^n$, e.g. a hyperplane through the origin, every $\mathbf{x} \in \mathbb{R}^n$ has a unique orthogonal projection on V. In fact, if $\dim V = k$ and the unit vectors $\mathbf{v}_1, \mathbf{v}_2, \ldots, \mathbf{v}_k$ constitute an *orthonormal basis* in V, we may always find an orthonormal basis in $\mathbb{R}^n$, given by

$$\mathbf{v}_1, \mathbf{v}_2, \ldots, \mathbf{v}_k, \mathbf{w}_{k+1}, \ldots, \mathbf{w}_n,$$

where $\mathbf{w}_{k+1}, \ldots, \mathbf{w}_n$ are suitable unit vectors. Thus, if

$$\mathbf{x} = \sum_{j=1}^{k} x_j \mathbf{v}_j + \sum_{j=k+1}^{n} x_j \mathbf{w}_j,$$

the projection of $\mathbf{x}$ on V is given by

$$P_V \mathbf{x} = \sum_{j=1}^{k} x_j \mathbf{v}_j.$$

On the other hand, the projection $P_V \mathbf{x}$ can be characterized through the following property, which does not involve a basis in $\mathbb{R}^n$: $P_V \mathbf{x}$ is *the point in V that minimizes the distance from* $\mathbf{x}$, that is

$$|P_V \mathbf{x} - \mathbf{x}| = \inf_{\mathbf{y} \in V} |\mathbf{y} - \mathbf{x}|. \tag{7.10}$$

In fact, if $\mathbf{y} = \sum_{j=1}^{k} y_j \mathbf{v}_j$, we have

$$|\mathbf{y} - \mathbf{x}|^2 = \sum_{j=1}^{k}(y_j - x_j)^2 + \sum_{j=k+1}^{n} x_j^2 \geq \sum_{j=k+1}^{n} x_j^2 = |P_V\mathbf{x} - \mathbf{x}|^2.$$

In this case, the "infimum" in (7.10) is actually a "minimum".

The uniqueness of $P_V\mathbf{x}$ follows from the fact that, if $\mathbf{y}^* \in V$ and

$$|\mathbf{y}^* - \mathbf{x}| = |P_V\mathbf{x} - \mathbf{x}|,$$

then we must have

$$\sum_{j=1}^{k}(y_j^* - x_j)^2 = 0,$$

whence $y_j^* = x_j$ for $j = 1, \ldots, k$, and therefore $\mathbf{y}^* = P_V\mathbf{x}$. Since

$$(\mathbf{x} - P_V\mathbf{x}) \perp \mathbf{v}, \qquad \forall \mathbf{v} \in V$$

every $\mathbf{x} \in \mathbb{R}^n$ may be written in a unique way in the form

$$\mathbf{x} = \mathbf{y} + \mathbf{z}$$

with $\mathbf{y} \in V$ and $\mathbf{z} \in V^\perp$, where $V^\perp$ denotes the subspace of the vectors orthogonal to V.

Then, we say that $\mathbb{R}^n$ is *direct sum* of the subspaces V and $V^\perp$ and we write

$$\mathbb{R}^n = V \oplus V^\perp.$$

Finally,

$$|\mathbf{x}|^2 = |\mathbf{y}|^2 + |\mathbf{z}|^2$$

which is the Pythagoras' Theorem in $\mathbb{R}^n$.

We may extend all the above consideration to infinite-dimensional Hilbert spaces H, if we consider **closed subspaces** V of H. Here *closed* means with respect to the convergence induced by the norm. More precisely, a subset $U \subset H$ is closed in H if it contains all the limit points of sequences in U. Observe that if V has *finite dimension* k, it is automatically closed, since it is isomorphic to $\mathbb{R}^k$ (or $\mathbb{C}^k$). Also, a closed subspace of a Hilbert space is a Hilbert space as well, with respect to the inner product in H.

Unless stated explicitly, **from now on we consider Hilbert spaces over $\mathbb{R}$** (real Hilbert spaces), endowed with inner product $(\cdot, \cdot)$ and induced norm $\|\cdot\|$.

Theorem 7.8 (Projection Theorem). *Let V be a closed subspace of a Hilbert space H. Then, for every $x \in H$, there exists a unique element $P_V x \in V$ such that*

$$\|P_V x - x\| = \inf_{v \in V} \|v - x\|. \tag{7.11}$$

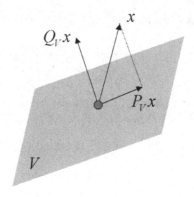

Fig. 7.1. Projection Theorem

Moreover, the following properties hold:

1. $P_V x = x$ *if and only if* $x \in V$.
2. *Let* $Q_V x = x - P_V x$. *Then* $Q_V x \in V^\perp$ *and*

$$\|x\|^2 = \|P_V x\|^2 + \|Q_V x\|^2.$$

Proof. Let

$$d = \inf_{v \in V} \|v - x\|.$$

By the definition of least upper bound, we may select a sequence $\{v_m\} \subset V$, such that $\|v_m - x\| \to d$ as $m \to \infty$. In fact, for every integer $m \geq 1$ there exists $v_m \in V$ such that

$$d \leq \|v_m - x\| < d + \frac{1}{m}. \tag{7.12}$$

Letting $m \to \infty$ in (7.12), we get $\|v_m - x\| \to d$.

We now show that $\{v_m\}$ is a Cauchy sequence. In fact, using the parallelogram law for the vectors $v_k - x$ and $v_m - x$, we obtain

$$\|v_k + v_m - 2x\|^2 + \|v_k - v_m\|^2 = 2\|v_k - x\|^2 + 2\|v_m - x\|^2. \tag{7.13}$$

Since $\frac{v_k + v_m}{2} \in V$, we may write

$$\|v_k + v_m - 2x\|^2 = 4 \cdot \left\|\frac{v_k + v_m}{2} - x\right\|^2 \geq 4d^2$$

whence, from (7.13):

$$\|v_k - v_m\|^2 = 2\|v_k - x\|^2 + 2\|v_m - x\|^2 - \|v_k + v_m - 2x\|^2$$
$$\leq 2\|v_k - x\|^2 + 2\|v_m - x\|^2 - 4d^2.$$

Letting $k, m \to \infty$, the right hand side goes to zero and therefore

$$\|v_k - v_m\| \to 0$$

as well. This proves that $\{v_m\}$ is a Cauchy sequence.

Since H is complete, v_m converges to an element $w \in H$ which belongs to V, because V *is closed*. Using the norm continuity (Proposition 7.1) we deduce

$$\|v_m - x\| \to \|w - x\| = d$$

so that w realizes the minimum distance from x among the elements in V.

We have to prove the uniqueness of w. Suppose $\bar{w} \in V$ is another element such that $\|\bar{w} - x\| = d$. The parallelogram law, applied to the vectors $w - x$ and $\bar{w} - x$, yields

$$\|w - \bar{w}\|^2 = 2\|w - x\|^2 + 2\|\bar{w} - x\|^2 - 4\left\|\frac{w + \bar{w}}{2} - x\right\|^2$$

$$\leq 2d^2 + 2d^2 - 4d^2 = 0$$

whence $w = \bar{w}$.

We have proved that there exists a unique element $w = P_V x \in V$ such that

$$\|x - P_V x\| = d.$$

To prove 1, observe that, since V is closed, $x \in V$ if and only if $d = 0$, which means $x = P_V x$.

To show 2, let $Q_V x = x - P_V x$, $v \in V$ and $t \in \mathbb{R}$. Since $P_V x + tv \in V$ for every t, we have:

$$d^2 \leq \|x - (P_V x + tv)\|^2 = \|Q_V x - tv\|^2$$

$$= \|Q_V x\|^2 - 2t\,(Q_V x, v) + t^2\|v\|^2$$

$$= d^2 - 2t\,(Q_V x, v) + t^2\|v\|^2.$$

Erasing d^2 and dividing by $t > 0$, we get

$$(Q_V x, v) \leq \frac{t}{2}\|v\|^2$$

which forces $(Q_V x, v) \leq 0$; dividing by $t < 0$ we get

$$(Q_V x, v) \geq \frac{t}{2}\|v\|^2$$

which forces $(Q_V x, v) \geq 0$. Thus $(Q_V x, v) = 0$ which means $Q_V x \in V^\perp$ and implies that

$$\|x\|^2 = \|P_V x + Q_V x\|^2 = \|P_V x\|^2 + \|Q_V x\|^2,$$

concluding the proof.   □

The elements $P_V x$, $Q_V x$ are called **orthogonal projections** of x **on** V and $V^\perp$, respectively. The least upper bound in (7.11) is actually a minimum.

Moreover thanks to properties 1, 2, we say that H *is direct sum of* V *and* $V^{\perp}$:

$$H = V \oplus V^{\perp}.$$

Note that
$$V^{\perp} = \{0\} \qquad \text{if and only if} \qquad V = H.$$

Remark 7.5. Another characterization of $P_V x$ is the following (see Exercise 7.14): $u = P_V x$ *if and only if*

$$\begin{cases} 1. \ u \in V \\ 2. \ (x - u, v) = 0, \ \forall v \in V. \end{cases}$$

Remark 7.6. It is useful to point out that, even if V is *not a closed* subspace of H, the subspace $V^{\perp}$ is *always closed*. In fact, if $y_n \to y$ and $\{y_n\} \subset V^{\perp}$, we have, for every $x \in V$,

$$(y, x) = \lim_{n \to \infty} (y_n, x) = 0$$

whence $y \in V^{\perp}$.

Example 7.7. Let $\Omega \subset \mathbb{R}^n$ be a set of finite measure. Consider in $L^2(\Omega)$ the one-dimensional subspace V of constant functions (a basis is given by $f \equiv 1$, for instance). Since it is finite-dimensional, V is closed in $L^2(\Omega)$. Given $f \in L^2(\Omega)$, to find the projection $P_V f$, we solve the minimization problem

$$\min_{\lambda \in \mathbb{R}} \int_{\Omega} (f - \lambda)^2.$$

Since

$$\int_{\Omega} (f - \lambda)^2 = \int_{\Omega} f^2 - 2\lambda \int_{\Omega} f + \lambda^2 |\Omega|,$$

we see that the minimizer is

$$\lambda = \frac{1}{|\Omega|} \int_{\Omega} f.$$

Therefore

$$P_V f = \frac{1}{|\Omega|} \int_{\Omega} f \quad \text{and} \quad Q_V f = f - \frac{1}{|\Omega|} \int_{\Omega} f.$$

Thus, the subspace $V^{\perp}$ is given by the functions $g \in L^2(\Omega)$ with *zero mean value*. In fact these functions are orthogonal to $f \equiv 1$:

$$(g, 1)_0 = \int_{\Omega} g = 0.$$

7.2.4 Orthonormal bases

A Hilbert space is said to be **separable** when there exists a *countable dense* subset of H. An *orthonormal basis* in a separable *Hilbert space* H is a sequence $\{w_k\}_{k\geq 1} \subset H$ such that[6]

$$\begin{cases} (w_k, w_j) = \delta_{kj} & k, j \geq 1 \\ \|w_k\| = 1 & k \geq 1 \end{cases}$$

and *every* $x \in H$ *may be expanded in the form*

$$x = \sum_{k=1}^{\infty} (x, w_k)\, w_k. \tag{7.14}$$

The series (7.14) is called **generalized Fourier series** and the numbers $c_k = (x, w_k)$ are the *Fourier coefficients* of x with respect to the basis $\{w_k\}$. Moreover (Pythagoras again!):

$$\|x\|^2 = \sum_{k=1}^{\infty} (x, w_k)^2.$$

Given an orthonormal basis $\{w_k\}_{k\geq 1}$, the projection of $x \in H$ on the subspace V spanned by, say, $w_1, \ldots, w_N$ is given by

$$P_V x = \sum_{k=1}^{N} (x, w_k)\, w_k.$$

An example of separable Hilbert space is $L^2(\Omega)$, $\Omega \subseteq \mathbb{R}^n$. In particular, the set of functions

$$\frac{1}{\sqrt{2\pi}}, \frac{\cos x}{\sqrt{\pi}}, \frac{\sin x}{\sqrt{\pi}}, \frac{\cos 2x}{\sqrt{\pi}}, \frac{\sin 2x}{\sqrt{\pi}}, \ldots, \frac{\cos mx}{\sqrt{\pi}}, \frac{\sin mx}{\sqrt{\pi}}, \ldots$$

constitutes an orthonormal basis in $L^2(0, 2\pi)$ (see Appendix A).

It turns out that:

Proposition 7.2. *Every separable Hilbert space H admits an orthonormal basis.*

Proof (sketch). Let $\{z_k\}_{k\geq 1}$ be dense in H. Disregarding, if necessary, those elements which are spanned by other elements in the sequence, we may assume that $\{z_k\}_{k\geq 1}$ constitutes an *independent set*, i.e. every finite subset of $\{z_k\}_{k\geq 1}$ is composed by independent elements.

Then, an orthonormal basis $\{w_k\}_{k\geq 1}$ is obtained by applying to $\{z_k\}_{k\geq 1}$ the following so called *Gram-Schmidt process*. First, construct by induction a

---

[6] δ_{jk} is the Kronecker symbol.

sequence $\{\tilde{w}\}_{k\geq 1}$ as follows. Let $\tilde{w}_1 = z_1$. Once $\tilde{w}_{k-1}$ is known, we construct $\tilde{w}_k$ by subtracting from z_k its components with respect to $\tilde{w}_1, \ldots, \tilde{w}_{k-1}$:

$$\tilde{w}_k = z_k - \frac{(z_k, \tilde{w}_{k-1})}{\|\tilde{w}_{k-1}\|^2}\tilde{w}_{k-1} - \cdots - \frac{(z_k, \tilde{w}_1)}{\|\tilde{w}_1\|^2}\tilde{w}_1.$$

In this way, $\tilde{w}_k$ is orthogonal to $\tilde{w}_1, \ldots, \tilde{w}_{k-1}$. Finally, set $w_k = \tilde{w}_k / \|\tilde{w}_{k-1}\|$. Since $\{z_k\}_{k\geq 1}$ is dense in H, then $\{w_k\}_{k\geq 1}$ is dense in H as well. Thus $\{w_k\}_{k\geq 1}$ is an orthonormal basis. $\qquad\square$

In the applications, orthonormal bases arise from solving particular boundary value problems, often in relation to the separation of variables method. Typical examples come from the vibrations of a non homogeneous string or from diffusion in a rod with non constant thermal properties c_v, ρ, κ. The first example leads to the wave equation

$$\rho(x)\, u_{tt} - \tau u_{xx} = 0.$$

Separating variables $(u(x,t) = v(x)z(t))$, we find for the spatial factor the equation

$$\tau v'' + \lambda \rho v = 0.$$

In the second example we are led to

$$(\kappa v')' + \lambda c_v \rho v = 0.$$

These equations are particular cases of a general class of ordinary differential equations of the form

$$(pu')' + qu + \lambda w u = 0 \qquad (7.15)$$

called *Sturm-Liouville* equations. Usually one looks for solutions of (7.15) in an interval (a, b), $-\infty \leq a < b \leq +\infty$, satisfying suitable conditions at the end points. The natural assumptions on p and q are $p \neq 0$ in (a, b) and p, q, p^{-1} locally integrable in (a, b). The function w plays the role of a *weight function*, continuous in $[a, b]$ and positive in (a, b).

In general, the resulting boundary value problem has non trivial solutions only for particular values of λ, called *eigenvalues*. The corresponding solutions are called *eigenfunctions* and it turns out that, when suitably normalized, they constitute an orthonormal basis in the Hilbert space $L_w^2(a, b)$, the set of Lebesgue measurable functions in (a, b) such that

$$\|u\|_{L_w^2}^2 = \int_a^b u^2(x)\, w(x)\, dx < \infty,$$

endowed with the inner product

$$(u, v)_{L_w^2} = \int_a^b u(x)\, v(x)\, w(x)\, dx.$$

We list below some examples.

Example 7.8 (Legendre polynomials). Consider the problem

$$\left(\left(1-x^2\right)u'\right)' + \lambda u = 0 \qquad \text{in } (-1,1)$$

with weighted Neumann conditions

$$\left(1-x^2\right)u'(x) \to 0 \qquad \text{as } x \to \pm 1.$$

The differential equation is known as *Legendre's* equation. The eigenvalues are $\lambda_n = n(n+1)$, $n = 0,1,2,\ldots$ The corresponding eigenfunctions are the *Legendre polynomials*, defined by $L_0(x) = 1, L_1(x) = x$,

$$(n+1)L_{n+1} = (2n+1)xL_n - nL_{n-1} \qquad (n > 1)$$

or by *Rodrigues' formula*

$$L_n(x) = \frac{1}{2^n n!}\frac{d^n}{dx^n}\left(x^2-1\right)^n \qquad (n \geq 0).$$

For instance, $L_2(x) = (3x^2 - 1)/2$, $L_3(x) = (5x^3 - 3x)/2$. The normalized polynomials

$$\sqrt{\frac{2n+1}{2}}L_n$$

constitute an orthonormal basis in $L^2(-1,1)$ (here $w(x) \equiv 1$). Every function $f \in L^2(-1,1)$ has an expansion

$$f(x) = \sum_{n=0}^{\infty} f_n L_n(x)$$

where $f_n = \frac{2n+1}{2}\int_{-1}^{1} f(x)L_n(x)\,dx$, with convergence in $L^2(-1,1)$.

Example 7.9 (Hermite polynomials). Consider the problem

$$\begin{cases} u'' - 2xu' + 2\lambda u = 0 & \text{in } (-\infty, +\infty) \\ e^{-x^2/2}u(x) \to 0 & \text{as } x \to \pm\infty. \end{cases}$$

The differential equation is known as *Hermite's* equation (see Exercise 7.16) and may be written in the form (7.15):

$$(e^{-x^2}u')' + 2\lambda e^{-x^2}u = 0$$

which shows the proper weight function $w(x) = e^{-x^2}$. The eigenvalues are $\lambda_n = n$, $n = 0,1,2,\ldots$. The corresponding eigenfunctions are the *Hermite polynomials* defined by *Rodrigues' formula*

$$H_n(x) = (-1)^n e^{x^2}\frac{d^n}{dx^n}e^{-x^2} \qquad (n \geq 0).$$

For instance

$$H_0(x) = 1, \quad H_1(x) = 2x, \quad H_2(x) = 4x^2 - 2, \quad H_3(x) = 8x^3 - 12x.$$

The normalized polynomials $\pi^{-1/4}(2^n n!)^{-1/2} H_n$ constitute an orthonormal basis in $L_w^2(\mathbb{R})$, with $w(x) = e^{-x^2}$. Every $f \in L_w^2(\mathbb{R})$ has an expansion

$$f(x) = \sum_{n=0}^{\infty} f_n H_n(x)$$

where $f_n = [\pi^{1/2} 2^n n!]^{-1} \int_{\mathbb{R}} f(x) H_n(x) e^{-x^2} dx$, with convergence in $L_w^2(\mathbb{R})$.

Example 7.10 (Bessel functions). After separating variables in the model for the vibration of a circular membrane the following *parametric Bessel equation of order* p arises:

$$x^2 u'' + xu' + \left(\lambda x^2 - p^2\right) u = 0 \qquad x \in (0, a) \tag{7.16}$$

where $p \geq 0$, $\lambda \geq 0$, with the boundary conditions

$$u(0) \text{ finite}, \quad u(a) = 0. \tag{7.17}$$

Equation (7.16) may be written in Sturm-Liouville form as

$$(xu')' + \left(\lambda x - \frac{p^2}{x}\right) u = 0$$

which shows the proper weight function $w(x) = x$. The simple rescaling $z = \sqrt{\lambda} x$ reduces (7.16) to the *Bessel equation of order* p

$$z^2 \frac{d^2 u}{dz^2} + z \frac{du}{dz} + \left(z^2 - p^2\right) u = 0 \tag{7.18}$$

where the dependence on the parameter λ is removed. The only bounded solutions of (7.18) are the *Bessel functions of first kind and order* p, given by

$$J_p(z) = \sum_{k=0}^{\infty} \frac{(-1)^k}{\Gamma(k+1)\Gamma(k+p+1)} \left(\frac{z}{2}\right)^{p+2k}$$

where

$$\Gamma(s) = \int_0^{\infty} e^{-t} t^{s-1} dt \tag{7.19}$$

is the Euler Γ-function. In particular, if $p = n \geq 0$, integer:

$$J_n(z) = \sum_{k=0}^{\infty} \frac{(-1)^k}{k!(k+n)!} \left(\frac{z}{2}\right)^{n+2k}.$$

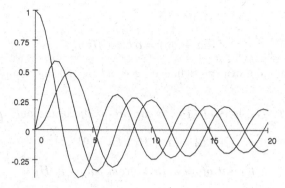

Fig. 7.2. Graphs of J_0, J_1 and J_2

For every p, there exists an infinite, increasing sequence $\{\alpha_{pj}\}_{j\geq 1}$ of positive zeroes of J_p

$$J_p(\alpha_{pj}) = 0 \qquad (j = 1, 2, \ldots).$$

Then, the eigenvalues of problem (7.16), (7.17) are given by $\lambda_{pj} = \left(\dfrac{\alpha_{pj}}{a}\right)^2$, with corresponding eigenfunctions $u_{pj}(x) = J_p\left(\dfrac{\alpha_{pj}}{a}x\right)$. The normalized eigenfunctions

$$\frac{\sqrt{2}}{aJ_{p+1}(\alpha_{pj})} J_p\left(\frac{\alpha_{pj}}{a}x\right)$$

constitute an orthonormal basis in $L_w^2(0, a)$, with $w(x) = x$. Every function $f \in L_w^2(0, a)$ has an expansion in *Fourier-Bessel series*

$$f(x) = \sum_{j=1}^{\infty} f_j J_p\left(\frac{\alpha_{pj}}{a}x\right),$$

where

$$f_j = \frac{2}{a^2 J_{p+1}^2(\alpha_{pj})} \int_0^a xf(x) J_p\left(\frac{\alpha_{pj}}{a}x\right) dx,$$

convergent in $L_w^2(0, a)$.

7.3 Linear Operators and Duality

7.3.1 Linear operators

Let H_1 and H_2 be Hilbert spaces. A **linear operator from H_1 into H_2** is a function

$$L : H_1 \to H_2$$

such that[7], $\forall \alpha, \beta \in \mathbb{R}$ and $\forall x, y \in H_1$

$$L(\alpha x + \beta y) = \alpha Lx + \beta Ly.$$

For every linear operator we define its *Kernel*, $\mathcal{N}(L)$ and *Range*, $\mathcal{R}(L)$, as follows:

Definition 7.7. *The **kernel** of L, is the pre-image of the null vector in H_2*

$$\mathcal{N}(L) = \{x \in H_1 : Lx = 0\}.$$

*The **range** of L is the set of all outputs from points in H_1*

$$\mathcal{R}(L) = \{y \in H_2 : \exists x \in H_1, Lx = y\}.$$

$\mathcal{N}(L)$ and $\mathcal{R}(L)$ are linear subspaces of H_1 and H_2, respectively.

Our main objects will be linear bounded operators.

Definition 7.8. *A linear operator $L : H_1 \to H_2$ is **bounded** if there exists a number C such that*

$$\|Lx\|_{H_2} \le C \|x\|_{H_1}, \qquad \forall x \in H_1. \tag{7.20}$$

The number C controls the expansion rate operated by L on the elements of H_1. In particular, if $C < 1$, L contracts the sizes of the vectors in H_1.

If $x \ne 0$, using the linearity of L, we may write (7.20) in the form

$$\left\| L \left(\frac{x}{\|x\|_{H_1}} \right) \right\|_{H_2} \le C$$

which is equivalent to

$$\sup_{\|x\|_{H_1}=1} \|Lx\|_{H_2} = K < \infty, \tag{7.21}$$

since $x / \|x\|_{H_1}$ is a unit vector in H_1. Clearly $K \le C$.

Proposition 7.3. *A linear operator $L : H_1 \to H_2$ is bounded if and only if it is continuous.*

Proof. Let L be bounded. From (7.20) we have, $\forall x, x_0 \in H_1$,

$$\|L(x - x_0)\|_{H_2} \le C \|x - x_0\|_{H_1}$$

so that, if $\|x - x_0\|_{H_1} \to 0$, also $\|Lx - Lx_0\|_{H_2} = \|L(x - x_0)\|_{H_2} \to 0$. This shows the continuity of L.

---

[7] Notation: if L is linear, when no confusion arises, we may write Lx instead of $L(x)$.

Let L be continuous. In particular, L is continuous at $x = 0$ so that there exists δ such that

$$\|Lx\|_{H_2} \leq 1 \quad \text{if } \|x\|_{H_1} \leq \delta.$$

Choose now $y \in H_1$ with $\|y\|_{H_1} = 1$ and let $z = \delta y$. We have $\|z\|_{H_1} = \delta$ which implies

$$\delta \|Ly\|_{H_2} = \|Lz\|_{H_2} \leq 1$$

or

$$\|Ly\|_{H_2} \leq \frac{1}{\delta}$$

and (7.21) holds with $K \leq C = 1/\delta$.   □

Given two Hilbert spaces H_1 and H_2, we denote by

$$\mathcal{L}(H_1, H_2)$$

the *family of all linear bounded operators from H_1 into H_2*. If $H_1 = H_2$ we simply write $\mathcal{L}(H)$.

$\mathcal{L}(H_1, H_2)$ becomes a linear space if we define, for $x \in H_1$ and $\lambda \in \mathbb{R}$,

$$(G + L)(x) = Gx + Lx$$
$$(\lambda L)x = \lambda Lx$$

for every $L, G \in \mathcal{L}(H_1, H_1)$ Also, we may use the number K in (7.21) as a norm in $\mathcal{L}(H_1, H_2)$

$$\|L\|_{\mathcal{L}(H_1, H_2)} = \sup_{\|x\|_{H_1} = 1} \|Lx\|_{H_2}. \tag{7.22}$$

When no confusion arises we will write simply $\|L\|$ instead of $\|L\|_{\mathcal{L}(H_1, H_2)}$. Thus, for every $L \in \mathcal{L}(H_1, H_2)$, we have

$$\|Lx\|_{H_2} \leq \|L\| \, \|x\|_{H_1}.$$

The resulting space is complete, so that:

Proposition 7.4. *Endowed with the norm (7.22), $\mathcal{L}(H_1, H_2)$ is a Banach space.*

Example 7.11. Let $\mathbf{A}$ be an $m \times n$ real matrix. The map

$$L : \mathbf{x} \longmapsto \mathbf{Ax}$$

is a linear operator from $\mathbb{R}^n$ into $\mathbb{R}^m$ (see Exercise 7.1).

Example 7.12. Let V be a *closed subspace* of a Hilbert space H. The projections

$$x \longmapsto P_V x, \qquad x \longmapsto Q_V x,$$

defined in Theorem 7.8, are bounded linear operators from H into H. In fact, from $\|x\|^2 = \|P_V x\|^2 + \|Q_V x\|^2$, it follows immediately that

$$\|P_V x\| \le \|x\|, \qquad \|Q_V x\| \le \|x\|$$

so that (7.20) holds with $C = 1$. Since $P_V x = x$ when $x \in V$ and $Q_V x = x$ when $x \in V^\perp$, it follows that $\|P_V\| = \|Q_V\| = 1$. Finally, observe that

$$\mathcal{N}(P_V) = \mathcal{R}(Q_V) = V^\perp \quad \text{and} \quad \mathcal{N}(Q_V) = \mathcal{R}(P_V) = V.$$

Example 7.13. Let V and H be Hilbert spaces with[8] $V \subset H$. Considering an element in V as an element of H, we define the operator $I_{V \to H} : V \to H$,

$$I_{V \to H}(u) = u,$$

which is called *embedding of V into H*. $I_{V \to H}$ is clearly a linear operator and it is also bounded if there exists a constant C such that

$$\|u\|_H \le C \|u\|_V, \qquad \text{for every } u \in V.$$

In this case, we say that V *is continuously embedded in H* and we write

$$V \hookrightarrow H.$$

For instance, $H^1_{per}(0, 2\pi) \hookrightarrow L^2(0, 2\pi)$.

7.3.2 Functionals and dual space

When $H_2 = \mathbb{R}$ (or $\mathbb{C}$, for complex Hilbert spaces), a linear operator $L : H \to \mathbb{R}$ takes the name of **functional**.

Definition 7.9. *The collection of all bounded linear functionals on a Hilbert space H is called **dual space** of H and denoted by H^* (instead of $\mathcal{L}(H, \mathbb{R})$).*

Example 7.14. Let $H = L^2(\Omega)$, $\Omega \subseteq \mathbb{R}^n$ and fix $g \in L^2(\Omega)$. The functional defined by

$$L_g : f \longmapsto \int_\Omega fg$$

is linear and bounded (see Exercise 7.2).

Example 7.15. A continuous functional is induced by the inner product with a fixed element in $L^2(\Omega)$: let H be a Hilbert space, for fixed $y \in H$, the functional

$$L_1 : x \longmapsto (x, y)$$

is continuous (see Exercise 7.3).

---

[8] The inner products in V and H may be different.

The possibility to identify the dual space of a Hilbert space H is crucial in many instances. It's easy to show that the inner product with a fixed element y in H defines an element of H^*, whose norm is exactly $\|y\|$. From Linear Algebra it is well known that *all* linear functionals in a finite-dimensional space can be represented in that way. Precisely, if L is linear in $\mathbb{R}^n$, there exists a vector $\mathbf{a} \in \mathbb{R}^n$ such that, for every $\mathbf{h} \in \mathbb{R}^n$,

$$L\mathbf{h} = \mathbf{a} \cdot \mathbf{h}$$

and $\|L\| = |\mathbf{a}|$. The following theorem says that an analogous result holds in Hilbert spaces.

Theorem 7.9 (Riesz's Representation Theorem). *Let H be a Hilbert space. For every $L \in H^*$ there exists a unique $u_L \in H$ such that:*

$$Lx = (u_L, x) \qquad \text{for every } x \in H.$$

Moreover $\|L\| = \|u_L\|$.

Proof. Let $\mathcal{N}$ be the kernel of L. If $\mathcal{N} = H$, then L is the *null operator* and $u_L = 0$. If $\mathcal{N} \subset H$, then $\mathcal{N}$ is a *closed* subspace of H. In fact, if $\{x_n\} \subset \mathcal{N}$ and $x_n \to x$, then $0 = Lx_n \to Lx$ so that $x \in \mathcal{N}$; thus $\mathcal{N}$ contains all its limit points and therefore is closed.

Then, by the Projection Theorem, there exists $z \in \mathcal{N}^\perp$, $z \neq 0$. Thus $Lz \neq 0$ and, given any $x \in H$, the element

$$w = x - \frac{Lx}{Lz} z$$

belongs to $\mathcal{N}$. In fact

$$Lw = L\left(x - \frac{Lx}{Lz} z \right) = Lx - \frac{Lx}{Lz} Lz = 0.$$

Since $z \in \mathcal{N}^\perp$, we have

$$0 = (z, w) = (z, x) - \frac{Lx}{Lz} \|z\|^2$$

which entails

$$Lx = \frac{L(z)}{\|z\|^2} (z, x).$$

Therefore if $u_L = L(z) \|z\|^{-2} z$, then $Lx = (u_L, x)$.

For the uniqueness, observe that, if $v \in H$ and

$$Lx = (v, x) \quad \text{for every } x \in H,$$

subtracting this equation from $Lx = (u_L, x)$, we infer

$$(u_L - v, x) = 0 \qquad \text{for every } x \in H$$

which forces $v = u_L$.

To show $\|L\| = \|u_L\|$, use Schwarz's inequality

$$|(u_L, x)| \leq \|x\| \, \|u_L\|$$

to get

$$\|L\| = \sup_{\|x\|=1} |Lx| = \sup_{\|x\|=1} |(u_L, x)| \leq \|u_L\|.$$

On the other hand,

$$\|u_L\|^2 = (u_L, u_L) = Lu_L \leq \|L\| \, \|u_L\|$$

whence

$$\|u_L\| \leq \|L\|.$$

Thus $\|L\| = \|u_L\|$. $\square$

The Riesz's map $R : H^* \to H$ given by

$$L \longmapsto u_L$$

is a *canonical isometry*, since it preserves the norm:

$$\|L\| = \|u_L\|.$$

We say that u_L is *the Riesz element associated with* L, with respect to the scalar product $(\cdot, \cdot)$. Moreover, H^* endowed with the inner product

$$(L_1, L_2)_{H^*} = (u_{L_1}, u_{L_2})$$

is clearly a Hilbert space. Thus, in the end, the Representation Theorem allows the **identification of a Hilbert space with its dual**.

Typically, $L^2(\Omega)$ or l_2 are identified with their duals.

Remark 7.7. *Warning*: there are situations in which the above canonical identification requires some care. A typical case occurs when dealing with a pair of Hilbert spaces V, H such that

$$V \hookrightarrow H \quad \text{and} \quad H^* \hookrightarrow V^*.$$

In this conditions it is possible to identify H and H^* and write

$$V \hookrightarrow H \hookrightarrow V^*,$$

but at this point the identification of V with V^* is forbidden, since it would give rise to nonsense!

Remark 7.8. A few words about **notations**. The symbol $(\cdot, \cdot)$ or $(\cdot, \cdot)_H$ denotes the inner product in a Hilbert space H. Let now $L \in H^*$. For the *action* of the functional L on an element $x \in H$ we used the symbol Lx. Sometimes, when it is useful or necessary to emphasize the *duality (or pairing)* between H and H^*, we can use notations $\langle L, x \rangle_*$ or $_{H^*}\langle L, x \rangle_H$.

7.4 Abstract Variational Problems

7.4.1 Bilinear forms and the Lax-Milgram Theorem

In the variational formulation of boundary value problems a key role is played by *bilinear forms*. Given two linear spaces V_1, V_2, a **bilinear form in** $V_1 \times V_2$ is a function

$$a : V_1 \times V_2 \to \mathbb{R}$$

satisfying the following properties:

i) For every $y \in V_2$, the function

$$x \longmapsto a(x, y)$$

is linear in V_1.

ii) For every $x \in V_1$, the function

$$y \longmapsto a(x, y)$$

is linear in V_2.

When $V_1 = V_2$, we simply say that a is a *bilinear form in* V.

Remark 7.9. In complex inner product spaces we define *sesquilinear forms*, instead of bilinear forms, replacing ii) by:

ii$_{bis}$) for every $x \in V_1$, the function

$$y \longmapsto a(x, y)$$

is *anti-linear*[9] in V_2.

Here are some examples.

• A typical example of bilinear form in a Hilbert space is its inner product.

• The formula

$$a\,(u, v) = \int_a^b (p(x)u'v' + q(x)u'v + r(x)uv)\ dx$$

where p, q, r are bounded functions, defines a bilinear form in $C^1\left([a, b]\right)$.

More generally, if Ω is a bounded domain in $\mathbb{R}^n$,

$$a(u,v) = \int_\Omega (\alpha\, \nabla u \cdot \nabla v + u\mathbf{b}\,(\mathbf{x}) \cdot \nabla v + a_0\,(\mathbf{x})\, uv)\ dx \qquad (\alpha > 0),$$

or

$$a(u,v) = \int_\Omega \alpha\, \nabla u \cdot \nabla v\ dx + \int_{\partial\Omega} huv\ d\sigma \qquad (\alpha > 0),$$

($\mathbf{b}$, a_0, h bounded) are bilinear forms in $C^1\left(\overline{\Omega}\right)$.

---

[9] That is $a\,(x, \alpha y + \beta z) = \overline{\alpha}a\,(x, y) + \overline{\beta}a\,(x, z)$.

- A bilinear form in $C^2\left(\overline{\Omega}\right)$ involving higher order derivatives is

$$a\left(u,v\right) = \int_{\Omega} \Delta u\, \Delta v\, d\mathbf{x}.$$

Let V be a Hilbert space, a be a bilinear form in V and $F \in V^*$. Consider the following problem, called *abstract variational problem*:

$$\begin{cases} \text{Find } u \in V \\ \qquad \text{such that} \\ a\left(u,v\right) = \langle F,v \rangle_* \quad \forall v \in V. \end{cases} \qquad (7.23)$$

As we shall see, many boundary values problems can be recast in this form. The fundamental result is:

Theorem 7.10 (Lax-Milgram). *Let V be a real Hilbert space endowed with inner product $(\cdot,\cdot)$ and norm $\|\cdot\|$. Let $a = a\left(u,v\right)$ be a bilinear form in V. If:*

i) *a is **continuous**, i.e. there exists a constant M such that*

$$|a(u,v)| \le M \, \|u\| \, \|v\|, \qquad \forall u,v \in V;$$

ii) *a is $V-$**coercive**, i.e. there exists a constant $\alpha > 0$ such that*

$$a(v,v) \ge \alpha \, \|v\|^2, \qquad \forall v \in V, \qquad (7.24)$$

then there exists a unique solution $\overline{u} \in V$ of problem (7.23). Moreover, the following stability estimate holds:

$$\|\overline{u}\| \le \frac{1}{\alpha} \, \|F\|_{V^*} . \qquad (7.25)$$

Remark 7.10. The coerciveness inequality (7.24) may be considered as an abstract version of the *energy* or *integral estimates* we met in the previous chapters. Usually, it is the key estimate to prove in order to apply Theorem 7.10.

Remark 7.11. Inequality (7.25) is called *stability estimate* for the following reason. The functional F, element of V^*, encodes the "data" of the problem (7.23). Since for every F there is a unique solution $u(F)$, the map

$$F \longmapsto u(F)$$

is a well defined *function* from V^* onto V. Also, everything here has a linear nature, so that the solution map is linear as well. To check it, let $\lambda, \mu \in \mathbb{R}$, F_1, $F_2 \in V^*$ and u_1, u_2 the corresponding solutions. The bilinearity of a, gives

$$\begin{aligned} a(\lambda u_1 + \mu u_2, v) &= \lambda a(u_1,v) + \mu a(u_2,v) = \\ &= \lambda F_1 v + \mu F_2 v. \end{aligned}$$

Therefore, the same linear combination of the solutions corresponds to a linear combination of the data; this expresses the *principle of superposition* for problem (7.23). Applying now (7.25) to $u_1 - u_2$, we obtain

$$\|u_1 - u_2\| \leq \frac{1}{\alpha} \|F_1 - F_2\|_{V^*}.$$

Thus, close data imply close solutions. The stability constant $1/\alpha$ plays an important role, since it controls the norm-variation of the solutions in terms of the variations on the data, measured by $\|F_1 - F_2\|_{V^*}$. This entails, in particular, that the more the coerciveness constant α is large, the more "stable" is the solution.

Proof (Theorem 7.10). We split it into several steps.

1. *Reformulation of problem* (7.23). For every fixed $u \in V$, by the continuity of a, the linear map

$$v \mapsto a\,(u,v)$$

is bounded in V and therefore it defines an element of V^*. From Riesz's Representation Theorem, there exists a unique $A\,[u] \in V$ such that

$$a\,(u,v) = (A[u],v)\,; \qquad \forall v \in V. \tag{7.26}$$

Since $F \in V^*$ as well, there exists a unique $z_F \in V$ such that

$$Fv = (z_F,v) \qquad \forall v \in V$$

and moreover $\|F\|_{V^*} = \|z_F\|$. Then, problem (7.23) can be recast in the following way:

$$\begin{cases} \qquad \text{Find } u \in V \\ \qquad \text{such that} \\ (A\,[u]\,,v) = (z_F,v)\,, \quad \forall v \in V \end{cases}$$

which, in turn, is equivalent to **finding** u **such that**

$$A\,[u] = z_F. \tag{7.27}$$

We want to show that (7.27) has exactly one solution. To do this we show that

$$A : V \to V$$

is a *linear, continuous, one-to-one, surjective* map.

2. *Linearity and continuity of A.* We repeatedly use the definition of A and the bilinearity of a. To show linearity, we write, for every $u_1, u_2, v \in V$ and $\lambda_1, \lambda_2 \in \mathbb{R}$,

$$(A\,[\lambda_1 u_1 + \lambda_2 u_2]\,,v) = a\,(\lambda_1 u_1 + \lambda_2 u_2, v) = \lambda_1 a\,(u_1, v) + \lambda_2 a\,(u_2, v)$$
$$= \lambda_1\,(A\,[u_1]\,,v) + \lambda_2\,(A\,[u_2]\,,v) = (\lambda_1 A\,[u_1] + \lambda_2 A\,[u_2]\,,v)$$

whence

$$A \left[\lambda_1 u_1 + \lambda_2 u_2\right] = \lambda_1 A \left[u_1\right] + \lambda_2 A \left[u_2\right].$$

Thus A is linear and we may write Au instead of $A \left[u\right]$. For the continuity, observe that

$$\|Au\|^2 = (Au, Au) = a(u, Au)$$
$$\leq M \|u\| \|Au\|$$

whence

$$\|Au\| \leq M \|u\|.$$

3. *A is one-to-one and has closed range*, i.e.

$$\mathcal{N}(A) = \{0\} \quad \text{and} \quad \mathcal{R}(A) \text{ is a closed subspace of } V.$$

In fact, the coercivity of a yields

$$\alpha \|u\|^2 \leq a(u, u) = (Au, u) \leq \|Au\| \|u\|$$

whence

$$\|u\| \leq \frac{1}{\alpha} \|Au\|. \tag{7.28}$$

Thus, $Au = 0$ implies $u = 0$ and hence $\mathcal{N}(A) = \{0\}$. To prove that $\mathcal{R}(A)$ is closed we have to consider a sequence $\{y_m\} \subset \mathcal{R}(A)$ such that

$$y_m \to y \in V$$

as $m \to \infty$, and show that $y \in \mathcal{R}(A)$. Since $y_m \in \mathcal{R}(A)$, there exists u_m such that $Au_m = y_m$. From (7.28) we infer

$$\|u_k - u_m\| \leq \frac{1}{\alpha} \|y_k - y_m\|$$

and therefore, since $\{y_m\}$ is convergent, $\{u_m\}$ is a Cauchy sequence. Since V is complete, there exists $u \in V$ such that

$$u_m \to u$$

as $m \to \infty$ and the continuity of A yields $y_m = Au_m \to Au$. Thus $Au = y$, so that $y \in \mathcal{R}(A)$ and $\mathcal{R}(A)$ is closed.

4. *A is surjective, that is* $\mathcal{R}(A) = V$. Suppose $\mathcal{R}(A) \subset V$. Since $\mathcal{R}(A)$ is a closed subspace, by the Projection Theorem there exists $z \neq 0$, $z \in \mathcal{R}(A)^{\perp}$. In particular, this implies

$$0 = (Az, z) = a(z, z) \geq \alpha \|z\|^2$$

whence $z = 0$. That is a contradiction. Therefore $\mathcal{R}(A) = V$.

5. *Solution of problem* (7.23). Since A is one-to-one and $\mathcal{R}(A) = V$, there exists exactly one solution $\bar{u} \in V$ of equation

$$Au = z_F.$$

From point **1**, $\bar{u}$ is the unique solution of problem (7.23) as well.

6. *Stability estimate.* From (7.28) with $u = \bar{u}$, we obtain

$$\|\bar{u}\| \le \frac{1}{\alpha} \|A\bar{u}\| = \frac{1}{\alpha} \|z_F\| = \frac{1}{\alpha} \|F\|_{V^*}$$

and the proof is complete. $\qquad\qquad\square$

7.4.2 Minimization of quadratic functionals

When a is *symmetric*, i.e. if

$$a(u, v) = a(v, u) \qquad \forall u, v \in V,$$

the abstract variational problem (7.23) is equivalent to a *minimization* problem. In fact, consider the quadratic functional

$$E(v) = \frac{1}{2} a(v, v) - \langle F, v \rangle_*.$$

We have:

Theorem 7.11. *Let a be symmetric. Then $\bar{u}$ is solution of problem (7.23) if and only if $\bar{u}$ is a minimizer of E, that is*

$$E(\bar{u}) = \min_{v \in V} E(v).$$

Proof. For every $\varepsilon \in \mathbb{R}$ and every "variation" $v \in V$ we have

$$E(\bar{u} + \varepsilon v) - E(\bar{u})$$
$$= \left\{ \frac{1}{2} a(\bar{u} + \varepsilon v, \bar{u} + \varepsilon v) - \langle F, \bar{u} + \varepsilon v \rangle_* \right\} - \left\{ \frac{1}{2} a(\bar{u}, \bar{u}) - \langle F, \bar{u} \rangle_* \right\}$$
$$= \varepsilon \{ a(\bar{u}, v) - \langle F, v \rangle_* \} + \frac{1}{2} \varepsilon^2 a(v, v).$$

Now, if $\bar{u}$ is the solution of problem (7.23), then $a(\bar{u}, v) - \langle F, v \rangle_* = 0$. Therefore

$$E(\bar{u} + \varepsilon v) - E(\bar{u}) = \frac{1}{2} \varepsilon^2 a(v, v) \ge 0$$

so that $\bar{u}$ minimizes E. On the other hand, if $\bar{u}$ is a minimizer of E, then

$$E(\bar{u} + \varepsilon v) - E(\bar{u}) \ge 0,$$

which entails

$$\varepsilon \left\{ a \left(\overline{u}, v \right) - \langle F, v \rangle_* \right\} + \frac{1}{2} \varepsilon^2 a \left(v, v \right) \geq 0.$$

This inequality forces (why?)

$$a \left(\overline{u}, v \right) - \langle F, v \rangle_* = 0 \qquad \forall v \in V \tag{7.29}$$

and $\overline{u}$ is a solution of problem (7.23)). □

Letting $\varphi \left(\varepsilon \right) = E \left(\overline{u} + \varepsilon v \right)$, from the above calculations we have

$$\varphi' (0) = a \left(\overline{u}, v \right) - \langle F, v \rangle_* .$$

Thus, the linear functional

$$v \longmapsto a \left(\overline{u}, v \right) - \langle F, v \rangle_*$$

appears as the **derivative of** E at $\overline{u}$ along the direction v and we write

$$E' \left(\overline{u} \right) v = a \left(\overline{u}, v \right) - \langle F, v \rangle_* . \tag{7.30}$$

In Calculus of Variation E' is called **first variation** and denoted by δE.

If a is symmetric, the *variational equation*

$$E' \left(u \right) v = a \left(u, v \right) - \langle F, v \rangle_* = 0, \qquad \forall v \in V \tag{7.31}$$

is called **Euler equation** for the functional E. This equation constitutes an abstract version of the *principle of virtual work*, while E often represents an "energy".

Remark 7.12. A bilinear form a, symmetric and coercive, induces in V the inner product

$$\left(u, v \right)_a = a \left(u, v \right).$$

In this case, existence, uniqueness and stability for problem (7.23) follow directly from Riesz's Representation Theorem. In particular, *there exists a unique minimizer $\overline{u}$ of E.*

7.4.3 Approximation and Galerkin method

The solution u of the abstract variational problem (7.23), satisfies the equation

$$a \left(u, v \right) = \langle F, v \rangle_* \tag{7.32}$$

for *every* v in the Hilbert space V. In concrete applications, it is important to compute approximate solutions with a given degree of accuracy and the infinite dimension of V is the main obstacle. Often, however, V may be written as a *union of finite-dimensional subspaces*, so that, in principle, it could be reasonable to obtain approximate solutions by "projecting" equation (7.32) on those subspaces. This is the idea of **Galerkin's method**. In principle, the higher the dimension of the subspace the better should be the degree of approximation. More precisely, the idea is to construct a sequence $\{V_k\}$ of subspaces of V with the following properties:

a) every V_k is *finite-dimensional*: $\dim V_k = k$;

b) $V_k \subset V_{k+1}$ (actually, not strictly necessary);

c) $\overline{\cup V_k} = V$.

To realize the projection, assume that the vectors $\psi_1, \psi_2, \ldots, \psi_k$ span V_k. Then, we look for an approximation of the solution u in the form

$$u_k = \sum_{j=1}^{k} c_j \psi_j, \tag{7.33}$$

by solving the *projected* problem

$$a(u_k, v) = \langle F, v \rangle_* \qquad \forall v \in V_k. \tag{7.34}$$

Since $\{\psi_1, \psi_2, \ldots, \psi_k\}$ constitutes a basis in V_k, (7.34) amounts to requiring

$$a(u_k, \psi_r) = \langle F, \psi_r \rangle_* \qquad r = 1, \ldots, k. \tag{7.35}$$

Substituting (7.33) into (7.35), we obtain the k linear algebraic equations

$$\sum_{j=1}^{k} c_j a(\psi_j, \psi_r) = \langle F, \psi_r \rangle_* \qquad r = 1, 2, \ldots, k \tag{7.36}$$

for the unknown coefficients $c_1, c_2, \ldots, c_k$. Introducing the vectors

$$\mathbf{c} = \begin{pmatrix} c_1 \\ c_2 \\ \vdots \\ c_k \end{pmatrix}, \quad \mathbf{F} = \begin{pmatrix} \langle F, \psi_1 \rangle_* \\ \langle F, \psi_2 \rangle_* \\ \vdots \\ \langle F, \psi_k \rangle_* \end{pmatrix}$$

and the matrix $\mathbf{A} = (a_{rj})$, with entries

$$a_{rj} = a(\psi_j, \psi_r), \qquad j, r = 1, \ldots, k,$$

we may write (7.36) in the compact form

$$\mathbf{Ac} = \mathbf{F}. \tag{7.37}$$

The matrix $\mathbf{A}$ is called *stiffness matrix* and clearly plays a key role in the numerical analysis of the problem.

If the bilinear form a is coercive, $\mathbf{A}$ is *strictly positive*. In fact, let $\boldsymbol{\xi} \in \mathbb{R}^k$. Then, by linearity and coercivity

$$\mathbf{A}\boldsymbol{\xi} \cdot \boldsymbol{\xi} = \sum_{r,j=1}^{k} a_{rj}\xi_r\xi_j = \sum_{r,j=1}^{k} a\left(\psi_j, \psi_r\right)\xi_r\xi_j$$

$$= \sum_{r,j=1}^{k} a\left(\xi_j\psi_j, \xi_r\psi_r\right) = a\left(\sum_{j=1}^{k}\xi_j\psi_j, \sum_{r=1}^{k}\xi_r\psi_r\right)$$

$$\geq \alpha \left\|\mathbf{v}\right\|^2$$

where

$$\mathbf{v} = \sum_{j=1}^{k}\xi_j\psi_j \in V_k.$$

Since $\{\psi_1, \psi_2, \ldots, \psi_k\}$ is a basis in V_k, we have $\mathbf{v} = \mathbf{0}$ if and only if $\boldsymbol{\xi} = \mathbf{0}$. Therefore $\mathbf{A}$ is strictly positive and, in particular, non singular.

Thus, for each $k \geq 1$, there exists a unique solution $u_k \in V_k$ of (7.37). We want to show that $u_k \to u$, as $k \to \infty$, i.e. the *convergence of the method*, and give a control of the approximation error.

For this purpose, we prove the following lemma, which also shows the role of the continuity and the coercivity constants (M and α, respectively) of the bilinear form a.

Lemma 7.1 (Céa). *Assume that the hypotheses of the Lax-Milgram Theorem hold and let u be the solution of problem* (7.23). *If u_k is the solution of problem* (7.35), *then*

$$\|u - u_k\| \leq \frac{M}{\alpha} \inf_{v \in V_k} \|u - v\|. \tag{7.38}$$

Proof. We have

$$a\left(u_k, v\right) = \langle F, v \rangle_*, \qquad \forall v \in V_k$$

and

$$a\left(u, v\right) = \langle F, v \rangle_*, \qquad \forall v \in V_k.$$

Subtracting the two equations we obtain

$$a\left(u - u_k, v\right) = 0, \qquad \forall v \in V_k.$$

In particular, since $v - u_k \in V_k$, we have

$$a\left(u - u_k, v - u_k\right) = 0, \qquad \forall v \in V_k$$

which implies

$$a\left(u - u_k, u - u_k\right) = a\left(u - u_k, u - v\right) + a\left(u - u_k, v - u_k\right)$$
$$= a\left(u - u_k, u - v\right).$$

Then, by the coercivity of a,

$$\alpha \left\| u - u_k \right\|^2 \le a\left(u - u_k, u - u_k\right) \le M \left\| u - u_k \right\| \left\| u - v \right\|$$

whence,

$$\left\| u - u_k \right\| \le \frac{M}{\alpha} \left\| u - v \right\| . \tag{7.39}$$

This inequality holds for every $v \in V_k$, with $\frac{M}{\alpha}$ independent of k. Therefore (7.39) still holds if we take in the right hand side the infimum over all $v \in V_k$. $\square$

Convergence of Galerkin's method. Since we have assumed that

$$\overline{\cup V_k} = V,$$

there exists a sequence $\{w_k\} \subset V_k$ such that $w_k \to u$ as $k \to \infty$. Céa's Lemma gives, for every k

$$\left\| u - u_k \right\| \le \frac{M}{\alpha} \inf_{v \in V_k} \left\| u - v \right\| \le \frac{M}{\alpha} \left\| u - w_k \right\|$$

whence, as $k \to +\infty$

$$\left\| u - u_k \right\| \to 0.$$

7.5 Distributions and Functions

7.5.1 Preliminary concepts

We have seen in Section 3.3.3 the concept of *Dirac measure* arising in connection with the fundamental solutions of the diffusion and the wave equations. Another interesting situation is the following, where the Dirac measure models a mechanical impulse.

Consider a mass m moving along the x−axis with constant speed $v\mathbf{i}$ (see Fig. 7.3). At time $t = t_0$ an *elastic* collision with a vertical wall occurs. After the collision, the mass moves with opposite speed $-v\mathbf{i}$. If v_2, v_1 denote the scalar speeds at times t_1, t_2, $t_1 < t_2$, by the laws of mechanics we should have

$$m(v_2 - v_1) = \int_{t_1}^{t_2} F\left(t\right) dt,$$

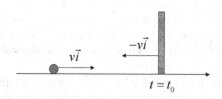

Fig. 7.3. Elastic collision at time $t = t_0$

where F denotes the intensity of the force acting on m. When $t_1 < t_2 < t_0$ or $t_0 < t_1 < t_2$, then $v_2 = v_1 = v$ or $v_2 = v_1 = -v$ and therefore $F = 0$: no force is acting on m before and after the collision. However, if $t_1 < t_0 < t_2$, the left hand side is equal to $2mv \neq 0$. If we insist to model the intensity of the force by a function F, the integral in the right hand side is zero and we obtain a contradiction.

Indeed, in this case, F is a force concentrated at time t_0, of intensity $2mv$, that is

$$F(t) = 2mv \, \delta(t - t_0).$$

In this chapter we see how the Dirac delta is perfectly included in the theory of *distributions or Schwartz generalized functions*. The key idea in this theory is to describe a mathematical object through its action on smooth test functions φ, with compact support. In the case of the Dirac δ, such action is expressed by the formula (see Definition 3.2)

$$\int \delta(x) \, \varphi(x) \, dx = \varphi(0)$$

where, we recall, the integral symbol is purely formal. As we shall shortly see, the appropriate notation is $\langle \delta, \varphi \rangle = \varphi(0)$.

Of course, by a principle of coherence, among the *generalized functions* we should be able to recover the *usual* functions of Analysis. This fact implies that the choice of the test functions cannot be arbitrary. In fact, let $\Omega \subseteq \mathbb{R}^n$ be a domain and take for instance a function $u \in L^2(\Omega)$. Usually u is described *pointwise* by the law

$$\mathbf{x} \longmapsto u(\mathbf{x}).$$

There is however another way to think of u, that is through its *action* on a test function φ, described by the linear functional

$$I_u : \varphi \longmapsto (u, \varphi)_{L^2(\Omega)} = \int_\Omega u\varphi \, d\mathbf{x}.$$

We ask: is it possible to reconstruct u from the knowledge of $I_u(\varphi)$, when φ varies on a set of *nice* functions?

Certainly this is impossible if we use only a restricted set of *test* functions. However, it is possible to recover u from the value of $I_u(\varphi)$, when φ varies in a **dense** set in $L^2(\Omega)$. In the next subsection we construct this set and show how the identification works.

Let us note, however, that the main purpose of introducing the Schwartz distributions is not restricted to a mere extension of the notion of function but it relies on the possibility of broadening the domain of *calculus* in a significant way, opening the door to an enormous amount of new applications. Here the key idea is to use integration by parts to carry the derivatives onto the test functions.

In the first part of this chapter we give the basic concepts of the theory of Schwartz distributions, mainly finalized to the introduction of Sobolev spaces. The basic reference is [35].

7.5.2 Test functions

Recall that, given a continuous function v, defined in a domain $\Omega \subseteq \mathbb{R}^n$, the *support of v* is given by *the closure of the set of points where v is different from zero*

$$\text{supp}(v) = \text{closure of } \{\mathbf{x} \in \mathbb{R}^n : v(\mathbf{x}) \neq 0\}.$$

Actually, the support or, better, the *essential support*, is defined also for measurable functions, not necessarily continuous in Ω. Namely, let Z be the union of the open sets on which $v = 0$ a.e. Then, the closure of $\Omega \backslash Z$ is called *the essential support of v* and we use the same symbol $\text{supp}(v)$ to denote it.

We say that v is *compactly supported* in Ω, if $\text{supp}(v)$ is a *compact* subset of Ω.

Definition 7.10. *Denote by $C_0^\infty(\Omega)$ the set of functions belonging to $C^\infty(\Omega)$, compactly supported in Ω. We call* **test functions** *the elements of $C_0^\infty(\Omega)$.*

The elements of $L^p(\Omega), 1 \leq p < \infty$, can be approximate in L^p norm by functions in $C_0^\infty(\Omega)$. Precisely, the following important theorm holds.

Theorem 7.12. $C_0^\infty(\Omega)$ *is dense in $L^p(\Omega)$ forevery $1 \leq p < \infty$. That is, given $f \in L^p(\Omega)$ there exists a sequence $\{f_k\} \subset C_0^\infty(\Omega)$ such that $\|f_k - f\|_{L^p(\Omega)} \to 0$ as $k \to +\infty$.*

We now go back to our identification problem, in Section 7.5.1. Assume that $u_1, u_2 \in L^2(\Omega)$ and that

$$I_{u_1}(\varphi) = \int_\Omega u_1 \varphi \, d\mathbf{x} = \int_\Omega u_2 \varphi \, d\mathbf{x} = I_{u_2}(\varphi)$$

for *every* $\varphi \in C_0^\infty$. Then

$$\int_\Omega (u_1 - u_2)\varphi \, d\mathbf{x} = 0 \tag{7.40}$$

for *every* $\varphi \in C_0^\infty(\Omega)$. Now, given $\psi \in L^2(\Omega)$, by the density of $C_0^\infty(\Omega)$ in $L^2(\Omega)$, there exists a sequence of test functions $\{\varphi_k\}$ such that $\|\varphi_k - \psi\|_0 \to 0$ as $k \to \infty$. Then[10],

$$0 = \int_\Omega (u - v)\varphi_k \, d\mathbf{x} \to \int_\Omega (u - v)\psi \, d\mathbf{x}$$

so that (7.40) holds *for every* $\psi \in L^2(\Omega)$.

Choosing $\varphi = u_1 - u_2$ in (7.40), we infer

$$\int_\Omega (u_1 - u_2)^2 d\mathbf{x} = 0$$

---

[10] From

$$\left| \int_\Omega (u - v)(\varphi_k - \psi) \, d\mathbf{x} \right| \leq \|u - v\|_0 \|\varphi_k - \psi\|_0.$$

which implies $u_1 = u_2$ a.e. in Ω. Thus, the value of I_u on $C_0^\infty (\Omega)$ **identifies uniquely** u in $L^2 (\Omega)$. In other words we can **identify** u with the functional I_u.

7.5.3 Distributions

We now endow $C_0^\infty (\Omega)$ with a suitable notion of convergence. Recall that the symbol

$$D^\alpha = \frac{\partial^{\alpha_1}}{\partial x_1^{\alpha_1}} \cdots \frac{\partial^{\alpha_n}}{\partial x_n^{\alpha_n}}, \qquad \alpha = (\alpha_1, \ldots, \alpha_n),$$

denotes a derivative of order $|\alpha| = \alpha_1 + \ldots + \alpha_n$.

Definition 7.11. *Let* $\{\varphi_k\} \subset C_0^\infty (\Omega)$ *and* $\varphi \in C_0^\infty (\Omega)$*. We say that*

$$\varphi_k \to \varphi \qquad in\ C_0^\infty (\Omega) \qquad as\ k \to +\infty$$

if the following conditions are fulfilled

1. $D^\alpha \varphi_k \to D^\alpha \varphi$ *uniformly in* Ω, $\forall \alpha = (\alpha_1, \ldots, \alpha_n)$;
2. *there exists a compact set* $K \subset \Omega$ *containing the support of every* φ_k.

It is possible to show that the limit so defined is *unique*. The space $C_0^\infty (\Omega)$ is denoted by $\mathcal{D} (\Omega)$, when endowed with the above notion of convergence.

Following the discussion in the first section, we focus on the linear functionals in $\mathcal{D} (\Omega)$. If L is one of those, we shall use the *bracket* (or *pairing*) $\langle L, \varphi \rangle$ to denote the action of L on a test function φ.

We say that linear functional

$$L : \mathcal{D} (\Omega) \to \mathbb{R}$$

is *continuous* in $\mathcal{D} (\Omega)$ if

$$\langle L, \varphi_k \rangle \to \langle L, \varphi \rangle, \qquad \text{whenever } \varphi_k \to \varphi \text{ in } \mathcal{D} (\Omega). \qquad (7.41)$$

Note that, given the linearity of L, it would be enough to check (7.41) in the case $\varphi = 0$.

Definition 7.12. *A **distribution** in* Ω *is a linear continuous functional in* $\mathcal{D} (\Omega)$*. The set of distributions is denoted by* $\mathcal{D}' (\Omega)$.

Two distributions F and G coincide when their action on every test function is the same, i.e. if

$$\langle F, \varphi \rangle = \langle G, \varphi \rangle, \qquad \forall \varphi \in \mathcal{D} (\Omega).$$

To every $u \in L^2 (\Omega)$ corresponds the functional I_u whose action on φ is

$$\langle I_u, \varphi \rangle = \int_\Omega u\varphi \, dx,$$

which is certainly continuous in $\mathcal{D}(\Omega)$. Therefore I_u is a distribution in $\mathcal{D}'(\Omega)$ and we have seen at the end of Section 7.5.1 that I_u may be identified with u.

Thus, the notion of distribution generalizes the notion of function (in $L^2(\Omega)$) and the pairing $\langle \cdot, \cdot \rangle$ between $\mathcal{D}(\Omega)$ and $\mathcal{D}'(\Omega)$ generalizes the inner product in $L^2(\Omega)$.

The same arguments works for every function $u \in L^1_{loc}(\Omega)$, that is if u is integrable on every compact subset of Ω.

On the other hand, if $u \notin L^1_{loc}$, u **cannot** represent a distribution. A typical example is $u(x) = 1/x$ which does not belongs to $L^1_{loc}(\mathbb{R})$.

Example 7.16 (Dirac delta). The *Dirac delta* at the point $\mathbf{y} \in \mathbb{R}^n$, i.e. $\delta_{\mathbf{y}} : \mathcal{D}(\mathbb{R}^n) \to \mathbb{R}$, whose action is

$$\langle \delta_{\mathbf{y}}, \varphi \rangle = \varphi(\mathbf{y}),$$

is a distribution $\mathcal{D}'(\mathbb{R}^n)$, as it is easy to check. See the Exercise 7.5 to see different approximation of the Dirac delta.

$\mathcal{D}'(\Omega)$ is a linear space. Indeed if α, β are real (or complex) scalars, $\varphi \in \mathcal{D}(\Omega)$ and $L_1, L_2 \in \mathcal{D}'(\Omega)$, we define $\alpha L_1 + \beta L_2 \in \mathcal{D}'(\Omega)$ by means of the formula

$$\langle \alpha L_1 + \beta L_2, \varphi \rangle = \alpha \langle L_1, \varphi \rangle + \beta \langle L_2, \varphi \rangle.$$

In $\mathcal{D}'(\Omega)$ we may introduce a notion of (weak) convergence: $\{L_k\}$ *converges to L as $k \to \infty$ in $\mathcal{D}'(\Omega)$ if*

$$\langle L_k, \varphi \rangle \to \langle L, \varphi \rangle, \qquad \forall \varphi \in \mathcal{D}(\Omega).$$

If $1 \leq p \leq \infty$, we have the **continuous embeddings**:

$$L^p(\Omega) \hookrightarrow L^1_{loc}(\Omega) \hookrightarrow \mathcal{D}'(\Omega).$$

This means that, if $u_k \to u$ in $L^p(\Omega)$ or in $L^1_{loc}(\Omega)$, then[11] $u_k \to u$ in $\mathcal{D}'(\Omega)$ as well.

With respect to this convergence, $\mathcal{D}'(\Omega)$ possesses a *completeness* property that may be used to construct a distribution or to recognize that some linear functional in $\mathcal{D}(\Omega)$ is a distribution. Precisely, one can prove the following result.

---

[11] For instance, let $\varphi \in \mathcal{D}(\Omega)$. We have, by Hölder's inequality:

$$\left| \int_{\Omega} (u_k - u) \varphi \, dx \right| \leq \|u_k - u\|_{L^p(\Omega)} \|\varphi\|_{L^q(\Omega)}$$

where $q = p/(p-1)$. Then, if $\|u_k - u\|_{L^p(\Omega)} \to 0$, also $\int_{\Omega} (u_k - u) \varphi \, d\mathbf{x} \to 0$, showing the convergence of $\{u_k\}$ in $\mathcal{D}'(\Omega)$.

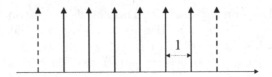

Fig. 7.4. A train of impulses

Proposition 7.5. *Let $\{F_k\} \subset \mathcal{D}'(\Omega)$ such that*

$$\lim_{k \to \infty} \langle F_k, \varphi \rangle$$

exists and is finite for all $\varphi \in \mathcal{D}(\Omega)$. Call $F(\varphi)$ this limit. Then, $F \in \mathcal{D}'(\Omega)$ and $F_k \to F$ in $\mathcal{D}'(\Omega)$.

In particular, if the numerical series

$$\sum_{k=1}^{\infty} \langle F_k, \varphi \rangle$$

converges for all $\varphi \in \mathcal{D}(\Omega)$, then $\sum_{k=1}^{\infty} F_k = F \in \mathcal{D}'(\Omega)$.

Example 7.17 (Dirac comb). For every $\varphi \in \mathcal{D}(\mathbb{R})$, the numerical series

$$\sum_{k=-\infty}^{\infty} \langle \delta(x-k), \varphi \rangle = \sum_{k=-\infty}^{\infty} \varphi(k)$$

is convergent, since only a finite number of terms is different from zero[12]. From Proposition 7.5, we deduce that the series

$$comb(x) = \sum_{k=-\infty}^{\infty} \delta(x-k) \tag{7.42}$$

is convergent in $\mathcal{D}'(\mathbb{R})$ and its sum is a distribution called **Dirac comb**. This name is due to the fact it models a train of impulses concentrated at the integers (see Fig. 7.4, using some ... fantasy).

7.5.4 Calculus

The derivative in the sense of distributions

A central concept in the theory of the Schwartz distributions is the notion of *weak* or *distributional derivative*. Clearly we have to weaken the classical definition, since, for instance, we are going to define the derivative for a function $u \in L_{loc}^1$, which may be quite irregular.

---

[12] Only a finite number of integers k belongs to the support of φ.

The idea is to carry the derivative onto the test functions, as if we were using the integration by parts formula.

Let us start from a function $u \in C^1(\Omega)$. If $\varphi \in \mathcal{D}(\Omega)$, denoting by $\boldsymbol{\nu} = (\nu_1, \ldots, \nu_n)$ the outward normal unit vector to $\partial\Omega$, we have

$$\int_\Omega \varphi \partial_{x_i} u \, d\mathbf{x} = \int_{\partial\Omega} \varphi u \, \nu_i \, d\mathbf{x} - \int_\Omega u \partial_{x_i} \varphi \, d\mathbf{x}$$

$$= -\int_\Omega u \partial_{x_i} \varphi \, d\mathbf{x}$$

since $\varphi = 0$ on $\partial\Omega$. The equation

$$\int_\Omega \varphi \, \partial_{x_i} u \, d\mathbf{x} = -\int_\Omega u \, \partial_{x_i} \varphi \, d\mathbf{x},$$

interpreted in $\mathcal{D}'(\Omega)$, becomes

$$\langle \partial_{x_i} u, \varphi \rangle = -\langle u, \partial_{x_i} \varphi \rangle. \tag{7.43}$$

Formula (7.43) shows that the action of $\partial_{x_i} u$ on the test function φ equals the action of u on the test function $-\partial_{x_i}\varphi$. On the other hand, formula (7.43) makes perfect sense if we replace u by any $F \in \mathcal{D}'(\Omega)$ and it is not difficult to check that it defines a continuous linear functional in $\mathcal{D}(\Omega)$. This leads to the following fundamental notion:

Definition 7.13. *Let $F \in \mathcal{D}'(\Omega)$. The derivative $\partial_{x_i} F$ is the distribution defined by the formula*

$$\langle \partial_{x_i} F, \varphi \rangle = -\langle F, \partial_{x_i} \varphi \rangle, \qquad \forall \varphi \in \mathcal{D}(\Omega).$$

From (7.43), if $u \in C^1(\Omega)$ its derivatives in the sense of distributions of u coincide with the classical ones. This is the reason we keep the same notations in the two cases.

Note that the derivative of a distribution is **always defined**! Moreover, since any derivative of a distribution is a distribution, we deduce the convenient fact that **every distribution possesses derivatives of any order** (in $\mathcal{D}'(\Omega)$):

$$\langle D^\alpha F_k, \varphi \rangle = (-1)^{|\alpha|} \langle F_k, D^\alpha \varphi \rangle.$$

For example, the second order derivative

$$\partial_{x_i x_k} F = \partial_{x_i} \left(\partial_{x_k} F \right)$$

is defined by

$$\langle \partial_{x_i x_k} F, \varphi \rangle = \langle u, \partial_{x_i x_k} \varphi \rangle. \tag{7.44}$$

Not only. Since φ is smooth, then $\partial_{x_i x_k} \varphi = \partial_{x_k x_i} \varphi$ so that (7.44) yields

$$\partial_{x_i x_k} F = \partial_{x_k x_i} F.$$

Thus, for **all** $F \in \mathcal{D}'(\Omega)$ we may always reverse the order of differentiation *without any restriction*.

Example 7.18. Let $u(x) = \mathcal{H}(x)$, the Heaviside function. In $\mathcal{D}'(\mathbb{R})$ we have $\mathcal{H}' = \delta$. In fact, let $\varphi \in \mathcal{D}(\mathbb{R})$. By definition,

$$\langle \mathcal{H}', \varphi \rangle = -\langle \mathcal{H}, \varphi' \rangle.$$

On the other hand, $\mathcal{H} \in L^1_{loc}(\mathbb{R})$, hence

$$\langle \mathcal{H}, \varphi' \rangle = \int_{\mathbb{R}} \mathcal{H}(x) \varphi'(x) \, dx = \int_0^\infty \varphi'(x) \, dx = -\varphi(0)$$

whence

$$\langle \mathcal{H}', \varphi \rangle = \varphi(0) = \langle \delta, \varphi \rangle$$

or $\mathcal{H}' = \delta$.

Another aspect of the idyllic relationship between calculus and distributions is given by the following theorem, which expresses the continuity in $\mathcal{D}'(\Omega)$ of every derivative D^α.

Proposition 7.6. If $F_k \to F$ in $\mathcal{D}'(\Omega)$ then, $D^\alpha F_k \to D^\alpha F$ in $\mathcal{D}'(\Omega)$ for any multi-index α.

Proof. $F_k \to F$ in $\mathcal{D}'(\Omega)$ means that $\langle F_k, \varphi \rangle \to \langle F, \varphi \rangle$, $\forall \varphi \in \mathcal{D}(\Omega)$. In particular, since $D^\alpha \varphi \in \mathcal{D}(\Omega)$,

$$\langle D^\alpha F_k, \varphi \rangle = (-1)^{|\alpha|} \langle F_k, D^\alpha \varphi \rangle \to (-1)^{|\alpha|} \langle F, D^\alpha \varphi \rangle = \langle D^\alpha F, \varphi \rangle. \qquad \square$$

As a consequence, if $\sum_{k=1}^\infty F_k = F$ in $\mathcal{D}'(\Omega)$, then

$$\sum_{k=1}^\infty D^\alpha F_k = D^\alpha F \qquad \text{in } \mathcal{D}'(\Omega).$$

Thus, term by term differentiation is **always** permitted in $\mathcal{D}'(\Omega)$.

The following proposition expresses a well known fact for functions, we omit the rather difficult proof.

Proposition 7.7. Let Ω be a domain in $\mathbb{R}^n$. If $F \in \mathcal{D}'(\Omega)$ and $\partial_{x_j} F = 0$ for every $j = 1, \ldots, n$, then F is a constant function.

Gradient, divergence, laplacian

There is no problem to define *vector valued distributions*. The space of test functions is $\mathcal{D}(\Omega; \mathbb{R}^n)$, i.e. the set of vectors $\varphi = (\varphi_1, \ldots, \varphi_n)$ whose components belong to $\mathcal{D}(\Omega)$.

A distribution $\mathbf{F} \in \mathcal{D}'(\Omega; \mathbb{R}^n)$ is given by $\mathbf{F} = (F_1, \ldots, F_n)$ with $F_j \in \mathcal{D}'(\Omega)$, $j = 1, \ldots, n$. The pairing between $\mathcal{D}(\Omega; \mathbb{R}^n)$ and $\mathcal{D}'(\Omega; \mathbb{R}^n)$ is defined by

$$\langle \mathbf{F}, \varphi \rangle = \sum_{i=1}^n \langle F_i, \varphi_i \rangle. \tag{7.45}$$

• The gradient of $F \in \mathcal{D}'\left(\Omega\right)$, $\Omega \subset \mathbb{R}^n$, is simply

$$\nabla F = \left(\partial_{x_1} F, \partial_{x_2} F, \ldots, \partial_{x_n} F\right).$$

Clearly $\nabla F \in \mathcal{D}'\left(\Omega; \mathbb{R}^n\right)$. If $\varphi \in \mathcal{D}\left(\Omega; \mathbb{R}^n\right)$, we have

$$\langle \nabla F, \varphi \rangle = \sum_{i=1}^{n} \langle \partial_{x_i} F, \varphi_i \rangle = -\sum_{i=1}^{n} \langle F, \partial_{x_i} \varphi_i \rangle = -\langle F, \mathrm{div}\varphi \rangle$$

whence

$$\langle \nabla F, \varphi \rangle = -\langle F, \mathrm{div}\varphi \rangle \tag{7.46}$$

which shows the action of ∇F on φ.

• For $\mathbf{F} \in \mathcal{D}'\left(\Omega; \mathbb{R}^n\right)$, we set

$$\mathrm{div}\mathbf{F} = \sum_{i=1}^{n} \partial_{x_i} F_i.$$

Clearly $\mathrm{div}\mathbf{F} \in \mathcal{D}'\left(\Omega\right)$. If $\varphi \in \mathcal{D}\left(\Omega\right)$, then

$$\langle \mathrm{div}\mathbf{F}, \varphi \rangle = \langle \sum_{1=1}^{n} \partial_{x_i} F_i, \varphi \rangle = -\sum_{1=1}^{n} \langle F_i, \partial_{x_i} \varphi \rangle = -\langle \mathbf{F}, \nabla\varphi \rangle$$

whence

$$\langle \mathrm{div}\mathbf{F}, \varphi \rangle = -\langle \mathbf{F}, \nabla\varphi \rangle. \tag{7.47}$$

• The Laplace operator is defined in $\mathcal{D}'\left(\Omega\right)$ by

$$\Delta F = \sum_{i=1}^{n} \partial_{x_i x_i} F.$$

If $\varphi \in \mathcal{D}\left(\Omega\right)$, then

$$\langle \Delta F, \varphi \rangle = \langle F, \Delta\varphi \rangle.$$

Using (7.46), (7.47) we get

$$\langle \Delta F, \varphi \rangle = \langle F, \mathrm{div}\nabla\varphi \rangle = -\langle \nabla F, \nabla\varphi \rangle = \langle \mathrm{div}\nabla F, \varphi \rangle$$

whence $\Delta = \mathrm{div}\nabla$ also in $\mathcal{D}'\left(\Omega\right)$.

Example 7.19. Consider the **fundamental solution** for the Laplace operator in $\mathbb{R}^3$

$$u\left(\mathbf{x}\right) = \frac{1}{4\pi} \frac{1}{|\mathbf{x}|}.$$

Observe that $u \in L^1_{loc}\left(\mathbb{R}^3\right)$ so that $u \in \mathcal{D}'\left(\mathbb{R}^3\right)$. We want to show that, in $\mathcal{D}'\left(\mathbb{R}^3\right)$,

$$-\Delta u = \delta. \tag{7.48}$$

First of all, if $\Omega \subset \mathbb{R}^3$ and $\mathbf{0} \notin \Omega$, we know that u is *harmonic in* Ω, that is

$$\Delta u = 0 \qquad \text{in } \Omega$$

in the classical sense and therefore also in $\mathcal{D}'(\mathbb{R}^3)$. Thus, let $\varphi \in \mathcal{D}(\mathbb{R}^3)$ with $\mathbf{0} \in \text{supp}(\varphi)$. We have, since $u \in L^1_{loc}(\mathbb{R}^3)$

$$\langle \Delta u, \varphi \rangle = \langle u, \Delta \varphi \rangle = \frac{1}{4\pi} \int_{\mathbb{R}^3} \frac{1}{|\mathbf{x}|} \Delta \varphi(\mathbf{x}) \, d\mathbf{x}. \tag{7.49}$$

We would like to carry the laplacian onto $1/|\mathbf{x}|$. However, this cannot be done directly, since the integrand is not continuous at $\mathbf{0}$. Therefore we exclude a small sphere $B_r = B_r(\mathbf{0})$ from our integration region and write

$$\int_{\mathbb{R}^3} \frac{1}{|\mathbf{x}|} \Delta \varphi(\mathbf{x}) \, d\mathbf{x} = \lim_{r \to 0} \int_{B_R \backslash B_r} \frac{1}{|\mathbf{x}|} \Delta \varphi(\mathbf{x}) \, d\mathbf{x} \tag{7.50}$$

where $B_R = B_R(\mathbf{0})$ is a sphere containing the support of φ. An integration by parts in the ring $C_{R,r} = B_R \backslash B_r$ yields[13]

$$\int_{B_R \backslash B_r} \frac{1}{|\mathbf{x}|} \Delta \varphi(\mathbf{x}) \, d\mathbf{x} = \int_{\partial B_r} \frac{1}{r} \partial_\nu \varphi(\mathbf{x}) \, d\sigma - \int_{C_{R,r}} \nabla \left(\frac{1}{|\mathbf{x}|} \right) \cdot \nabla \varphi(\mathbf{x}) \, d\mathbf{x}$$

where $\nu = -\frac{\mathbf{x}}{|\mathbf{x}|}$ is the *outward* normal unit vector on ∂B_r. Integrating once more by parts the last integral, we obtain:

$$\int_{B_R \backslash B_r} \nabla \left(\frac{1}{|\mathbf{x}|} \right) \cdot \nabla \varphi(\mathbf{x}) \, d\mathbf{x} = \int_{\partial B_r} \partial_\nu \left(\frac{1}{|\mathbf{x}|} \right) \varphi(\mathbf{x}) \, d\sigma - \int_{C_{R,r}} \Delta \left(\frac{1}{|\mathbf{x}|} \right) \varphi(\mathbf{x}) \, d\mathbf{x}$$

$$= \int_{\partial B_r} \partial_\nu \left(\frac{1}{|\mathbf{x}|} \right) \varphi(\mathbf{x}) \, d\sigma,$$

since $\Delta \left(\frac{1}{|\mathbf{x}|} \right) = 0$ inside the ring $C_{R,r}$. From the above computations we infer

$$\int_{B_R \backslash B_r} \frac{1}{|\mathbf{x}|} \Delta \varphi(\mathbf{x}) \, d\mathbf{x} = \int_{\partial B_r} \frac{1}{r} \partial_\nu \varphi(\mathbf{x}) \, d\sigma - \int_{\partial B_r} \partial_\nu \left(\frac{1}{|\mathbf{x}|} \right) \varphi(\mathbf{x}) \, d\sigma. \tag{7.51}$$

We have:

$$\frac{1}{r} \left| \int_{\partial B_r} \partial_\nu \varphi(\mathbf{x}) \, d\sigma \right| \leq \frac{1}{r} \int_{\partial B_r} |\partial_\nu \varphi(\mathbf{x})| \, d\sigma \leq 4\pi r \max_{\mathbb{R}^3} |\nabla \varphi|$$

and therefore

$$\lim_{r \to 0} \int_{\partial B_r} \frac{1}{r} \partial_\nu \varphi(\mathbf{x}) \, d\sigma = 0.$$

---

[13] Recall that $\varphi = 0$ and $\nabla \varphi = \mathbf{0}$ on ∂B_R.

Moreover, since

$$\partial_\nu \left(\frac{1}{|\mathbf{x}|}\right) = \nabla \left(\frac{1}{|\mathbf{x}|}\right) \cdot \left(-\frac{\mathbf{x}}{|\mathbf{x}|}\right) = \left(-\frac{\mathbf{x}}{|\mathbf{x}|^3}\right) \cdot \left(-\frac{\mathbf{x}}{|\mathbf{x}|}\right) = \frac{1}{|\mathbf{x}|^2},$$

we may write

$$\int_{\partial B_r} \partial_\nu \left(\frac{1}{|\mathbf{x}|}\right) \varphi(\mathbf{x})\, d\sigma = 4\pi \frac{1}{4\pi r^2} \int_{\partial B_r} \varphi(\mathbf{x})\, d\sigma \to 4\pi\varphi(\mathbf{0}).$$

Thus, from (7.51) we get

$$\lim_{r \to 0} \int_{B_R \setminus B_r} \frac{1}{|\mathbf{x}|} \Delta\varphi(\mathbf{x})\, d\mathbf{x} = -4\pi\varphi(\mathbf{0})$$

and finally (7.49) yields

$$\langle \Delta u, \varphi \rangle = -\varphi(\mathbf{0}) = -\langle \delta, \varphi \rangle$$

whence $-\Delta u = \delta$.

Multiplication. Leibniz rule

Let us analyze the multiplication between two distributions. Does it make any sense to define, for instance, the product $\delta \cdot \delta = \delta^2$ as a distribution in $\mathcal{D}'(\mathbb{R})$?

Things are not so smooth. An idea for defining δ^2 may be the following: take a sequence $\{u_k\}$ of functions in $L^1_{loc}(\mathbb{R})$ such that $u_k \to \delta$ in $\mathcal{D}'$, compute u_k^2 and set

$$\delta^2 = \lim_{k \to \infty} u_k^2 \qquad \text{in } \mathcal{D}'.$$

Since we may approximate δ in $\mathcal{D}'$ in many ways, it is necessary that the definition *does not depend* on the approximating sequence. In other words, to compute δ^2 we must be free to choose any approximating sequence. However, this is illusory. Indeed choose

$$u_k = k\chi_{[0,1/k]}.$$

We have $u_k \to \delta$ in $\mathcal{D}'(\mathbb{R})$ but, if $\varphi \in D(\mathbb{R})$, by the Mean Value Theorem we have

$$\int_{\mathbb{R}} u_k^2 \varphi = k^2 \int_0^{1/k} \varphi = k\varphi(x_k)$$

for some $x_k \in [0, 1/k]$. Now, if $\varphi(x_k) > 0$, say, we deduce that

$$\int_{\mathbb{R}} u_k^2 \varphi \to +\infty, \qquad k \to +\infty$$

so that $\{u_k^2\}$ *does not converge* in $\mathcal{D}'(\mathbb{R})$.

The method does not work and it seems that there is no other reasonable way to define δ^2. Thus, we simply give up defining δ^2 as a distribution or, in general, the product of a pair of distributions. However, if $F \in \mathcal{D}'(\Omega)$ and $u \in C^\infty(\Omega)$, we may define the product uF by the formula

$$\langle uF, \varphi \rangle = \langle F, u\varphi \rangle, \qquad \forall \varphi \in \mathcal{D}(\Omega).$$

First of all, this makes sense since $u\varphi \in \mathcal{D}(\Omega)$. Also, if $\varphi_k \to \varphi$ in $\mathcal{D}(\Omega)$, then $u\varphi_k \to u\varphi$ in $\mathcal{D}(\Omega)$ and

$$\langle uF, \varphi_k \rangle = \langle F, u\varphi_k \rangle \to \langle F, u\varphi \rangle = \langle uF, \varphi \rangle.$$

so that uF is a well defined element of $\mathcal{D}'(\Omega)$.

Example 7.20. Let $u \in C^\infty(\mathbb{R})$. We have

$$u\delta = u(0)\delta.$$

Indeed, if $\varphi \in \mathcal{D}(\mathbb{R})$,

$$\langle u\delta, \varphi \rangle = \langle \delta, u\varphi \rangle = u(0)\varphi(0) = \langle u(0)\delta, \varphi \rangle.$$

Note that the product $u\delta$ makes sense even if u is only continuous. In particular

$$x\delta = 0.$$

The *Leibniz* rule holds: let $F \in \mathcal{D}'(\Omega)$ and $u \in C^\infty(\Omega)$; then

$$\partial_{x_i}(uF) = u\,\partial_{x_i}F + \partial_{x_i}u\,F. \tag{7.52}$$

In fact, let $\varphi \in \mathcal{D}(\Omega)$; we have:

$$\langle \partial_{x_i}(uF), \varphi \rangle = -\langle uF, \partial_{x_i}\varphi \rangle = -\langle F, u\partial_{x_i}\varphi \rangle$$

while

$$\langle u\,\partial_{x_i}F + \partial_{x_i}u\,F, \varphi \rangle = \langle \partial_{x_i}F, u\varphi \rangle + \langle F, \varphi\partial_{x_i}u \rangle$$
$$= -\langle F, \partial_{x_i}(u\varphi) \rangle + \langle F, \varphi\partial_{x_i}u \rangle = \langle F, u\partial_{x_i}\varphi \rangle$$

and (7.52) follows.

Example 7.21. From $x\delta = 0$ and Leibniz formula we obtain

$$\delta + x\delta' = 0.$$

More generally,

$$x^m\delta^{(k)} = 0 \quad \text{in } \mathcal{D}'(\mathbb{R}), \quad \text{if } 0 \le k < m.$$

7.6 Sobolev Spaces

Sobolev spaces constitute one of the most relevant functional settings for the treatment of boundary value problems. Here, we will be mainly concerned with Sobolev spaces based on $L^2(\Omega)$, developing only the theoretical elements we will need in the sequel[14]. These are spaces of square summable functions on a domain Ω in $\mathbb{R}^n$ whose distributional derivatives up to a certain level are still square summable functions. We shall assume that Ω is a bounded domain whose boundary $\partial\Omega$ is sufficiently regular (*Lipschitz domains*), that is at *almost* every point $\mathbf{p} \in \partial\Omega$ the *tangent plane* and *the interior* or *exterior normals* are well defined. Typical admissible domains are polygons, circular sectors in $\mathbb{R}^2$, polyhedra, cones, cilinders, quite important in several concrete applications and in the numerical approximation techniques.

7.6.1 The space $H^1(\Omega)$

In our context, one of the most important space is the space of functions in $L^2(\Omega)$ whose first partial distibutional derivatives are still functions in $L^2(\Omega)$. To denote this space one uses the symbol[15] $H^1(\Omega)$; thus:

$$H^1(\Omega) = \{v \in L^2(\Omega) : \nabla v \in L^2(\Omega; \mathbb{R}^n)\}.$$

In other words, if $v \in H^1(\Omega)$, every partial derivative $\partial_{x_i} v$ is a function $v_i \in L^2(\Omega)$. This means that

$$\langle \partial_{x_i} v, \varphi \rangle = -(v, \partial_{x_i}\varphi)_0 = (v_i, \varphi)_0, \qquad \forall \varphi \in \mathcal{D}(\Omega)$$

or, more explicitly,

$$\int_\Omega v(\mathbf{x})\, \partial_{x_i}\varphi(\mathbf{x})\, d\mathbf{x} = -\int_\Omega v_i(\mathbf{x})\, \varphi(\mathbf{x})\, d\mathbf{x}, \qquad \forall \varphi \in \mathcal{D}(\Omega).$$

In many applied situations, the Dirichlet integral

$$\int_\Omega |\nabla v|^2$$

represents an energy. The functions in $H^1(\Omega)$ are therefore associated with *configurations having finite energy*. We have:

Theorem 7.13. $H^1(\Omega)$ *is a Hilbert space, continuously embedded in* $L^2(\Omega)$. *The gradient operator is continuous from* $H^1(\Omega)$ *into* $L^2(\Omega; \mathbb{R}^n)$.

---

[14] We omit the most technical proofs, that can be found, for instance, in the classical books of *Adams* [28], or *Maz'ya* [31].

[15] Also $H^{1,2}(\Omega)$ or $W^{1,2}(\Omega)$ are used.

The inner product and the norm in $H^1(\Omega)$ are given, respectively, by

$$(u, v)_{H^1(\Omega)} = \int_\Omega uv \, dx + \int_\Omega \nabla u \cdot \nabla v \, dx.$$

and

$$\|u\|_{H^1(\Omega)}^2 = \int_\Omega u^2 dx + \int_\Omega |\nabla u|^2 \, dx.$$

Remark 7.13. If no confusion arises, the symbol $(u, v)_{1,2}$ can be used instead of $(u, v)_{H^1(\Omega)}$ and $\|u\|_{1,2}$ instead of $\|u\|_{H^1(\Omega)}$. The numbers $1, 2$ in the symbol $\|\cdot\|_{1,2}$ stay for "*first* derivatives in L^2".

Example 7.22. Let $\Omega = B_1(0) = \{\mathbf{x} \in \mathbb{R}^2 : |\mathbf{x}| < 1\}$ and

$$u(\mathbf{x}) = (-\log |\mathbf{x}|)^a, \qquad \mathbf{x} \neq \mathbf{0}.$$

We have, using polar coordinates,

$$\int_{B_1(0)} u^2 = 2\pi \int_0^1 (-\log r)^{2a} \, r dr < \infty, \qquad \text{for every } a \in \mathbb{R},$$

so that $u \in L^2(B_1(0))$ for every $a \in \mathbb{R}$. Moreover:

$$u_{x_i} = -ax_i |\mathbf{x}|^{-2} (-\log |\mathbf{x}|)^{a-1}, \, i = 1, 2,$$

and therefore

$$|\nabla u| = \left| a (-\log |\mathbf{x}|)^{a-1} \right| |\mathbf{x}|^{-1}.$$

Thus, using polar coordinates, we get

$$\int_{B_1(0)} |\nabla u|^2 = 2\pi a^2 \int_0^1 |\log r|^{2a-2} \, r^{-1} dr.$$

This integral is finite only if $2 - 2a > 1$ or $a < 1/2$. In particular, ∇u represents the gradient of u in the distribution sense as well. We conclude that $u \in H^1(B_1(0))$ only if $a < 1/2$.

We point out that when $a > 0$, u is **unbounded** near $\mathbf{0}$.

We have affirmed that the Sobolev spaces constitute an adequate functional setting to solve boundary value problems. This point requires some observations. When we write $f \in L^2(\Omega)$, we may think of a single function

$$f : \Omega \to \mathbb{R} \text{ (or } \mathbb{C}),$$

square summable in the Lebesgue sense. However, if we want to exploit the Hilbert space structure of $L^2(\Omega)$, we need to identify two functions when they are equal a.e. in Ω. Adopting this point of view, each element in $L^2(\Omega)$ is actually an *equivalence class* of which f is a *representative*. The drawback here is that it does not make sense anymore to compute the *value of f at a single point*, since a point is a set with measure zero!

The same considerations hold for "functions" in $H^1(\Omega)$, since

$$H^1(\Omega) \subset L^2(\Omega).$$

On the other hand, if we deal with a boundary value problem, it is clear that *we would like to compute the solution at any point in Ω*!

Even more important is the question of the *trace of a function on the boundary of a domain*. By *trace* of f on $\partial\Omega$ we mean the restriction of f to $\partial\Omega$. In a Dirichlet or Neumann problem we assign precisely the trace of the solution or of its normal derivative on $\partial\Omega$, which is a set with measure zero. Does this make any sense if $u \in H^1(\Omega)$?

There are two cases, in which the trace problem may be solved quite simply: the one-dimensional case and the case of functions with zero trace. These cases will be enough for our purposes. We start with the first case.

• *Characterization of $H^1(a, b)$.* As Example 7.22 shows, a function in $H^1(\Omega)$ may be unbounded. In dimension $n = 1$ this cannot occur. In fact, the elements in $H^1(a, b)$ are continuous functions[16] in $[a, b]$.

Moreover, the fundamental theorem of integral calculus holds. Precisely, we have:

Proposition 7.8. *Let $u \in L^2(a, b)$. Then $u \in H^1(a, b)$ if and only if u is continuous in $[a, b]$ and there exists $w \in L^2(a, b)$ such that*

$$u(y) = u(x) + \int_x^y w(s)\, ds, \qquad \forall x, y \in [a, b]. \tag{7.53}$$

Moreover $u' = w$ (both a.e. and in the sense of distribution).

Proof. Assume that u is continuous in $[a, b]$ and that (7.53) holds with $w \in L^2(a, b)$. Choose $x = a$. Replacing, if necessary, u by $u - u(a)$, we may assume $u(a) = 0$, so that

$$u(y) = \int_a^y w(s)\, ds, \qquad \forall x, y \in [a, b].$$

Let $v \in \mathcal{D}(a, b)$. We have:

$$\langle u', v \rangle = -\langle u, v' \rangle = -\int_a^b u(s) v'(s)\, ds = -\int_a^b \left[\int_a^s w(t)\, dt \right] v'(s)\, ds =$$

(exchanging the order of integration)

$$= -\int_a^b \left[\int_t^b v'(s)\, ds \right] w(t)\, dt = \int_a^b v(t) w(t)\, dt = \langle w, v \rangle.$$

---

[16] Rigorously: every *equivalence class* in $H^1(a, b)$ *has a continuous representative in* $[a, b]$.

Thus $u' = w$ in $\mathcal{D}'(a,b)$ and therefore $u \in H^1(a,b)$. From the Lebesgue Differentiation Theorem we deduce that $u' = w$ a.e. as well.

Viceversa, let $u \in H^1(a,b)$. Define

$$v(x) = \int_c^x u'(s)\,ds, \qquad c, x \in [a,b]. \tag{7.54}$$

The function v is continuous in $[a,b]$ and the above proof shows that $v' = u'$ in $\mathcal{D}'(a,b)$. Then (Proposition 7.7) $u = v + C$, $C \in \mathbb{R}$ and therefore u is continuous in $[a,b]$ as well. Moreover, (7.54) yields

$$u(y) - u(x) = v(y) - v(x) = \int_x^y u'(s)\,ds$$

which is (7.53). $\square$

Since a function $u \in H^1(a,b)$ is continuous in $[a,b]$, the value $u(x_0)$ at every point $x_0 \in [a,b]$ makes perfect sense. In particular *the trace of u* at the end points of the interval is given by the values $u(a)$ and $u(b)$.

7.6.2 The spaces $H_0^1(\Omega)$ and $H_\Gamma^1(\Omega)$

Let $\Omega \subseteq \mathbb{R}^n$. We introduce an important subspace of $H^1(\Omega)$.

Definition 7.14. *We denote by $H_0^1(\Omega)$ the closure of $\mathcal{D}(\Omega)$ in $H^1(\Omega)$.*

Thus $u \in H_0^1(\Omega)$ if and only if there exists a sequence $\{\varphi_k\} \subset \mathcal{D}(\Omega)$ such that $\varphi_k \to u$ in $H^1(\Omega)$, i.e. both $\|\varphi_k - u\|_0 \to 0$ and $\|\nabla\varphi_k - \nabla u\|_0 \to 0$ as $k \to \infty$.

Since the test functions in $\mathcal{D}(\Omega)$ have zero trace on $\partial\Omega$, every $u \in H_0^1(\Omega)$ "inherits" this property and it is reasonable to consider the elements $H_0^1(\Omega)$ as the functions in $H^1(\Omega)$ with *zero trace on $\partial\Omega$*. Clearly, $H_0^1(\Omega)$ is a Hilbert subspace of $H^1(\Omega)$.

An important property that holds in $H_0^1(\Omega)$, particularly useful in the solution of boundary value problems, is expressed by the following **Poincaré's inequality**.

Theorem 7.14. *Let $\Omega \subset \mathbb{R}^n$ be a bounded domain. There exists a positive constant C_P (Poincaré constant) such that, for every $u \in H_0^1(\Omega)$,*

$$\|u\|_0 \le C_P \|\nabla u\|_0. \tag{7.55}$$

Proof. We use a strategy which is rather common for proving formulas in $H_0^1(\Omega)$. First, we prove the formula for $v \in \mathcal{D}(\Omega)$; then, if $u \in H_0^1(\Omega)$, select a sequence $\{v_k\} \subset \mathcal{D}(\Omega)$ converging to u in norm $\|\cdot\|_{1,2}$ as $k \to \infty$, that is

$$\|v_k - u\|_0 \to 0, \qquad \|\nabla v_k - \nabla u\|_0 \to 0.$$

In particular
$$\|v_k\|_0 \to \|u\|_0, \qquad \|\nabla v_k\|_0 \to \|\nabla u\|_0.$$
Since (7.55) holds for every v_k, we have $\|v_k\|_0 \le C_P \|\nabla v_k\|_0$. Letting $k \to \infty$ we obtain (7.55) for u. Thus, it is enough to prove (7.55) for $v \in \mathcal{D}(\Omega)$. To this purpose, from the Gauss Divergence Theorem, we may write

$$\int_\Omega \operatorname{div}(v^2 \mathbf{x})\, d\mathbf{x} = 0 \qquad (7.56)$$

since $v = 0$ on $\partial\Omega$. Now,

$$\operatorname{div}(v^2 \mathbf{x}) = 2v\nabla v \cdot \mathbf{x} + nv^2$$

so that (7.56) yields

$$\int_\Omega v^2 d\mathbf{x} = -\frac{2}{n}\int_\Omega v\nabla v \cdot \mathbf{x}\, d\mathbf{x}.$$

Since Ω is bounded, we have $\max_{\mathbf{x}\in\Omega} |\mathbf{x}| = M < \infty$; therefore, using Schwarz's inequality, we get

$$\int_\Omega v^2 d\mathbf{x} = \frac{2}{n}\left|\int_\Omega v\nabla v \cdot \mathbf{x}\, d\mathbf{x}\right| \le \frac{2M}{n}\left(\int_\Omega v^2 d\mathbf{x}\right)^{1/2}\left(\int_\Omega |\nabla v|^2 d\mathbf{x}\right)^{1/2}.$$

Simplyfying, it follows that

$$\|v\|_0 \le C_P \|\nabla v\|_0$$

with $C_P = 2M/n$. $\qquad\qquad\square$

Inequality (7.55) implies that in $H_0^1(\Omega)$ the norm $\|u\|_{1,2}$ is equivalent to $\|\nabla u\|_0$. Indeed

$$\|u\|_{1,2} = \sqrt{\|u\|_0^2 + \|\nabla u\|_0^2}$$

and from (7.55),

$$\|\nabla u\|_0 \le \|u\|_{1,2} \le \sqrt{C_P^2 + 1}\,\|\nabla u\|_0.$$

Unless explicitly stated, **we will choose in $H_0^1(\Omega)$**

$$(u,v)_1 = (\nabla u, \nabla v)_0 \qquad \text{and} \qquad \|u\|_1 = \|\nabla u\|_0$$

as inner product and norm, respectively.

In similar way, we may define the space of functions $u \in H^1(\Omega)$ vanishing on a relatively open subset $\Gamma \subset \partial\Omega$.

Let $V_{0,\Gamma}$ be the set of functions in $\mathcal{D}(\overline{\Omega})$ vanishing in *a neighborhood* of $\overline{\Gamma}$. Then, we introduce the following set:

Definition 7.15. *$H_{0,\Gamma}^1(\Omega)$ is the closure of $V_{0,\Gamma}$ in $H^1(\Omega)$.*

Remark 7.14. It is possible to prove that the Poincaré inequality holds in $H_{0,\Gamma}^1(\Omega)$.

7.6.3 The dual of $H_0^1(\Omega)$

In the applications of the Lax-Milgram Theorem to boundary value problems, one often needs to identify a given functional F as an element of the dual of $H_0^1(\Omega)$ or, in other words, if F is a linear and continuous functional on $H_0^1(\Omega)$. The dual of $H_0^1(\Omega)$ is denoted by a special symbol.

Definition 7.16. *We denote by* $H^{-1}(\Omega)$ *the dual of* $H_0^1(\Omega)$ *with the norm*

$$\|F\|_{-1} = \sup\left\{|Fv| : v \in H_0^1(\Omega),\ \|v\|_1 \leq 1\right\}.$$

The first thing to observe is that, since $\mathcal{D}(\Omega)$ is dense (by definition) and continuously embedded in $H_0^1(\Omega)$, $H^{-1}(\Omega)$ is a *space of distributions*. This means two things:

a) if $F \in H^{-1}(\Omega)$, its restriction to $\mathcal{D}(\Omega)$ is a distribution;

b) if $F, G \in H^{-1}(\Omega)$ and $F\varphi = G\varphi$ for every $\varphi \in \mathcal{D}(\Omega)$, then $F = G$.

To prove a) it is enough to note that if $\varphi_k \to \varphi$ in $\mathcal{D}(\Omega)$, then $\varphi_k \to \varphi$ in $H_0^1(\Omega)$ as well, and therefore $F\varphi_k \to F\varphi$. Thus $F \in \mathcal{D}'(\Omega)$.

To prove b), let $u \in H_0^1(\Omega)$ and $\varphi_k \to u$ in $H_0^1(\Omega)$, with $\varphi_k \in \mathcal{D}(\Omega)$. Then, since $F\varphi_k = G\varphi_k$ we may write

$$Fu = \lim_{k \to +\infty} F\varphi_k = \lim_{k \to +\infty} G\varphi_k = Gu$$

whence $F = G$.

Thus, $H^{-1}(\Omega)$ is in *one-to-one* correspondence with a subspace of $\mathcal{D}'(\Omega)$ and in this sense we will write

$$H^{-1}(\Omega) \subset \mathcal{D}'(\Omega).$$

Which distributions belong to $H^{-1}(\Omega)$? The following theorem gives a satisfactory answer.

Theorem.7.15. $H^{-1}(\Omega)$ *is the set of distributions of the form*

$$F = f_0 + \mathrm{div}\mathbf{f} \tag{7.57}$$

where $f_0 \in L^2(\Omega)$ *and* $\mathbf{f} = (f_1, \ldots, f_n) \in L^2(\Omega; \mathbb{R}^n)$. *Moreover:*

$$\|F\|_{-1} \leq (1 + C_P)\left\{\|f_0\|_0 + \|\mathbf{f}\|_0\right\}. \tag{7.58}$$

Proof. Let $F \in H^{-1}(\Omega)$. From Riesz's Representation Theorem, there exists a unique $u \in H_0^1(\Omega)$ such that

$$(u, v)_1 = Fv \qquad \forall v \in H_0^1(\Omega).$$

Since

$$(u, v)_1 = (\nabla u, \nabla v) = -\langle \mathrm{div}\nabla u, v \rangle$$

in $\mathcal{D}'(\Omega)$, it follows that (7.57) holds with $f_0 = 0$ and $\mathbf{f} = -\nabla u$. Moreover, $\|u\|_1 = \|F\|_{-1}$.

Viceversa, let $F = f_0 + \mathrm{div}\ \mathbf{f}$, with $f_0 \in L^2(\Omega)$ and $\mathbf{f} = (f_1, \ldots, f_n) \in L^2(\Omega; \mathbb{R}^n)$. Then $F \in \mathcal{D}'(\Omega)$ and, letting $Fv = \langle F, v \rangle$, we have;

$$Fv = \int_\Omega f_0 v\ d\mathbf{x} + \int_\Omega \mathbf{f} \cdot \nabla v\ d\mathbf{x} \qquad \forall v \in \mathcal{D}(\Omega).$$

From the Schwarz and Poincaré inequalities, we have

$$|Fv| \leq (C_P + 1) \{\|f_0\|_0 + \|\mathbf{f}\|_0\} \|v\|_1. \tag{7.59}$$

Thus, F is continuous in the H_0^1−norm. It remains to show that F has a unique continuous extension to all $H_0^1(\Omega)$. Take $u \in H_0^1(\Omega)$ and $\{v_k\} \subset \mathcal{D}(\Omega)$ such that $\|v_k - u\|_1 \to 0$. Then, (7.59) yields

$$|Fv_k - Fv_h| \leq (1 + C_P) \{\|f_0\|_0 + \|\mathbf{f}\|_0\} \|v_k - v_h\|_1.$$

Therefore $\{Fv_k\}$ is a Cauchy sequence in $\mathbb{R}$ and converges to a limit which is independent of the sequence approximating u (why?) and which we may denote by Fu. Finally, since

$$|Fu| = \lim_{k \to \infty} |Fv_k| \quad \text{and} \quad \|u\|_1 = \lim_{k \to \infty} \|v_k\|_1,$$

from (7.59) we get:

$$|Fu| \leq (1 + C_P) \{\|f_0\|_0 + \|\mathbf{f}\|_0\} \|u\|_1$$

showing that $F \in H^{-1}(\Omega)$. $\qquad\qquad\qquad\qquad\qquad\qquad\qquad$ □

Theorem 7.15 says that the elements of $H^{-1}(\Omega)$ are represented by a linear combination of functions in $L^2(\Omega)$ and their first derivatives (in the sense of distributions). In particular, $L^2(\Omega) \hookrightarrow H^{-1}(\Omega)$.

Example 7.23. If $n = 1$, the Dirac δ belongs to $H^{-1}(-a, a)$. Indeed, we have $\delta = \mathcal{H}'$ where $\mathcal{H}$ is the Heaviside function, and $\mathcal{H} \in L^2(-a, a)$.

However, if $n \geq 2$ and $\mathbf{0} \in \Omega$, $\delta \notin H^{-1}(\Omega)$. For instance, let $n = 2$ and $\Omega = B_1(\mathbf{0})$. Assume $\delta \in H^{-1}(\Omega)$. Then we may write

$$\delta = f_0 + \mathrm{div}\ \mathbf{f}$$

for some $f_0 \in L^2(\Omega)$ and $\mathbf{f} \in L^2(\Omega; \mathbb{R}^2)$. Thus, for every $\varphi \in \mathcal{D}(\Omega)$,

$$\varphi(\mathbf{0}) = \langle \delta, \varphi \rangle = \langle f_0 + \mathrm{div}\ \mathbf{f}, \varphi \rangle = \int_\Omega [f_0 \varphi - \mathbf{f} \cdot \nabla \varphi]\ d\mathbf{x}.$$

From Schwarz's inequality, it follows that

$$|\varphi(\mathbf{0})|^2 \leq \left\{ \|f_0\|_0^2 + \|\mathbf{f}\|_0^2 \right\} \|\varphi\|_{1,2}^2$$

and, using the density of $\mathcal{D}(\Omega)$ in $H_0^1(\Omega)$, this estimate should hold for any $\varphi \in H_0^1(\Omega)$ as well. But this is impossible, since in $H_0^1(\Omega)$ there are functions which are unbounded near the origin, as we have seen in Example 7.22.

Example 7.24. Let Ω be a smooth, bounded domain in $\mathbb{R}^n$. Let $u = \chi_\Omega$ be its characteristic function. Since $\chi_\Omega \in L^2(\mathbb{R}^n)$, the distribution $\mathbf{F} = \nabla \chi_\Omega$ belongs to $H^{-1}(\mathbb{R}^n; \mathbb{R}^n)$. The support of $\mathbf{F} = \nabla \chi_\Omega$ coincides with $\partial\Omega$ and its action on a test $\varphi \in \mathcal{D}(\mathbb{R}^n; \mathbb{R}^n)$ is described by the following formula:

$$\langle \nabla \chi_\Omega, \varphi \rangle = - \int_{\mathbb{R}^n} \chi_\Omega \operatorname{div} \varphi \, dx = - \int_{\partial\Omega} \varphi \cdot \nu \, d\sigma.$$

We may regard $\mathbf{F}$ as a "delta uniformly distributed on $\partial\Omega$".

Remark 7.15. It is important to avoid confusion between $H^{-1}(\Omega)$ and $H^1(\Omega)^*$, the dual of $H^1(\Omega)$. Since, in general, $\mathcal{D}(\Omega)$ is **not dense** in $H^1(\Omega)$, the space $H^1(\Omega)^*$ is **not** a space of distributions. Indeed, although the restriction to $\mathcal{D}(\Omega)$ of every $T \in H^1(\Omega)^*$ is a distribution, this restriction does not identifies T. As a simple example, take $\mathbf{f} \in L^2(\Omega; \mathbb{R}^n)$ with $|\mathbf{f}| \geq c > 0$ a.e. and $\operatorname{div} \mathbf{f} = 0$. Define

$$T\varphi = \int_\Omega \mathbf{f} \cdot \nabla \varphi \, d\mathbf{x}.$$

Since $|T\varphi| \leq \|\mathbf{f}\|_0 \|\nabla \varphi\|_0$, we infer that $T \in H^1(\Omega)^*$. However, the restriction of T to $\mathcal{D}(\Omega)$ is the *null operator*, since in $\mathcal{D}'(\Omega)$ we have

$$\langle T, \varphi \rangle = - \langle \operatorname{div} \mathbf{f}, \varphi \rangle = 0 \qquad \forall \varphi \in \mathcal{D}(\Omega).$$

7.6.4 The spaces $H^m(\Omega)$ and $H_0^m(\Omega)$, $m > 1$

Involving higher order derivatives, we may construct new Sobolev spaces. Let N be the number of multi-indexes $\alpha = (\alpha_1, \dots, \alpha_n)$ such that $|\alpha| = \sum_{i=1}^n \alpha_i \leq m$.

Denote by $H^m(\Omega)$ the Sobolev space of the functions in $L^2(\Omega)$, whose *derivatives (in the sense of distributions) up to order m included, are functions in $L^2(\Omega)$.* Thus:

$$H^m(\Omega) = \left\{ v \in L^2(\Omega) : D^\alpha v \in L^2(\Omega), \quad \forall \alpha : |\alpha| \leq m \right\}.$$

We have.

Theorem 7.16. $H^m(\Omega)$ *is a Hilbert space, continuously embedded in $L^2(\Omega)$. The operators D^α, $|\alpha| \leq m$, are continuous from $H^m(\Omega)$ into $L^2(\Omega)$.*

The inner product and the norm in H^m are given, respectively, by

$$(u, v)_{H^m(\Omega)} = (u, v)_{m,2} = \sum_{|\alpha| \leq m} \int_\Omega D^\alpha u D^\alpha v \, d\mathbf{x}$$

and

$$\|u\|_{H^m(\Omega)}^2 = \|u\|_{m,2}^2 = \sum_{|\alpha| \leq m} \int_\Omega |D^\alpha u|^2 \, d\mathbf{x}.$$

If $u \in H^m(\Omega)$, any derivative of u of order k belongs to $H^{m-k}(\Omega)$; more generally, if $|\alpha| = k \le m$, then

$$D^\alpha u \in H^{m-k}(\Omega)$$

and the inclusion $H^m(\Omega) \subset H^{m-k}(\Omega)$, $k \ge 1$ is continuous.

Example 7.25. Let $B_1(0) \subset \mathbb{R}^3$ and consider $u(\mathbf{x}) = |\mathbf{x}|^{-a}$. It is easy to check that $u \in H^1(B_1(0))$ if $a < 1/2$. The second order derivatives of u are given by:

$$u_{x_i x_j} = a(a+2)x_i x_j |\mathbf{x}|^{-a-4} - a\delta_{ij} |\mathbf{x}|^{-a-2}.$$

Then

$$\left| u_{x_i x_j} \right| \le |a(a+2)| \, |\mathbf{x}|^{-a-2}$$

so that $u_{x_i x_j} \in L^2(B_1(0))$ if $2a + 4 < 3$, or $a < -\frac{1}{2}$. Thus $u \in H^2(B_1(0))$ if $a < -1/2$.

An important subspace of $H^m(\Omega)$ is the set of functions in $H^m(\Omega)$ vanishing on $\partial\Omega$ together with their normal derivatives up to order $m-1$ included. This space is denoted by $H_0^m(\Omega)$.

7.6.5 Calculus rules

Most calculus rules in H^m are formally similar to the classical ones, although their proofs are not so trivial. We list here a few of them.

Derivative of a product

Let $u \in H^1(\Omega)$ and $v \in \mathcal{D}(\Omega)$. Then $uv \in H^1(\Omega)$ and

$$\nabla(uv) = u\nabla v + v\nabla u. \tag{7.60}$$

Formula (7.60) holds if both $u, v \in H^1(\Omega)$ as well. In this case, however,

$$uv \in L^1(\Omega) \quad \text{and} \quad \nabla(uv) \in L^1(\Omega; \mathbb{R}^n).$$

Composition I

Let Ω' be a domain in $\mathbb{R}^n$. Let $u \in H^1(\Omega)$ and $g : \Omega' \to \Omega$ be one-to-one and Lipschitz. Then, the composition

$$u \circ g : \Omega' \to \mathbb{R}$$

belongs to $H^1(\Omega')$ and

$$\partial_{x_i}[u \circ g](\mathbf{x}) = \sum_{k=1}^{n} \partial_{x_k} u(g(\mathbf{x})) \partial_{x_i} g_k(\mathbf{x}) \tag{7.61}$$

both a.e. in Ω and in $\mathcal{D}'(\Omega)$. In particular, the Lipschitz change of variables $\mathbf{y} = g(\mathbf{x})$ transforms $H^1(\Omega)$ into $H^1(\Omega')$.

Composition II

Let $u \in H^1(\Omega)$ and $f : \mathbb{R} \rightarrow \mathbb{R}$ be Lipschitz. Then, the composition

$$f \circ u : \Omega \rightarrow \mathbb{R}$$

belongs to $H^1(\Omega)$ and

$$\partial_{x_i} [f \circ u](\mathbf{x}) = f'(u(\mathbf{x})) \, \partial_{x_i} u(\mathbf{x}) \qquad (7.62)$$

both a.e. in Ω and in $\mathcal{D}'(\Omega)$.

In particular, choosing respectively

$$f(t) = |t|, \quad f(t) = \max\{t, 0\} \quad \text{and} \quad f(t) = -\min\{t, 0\},$$

it follows that the following functions:

$$|u|, \quad u^+ = \max\{u, 0\}, \quad \text{and} \quad u^- = -\min\{u, 0\}$$

all belong to $H^1(\Omega)$. For these functions, (7.62) yields

$$\nabla u^+ = \begin{cases} \nabla u & \text{if } u > 0 \\ 0 & \text{if } u \leq 0 \end{cases}, \quad \nabla u^- = \begin{cases} 0 & \text{if } u \geq 0 \\ -\nabla u & \text{if } u < 0 \end{cases}$$

and $\nabla(|u|) = \nabla u^+ + \nabla u^-$, $\nabla u = \nabla u^+ - \nabla u^-$. As a consequence, if $u \in H^1(\Omega)$ is constant in a set $K \subseteq \Omega$, then $\nabla u = 0$ a.e. in K.

7.7 Exercises

7.1. Calculate the norm of the operator introduced in the Example 7.11 in the page 269.

7.2. Calculate the norm of the operator introduced in the Example 7.14 in the page 270.

7.3. Calculate the norm of the operator introduced in the Example 7.15 in the page 270.

7.4. Consider $H = L^2(0,1)$ and let $F : H \longrightarrow \mathbb{R}$ be the linear functional

$$F(u) = \int_0^{1/2} u(t) \, dt.$$

Prove that F is well-defined and that $F \in H'$; then, use the Riesz representation theorem to calculate $\|F\|_{H'}$.

7.5. Approximate in three different ways the Dirac delta function δ in $\mathcal{D}'(\mathbb{R}^n)$.

7.6. Given $u(x) = |x|$ and $S(x) = \text{sign}(x)$. Prove that $u' = S$ in $\mathcal{D}'(\mathbb{R})$ and calculate u'' in $\mathcal{D}'(\mathbb{R})$.

7.7. In distributional sense calculate the derivative of $y = \arctan x^{-1}$.

7.8. Find the second mixed derivative of the distribution $\mathcal{H}(x, y) = \mathcal{H}(x)\mathcal{H}(y)$ in $\mathcal{D}'(\mathbb{R}^2)$.

7.9. Project x^2 on the vector subspace V of $L^2(0, 1)$ spanned by the linear combinations of $v_1 = 1$ and $v_2 = x$. Verify that such projection coincides with the function of V with minimum distance from x^2.

7.10. In $L^2(-1, 1)$, calculate the projection of $x(t) = e^t$ on the subspace V spanned by t and t^2.

7.11. Solve the homogeneous wire problem with fixed boundaries at $x = \pm 1$ subject to pointwise forces at $x = \pm 1/2$:

$$\begin{cases} -u'' = \delta(x - 1/2) + \delta(x + 1/2) & \text{in } (-1, 1) \\ u(-1) = u(1) = 0. \end{cases}$$

7.12. Verify that the function

$$y(x) = \frac{\sinh\sqrt{k}(1 - |x|)}{2\sqrt{k}\cosh\sqrt{k}}$$

solves the elastic homogeneous wire problem with fixed boundaries at $x = \pm 1$, subject to a unit pointwise force at $x = 0$ and to an elastic force proportional to the displacement, with elastic constant $k > 0$:

$$\begin{cases} -u'' + ku = \delta & \text{in } (-1, 1) \\ u(-1) = u(1) = 0. \end{cases}$$

7.13 (Heisenberg Uncertainty Principle). Let $\psi \in C^1(\mathbb{R})$ such that $x[\psi(x)]^2 \to 0$ as $|x| \to \infty$ and $\int_{\mathbb{R}} [\psi(x)]^2 dx = 1$.
    Show that
$$1 \le 2\int_{\mathbb{R}} x^2 |\psi(x)|^2\, dx \int_{\mathbb{R}} |\psi'(x)|^2\, dx.$$

If ψ is a Schrödinger wave function, the first factor in the right hand side measures the spread of the density of a particle, while the second one measures the spread of its momentum).

7.14. Let H be a Hilbert space and V a closed subspace of H. Show that $u = P_V x$ if and only if

$$\begin{cases} 1. \ u \in V \\ 2. \ (x - u, v) = 0, \ \forall v \in V. \end{cases}$$

7.15. Let $f \in L^2(-1,1)$. Find the polynomial of degree $\leq n$ that gives the best approximation of f in the least squares sense, that is, the polynomial p that minimizes
$$\int_{-1}^{1}(f-q)^2$$
among all polynomials q with degree $\leq n$.

7.16 (Hermite's equation; the quantum mechanics harmonic oscillator). Consider the equation
$$w'' + \left(2\lambda + 1 - x^2\right)w = 0 \qquad x \in \mathbb{R} \qquad (7.63)$$
with $w(x) \to 0$ as $x \to \pm\infty$.

a) Show that the change of variables $z = we^{x^2/2}$ transforms (7.63) into Hermite's equation for z:
$$z'' - 2xz' + 2\lambda z = 0$$
with $e^{-x^2/2}z(x) \to 0$ as $x \to \pm\infty$.

b) Consider the Schrödinger wave equation for the harmonic oscillator
$$\psi'' + \frac{8\pi^2 m}{h^2}\left(E - 2\pi^2 m\nu^2 x^2\right)\psi = 0 \qquad x \in \mathbb{R}$$
where m is the mass of the particle, E is the total energy, h is the Plank constant and ν is the vibrational frequency. The physically admissible solutions are those satisfying the following conditions:
$$\psi \to 0 \ \text{ as } x \to \pm\infty \qquad \text{and} \qquad \|\psi\|_{L^2(\mathbb{R})} = 1.$$

Show that there is a solution if and only if
$$E = h\nu\left(n + \frac{1}{2}\right) \qquad n = 0,1,2....$$
and, for each n, the corresponding solution is given by
$$\psi_n(x) = k_n H_n\left(2\pi\sqrt{\nu m/hx}\right)\exp\left(-\frac{2\pi^2\nu m}{h}x^2\right)$$
where $k_n = \left(\dfrac{4\pi\nu m}{2^{2n}(n!)^2 h}\right)^{1/2}$ and H_n is the $n-th$ Hermite polynomial.

7.17. Using separation of variables, solve the following steady state diffusion problem in three dimensions (r, θ, φ spherical coordinates, $0 \leq \theta \leq 2\pi$, $0 \leq \varphi \leq \pi$):
$$\begin{cases} \Delta u = 0 & r < 1, 0 < \varphi < \pi \\ u(1,\varphi) = g(\varphi) & 0 \leq \varphi \leq \pi. \end{cases}$$

7.18. The vertical displacement u of a circular membrane of radius a satisfies the bidimensional wave equation $u_{tt} = \Delta u$, with boundary condition $u(a, \theta, t) = 0$. Supposing the membrane initially at rest, write a formal solution of the problem.

7.19. Let $\{x_k\} \subset \mathbb{R}$, $x_k \to +\infty$. Show that $\sum_{k=1}^{\infty} c_k \delta(x - x_k)$ converges in $\mathcal{D}'(\mathbb{R})$ for all $\{c_k\} \subset \mathbb{R}$.

7.20. Let $u(x) = |x|$ and $S(x) = \text{sign}(x)$. Prove that $u' = S$ in $\mathcal{D}'(\mathbb{R})$.

7.21. Prove that $x^m \delta^{(k)} = 0$ in $\mathcal{D}'(\mathbb{R})$, if $0 \le k < m$.

7.22. Let $n = 3$ and $\mathbf{F} \in \mathcal{D}'(\Omega; \mathbb{R}^3)$. Define curl $\mathbf{F} \in \mathcal{D}'(\Omega; \mathbb{R}^3)$ by the formula

$$\text{curl } \mathbf{F} = (\partial_{x_2} F_3 - \partial_{x_3} F_2, \partial_{x_3} F_1 - \partial_{x_1} F_3, \partial_{x_1} F_2 - \partial_{x_2} F_1).$$

Check that, for all $\varphi = (\varphi_1, \varphi_2, \varphi_3) \in \mathcal{D}(\Omega; \mathbb{R}^3)$,

$$\langle \text{curl } \mathbf{F}, \varphi \rangle = \langle \mathbf{F}, \text{curl } \varphi \rangle.$$

7.23. Show that if

$$u(x_1, x_2) = -\frac{1}{\pi} \ln(x_1^2 + x_2^2)$$

then

$$-\Delta u = \delta, \quad \text{in } \mathcal{D}'(\mathbb{R}^2).$$

8

Variational formulation of elliptic problems

8.1 Elliptic Equations

Poisson's equation $\Delta u = f$ is the simplest among the *elliptic equations*, according to the classification in Section 6.5.1, at least in dimension two. This type of equations plays an important role in the modelling of a large variety of phenomena, often of stationary nature. Typically, in drift, diffusion and reaction models, like those considered in the preceding part of this book, a stationary condition corresponds to a steady state, with no more dependence on time.

Elliptic equations appear in the theory of electrostatic and electromagnetic potentials or in the search of vibration modes of elastic structures as well (e.g. through the method of separation of variables for the wave equation).

Let us define precisely what we mean by *elliptic equation* in dimension n.

Let $\Omega \subseteq \mathbb{R}^n$ be a domain, $\mathbf{A}(\mathbf{x}) = (a_{ij}(\mathbf{x}))$ a square matrix of order n, $\mathbf{b}(\mathbf{x}) = (b_1(\mathbf{x}), ..., b_n(\mathbf{x}))$, $\mathbf{c}(\mathbf{x}) = (c_1(\mathbf{x}), ..., c_n(\mathbf{x}))$ vector fields in $\mathbb{R}^n$, $a_0 = a_0(\mathbf{x})$ and $f = f(\mathbf{x})$ real functions. An equation of the form

$$- \sum_{i,j=1}^{n} \partial_{x_i}\left(a_{ij}(\mathbf{x}) u_{x_j}\right) + \sum_{i=1}^{n} \partial_{x_i}(b_i(\mathbf{x})u) + \sum_{i=1}^{n} c_i(\mathbf{x})u_{x_i} + a_0(\mathbf{x}) u = f(\mathbf{x})$$

(8.1)

or

$$- \sum_{i,j=1}^{n} a_{ij}(\mathbf{x}) u_{x_i x_j} + \sum_{i=1}^{n} b_i(\mathbf{x}) u_{x_i} + a_0(\mathbf{x}) u = f(\mathbf{x}) \qquad (8.2)$$

is said to be **elliptic in** Ω if $\mathbf{A}$ is **positive** in Ω, i.e. if the following *ellipticity condition* holds:

$$\sum_{i,j=1}^{n} a_{ij}(\mathbf{x}) \xi_i \xi_j > 0, \qquad \forall \mathbf{x} \in \Omega, \forall \boldsymbol{\xi} \in \mathbb{R}^n, \boldsymbol{\xi} \neq \mathbf{0}.$$

Salsa S., Vegni F.M.G., Zaretti A., Zunino P.: *A Primer on PDEs. Models, Methods, Simulations.*
Unitext – La Matematica per il 3+2 65.
DOI 10.1007/978-88-470-2862-3_8, © Springer-Verlag Italia 2013

We say that (8.1) is in **divergence form** since it mat be written as

$$\underbrace{-\mathrm{div}(\mathbf{A}\,(\mathbf{x})\,\nabla u)}_{diffusion} + \underbrace{\mathrm{div}(\mathbf{b}(\mathbf{x})u) + \mathbf{c}\,(\mathbf{x})\cdot\nabla u}_{transport} + \underbrace{a_0\,(\mathbf{x})\,u}_{reaction} = \underbrace{f\,(\mathbf{x})}_{external\ source} \qquad (8.3)$$

which emphasizes the particular structure of the higher order terms. Usually, the first term models the diffusion in heterogeneous or anisotropic media, when the constitutive law for the flux function $\mathbf{q}$ is given by the Fourier or Fick law:

$$\mathbf{q} = -\mathbf{A}\nabla u.$$

Here u may represent a temperature or the concentration of a substance. Thus, the term $-\mathrm{div}(\mathbf{A}\nabla u)$ is associated with thermal or molecular diffusion. The matrix $\mathbf{A}$ is called *diffusion matrix*; the dependence of $\mathbf{A}$ on $\mathbf{x}$ denotes anisotropic diffusion.

The examples in Chapter 2 explain the meaning of the other terms in equation (8.3). In particular, $\mathrm{div}(\mathbf{b}u)$ models *convection or transport* and corresponds to a flux function given by

$$\mathbf{q} = \mathbf{b}u.$$

The vector $\mathbf{b}$ has the dimensions of a **velocity**. Think, for instance, of the fumes emitted by a factory installation, which diffuse and are transported by the wind. In this case $\mathbf{b}$ is the wind velocity. Note that, if $\mathrm{div}\mathbf{b} = 0$, then $\mathrm{div}(\mathbf{b}u)$ reduces to $\mathbf{b}\cdot\nabla u$ which is of the same form of the third term $\mathbf{c}\cdot\nabla u$.

The term $a_0 u$ models *reaction*. If u is the concentration of a substance, a_0 represents the rate of decomposition ($a_0 > 0$) or growth ($a_0 < 0$).

Finally, f represents an external action, distributed in Ω, e.g. the rate of heat per unit mass supplied by an external source.

If the entries a_{ij} of the matrix $\mathbf{A}$ and the component b_j of $\mathbf{b}$ are all differentiable, we may compute the divergence of both $\mathbf{A}\nabla u$ and $\mathbf{b}u$, and reduce (8.1) to the *non-divergence form*

$$-\sum_{i,j=1}^{n} a_{ij}\,(\mathbf{x})\,u_{x_i x_j} + \sum_{k=1}^{n}\tilde{b}_k\,(\mathbf{x})\,u_{x_k} + \tilde{c}\,(\mathbf{x})\,u = f\,(\mathbf{x})$$

where

$$\tilde{b}_k\,(\mathbf{x}) = \sum_{i=1}^{n}\partial_{x_i} a_{ik}\,(\mathbf{x}) + b_k\,(\mathbf{x}) + c_k\,(\mathbf{x}) \quad \text{and} \quad \tilde{c}\,(\mathbf{x}) = \mathrm{div}\mathbf{b}\,(\mathbf{x}) + a_0\,(\mathbf{x})\,.$$

However, when the a_{ij} or the b_j are *not differentiable*, we must keep the divergence form and interpret the differential equation (8.3) in a suitable weak sense.

A *non-divergence form equation* is also associated with diffusion phenomena through stochastic processes which generalize the Brownian motion, called *diffusion processes*. Thus, the steady state case is a solution of a non-divergence form equation.

We develop the basic theory of elliptic equations in divergence form, recasting the most common boundary value problems within the functional framework of the abstract variational problems of Section 7.4. This functional setting turns out to be very well adapted to implement numerical approximation procedures, such as Galerkin method.

8.2 The Poisson Problem

Assume we are given a domain $\Omega \subset \mathbb{R}^n$ and two real functions $a_0, f : \Omega \to \mathbb{R}$. We want to determine a function u satisfying the equation

$$-\Delta u + a_0 u = f \qquad \text{in } \Omega$$

and one of the usual boundary conditions on $\partial\Omega$.

The Poisson problem models a variety of stationary phenomena. For instance, if $n = 2$, u can be interpreted as the equilibrium position of a perfectly elastic membrane under a distributed charge f.

A Dirichlet condition ($u = g$ on $\partial\Omega$) means that the boundary of the membrane keeps an assigned position. A Robin condition ($\partial_\nu u + \beta u = g$ on $\partial\Omega$, $\beta > 0$) corresponds to a membrane whose boundary is subject to an elastic restoring force. When $\beta = 0$ (Neumann condition) the membrane boundary is free to move along a vertical guide.

When u describes the equilibrium concentration of a substance, a Dirichlet condition prescribes the boundary concentration, while a Neumann condition prescribes its flux through the boundary.

Let us examine what we mean by *solving* the above Poisson problem. The obvious part is the final goal: we want to show *existence, uniqueness and stability* of the solution; then, based on these results, we want to *compute* the solution by Numerical Analysis approximation methods.

Less obvious is the *meaning of solution*. In fact, in principle, every problem may be formulated in several ways and a different notion of solution is associated with each way. What is important in the applications is to select the "most efficient notion" for the problem under examination, where "efficiency" may stand for the best compromise between *simplicity of both formulation and theoretical treatment, sufficient flexibility and generality, adaptability to numerical methods.*

Here is a (non exhaustive!) list of various notions of solution for the Poisson problem.

Classical solutions are twice continuously differentiable functions; the differential equation and the boundary conditions are satisfied in the usual pointwise sense.

Strong solutions belong to the Sobolev space $H^2(\Omega)$. Thus, they possess derivatives in $L^2(\Omega)$ up to the second order, in the sense of distributions.

The differential equation is satisfied in the pointwise sense, a.e. with respect to the Lebesgue measure in Ω, while the boundary condition is satisfied in the sense of traces.

Distributional solutions belong to $L^1_{loc}(\Omega)$ and the equation holds in the sense of distributions, i.e.:

$$\int_\Omega \{-u\Delta\varphi + a_0(\mathbf{x})u\varphi\}\, d\mathbf{x} = \int_\Omega f\varphi d\mathbf{x}, \quad \forall\varphi \in \mathcal{D}(\Omega).$$

The boundary condition is satisfied in a very weak sense.

Weak or variational solutions belong to the Sobolev space $H^1(\Omega)$. The boundary value problem is recast within the framework of the abstract variational theory developed in Section 7.4. Often the new formulation represents a version of the *principle of virtual work*.

Clearly, all these notions of solution must be connected by a *coherence principle*, which may be stated as follows: if all the data (domain, boundary data, forcing terms) and the solution are C^∞, *all the above notions must be equivalent*. Thus, the *non-classical* notions constitute a generalization of the classical one.

An important task, with consequences in the error control in numerical methods, is to establish the optimal degree of regularity of a non-classical solution.

More precisely, let u be a non-classical solution of the Poisson problem. The question is: *how much does the regularity of the data a_0, f and of the domain Ω affect the regularity of the solution?*

An exhaustive answer requires rather complicated tools. In the sequel we shall indicate only the most relevant results.

The theory for classical and strong solutions is well established and can be found in specialized books (see the references). From the numerical point of view, the *method of finite differences* best fits the differential structure of the problem and aims at approximating classical solutions.

The distributional theory is well developed, is quite general, but is not the most appropriate framework for solving boundary value problems.

Indeed, the sense in which the boundary values are attained is one of the most delicate points, when one is willing to widen the notion of solution.

For our purposes, the most convenient notion of solution is the last one: it leads to a quite flexible formulation with a sufficiently high degree of generality and a basic theory solely relying on the Lax-Milgram Theorem (Section 7.4). Moreover, the analogy (and often the coincidence) with the principle of virtual work indicates a direct connection with the physical interpretation.

Finally, the variational formulation is the most natural one to implement the *Galerkin method (finite elements, spectral elements, etc...)*, widely used in the numerical approximation of the solutions of boundary value problems.

To present the main ideas behind the variational formulation, we start from one-dimensional problems with an equation slightly more general than Poisson's equation.

8.3 Diffusion, Drift and Reaction $(n = 1)$

8.3.1 The problem

We shall derive the variational formulation of the following problem:

$$\begin{cases} \underbrace{-(p(x)u')'}_{diffusion} + \underbrace{q(x)u'}_{transport} + \underbrace{r(x)u}_{reaction} = f(x), & \text{in the interval } (a,b) \\ \text{boundary conditions} & \text{at } x = a \text{ and } x = b. \end{cases} \tag{8.4}$$

We may interpret (8.4) as a stationary problem of *diffusion, drift and reaction*.

The main steps for the weak formulation are the following:

a) Select a space of *smooth test functions, adapted to the boundary conditions*.

b) Multiply the differential equation by a *test* function and integrate over (a,b).

c) Carry one of the derivatives in the divergence term onto the test function via an integration by parts, using the boundary conditions and obtaining an *integral equation*.

d) Interpret the integral equation as an *abstract variational problem* (Section 7.4) in a suitable Hilbert space. In general, this is a Sobolev space, given by the topological closure of the space of test functions.

8.3.2 Dirichlet conditions

We start by analyzing *homogeneous Dirichlet conditions*:

$$\begin{cases} -(p(x)u')' + q(x)u' + r(x)u = f(x), & \text{in } (a,b) \\ u(a) = u(b) = 0. \end{cases} \tag{8.5}$$

Assume first that $p \in C^1([a,b])$, with $p > 0$, and $q, r, f \in C([a,b])$.

Let $u \in C^2(a,b) \cap C([a,b])$ be a classical solution of (8.5). We select $C_0^1(a,b)$ as the space of test functions. These test functions have a continuous derivative and compact support in (a,b). In particular, *they vanish at the end points*.

Now we multiply the equation by an arbitrary $v \in C_0^1(a,b)$ and integrate over (a,b). We find

$$-\int_a^b (pu')'v\,dx + \int_a^b [qu' + ru]v\,dx = \int_a^b fv\,dx. \qquad (8.6)$$

Integrating by parts the first term and using $v(a) = v(b) = 0$, we get

$$-\int_a^b (pu')'v\,dx = \int_a^b pu'v'\,dx - [pu'v]_a^b = \int_a^b pu'v'\,dx.$$

From (8.6) we derive the integral equation

$$\int_a^b [pu'v' + qu'v + ruv]\,dx = \int_a^b fv\,dx, \qquad \forall v \in C_0^1(a,b). \qquad (8.7)$$

Thus, (8.5) implies (8.7).

On the other hand, assume that (8.7) is true. Integrating by parts in the reverse order, we recover (8.6), which can be written in the form

$$\int_a^b \{-(pu')' + q(x)u' + r(x)u - f(x)\}v\,dx = 0 \qquad \forall v \in C_0^1(a,b).$$

The arbitrariness of v entails[1]

$$-(p(x)u')' + q(x)u' + r(x)u - f(x) = 0 \quad \text{in } (a,b)$$

i.e. the original differential equation.

Thus, **for classical solutions, the two formulations** (8.5) **and** (8.7) **are equivalent**. Observe that equation (8.7):

- involves *only one derivative* of u;
- makes perfect sense even if p, q, r and f are merely locally integrable;
- transforms (8.5) into an integral equation, valid on an infinite-dimensional space of test functions.

These features lead to the following functional setting:

a) we enlarge the class of test functions to $H_0^1(a,b)$, which is the closure of $C_0^1(a,b)$ in H^1−norm;

b) we look for a solution belonging to $H_0^1(a,b)$, in which the homogeneous Dirichlet conditions are already included.

Thus, the **weak** or **variational** formulation of problem (8.5) is: Determine $u \in H_0^1(a,b)$ such that

$$\int_a^b \{pu'v' + qu'v + ruv\}\,dx = \int_a^b fv\,dx, \qquad \forall v \in H_0^1(a,b). \qquad (8.8)$$

---

[1] If $g \in C([a,b])$ and $\int_a^b gv\,dx = 0$ for every $v \in C_0^1(a,b)$, then $g \equiv 0$ (exercise).

If we introduce the bilinear form

$$B\left(u,v\right) = \int_a^b \{pu'v' + qu'v + ruv\}\, dx$$

and the linear functional

$$Lv = \int_a^b fv\, dx,$$

equation (8.8) can be recast as

$$B\left(u,v\right) = Lv, \quad \forall v \in H_0^1(a,b).$$

Then existence, uniqueness and stability follow from the Lax-Milgram Theorem 7.10, under rather natural hypotheses on $p, q, r,\ f$. Recall that, by Poincaré's inequality (7.55) we have

$$\|u\|_0 \le C_P \|u'\|_0,$$

so that we may choose in $H_0^1(a,b)$ the norm

$$\|u\|_1 = \|u'\|_0$$

equivalent to $\|u\|_{1,2} = \|u\|_0 + \|u'\|_0$.

Proposition 8.1. *Assume that* $p, q, q', r \in L^\infty(a,b)$ *and* $f \in L^2(a,b)$. *If*

$$p\left(x\right) \ge \alpha > 0 \quad and \quad -\frac{1}{2}q'\left(x\right) + r\left(x\right) \ge 0 \quad a.e.\ in\ (a,b), \qquad (8.9)$$

then (8.8) has a unique solution $u \in H_0^1(a,b)$. *Moreover*

$$\|u'\|_0 \le \frac{C_P}{\alpha} \|f\|_0. \qquad (8.10)$$

Proof. Let us check that the hypotheses of the Lax-Milgram Theorem hold, with $V = H_0^1(a,b)$.

Continuity of the bilinear form B. We have:

$$|B\left(u,v\right)| \le \int_a^b \{\|p\|_{L^\infty} |u'v'| + \|q\|_{L^\infty} |u'v| + \|r\|_{L^\infty} |uv|\}\, dx.$$

Using the Schwarz and Poincaré inequalities, we obtain

$$|B\left(u,v\right)| \le \|p\|_{L^\infty} \|u'\|_0 \|v'\|_0 + \|q\|_{L^\infty} \|u'\|_0 \|v\|_0 + \|r\|_{L^\infty} \|u\|_0 \|v\|_0$$
$$\le \left(\|p\|_{L^\infty} + C_P \|q\|_{L^\infty} + C_P^2 \|r\|_{L^\infty}\right) \|u'\|_0 \|v'\|_0$$

so that B is continuous in V.

Coerciveness of B. We may write:

$$B\left(u, u\right) = \int_a^b \left\{p(u')^2 + qu'u + ru^2\right\} dx$$

$$\geq \alpha \left\|u'\right\|_0^2 + \frac{1}{2} \int_a^b q\left(u^2\right)' dx + \int_a^b ru^2 dx$$

$$(\text{integrating by parts}) = \alpha \left\|u'\right\|_0^2 + \int_a^b \left\{-\frac{1}{2}q' + r\right\} u^2 dx$$

$$(\text{from } (8.9)) \geq \alpha \left\|u'\right\|_0^2$$

and therefore B is $V-$coercive.

Continuity of L in V. The Schwarz and Poincaré inequalities yield

$$|Lv| = \left|\int_a^b fv\, dx\right| \leq \left\|f\right\|_0 \left\|v\right\|_0 \leq C_P \left\|f\right\|_0 \left\|v'\right\|_0.$$

so that $\left\|L\right\|_{V^*} \leq C_P \left\|f\right\|_0$.

Then, the Lax-Milgram Theorem gives existence, uniqueness and the stability estimate (8.10). □

Remark 8.1. If $q = 0$, the bilinear form B is symmetric. From Theorem 7.11, in this case the weak solution minimizes in $H_0^1\left(a, b\right)$ the "energy functional"

$$J\left(u\right) = \int_a^b \left\{p\left(u'\right)^2 + ru^2 - 2fu\right\} dx.$$

Then, equation (8.8) coincides with the Euler equation of J:

$$J'\left(u\right) v = 0, \qquad \forall v \in H_0^1\left(a, b\right).$$

Remark 8.2. In the case of nonhomogeneous Dirichlet conditions, e.g. $u\left(a\right) = A$, $u\left(b\right) = B$, set $w = u - y$, where $y = y\left(x\right)$ is the straight line through the points (a, A), (b, B), given by

$$y\left(x\right) = A + \left(x - a\right) \frac{B - A}{b - a}.$$

Then, the variational problem for w is

$$\int_a^b [pw'v' + qw'v + rwv]dx = \int_a^b \left(Fv + Gv'\right) dx \qquad \forall v \in H_0^1(a, b) \qquad (8.11)$$

with

$$F\left(x\right) = f\left(x\right) + \frac{B - A}{b - a}q\left(x\right) - r(x)\left(A + \left(x - a\right)\frac{B - A}{b - a}\right)$$

and

$$G\left(x\right) = \frac{B - A}{b - a}p\left(x\right).$$

8.3.3 Neumann conditions

We now derive the weak formulation of the Neumann problem

$$\begin{cases} -(p(x)u')' + q(x)u' + r(x)u = f(x), & \text{in } (a,b) \\ -p(a)u'(a) = A, \quad p(b)u'(b) = B. \end{cases} \tag{8.12}$$

The boundary conditions prescribe the outward flux at the end points. This way of writing the Neumann conditions, with the presence of the factor p in front of the derivative, is naturally associated with the divergence structure of the diffusion term.

Again, assume first that $p \in C^1([a,b])$, with $p > 0$, and $q, r, f \in C^0([a,b])$. A classical solution u has a continuous derivative up to the end points so that $u \in C^2(a,b) \cap C^1([a,b])$.

As space of test functions, we choose $C^1([a,b])$. Multiplying the equation by an arbitrary $v \in C^1([a,b])$ and integrating over (a,b), we find again

$$-\int_a^b (pu')'v\,dx + \int_a^b [qu' + ru]v\,dx = \int_a^b fv\,dx. \tag{8.13}$$

Integrating by parts the first term and using the Neumann conditions, we get

$$-\int_a^b (pu')'v\,dx = \int_a^b pu'v'\,dx - [pu'v]_a^b = \int_a^b pu'v'\,dx - v(b)B - v(a)A.$$

Then (8.13) becomes

$$\int_a^b [pu'v' + qu'v + ruv]\,dx - v(b)B - v(a)A = \int_a^b fv\,dx, \tag{8.14}$$

for every $v \in C^1([a,b])$.

Thus, (8.12) implies (8.14). If the choice of the test functions is correct, we should be able to recover the classical formulation from (8.14).

Indeed, let us start recovering the differential equation. Since

$$C_0^1(a,b) \subset C^1([a,b]),$$

(8.14) clearly holds for every $v \in C_0^1(a,b)$. Then, (8.14) reduces to (8.7) and we deduce, as before,

$$-(pu')' + qu' + ru - f = 0, \qquad \text{in } (a,b). \tag{8.15}$$

Let us now use the test functions which *do not vanish at the end points*. Integrating by parts the first term in (8.14) we have:

$$\int_a^b pu'v'\,dx = -\int_a^b (pu')'v\,dx + p(b)v(b)u'(b) - p(a)v(a)u'(a).$$

Inserting this expression into (8.14) and taking into account (8.15) we find:

$$v\left(b\right)\left[p\left(b\right)u'\left(b\right)-B\right]-v\left(a\right)\left[p\left(a\right)u'\left(a\right)+A\right]=0.$$

The arbitrariness of the values $v\left(b\right)$ and $v\left(a\right)$ forces

$$p\left(b\right)u'\left(b\right)=B,\qquad -p\left(a\right)u'\left(a\right)=A,$$

recovering the Neumann conditions as well.

Thus, **for classical solutions, the two formulations** (8.12) **and** (8.14) **are equivalent**.

Enlarging the class of test functions to $H^1(a,b)$, which is the closure of $C^1\left(\left[a,b\right]\right)$ in H^1−norm, we may state the **weak** or **variational** formulation of problem (8.12) as follows:

Determine $u\in H^1(a,b)$ such that, $\forall v\in H^1(a,b)$,

$$\int_a^b\{pu'v'+qu'v+ruv\}\,dx=\int_a^b fv\,dx+v\left(b\right)B+v\left(b\right)A.\qquad(8.16)$$

We point out that the Neumann conditions are encoded in equation (8.16), rather than forced by the choice of the test functions, as in the Dirichlet problem.

Introducing the bilinear form

$$B\left(u,v\right)=\int_a^b\{pu'v'+qu'v+ruv\}\,dx$$

and the linear functional

$$Lv=\int_a^b fv\,dx+v\left(b\right)B+v\left(a\right)A,$$

equation (8.16) can be recast in the abstract form

$$B\left(u,v\right)=Lv,\quad\forall v\in H^1(a,b).$$

Again, existence, uniqueness and stability of a weak solution follow from the Lax-Milgram Theorem, under rather natural hypotheses on p,q,r,f.

Recall that if $v\in H^1(a,b)$, the Proposition 7.8 yields to the estimate

$$v(x)\le C^*\left\|v\right\|_{1,2}\qquad(8.17)$$

for every $x\in[a,b]$, with $C^*=\sqrt{2}\max\left\{(b-a)^{-1/2},(b-a)^{1/2}\right\}.$

Proposition 8.2. *Assume that:*

i) *$p, q, r \in L^\infty(a, b)$ and $f \in L^2(a, b)$;*

ii) *$p(x) \geq \alpha_0 > 0$, $r(x) \geq c_0 > 0$ a.e. in (a, b) and*

$$K_0 \equiv \min\{\alpha_0, c_0\} - \frac{1}{2}\|q\|_{L^\infty} > 0.$$

Then, (8.16) has a unique solution $u \in H^1(a, b)$. Furthermore

$$\|u\|_{1,2} \leq K_0^{-1}\left\{\|f\|_0 + C^*(|A| + |B|)\right\}. \tag{8.18}$$

Proof. Let us check that the hypotheses of the Lax-Milgram Theorem hold, with $V = H^1(a, b)$.

Continuity of the bilinear form B. We have:

$$|B(u, v)| \leq \int_a^b \left\{\|p\|_{L^\infty}|u'v'| + \|q\|_{L^\infty}|u'v| + \|r\|_{L^\infty}|uv|\right\} dx.$$

Using Schwarz's inequality, we easily get

$$|B(u, v)| \leq (\|p\|_{L^\infty} + \|q\|_{L^\infty} + \|r\|_{L^\infty})\|u\|_{1,2}\|v\|_{1,2}$$

so that B is continuous in V.

Coerciveness of B. We have

$$B(u, u) = \int_a^b \left\{p(u')^2 + qu'u + ru^2\right\} dx.$$

The Schwarz inequality gives

$$\left|\int_a^b qu'u\, dx\right| \leq \|q\|_{L^\infty}\|u'\|_0\|u\|_0 \leq \frac{1}{2}\|q\|_{L^\infty}\left\{\|u'\|_0^2 + \|u\|_0^2\right\}.$$

Then, by *ii*),

$$B(u, u) \geq \left(\alpha_0 - \frac{1}{2}\|q\|_{L^\infty}\right)\|u'\|_0^2 + \left(c_0 - \frac{1}{2}\|q\|_{L^\infty}\right)\|u\|_0^2 \geq K_0\|u\|_{1,2}^2$$

so that B is V-coercive.

Continuity of L in V. Schwarz's inequality and (8.17) yield

$$|Lv| \leq \|f\|_0\|v\|_0 + |v(b)B + v(a)A| \leq$$
$$\leq \{\|f\|_0 + C^*(|A| + |B|)\}\|v\|_{1,2}$$

whence $\|L\|_{V^*} \leq \|f\|_0 + C^*(|A| + |B|)$.

Then, the Lax-Milgram Theorem gives existence, uniqueness and the stability estimate (8.18). $\qquad\square$

Remark 8.3. In general, without the condition like $c(x) \geq c_0 > 0$ we cannot expect existence and uniqueness of a solution. In fact, suppose, $p = 1$, $q = r = 0$. The problem reduces to

$$\begin{cases} u'' = f & \text{in } (a,b) \\ -u'(a) = A, \quad u'(b) = B. \end{cases} \tag{8.19}$$

Hypothesis $ii)$ is not satisfied (since $r = 0$). If u is a solution of the problem and $k \in \mathbb{R}$, also $u + k$ is a solution of the same problem. We cannot expect uniqueness. Not even we may prescribe f, A, B arbitrarily, if we want that a solution exists. In fact, integrating the equation $u'' = f$ over (a,b), we deduce that the Neumann data and f must satisfy the *compatibility condition*

$$B + A = \int_a^b f(x)\, dx. \tag{8.20}$$

If (8.20) does not hold, *the problem has no solution*. Thus, to have existence and uniqueness we must require that:

a) (8.20) holds;
b) select a solution (e.g.) with zero mean value in (a,b).

Condition (8.20) is indeed quite natural. Problem (8.19), with $A = B = 0$, describes an elastic chord whose end points are free to slide along a vertical guide; in this case (8.20) becomes

$$\int_a^b f(x)\, dx = 0$$

which expresses the obvious fact that in equilibrium conditions the total distributed charge on the chord must vanish.

Remark 8.4. Notice that the Neumann condition is incorporated into the variational formulation of the problem and for this reason they are considered as *natural conditions*. On the opposite, a Dirichlet condition is prescribed at the beginning so that it constitutes a *forced condition*.

8.3.4 Robin and mixed conditions

Suppose that the boundary conditions in problem (8.12) are:

$$-p(a)\, u'(a) = A, \qquad p(b)\, u'(b) + hu(b) = B \quad (h > 0, \text{ constant})$$

where, for simplicity, the Robin condition is imposed at $x = b$ only. With small adjustments, we may repeat the same computations made for the Neumann conditions. The **weak formulation** is:

Determine $u \in H^1(a,b)$ such that, $\forall v \in H^1(a,b)$,

$$\int_a^b \{pu'v' + qu'v + ruv\}\, dx + hu(b)\, v(b) = \int_a^b fv\, dx + v(b)\, B + v(a)\, A. \tag{8.21}$$

Introducing the bilinear form

$$\tilde{B}(u,v) = \int_a^b \{pu'v' + qu'v + ruv\}\, dx + hu(b)\, v(b)$$

we may write our problem in the abstract form

$$\tilde{B}(u,v) = Lv \qquad \forall v \in H^1(a,b).$$

We have:

Proposition 8.3. *Assume that i) and ii) of Proposition 8.2 hold and that $h > 0$. Then, (8.21) has a unique solution $u \in H^1(a,b)$. Furthermore*

$$\|u\|_{1,2} \le K_0^{-1}\{\|f\|_0 + C^*(|A| + |B|)\}.$$

Proof. Let $V = H^1(a,b)$. Since

$$\tilde{B}(u,u) = B(u,u) + hu^2(b) \ge K_0 \|u\|_{1,2}^2$$

and

$$\left|\tilde{B}(u,v)\right| \le |B(u,v)| + h\,|u(b)\,v(b)|$$
$$\le \left(\|p\|_{L^\infty} + \|q\|_{L^\infty} + \|r\|_{L^\infty} + h(C^*)^2\right)\|u\|_{1,2}\,\|v\|_{1,2},$$

$\tilde{B}$ is continuous and V−coercive. The conclusion follows easily. $\qquad\square$

Mixed conditions. The weak formulation of mixed problems does not present particular difficulties. Suppose, for instance, we assign at the end points the conditions

$$u(a) = 0, \qquad p(b)\, u'(b) = B.$$

Thus, we have a mixed Dirichlet-Neumann problem. The only relevant observation is the choice of the functional setting. Since $u(a) = 0$, we have to choose $V = H^1_{0,a}$, the space of functions $v \in H^1(a,b)$, vanishing at $x = a$. The Poincaré inequality holds in $H^1_{0,a}$, so that we may choose $\|u'\|_0$ as the norm in $H^1_{0,a}$. Moreover, the following inequality

$$v(x) \le C^{**}\|v'\|_0 \tag{8.22}$$

holds[2] for every $x \in [a,b]$, with $C^{**} = (b-a)^{1/2}$.

The **weak formulation** is: *find $u \in H^1_{0,a}$ such that*

$$\int_a^b \{pu'v' + qu'v + ruv\}\, dx = \int_a^b fv\, dx + v(b)\, B, \qquad \forall v \in H^1_{0,a}. \tag{8.23}$$

---

[2] Since $v(a) = 0$, we have $v(x) = \int_a^x v'$ so that $|v(x)| \le \sqrt{b-a}\,\|v'\|_0$.

We have:

Proposition 8.4. *Assume that* i) *and* ii) *of Proposition 8.1 hold. Then,* (8.23) *has a unique solution* $u \in H_{0,a}^1$. *Furthermore*

$$\|u'\|_0 \leq K_0^{-1} \{C_P \|f\|_0 + C^{**} |B|\}.$$

We leave the proof as an exercise.

8.4 Variational Formulation of Poisson's Problem

Guided by the one-dimensional case, we now analyze the variational formulation of Poisson's problem in dimension $n > 1$.

8.4.1 The Dirichlet problem

Let $\Omega \subset \mathbb{R}^n$ be a *bounded domain*. We examine the following problem

$$\begin{cases} -\Delta u + a_0\left(\mathbf{x}\right) u = f & \text{in } \Omega \\ \qquad\qquad\quad u = 0 & \text{on } \partial\Omega. \end{cases} \tag{8.24}$$

To achieve a weak formulation, we first assume that a_0 and f are smooth and that $u \in C^2\left(\Omega\right) \cap C^0\left(\overline{\Omega}\right)$ is a classical solution of (8.24). We select $C_0^1\left(\Omega\right)$ as the space of test functions, having continuous first derivatives and compact support in Ω. In particular, *they vanish in a neighborhood of* $\partial\Omega$. Let $v \in C_0^1\left(\Omega\right)$ and multiply the Poisson equation by v. We get

$$\int_\Omega \{-\Delta u + a_0 u - f\} v \, d\mathbf{x} = 0. \tag{8.25}$$

Integrating by parts and using the boundary condition, we obtain

$$\int_\Omega \{\nabla u \cdot \nabla v + a_0 uv\} \, d\mathbf{x} = \int_\Omega fv \, d\mathbf{x}, \qquad \forall v \in C_0^1\left(\Omega\right). \tag{8.26}$$

Thus (8.24) implies (8.26).

On the other hand, assume (8.26) is true. Integrating by parts in the reverse order we return to (8.25), which entails $-\Delta u + a_0 u - f = 0$ in Ω.

Thus, **for classical solutions, the two formulations** (8.24) and (8.26) **are equivalent**.

Observe that (8.26) only involves first order derivatives of the solution and of the test function. Then, enlarging the space of test functions to $H_0^1\left(\Omega\right)$, closure of $C_0^1\left(\Omega\right)$ in the norm $\|u\|_1 = \|\nabla u\|_0$, we may state the **weak** formulation of problem (8.24) as follows:

Determine $u \in H_0^1\left(\Omega\right)$ such that

$$\int_\Omega \{\nabla u \cdot \nabla v + a_0 uv\} \, d\mathbf{x} = \int_\Omega fv \, d\mathbf{x}, \qquad \forall v \in H_0^1\left(\Omega\right). \tag{8.27}$$

Introducing the bilinear form

$$B\left(u,v\right) = \int_{\Omega} \left\{\nabla u \cdot \nabla v + a_0 uv\right\}\, d\mathbf{x}$$

and the linear functional

$$Lv = \int_{\Omega} fv\, d\mathbf{x},$$

equation (8.27) corresponds to the abstract variational problem

$$B\left(u,v\right) = Lv, \quad \forall v \in H_0^1\left(\Omega\right).$$

Then, the well-posedness of this problem follows from the the Lax-Milgram Theorem under the hypothesis $a_0 \geq 0$. Precisely:

Theorem 8.1. *Assume that $f \in L^2\left(\Omega\right)$ and that $0 \leq a_0\left(\mathbf{x}\right) \leq \gamma_0$ a.e. in Ω. Then, problem (8.27) has a unique solution $u \in H_0^1\left(\Omega\right)$. Moreover*

$$\|\nabla u\|_0 \leq \frac{C_P}{\alpha} \|f\|_0.$$

Proof. We check that the hypotheses of the Lax-Milgram Theorem hold, with $V = H_0^1(\Omega)$.

Continuity of the bilinear form B. The Schwarz and Poincaré inequalities yield:

$$|B\left(u,v\right)| \leq \|\nabla u\|_0 \|\nabla v\|_0 + \gamma_0 \|u\|_0 \|v\|_0$$
$$\leq \left(1 + C_P^2 \gamma_0\right) \|\nabla u\|_0 \|\nabla v\|_0$$

so that B is continuous in $H_0^1\left(\Omega\right)$.

Coerciveness of B. It follows from

$$B\left(u,u\right) = \int_{\Omega} |\nabla u|^2\, d\mathbf{x} + \int_{\Omega} a_0 u^2 d\mathbf{x} \geq \alpha \|\nabla u\|_0^2$$

since $a_0 \geq 0$.

Continuity of L. The Schwarz and Poincaré inequalities give

$$|Lv| = \left|\int_{\Omega} fv\, d\mathbf{x}\right| \leq \|f\|_0 \|v\|_0 \leq C_P \|f\|_0 \|\nabla v\|_0.$$

Hence $L \in H^{-1}\left(\Omega\right)$ and $\|L\|_{H^{-1}(\Omega)} \leq C_P \|f\|_0$. The conclusions follow from the Lax-Milgram Theorem. $\qquad\square$

Remark 8.5. Suppose that $c = 0$ and that u represents the equilibrium position of an elastic membrane. Then $B\left(u,v\right)$ represents the work done by the elastic internal forces, due to a *virtual displacement* v. On the other hand Lv expresses the work done by the external forces. The weak formulation (8.27) states that these two works balance, which constitutes a version of the *principle of virtual work*.

Furthermore, due to the symmetry of B, the solution u of the problem **minimizes in $H_0^1(\Omega)$ the Dirichlet functional**

$$E(u) = \underbrace{\int_\Omega |\nabla u|^2 \, d\mathbf{x}}_{\text{internal elastic energy}} - \underbrace{\int_\Omega fu \, d\mathbf{x}}_{\text{external potential energy}}$$

which represents the **total potential energy**. Equation (8.27) constitutes the Euler equation for E.

Thus, in agreement with the principle of virtual work, u *minimizes the potential energy among all the admissible configurations*.

Similar observations can be made for the other types of boundary conditions.

Non homogeneous Dirichlet conditions. A nonhomogeneous Dirichlet condition can be prescribed selecting $g \in H^1(\Omega)$ and asking that $u - g \in H_0^1(\Omega)$. In this way we ask in a weak sense that $u - g = 0$ on $\partial\Omega$. Then, setting

$$w = u - g$$

we are reduced to homogeneous boundary conditions. In fact, $w \in H_0^1(\Omega)$ and is a solution of the equation

$$\int_\Omega \{\nabla w \cdot \nabla v \, dx + a_0 wv\} \, dx = \int_\Omega Fv \, dx, \quad \forall v \in H_0^1(\Omega)$$

where $F = f - \nabla g - a_0 g \in L^2(\Omega)$. The Lax-Milgram Theorem yields existence, uniqueness and the stability estimate

$$\|\nabla w\|_0 \le C_P\{\|f\|_0 + (1 + a_0)\|g\|_{1,2}\}. \tag{8.28}$$

Thus, for u we can write

$$\|u\|_{1,2} \le (1 + C_P)\|\nabla w\|_0 + \|g\|_{1,2}$$
$$\le (1 + C_P) C_P\{\|f\|_0 + (1 + a_0)\|g\|_{1,2}\}.$$

8.4.2 Neumann, Robin and mixed problems

Let $\Omega \subset \mathbb{R}^n$ be a bounded, *Lipschitz* domain. We examine the following problem:

$$\begin{cases} -\Delta u + a_0(\mathbf{x}) u = f & \text{in } \Omega \\ \partial_\nu u = g & \text{on } \partial\Omega \end{cases} \tag{8.29}$$

where $\alpha > 0$ is constant and $\boldsymbol{\nu}$ denotes the outward normal unit vector to $\partial\Omega$. As usual, to derive a weak formulation, we first assume that a_0, f and g are smooth and that $u \in C^2(\Omega) \cap C^1(\overline{\Omega})$ is a classical solution of (8.29). We choose $C^1(\overline{\Omega})$ as the space of test functions, having continuous first derivatives up to $\partial\Omega$. Let $v \in C^1(\overline{\Omega})$, arbitrary, and multiply the Poisson equation by v. Integrating over Ω, we get

$$\int_\Omega \{-\Delta u + a_0 u\}\, v\, dx = \int_\Omega fv\, dx. \qquad (8.30)$$

An integration by parts gives

$$-\int_{\partial\Omega} \partial_\nu u\, v d\sigma + \int_\Omega \{\nabla u \cdot \nabla v + a_0 uv\}\, dx = \int_\Omega fv\, dx, \quad \forall v \in C^1\left(\overline{\Omega}\right). \qquad (8.31)$$

Using the Neumann condition we may write

$$\int_\Omega \{\nabla u \cdot \nabla v + a_0 uv\}\, dx = \int_\Omega fv\, dx + \int_{\partial\Omega} gv\, d\sigma \qquad \forall v \in C^1\left(\overline{\Omega}\right). \qquad (8.32)$$

Thus (8.29) implies (8.32).

On the other hand, suppose that (8.32) is true. Integrating by parts in the reverse order, we find

$$\int_\Omega \{-\Delta u + a_0 u - f\,\}\, v\, dx + \int_{\partial\Omega} \partial_\nu u\, v d\sigma = \int_{\partial\Omega} gv\, d\sigma, \qquad (8.33)$$

for every $\forall v \in C^1\left(\overline{\Omega}\right)$. Since $C_0^1\left(\Omega\right) \subset C^1\left(\overline{\Omega}\right)$ we may insert any $v \in C_0^1\left(\Omega\right)$ into (8.33), to get

$$\int_\Omega \{-\Delta u + a_0 u - f\,\}\, v\, dx = 0.$$

The arbitrariness of $v \in C_0^1\left(\Omega\right)$ entails $-\Delta u + a_0 u - f = 0$ in Ω. Therefore (8.33) becomes

$$\int_{\partial\Omega} \partial_\nu u\, v d\sigma = \int_{\partial\Omega} gv\, d\sigma \qquad \forall v \in C^1\left(\overline{\Omega}\right)$$

and the arbitrariness of $v \in C^1\left(\overline{\Omega}\right)$ forces $\partial_\nu u = g$, recovering the Neumann condition as well.

Thus, **for classical solutions, the two formulations** (8.29) and (8.32) **are equivalent**.

Recall now that $C^1\left(\overline{\Omega}\right)$ is dense in $H^1\left(\Omega\right)$, which therefore constitutes the natural Sobolev space for the Neumann problem. Then, enlarging the space of test functions to $H^1\left(\Omega\right)$, we may give the **weak** formulation of problem (8.29) as follows.

Determine $u \in H^1\left(\Omega\right)$ such that

$$\int_\Omega \{\nabla u \cdot \nabla v + a_0 uv\}\, dx = \int_\Omega fv dx + \int_{\partial\Omega} gv d\sigma, \qquad \forall v \in H^1\left(\Omega\right). \qquad (8.34)$$

Again we point out that the Neumann condition is encoded in (8.34) and not explicitly expressed as in the case of Dirichlet boundary conditions. Since we used the density of $C^1(\overline{\Omega})$ in $H^1(\Omega)$ and the trace of v on $\partial\Omega$, some regularity of the domain (Lipschitz is enough) is needed, even in the variational formulation. Introducing the bilinear form

$$B(u,v) = \int_\Omega \{\nabla u \cdot \nabla v + a_0 uv\}\, d\mathbf{x} \tag{8.35}$$

and the linear functional

$$Lv = \int_\Omega fv\, d\mathbf{x} + \int_{\partial\Omega} gv d\sigma, \tag{8.36}$$

(8.34) may be formulated as the abstract variational problem

$$B(u,v) = Lv, \quad \forall v \in H_0^1(\Omega).$$

We state the well-posedness of this problem under reasonable hypotheses on the data.

Theorem 8.2. Let $\Omega \subset \mathbb{R}^n$ be a bounded, Lipschitz domain, $f \in L^2(\Omega)$, $g \in L^2(\partial\Omega)$ and $0 < c_0 \le a_0(\mathbf{x}) \le \gamma_0$ a.e.in Ω.
   Then, problem (8.34) has a unique solution $u \in H^1(\Omega)$. Moreover,

$$\|u\|_{1,2} \le \frac{1}{\min\{\alpha, c_0\}} \left\{\|f\|_0 + \overline{C}\alpha \|g\|_{L^2(\partial\Omega)}\right\}.$$

Along the proof, we use the following (important) the *trace inequality*

$$\|v\|_{L^2(\partial\Omega)} \le \overline{C}(n,\Omega) \|v\|_{1,2}. \tag{8.37}$$

We don't give here formal details about the *trace* theory, although we've proved this inequality in dimension 1 (see [28] and [37]).

Proof. We check that the hypotheses of the Lax-Milgram Theorem hold, with $V = H^1(\Omega)$.

• *Continuity of the bilinear form B.* The Schwarz inequality yields:

$$|B(u,v)| \le \|\nabla u\|_0 \|\nabla v\|_0 + \gamma_0 \|u\|_0 \|v\|_0$$
$$\le (1+\gamma_0) \|u\|_{1,2} \|v\|_{1,2}$$

so that B is continuous in $H^1(\Omega)$.

• *Coerciveness of B.* It follows from

$$B(u,u) = \int_\Omega |\nabla u|^2\, d\mathbf{x} + \int_\Omega a_0 u^2 d\mathbf{x} \ge \min\{1, c_0\} \|u\|_{1,2}^2$$

since $a_0(\mathbf{x}) \ge c_0 > 0$ a.e. in Ω.

• *Continuity of L.* From Schwarz's inequality and (8.37) we get:

$$|Lv| \leq \left| \int_\Omega fv \, d\mathbf{x} \right| + \left| \int_{\partial\Omega} gv \, d\sigma \right| \leq \|f\|_0 \|v\|_0 + \|g\|_{L^2(\partial\Omega)} \|v\|_{L^2(\partial\Omega)}$$

$$\leq \left\{ \|f\|_0 + \overline{C} \|g\|_{L^2(\partial\Omega)} \right\} \|v\|_{1,2} .$$

Therefore L is continuous in $H^1(\Omega)$ with

$$\|L\|_{H^1(\Omega)^*} \leq \|f\|_{L^2(\Omega)} + \overline{C} \|g\|_{L^2(\partial\Omega)} .$$

The conclusion follows from the Lax-Milgram Theorem.   □

Remark 8.6. As in the one-dimensional case, without the condition $a_0(\mathbf{x}) \geq c_0 > 0$, neither the existence nor the uniqueness of a solution is guaranteed. Let, for example, $a_0 = 0$. Then two solutions of the same problem differ by a constant. A way to restore uniqueness is to select a solution with, e.g., zero mean value, that is

$$\int_\Omega u(\mathbf{x}) \, d\mathbf{x} = 0.$$

The existence of a solution requires the following *compatibility condition* on the data f and g

$$\int_\Omega f \, d\mathbf{x} + \alpha \int_{\partial\Omega} g \, d\sigma = 0, \tag{8.38}$$

obtained by substituting $v = 1$ into the equation

$$\int_\Omega \alpha \nabla u \cdot \nabla v \, d\mathbf{x} = \int_\Omega fv \, d\mathbf{x} + \alpha \int_{\partial\Omega} gv \, d\sigma.$$

Note that, since Ω is bounded, the function $v = 1$ belongs to $H^1(\Omega)$.

If $a_0 = 0$ and (8.38) does not hold, problem (8.29) has no solution. Viceversa, we shall see later that, if this condition is fulfilled, a solution exists.

If $g = 0$, (8.38) has a simple interpretation. Indeed problem (8.29) is a model for the equilibrium configuration of a membrane whose boundary is free to slide along a vertical guide. The compatibility condition $\int_\Omega f d\mathbf{x} = 0$ expresses the obvious fact that, at equilibrium, the resultant of the external loads must vanish.

8.5 Eigenvalues of the Laplace operator

8.5.1 Separation of variables revisited

Using the method of separation of variables, in the first part of the book, we have constructed solutions of boundary value problems by superposition of special solutions. However, explicit computations can be performed only when the geometry of the relevant domain is quite particular. What may we say in general? Let us consider an example from diffusion.

Suppose we have to solve the problem

$$
\begin{cases}
u_t = \Delta u & (x,y) \in \Omega,\ t > 0 \\
u(x,y,0) = g(x,y) & \cdot (x,y) \in \Omega \\
u(x,y,t) = 0 & (x,y) \in \partial\Omega,\ t > 0
\end{cases}
$$

where Ω is a bounded bi-dimensional domain. Let us look for solutions of the form

$$
u(x,y,t) = v(x,y)\,w(t).
$$

Substituting into the differential equation, with some elementary manipulations, we obtain

$$
\frac{w'(t)}{w(t)} = \frac{\Delta v(x,y)}{v(x,y)} = -\lambda,
$$

where λ is a constant, which leads to the two problems

$$
w' + \lambda w = 0 \qquad t > 0 \tag{8.39}
$$

and

$$
\begin{cases}
-\Delta v = \lambda v & \text{in } \Omega \\
v = 0 & \text{on } \partial\Omega.
\end{cases} \tag{8.40}
$$

A number λ such that there exists a non trivial solution v of (8.40) is called a *Dirichlet eigenvalue of the operator* $-\Delta$ *in* Ω and v is a corresponding *eigenfunction*. Now, the original problem can be solved if the following two properties hold:

a) There exists a sequence of (real) eigenvalues λ_k with corresponding eigenvectors u_k. Solving (8.39) for $\lambda = \lambda_k$ yields

$$
w_k(t) = ce^{-\lambda_k t} \qquad c \in \mathbb{R}.
$$

b) The initial data g can be expanded is series of eigenfunctions:

$$
u(x,y) = \sum g_k u_k(x,y).
$$

Then, the solution is given by

$$
u(x,y,t) = \sum g_k e^{-\lambda_k t} u_k(x,y)
$$

where the series converges in some suitable sense.

In particular, condition b) requires that the set of Dirichlet eigenfunctions of $-\Delta$ constitutes a basis in the space of initial data. Thus, for instance, we may consider the *Dirichlet eigenfunctions for the Laplace operator in a domain* Ω, i.e. the non trivial solutions of the problem

$$
\begin{cases}
-\Delta u = \lambda u & \text{in } \Omega \\
u = 0 & \text{on } \partial\Omega.
\end{cases} \tag{8.41}
$$

A weak solution of problem (8.41) is a function $u \in H_0^1(\Omega)$ such that

$$a(u, v) \equiv (\nabla u, \nabla v)_0 = \lambda (u, v)_0 \qquad \forall v \in H_0^1(\Omega).$$

The following theorem holds.

Theorem 8.3. *Let Ω be a bounded domain. Then, there exists in $L^2(\Omega)$ an orthonormal basis $\{u_k\}_{k \geq 1}$ consisting of Dirichlet eigenfunctions for the Laplace operator. The corresponding eigenvalues $\{\lambda_k\}_{k \geq 1}$ are all positive and may be arranged in an increasing sequence*

$$0 < \lambda_1 < \lambda_2 \leq \cdots \leq \lambda_k \leq \cdots$$

with $\lambda_k \to +\infty$.

The sequence $\{u_k / \sqrt{\lambda_k}\}_{k \geq 1}$ constitutes an orthonormal basis in $H_0^1(\Omega)$, with respect to the scalar product $(u, v)_1 = (\nabla u, \nabla v)_0$.

Remark 8.7. Let $u \in L^2(\Omega)$ and denote by $c_k = (u, u_k)_0$ the Fourier coefficients of u with respect to the orthonormal basis $\{u_k\}_{k \geq 1}$. Then we may write

$$u = \sum_{k=1}^{\infty} c_k u_k \qquad \text{and} \qquad \|u\|_0^2 = \sum_{k=1}^{\infty} c_k^2.$$

Note that

$$\|\nabla u_k\|_0^2 = (\nabla u_k, \nabla u_k)_0 = \lambda_k (u_k, u_k)_0 = \lambda_k.$$

Thus, $u \in H_0^1(\Omega)$ if and only if

$$\|\nabla u\|_0^2 = \sum_{k=1}^{\infty} \lambda_k c_k^2 < \infty. \qquad (8.42)$$

Moreover, (8.42) implies that, for every $u \in H_0^1(\Omega)$,

$$\|\nabla u\|_0^2 \geq \lambda_1 \sum_{k=1}^{\infty} c_k^2 = \lambda_1 \|u\|_0^2.$$

We deduce the following **variational principle for the first Dirichlet eigenvalue**:

$$\lambda_1 = \min \left\{ \frac{\int_\Omega |\nabla u|^2}{\int_\Omega u^2} : u \in H_0^1(\Omega), \, u \text{ non identically zero} \right\}. \qquad (8.43)$$

The quotient in (8.43) is called *Raiyeigh's quotient*.

Similar theorems hold for the other types of boundary value problems as well. For instance, the *Neumann eigenfunctions for the Laplace operator in Ω* are the non trivial solutions of the problem

$$\begin{cases} -\Delta u = \mu u & \text{in } \Omega \\ \partial_\nu u = 0 & \text{on } \partial\Omega. \end{cases}$$

The following theorem holds.

Theorem 8.4. *If Ω is a bounded Lipschitz domain, there exists in $L^2(\Omega)$ an orthonormal basis $\{u_k\}_{k\geq 1}$ consisting of Neumann eigenfunctions for the Laplace operator. The corresponding eigenvalues form a non decreasing sequence $\{\mu_k\}_{k\geq 1}$, with $\mu_1 = 0$ and $\mu_k \to +\infty$.*

Moreover, the sequence $\{u_k/\sqrt{\mu_k + 1}\}_{k\geq 1}$ constitutes an orthonormal basis in $H^1(\Omega)$, with respect to the scalar product $(u,v)_{1,2} = (u,v)_0 + (\nabla u, \nabla v)_0$.

8.5.2 An asymptotic stability result

The results in the last subsection may be used sometimes to prove the asymptotic stability of a steady state solution of an evolution equation as time $t \to +\infty$.

As an example consider the following problem for the heat equation. Suppose that $u \in C^{2,1}(\overline{\Omega} \times [0,+\infty))$ is the (unique) solution of

$$\begin{cases} u_t - \Delta u = f(\mathbf{x}) & \mathbf{x} \in \Omega, \, t > 0 \\ u(\mathbf{x},0) = u_0(\mathbf{x}) & \mathbf{x} \in \Omega \\ u(\sigma,t) = 0 & \sigma \in \partial\Omega, \, t > 0 \end{cases}$$

where Ω is a smooth, bounded domain. Denote by $u_\infty = u_\infty(\mathbf{x})$ the solution of the stationary problem

$$\begin{cases} -\Delta u_\infty = f & \text{in } \Omega \\ u_\infty = 0 & \text{on } \partial\Omega. \end{cases}$$

Proposition 8.5. *For $t \geq 0$, we have*

$$\|u(\cdot,t) - u_\infty\|_0 \leq e^{-\lambda_1 t}\{\|u_0\|_0 + C_P^2 \|f\|_0\} \tag{8.44}$$

where λ_1 is the first Dirichlet eigenvalue for the Laplace operator in Ω.

Proof. Set $g(\mathbf{x}) = u_0(\mathbf{x}) - u_\infty(\mathbf{x})$. The function $w(\mathbf{x},t) = u(\mathbf{x},t) - u_\infty(\mathbf{x})$ solves the problem

$$\begin{cases} w_t - \Delta w = 0 & \mathbf{x} \in \Omega, \, t > 0 \\ w(\mathbf{x},0) = g(\mathbf{x}) & \mathbf{x} \in \Omega \\ w(\sigma,t) = 0 & \sigma \in \partial\Omega, \, t > 0. \end{cases} \tag{8.45}$$

Let us use the method of separation of variables and look for solutions of the form $w(\mathbf{x},t) = v(\mathbf{x})\,z(t)$. We find

$$\frac{z'(t)}{z(t)} = \frac{\Delta v(\mathbf{x})}{v(\mathbf{x})} = -\lambda$$

with λ constant. Thus we are lead to the eigenvalue problem

$$\begin{cases} -\Delta v = \lambda v & \text{in } \Omega \\ \quad\ v = 0 & \text{on } \partial\Omega. \end{cases}$$

From Proposition 7.2, there exists in $L^2(\Omega)$ an orthonormal basis $\{u_k\}_{k\geq 1}$ consisting of eigenvectors, corresponding to a sequence of non decreasing eigenvalues $\{\lambda_k\}$, with $\lambda_1 > 0$ and $\lambda_k \to +\infty$. Then, if $g_k = (g, u_k)_0$, we can write

$$g = \sum_1^\infty g_k u_k \qquad \text{and} \qquad \|g\|_0^2 = \sum_{k=1}^\infty g_k^2.$$

As a consequence, we find $z_k(t) = e^{-\lambda_k t}$ and finally

$$w(\mathbf{x},t) = \sum_1^\infty e^{-\lambda_k t} g_k u_k(\mathbf{x}).$$

Thus,

$$\|u(\cdot,t) - u_\infty\|_0^2 = \|w(\cdot,t)\|_0^2$$
$$= \sum_{k=1}^\infty e^{-2\lambda_k t} g_k^2$$

and since $\lambda_k > \lambda_1$ for every k, we deduce that

$$\|u(\cdot,t) - u_\infty\|_0^2 \leq \sum_{k=1}^\infty e^{-2\lambda_1 t} g_k^2 = e^{-2\lambda_1 t} \|g\|_0^2.$$

Theorem 8.1 yields, in particular

$$\|u_\infty\|_0 \leq C_P^2 \|f\|_0,$$

and hence

$$\|g\|_0 \leq \|u_0\|_0 + \|u_\infty\|_0$$
$$\leq \|u_0\|_0 + C_P^2 \|f\|_0$$

giving (8.44). $\qquad\qquad\qquad\qquad\qquad\qquad\qquad\qquad\qquad\qquad\qquad\qquad\square$

Proposition 8.5 implies that the steady state u_∞ is *asymptotically stable* in $L^2(\Omega)$ −norm as $t \to +\infty$. The speed of convergence to the steady state is exponential and it is determined by the first eigenvalue λ_1.

8.6 Equations in Divergence Form

8.6.1 Basic assumptions

In this section we consider boundary value problems for elliptic operators with general diffusion and transport terms. Let $\Omega \subset \mathbb{R}^n$ be a **bounded domain** and set

$$\mathcal{L}u = -\text{div}\,(\mathbf{A}\,(\mathbf{x})\,\nabla u) + \text{div}(\mathbf{b}(\mathbf{x})u) + \mathbf{c}\,(\mathbf{x}) \cdot \nabla u + a_0\,(\mathbf{x})\,u \qquad (8.46)$$

where $\mathbf{A} = (a_{ij})_{i,j=1,\dots,n}$, $\mathbf{b} = (b_1, \dots, b_n)$, $\mathbf{c} = (c_1, \dots, c_n)$ and a_0 is a real function.

Throughout this section, we will assume that the following hypotheses hold.

1. The differential operator $\mathcal{L}$ is **uniformly elliptic**, i.e. there exist **positive numbers** α and M such that:

$$\alpha\,|\boldsymbol{\xi}|^2 \le \mathbf{A}\,(\mathbf{x})\,\boldsymbol{\xi} \cdot \boldsymbol{\xi} \le M\,|\boldsymbol{\xi}|^2, \quad \forall \boldsymbol{\xi} \in \mathbb{R}^n, \text{ a.e. in } \Omega. \qquad (8.47)$$

2. The coefficients $\mathbf{b}$, $\mathbf{c}$ and a_0 are all **bounded**:

$$|\mathbf{b}\,(\mathbf{x})| \le \beta, \quad |\mathbf{c}\,(\mathbf{x})| \le \gamma, \quad |a_0\,(\mathbf{x})| \le \gamma_0, \quad \text{a.e. in } \Omega. \qquad (8.48)$$

The uniform ellipticity condition (8.47) states that $\mathbf{A}$ is *positive definite matrix* in Ω; α is called *ellipticity constant*. We point out that at this level of generality, we allow discontinuities also of the diffusion matrix $\mathbf{A}$, of the transport coefficients $\mathbf{b}$ and $\mathbf{c}$, in addition to the reaction coefficient a_0.

We want to extend to these type of operators the theory developed so far. The uniform ellipticity is a necessary requirement. In this section, we indicate some sufficient conditions assuring the well-posedness of the usual boundary value problems, based on the use of the Lax-Milgram Theorem.

As in the preceding sections, we start from the homogeneous Dirichlet problem. Nonhomogeneous Dirichlet conditions can be treated as in Remark 8.2.

8.6.2 Dirichlet problem

Consider the problem

$$\begin{cases} \mathcal{L}u = f + \text{div}\,\mathbf{f} & \text{in } \Omega \\ u = 0 & \text{on } \partial\Omega \end{cases} \qquad (8.49)$$

where $f \in L^2\,(\Omega)$ and $\mathbf{f} \in L^2\,(\Omega; \mathbb{R}^n)$.

A comment on the right hand side of (8.49) is in order. We have denoted by $H^{-1}\,(\Omega)$ the dual of $H_0^1\,(\Omega)$. We know (Theorem 7.15) that every element $F \in H^{-1}\,(\Omega)$ can be identified with an element in $\mathcal{D}'\,(\Omega)$ of the form

$$F = f + \text{div}\,\mathbf{f}.$$

Moreover

$$\|F\|_{H^{-1}(\Omega)} \le \|f\|_0 + \|\mathbf{f}\|_0 . \tag{8.50}$$

Thus, the right hand side of (8.49) represents a generic element of $H^{-1}(\Omega)$.

To derive a variational formulation of (8.49), we first assume that all the coefficients and the data f, $\mathbf{f}$ are smooth. Then, we multiply the equation by a test function $v \in C_0^1(\Omega)$ and integrate over Ω:

$$\int_\Omega [-\mathrm{div}(\mathbf{A}\nabla u - \mathbf{b}u)\, v]\, dx + \int_\Omega [\mathbf{c} \cdot \nabla u + a_0 u]\, v\, dx = \int_\Omega [f + \mathrm{div} \mathbf{f}]\, v dx.$$

Integrating by parts, we find, since $v = 0$ on $\partial\Omega$

$$\int_\Omega [-\mathrm{div}(\mathbf{A}\nabla u - \mathbf{b}u)\, v]\, dx = \int_\Omega [\mathbf{A}\nabla u \cdot \nabla v \; - \mathbf{b}u \cdot \nabla v]\, dx$$

and

$$\int_\Omega v \, \mathrm{div}\, \mathbf{f} \; dx = - \int_\Omega \mathbf{f} \cdot \nabla v \; dx.$$

Thus, the resulting equation is

$$\int_\Omega \{\mathbf{A}\nabla u \cdot \nabla v - \mathbf{b}u \cdot \nabla v + c v \cdot \nabla u \; + a_0 u v\}\, dx = \int_\Omega \{f v - \mathbf{f} \cdot \nabla v \;\}\, dx \tag{8.51}$$

for every $v \in C_0^1(\Omega)$.

It is not difficult to check that **for classical solutions, the two formulations (8.49) and (8.51) are equivalent.**

We now enlarge the space of test functions to $H_0^1(\Omega)$ and introduce the bilinear form

$$B(u,v) = \int_\Omega \{\mathbf{A}\nabla u \cdot \nabla v - \mathbf{b}u \cdot \nabla v + c v \cdot \nabla u \; + a_0 u v\}\, dx$$

and the linear functional

$$Fv = \int_\Omega \{f v - \mathbf{f} \cdot \nabla v \;\}\, dx.$$

Then, the **weak formulation** of problem (8.49) is the following:

Determine $u \in H_0^1(\Omega)$ such that

$$B(u,v) = Fv, \qquad \forall v \in H_0^1(\Omega). \tag{8.52}$$

A set of hypotheses that ensure the well-posedness of the problem is indicated in the following proposition.

Proposition 8.6. *Assume that hypotheses (8.47) and (8.48) hold and that* $f \in L^2(\Omega)$, $\mathbf{f} \in L^2(\Omega; \mathbb{R}^n)$. *Then if* $\mathbf{b}$ *and* $\mathbf{c}$ *have Lipschitz components and*

$$\frac{1}{2}\operatorname{div}(\mathbf{b} - \mathbf{c}) + a_0 \geq 0, \text{ a.e. in } \Omega, \tag{8.53}$$

problem (8.52) has a unique solution. Moreover, the following stability estimate holds:

$$\|u\|_1 \leq \frac{1}{\alpha}\{\|f\|_0 + \|\mathbf{f}\|_0\}. \tag{8.54}$$

Proof. We apply the Lax-Milgram Theorem with $V = H_0^1(\Omega)$. The continuity of B in V follows easily. In fact, the Schwarz inequality and the bounds in (8.48) give:

$$\left|\int_\Omega \mathbf{A}\nabla u \cdot \nabla v \, d\mathbf{x}\right| \leq \int_\Omega \sum_{i,j=1}^n \left|a_{ij}\partial_{x_i}u \, \partial_{x_j}v\right| d\mathbf{x}$$

$$\leq M\int_\Omega |\nabla u|\,|\nabla v|\, d\mathbf{x} \leq M\,\|\nabla u\|_0\,\|\nabla v\|_0\,.$$

Moreover, using Poincaré's inequality as well,

$$\left|\int_\Omega [\mathbf{b}u \cdot \nabla v - \mathbf{c}v \cdot \nabla u]\ d\mathbf{x}\right| \leq (\beta + \gamma)C_P\,\|\nabla u\|_0\,\|\nabla v\|_0$$

and

$$\left|\int_\Omega a_0 uv\ d\mathbf{x}\right| \leq \gamma_0\int_\Omega |u|\,|v|\, d\mathbf{x} \leq \gamma_0 C_P^2\,\|\nabla u\|_0\,\|\nabla v\|_0\,.$$

Thus, we can write

$$|B(u,v)| \leq \left(M + (\beta + \gamma)C_p + \gamma C_p^2\right)\|\nabla u\|_0\,\|\nabla v\|_0$$

which shows the continuity of B. Let us analyze the coerciveness of B. We have:

$$B(u,u) = \int_\Omega \left\{\mathbf{A}\nabla u \cdot \nabla u - (\mathbf{b} - \mathbf{c})u \cdot \nabla u + a_0 u^2\right\} d\mathbf{x}.$$

Observe that, since $u = 0$ on $\partial\Omega$, integrating by parts we obtain

$$\int_\Omega (\mathbf{b} - \mathbf{c})u \cdot \nabla u\ d\mathbf{x} = \frac{1}{2}\int_\Omega (\mathbf{b} - \mathbf{c}) \cdot \nabla u^2 d\mathbf{x} = -\frac{1}{2}\int_\Omega \operatorname{div}(\mathbf{b} - \mathbf{c})\ u^2 d\mathbf{x}.$$

Therefore, from (8.47) and (8.53), it follows that

$$B(u,u) \geq \alpha\int_\Omega |\nabla u|^2\, d\mathbf{x} + \int_\Omega \left[\frac{1}{2}\operatorname{div}(\mathbf{b} - \mathbf{c}) + a_0\right]u^2 d\mathbf{x} \geq \alpha\,\|\nabla u\|_0^2$$

so that B is V-coercive. Since we already know that $F \in H^{-1}(\Omega)$, the Lax-Milgram Theorem and (8.50) give existence, uniqueness and the stability estimate (8.54). □

Remark 8.8. If $\mathbf{A}$ is symmetric and $\mathbf{b} = \mathbf{c} = \mathbf{0}$, the solution u is a minimizer in $H_0^1(\Omega)$ for the "energy" functional

$$E(u) = \int_\Omega \left\{ \mathbf{A}\nabla u \cdot \nabla u + cu^2 - fu \right\} d\mathbf{x}.$$

The equation (8.52) constitutes the Euler equation for E.

In general we cannot prove that the bilinear form B is coercive. What we may affirm is that B is **weakly coercive**, i.e. *there exists $\lambda_0 \in \mathbb{R}$ such that:*

$$\tilde{B}(u,v) = B(u,v) + \lambda_0 (u,v)_0 \equiv B(u,v) + \lambda_0 \int_\Omega uv \, d\mathbf{x}$$

is coercive. In fact, from the elementary inequality

$$|ab| \le \varepsilon a^2 + \frac{1}{4\varepsilon} b^2, \qquad \forall \varepsilon > 0,$$

we get

$$\left| \int_\Omega (\mathbf{b} - \mathbf{c}) u \cdot \nabla u \, d\mathbf{x} \right| \le (\beta + \gamma) \int_\Omega |u \cdot \nabla u| \, d\mathbf{x} \le \varepsilon \|\nabla u\|_0^2 + \frac{(\beta + \gamma)^2}{4\varepsilon} \|u\|_0^2.$$

Therefore:

$$\tilde{B}(u,u) \ge \alpha \|\nabla u\|_0^2 + \lambda_0 \|u\|_0^2 - \varepsilon \|\nabla u\|_0^2 - \left(\frac{(\beta + \gamma)^2}{4\varepsilon} + \gamma \right) \|u\|_0^2. \quad (8.55)$$

If we choose $\varepsilon = \alpha/2$ and $\lambda_0 = (\beta + \gamma)^2/4\varepsilon + \gamma$, we obtain

$$\tilde{B}(u,u) \ge \frac{\alpha}{2} \|\nabla u\|_0^2$$

which shows the coerciveness of $\tilde{B}$. We will use this condition in the next chapter.

8.6.3 Neumann problem

Let Ω be a bounded, *Lipschitz* domain. The Neumann condition for an operator in the divergence form (8.46) assigns on $\partial\Omega$ the flux naturally associated with the operator. This flux is composed by two terms: $\mathbf{A}\nabla u \cdot \boldsymbol{\nu}$, due to the diffusion term $-\text{div}\mathbf{A}\nabla u$, and $-\mathbf{b}u \cdot \boldsymbol{\nu}$, due to the convective term $\text{div}(\mathbf{b}u)$, where $\boldsymbol{\nu}$ is the outward unit normal on $\partial\Omega$. We set

$$\partial_\nu^{\mathcal{L}} u \equiv (\mathbf{A}\nabla u - \mathbf{b}u) \cdot \boldsymbol{\nu} = \sum_{i,j=1}^n a_{ij} \partial_{x_j} u \, \nu_i - u \sum_j b_j \nu_j.$$

We call $\partial_\nu^{\mathcal{L}} u$ *conormal derivative* of u. Thus, the correct Neumann problem is:

$$\begin{cases} \mathcal{L}u = f & \text{in } \Omega \\ \partial_\nu^{\mathcal{L}} u = g & \text{on } \partial\Omega \end{cases} \tag{8.56}$$

with $f \in L^2(\Omega)$ and $g \in L^2(\partial\Omega)$. The variational formulation of problem (8.56) may be obtained by the usual integration by parts technique. It is enough to note, that, multiplying the differential equation $\mathcal{L}u = f$ by a test function $v \in H^1(\Omega)$ and using the Neumann condition, we get, formally

$$\int_\Omega \{(\mathbf{A}\nabla u - \mathbf{b}u)\nabla v + (\mathbf{c}\cdot\nabla u)v + a_0 uv\}\, dx = \int_\Omega fv\, dx + \int_{\partial\Omega} gv\, d\sigma.$$

Introducing the bilinear form

$$B(u,v) = \int_\Omega \{(\mathbf{A}\nabla u - \mathbf{b}u)\nabla v + (\mathbf{c}\cdot\nabla u)v + a_0 uv\}\, dx \tag{8.57}$$

and the linear functional

$$Fv = \int_\Omega fv\, dx + \int_{\partial\Omega} gv\, d\sigma,$$

we are led to the following **weak formulation**, that can be easily checked to be equivalent to the original problem, when all the data are smooth.

Determine $u \in H^1(\Omega)$ such that

$$B(u,v) = Fv, \quad \forall v \in H^1(\Omega). \tag{8.58}$$

If the size of $\mathbf{b} - \mathbf{c}$ is small enough, problem (8.58) is well-posed, as the following proposition shows.

Proposition 8.7. *Assume that hypotheses (8.47) and (8.48) hold and that $f \in L^2(\Omega)$, $g \in L^2(\partial\Omega)$. If $a_0(\mathbf{x}) \geq c_0 > 0$ a.e. in Ω and*

$$\alpha_0 \equiv \min\{\alpha - (\beta + \gamma)/2, c_0 - (\beta + \gamma)/2\} > 0, \tag{8.59}$$

then, problem (8.58) has a unique solution. Moreover, the following stability estimate holds:

$$\|u\|_{1,2} \leq \frac{1}{\alpha_0}\left\{\|f\|_0 + \overline{C}(n,\Omega)\|g\|_{L^2(\partial\Omega)}\right\}.$$

Proof (sketch). Since

$$|B(u,v)| \leq (M + \beta + \gamma + \gamma_0)\|u\|_{1,2}\|v\|_{1,2}$$

B is continuous in $H^1(\Omega)$. Moreover, we may write

$$B(u,u) \geq \alpha\int_\Omega |\nabla u|^2\, dx - \left|\int_\Omega [(\mathbf{b}-\mathbf{c})\cdot\nabla u]\, u\, dx\right| + \int_\Omega a_0 u^2 dx.$$

From Schwarz's inequality and the inequality $2ab \leq a^2 + b^2$, we obtain

$$\left| \int_\Omega [(\mathbf{b} - \mathbf{c}) \cdot \nabla u] \; u \; dx \right| \leq (\beta + \gamma) \left\| \nabla u \right\|_0 \left\| u \right\|_0 \leq \frac{(\beta + \gamma)}{2} \left\| u \right\|_{1,2}^2 .$$

Thus, if (8.59) holds, we get $B(u, u) \geq \alpha_0 \left\| u \right\|_{1,2}^2$ and therefore B is coercive. Finally, using the trace inequality (8.37), it is not difficult to check that $F \in H^1(\Omega)^*$, with

$$\left\| F \right\|_{H^1(\Omega)^*} \leq \left\| f \right\|_0 + \overline{C}(n, \Omega) \left\| g \right\|_{L^2(\partial\Omega)} . \qquad \square$$

8.6.4 Robin and mixed problems

The variational formulation of the problem

$$\begin{cases} \mathcal{L}u = f & \text{in } \Omega \\ \partial_\nu^{\mathcal{L}} u + hu = g & \text{on } \partial\Omega \end{cases} \qquad (8.60)$$

is obtained by replacing the bilinear form B in problem (8.58), by

$$\tilde{B}(u, v) = B(u, v) + \int_{\partial\Omega} huv \; d\sigma.$$

If $0 \leq h(\mathbf{x}) \leq h_0$ a.e. on $\partial\Omega$, Proposition 8.7 still holds for problem (8.60).

As for the Neumann problem, in general the bilinear form B is only *weakly coercive*.

Mixed Dirichlet-Neumann problem. Let Γ_D be a non empty relatively open subset of $\partial\Omega$ and $\Gamma_N = \partial\Omega \backslash \Gamma_D$. Consider the mixed problem

$$\begin{cases} \mathcal{L}u = f & \text{in } \Omega \\ u = 0 & \text{on } \Gamma_D \\ \partial_\nu^{\mathcal{L}} u = g & \text{on } \Gamma_N. \end{cases}$$

As in Section 8.4.2, the correct functional setting is $H_{0,\Gamma_D}^1(\Omega)$ with the norm $\left\| u \right\|_{H_{0,\Gamma_D}^1(\Omega)} = \left\| \nabla u \right\|_0$. Introducing the linear functional

$$Fv = \int_\Omega fv \; d\mathbf{x} + \int_{\Gamma_N} gv \; d\sigma,$$

the **variational formulation** is the following: *Determine* $u \in H_{0,\Gamma_D}^1(\Omega)$ *such that*

$$B(u, v) = Fv, \qquad \forall v \in H_{0,\Gamma_D}^1(\Omega). \qquad (8.61)$$

Proceeding as in Proposition 8.6, we may prove the following result.

Proposition 8.8. *Assume that hypotheses (8.47) and (8.48) hold and that* $f \in L^2(\Omega)$, $g \in L^2(\Gamma_N)$. *If* **b** *and* **c** *have Lipschitz components and*

$$(\mathbf{b} - \mathbf{c}) \cdot \boldsymbol{\nu} \leq 0 \quad \text{a.e. on } \Gamma_N, \quad \frac{1}{2}\operatorname{div}(\mathbf{b} - \mathbf{c}) + a_0 \geq 0, \quad \text{a.e. in } \Omega,$$

then problem (8.61) has a unique solution $u \in H^1_{0,\Gamma_D}(\Omega)$. *Moreover, the following stability estimate holds:*

$$\|u\|_1 \leq \frac{1}{\alpha}\left\{\|f\|_0 + \overline{C}\|g\|_{L^2(\partial\Omega)}\right\}.$$

8.7 A Control Problem

Control problems are more and more important in modern technology. We give here an application of the variational theory we have developed so far, to a fairly simple temperature control problem.

8.7.1 Structure of the problem

Suppose that the temperature u of a homogeneous body, occupying a smooth bounded domain $\Omega \subset \mathbb{R}^3$, satisfies the following stationary conditions:

$$\begin{cases} \mathcal{L}u \equiv -\Delta u + \operatorname{div}(\mathbf{b}u) = z & \text{in } \Omega \\ u = 0 & \text{on } \partial\Omega \end{cases} \tag{8.62}$$

where $\mathbf{b} \in C^1(\overline{\Omega}; \mathbb{R}^3)$ is given, with $\operatorname{div}\mathbf{b} \geq 0$ in Ω.

In (8.62) we distinguish two types of dependent variables: the **control** variable z, that we take in $H = L^2(\Omega)$, and the **state** variable u.

Coherently, (8.62) is called the **state system**. Given a control z, from Proposition 8.6, (8.62) has a unique weak solution

$$u[z] \in V = H^1_0(\Omega).$$

Thus, setting

$$a(u, v) = \int_\Omega (\nabla u \cdot \nabla v - u\mathbf{b} \cdot \nabla v)\, d\mathbf{x},$$

$u[z]$ satisfies the **state equation**

$$a(u[z], v) = (z, v)_0 \qquad \forall v \in V \tag{8.63}$$

and

$$\|u[z]\|_1 \leq \|z\|_0. \tag{8.64}$$

Our problem is to choose the source term z in order to minimize the "distance" of u from a given target state u_d.

Of course there are many ways to measure the distance of u from u_d. If we are interested in a distance which involves u and u_d over an open subset $\Omega_0 \subseteq \Omega$, a reasonable choice may be

$$J(u, z) = \frac{1}{2} \int_{\Omega_0} (u - u_d)^2 \, d\mathbf{x} + \frac{\beta}{2} \int_{\Omega} z^2 d\mathbf{x} \tag{8.65}$$

where $\beta > 0$.

$J(u, z)$ is called **cost functional** or **performance index**. The second term in (8.65) is called *penalization term*; its role is, on one hand, to avoid using "too large" controls in the minimization of J, on the other hand, to assure coerciveness for J, as we shall see later on.

Summarizing, we may write our control problem in the following way:

Find $(u^*, z^*) \in H \times V$, such that

$$\begin{cases} J(u^*, z^*) = \min_{(u,z) \in V \times H} J(u, z) \\[2mm] \textit{under the conditions} \\[2mm] \mathcal{L}u = z \ \textit{in } \Omega, \ u = 0 \ \textit{on } \partial\Omega. \end{cases} \tag{8.66}$$

If (u^*, z^*) is a minimizing pair, u^* and z^* are called **optimal state and optimal control**, respectively.

Remark 8.9. When the control z is defined in an open subset Ω_0 of Ω, we say that it is a *distributed control*. In some cases, z may be defined only on $\partial\Omega$ and then is called *boundary control*.

Similarly, when the cost functional (8.65) involves the observation of u in $\Omega_0 \subseteq \Omega$, we say that the observation is *distributed*. On the other hand, one may observe u or $\partial_\nu u$ on $\Gamma \subseteq \partial\Omega$. These cases correspond to *boundary observations* and the cost functional has to take an appropriate form.

The main questions to face in a control problem are:

- establish existence and/or uniqueness of an optimal pair (u^*, z^*);
- derive necessary and/or sufficient optimality conditions;
- construct algorithms for the numerical approximation of (u^*, z^*).

8.7.2 Existence and uniqueness of an optimal pair

Given $z \in H$, we may substitute into J the unique solution $u = u[z]$ of (8.63) to get the functional

$$\tilde{J}(z) = J(u[z], z) = \frac{1}{2} \int_{\Omega_0} (u[z] - u_d)^2 \, d\mathbf{x} + \frac{\beta}{2} \int_{\Omega} z^2 d\mathbf{x},$$

depending only on z. Thus, our minimization problem (8.66) is reduced to find an optimal control $z^* \in H$ such that

$$\tilde{J}(z^*) = \min_{z \in H} \tilde{J}(z). \tag{8.67}$$

Once z^* is known, the optimal state is given by $u^* = u[z^*]$.

The strategy to prove existence and uniqueness of an optimal control is to use the relationship between minimization of quadratic functionals and abstract variational problems corresponding to symmetric bilinear forms, expressed in Theorem 7.11. The key point is to write $\tilde{J}(z)$ in the following way:

$$\tilde{J}(z) = \frac{1}{2}b(z,z) + Lz + q \tag{8.68}$$

where $q \in \mathbb{R}$ (irrelevant in the optimization) and:

• $b(z,w)$ is a bilinear form in H, *symmetric, continuous and H−coercive*;

• L is a *linear, continuous functional* in H.

Then, by Theorem 7.11, there exists a unique minimizer $z^* \in H$. Moreover z^* is the minimizer if and only if z^* satisfies the Euler equation (see (7.31))

$$\tilde{J}'(z^*)w = b(z^*,w) - Lw = 0 \qquad \forall w \in H. \tag{8.69}$$

This procedure yields the following result.

Theorem 8.5. *There exists a unique optimal control $z^* \in H$. Moreover, z^* is optimal if and only if the following Euler equation holds ($u^* = u[z^*]$):*

$$\tilde{J}'(z^*)w = \int_{\Omega_0} (u^* - u_d)u[w]\ d\mathbf{x} + \beta \int_{\Omega} z^*w = 0 \qquad \forall w \in H. \tag{8.70}$$

Proof. According to the above strategy, we write $\tilde{J}(z)$ in the form (8.68).

First note that the map $z \mapsto u[z]$ is *linear*. In fact, if $\alpha_1, \alpha_2 \in \mathbb{R}$, then $u[\alpha_1 z_1 + \alpha_2 z_2]$ is the solution of $\mathcal{L}u[\alpha_1 z_1 + \alpha_2 z_2] = \alpha_1 z_1 + \alpha_2 z_2 u_1$. Since $\mathcal{L}$ is linear,

$$\mathcal{L}(\alpha_1 u[z_1] + \alpha_2 u[z_2]) = \alpha_1 \mathcal{L}u[z_1] + \alpha_2 \mathcal{L}u[z_2] = \alpha_1 z_1 + \alpha_2 z_2$$

and therefore, by uniqueness, $u[\alpha_1 z_1 + \alpha_2 z_2] = \alpha_1 u[z_1] + \alpha_2 u[z_2]$.

As a consequence,

$$b(z,w) = \int_{\Omega_0} u[z]u[w]\ d\mathbf{x} + \beta \int_{\Omega} zw \tag{8.71}$$

is a bilinear form and

$$Lw = \int_{\Omega_0} u[w]u_d\ d\mathbf{x} \tag{8.72}$$

is a linear functional in H.

Moreover, b is symmetric (obvious), continuous and $H-$coercive. In fact, from (8.64) and the Schwarz and Poincaré inequalities, we have, since $\Omega_0 \subseteq \Omega$,

$$|b(z,w)| \leq \|u[z]\|_{L^2(\Omega_0)} \|u[w]\|_{L^2(\Omega_0)} + \beta \|z\|_0 \|w\|_0$$
$$\leq (C_P^2 + \beta)\|z\|_0 \|w\|_0$$

which gives the continuity of b. The $H-$coerciveness of b follows from

$$b(z,z) = \int_{\Omega_0} u^2[z]\,d\mathbf{x} + \beta \int_\Omega z^2 \geq \beta \|z\|_0^2.$$

Finally, from (8.64) and Poincaré's inequality,

$$|Lw| \leq \|u_d\|_{L^2(\Omega_0)} \|u[w]\|_{L^2(\Omega_0)} \leq C_P \|u_d\|_0 \|w\|_0,$$

and we deduce that L is continuous in H.

Now, if we set: $q = \int_{\Omega_0} u_d^2\,d\mathbf{x}$, it is easy to check that

$$\tilde{J}(z) = \frac{1}{2}b(z,z) - Lz + q.$$

Then, Theorem 7.11 yields existence and uniqueness of the optimal control and Euler equation (8.69) translates into (8.70) after simple computations. □

8.7.3 Lagrange multipliers and optimality conditions

The Euler equation (8.70) gives a characterization of the optimal control z^* but it is not suitable for its computation.

To obtain more manageable optimality conditions, let us change point of view by regarding the state equation $\mathcal{L}u[z] = -\Delta u + \mathrm{div}(\mathbf{b}u) = z$, with $u = 0$ on $\partial\Omega$, as a *constraint* for our minimization problem. Then, the key idea is to introduce a *multiplier* $p \in V$, to be chosen suitably later on, and *formally*[3] write $\tilde{J}(z)$ in the augmented form

$$\frac{1}{2}\int_{\Omega_0}(u[z] - u_d)^2\,d\mathbf{x} + \frac{\beta}{2}\int_\Omega z^2 d\mathbf{x} + \int_\Omega p[z - \mathcal{L}u[z]]\,d\mathbf{x}. \qquad (8.73)$$

In fact, we have just added zero. Since $z \longmapsto u[z]$ is a linear map,

$$\tilde{L}z = \int_\Omega p(z - \mathcal{L}u[z])\,d\mathbf{x}$$

is a linear functional in H and therefore Theorem 8.5 yields the Euler equation

$$\tilde{J}'(z^*)w = \int_{\Omega_0}(u^* - u_d)\,u[w]\,d\mathbf{x} + \int_\Omega (p + \beta z^*)w\,d\mathbf{x} - \int_\Omega p\,\mathcal{L}u[w]\,d\mathbf{x} = 0 \qquad (8.74)$$

---

[3] It can be proved that actually u belongs to $H^2(\Omega)$ so that $\mathcal{L}u$ is well defined as a function in $L^2(\Omega)$ and the integral makes sense.

for every $w \in H$. Now we integrate twice by parts the last term, recalling that $u[w] = 0$ on $\partial\Omega$. We find

$$\int_\Omega p\mathcal{L}u[w]\,d\mathbf{x} = \int_{\partial\Omega} p\left(-\partial_\nu u[w] + (\mathbf{b}\cdot\nu)u[w]\right)d\sigma$$
$$+ \int_\Omega (-\Delta p - \mathbf{b}\cdot\nabla p)\,u[w]\,d\mathbf{x}$$
$$= -\int_{\partial\Omega} p\,\partial_\nu u[w]\,d\sigma + \int_\Omega \mathcal{L}^*p\,u[w]\,d\mathbf{x},$$

where the operator $\mathcal{L}^* = -\Delta - \mathbf{b}\cdot\nabla$ is the formal adjoint of $\mathcal{L}$.

Now we choose the multiplier: let p^* be the solution of the following **adjoint** problem

$$\begin{cases} \mathcal{L}^*p = (u^* - u_d)\,\chi_{\Omega_0} & \text{in } \Omega \\ p = 0 & \text{on } \partial\Omega. \end{cases} \tag{8.75}$$

Using (8.75), the Euler equation (8.74) becomes

$$\tilde{J}'(z^*)\,w = \int_\Omega (p^* + \beta z^*)w\,d\mathbf{x} = 0 \qquad \forall w \in H, \tag{8.76}$$

equivalent to $p^* + \beta z^* = 0$.

Summarizing, we have proved the following result:

Theorem 8.6. *The control z^* and the state $u^* = u(z^*)$ are optimal if and only if there exists a multiplier $p^* \in V$ such that z^*, u^* and p^* satisfy the following optimality conditions*

$$\begin{cases} \mathcal{E}u^* = -\Delta u^* + \operatorname{div}(\mathbf{b}u^*) = z^* & \text{in } \Omega,\ u^* = 0 \text{ on } \partial\Omega \\ \mathcal{E}^*p^* = -\Delta p^* - \mathbf{b}\cdot\nabla p^* = (u^* - u_d)\,\chi_{\Omega_0} & \text{in } \Omega,\ p^* = 0 \text{ on } \partial\Omega \\ p^* + \beta z^* = 0. & \text{(Euler equation).} \end{cases}$$

Remark 8.10. The optimal multiplier p^* is also called **adjoint state**.

Remark 8.11. We may generate the state and the adjoint equations in weak form, introducing the *Lagrangian* $\mathcal{L} = \mathcal{L}(u, z, p)$, given by

$$\mathcal{L}(u, z, p) = J(u, z) - a(u, p) + (z, p)_0.$$

Notice that $\mathcal{L}$ is linear in p, therefore[4]

$$\mathcal{L}'_p(u^*, z^*, p^*)\,v = -a(u^*, v) + (z^*, v)_0 = 0$$

---

[4] $\mathcal{L}'_p$, $\mathcal{L}'_z$ and $\mathcal{L}'_u$ denote the derivatives of the quadratic functional $\mathcal{L}$ with respect to p, z, u, respectively.

corresponds to the state equation. Moreover

$$L'_u\left(u^*, z^*, p^*\right)\varphi = J'_u\left(u^*, z^*\right)\varphi - a(\varphi, p^*)$$
$$= \left(u^* - u_d, \varphi\right)_{L^2(\Omega_0)} - a^*(p^*, \varphi) = 0$$

generates the adjoint equation, while

$$L'_z\left(u^*, z^*, p^*\right)w = \beta\left(w, z^*\right)_0 + (w, p^*)_0 = 0$$

constitutes Euler equation.

Remark 8.12. It is interesting to examine the behavior of $\tilde{J}\left(z^*\right)$ as $\beta \to 0$. In our case it is possible to show that $\tilde{J}\left(z^*\right) \to 0$ as $\beta \to 0$.

Remark 8.13 (An iterative algorithm). From Euler equation (8.76) and the Riesz Representation Theorem, we infer that

$$p^* + \beta z^* \text{ is the Riesz element associated with } \tilde{J}'\left(z^*\right),$$

called the **gradient of J at z^*** and denoted by the usual symbol $\nabla J\left(z^*\right)$ or by $\delta z\left(z^*, p^*\right)$. Thus, we have

$$\nabla J\left(z^*\right) = p^* + \beta z^*.$$

It turns out that $-\nabla J\left(z^*\right)$ plays the role of the *steepest descent* direction *for* J, as in the finite-dimensional case. This suggests an iterative procedure to compute a sequence of controls $\{z_k\}_{k \geq 0}$, convergent to the optimal one.

Select an initial control z_0. If z_k is known $(k \geq 0)$, then z_{k+1} is computed according to the following scheme.

1. Solve the state equation $a\left(u_k, v\right) = \left(z_k, v\right)_0, \forall v \in V.$

2. Knowing u_k, solve the adjoint equation

$$a^*\left(p_k, \varphi\right) = \left(u_k - u_d, \varphi\right)_{L^2(\Omega_0)} \qquad \forall \varphi \in V.$$

3. Set

$$z_{k+1} = z_k - \tau_k \nabla J\left(z_k\right) \tag{8.77}$$

and select the *relaxation parameter* τ_k in order to assure that

$$J\left(z_{k+1}\right) < J\left(z_k\right). \tag{8.78}$$

Clearly, (8.78) implies the convergence of the sequence $\{J\left(z_k\right)\}$, though in general not to zero. Concerning the choice of the relaxation parameter, there are several possibilities. For instance, if $\beta \ll 1$, we know that the optimal value $J\left(z^*\right)$ is close to zero and then we may chose

$$\tau_k = J\left(z_k\right)\left|\nabla J\left(z_k\right)\right|^{-2}.$$

With this choice, (8.77) is a Newton type method:

$$z_{k+1} = z_k - \frac{\nabla J\left(z_k\right)}{\left|\nabla J\left(z_k\right)\right|^2} J\left(z_k\right).$$

8.8 Numerical methods

8.8.1 The finite element method in one space dimension

We have introduced in the previous chapter the Galerkin method, that is an approximation method of the abstract variational problem $B(u, v) = Lv$, based on a sequence of finite-dimensional subspaces $V_k \subset V$. The finite element method is characterized by a particular definition of the subspaces V_k. We will develop here the main lines of this method applied to the discretization of second order boundary value problems on the interval (x_a, x_b), where in general the space V can be identified with $H^1(x_a, x_b)$.

These preliminary considerations already show that the finite element method relies on completely different theoretical foundations than finite difference schemes. While the latter provide a nodal based approximation, without guarantees on the behavior of the numerical solution between the nodes, the approximation properties of finite element method can be defined and analyzed at any point on the domain. Anyway, for simple problems as the ones that we will address in this section, the two families of methods share some similarities.

For the definition of subspaces V_k we start from a uniform partition of (x_a, x_b), in k subintervals

$$[x_a, x_b] = \bigcup_{i=1}^{k} [x_{i-1}, x_i], \quad x_i = h \cdot i, \quad h = (x_b - x_a)/k.$$

In this context, we say that x_i are the **vertexes** of the partition. The space of **finite elements of degree** r on this partition is defied as the collection of continuous functions whose restriction on each interval $[x_{i-1}, x_i]$ is a polynomial function of degree less or equal to r, namely

$$X_k^r = \{v_k \in C^0(x_a, x_b) \ : \ v_k|_{(x_{i-1}, x_i)} \in \mathbb{P}^r(x_{i-1}, x_i), \text{ for } i = 1, \dots, k\}$$

where the notation $\mathbb{P}^r(x_{i-1}, x_i)$ stands for the space of polynomial functions of degree less or equal to r on (x_{i-1}, x_i). Clearly, $X_k^r(x_a, x_b)$ is a finite-dimensional function space where $\dim(X_k^r) = N_k^r$ is affected by both parameters k and r. Then, the finite element method consists to determine $u_k \in X_k^r(x_a, x_b)$ such that,

$$B(u_k, v_k) = Lv_k, \quad \forall v_k \in V_k.$$

Now, we aim to reformulate the previous discrete problem as a linear system of algebraic equations, that could then be solved by a suitable algorithm. From Chapter 7, we remind that the Galerkin method is equivalent to determine a vector of coefficients, $\mathbf{c} = \{c_j\}_{j=1}^{N_k^r}$, such that $\mathbf{A}_{kr}\mathbf{c} = \mathbf{F}_{kr}$ with $\mathbf{A}_{kr} \in \mathbb{R}^{N_k^r \times N_k^r}$, $\mathbf{F}_{kr} \in \mathbb{R}^{N_k^r}$ where,

$$u_k = \sum_{j=1}^{N_k^r} c_j \psi_j, \quad A_{kr,ij} = B(\psi_j, \psi_i), \quad F_{kr,i} = L\psi_i.$$

According to this equivalent problem formulation, we observe that a suitable basis of the finite element space $X_k^r(x_a, x_b)$ should satisfy the following criteria:

a) The support of each basis element, namely ψ_i, should be small so as to $\mathrm{supp}(\psi_i) \cap \mathrm{supp}(\psi_j) = \emptyset$ for as many indexes as possible. By this way, we a priori know that the matrix coefficient $\mathbf{A}_{kr}$ vanishes because $A_{kr,ij} = a(\psi_j, \psi_i) = 0$. On the one hand, this strategy reduces the computational cost for building A_{kr}. On the other hand, since A_{kr} is likely to be a narrow banded matrix, it will be possible to use efficient algorithms to solve the corresponding linear system.

b) We look for a Lagrangian basis. Namely, provided that each basis element refers ψ_i to a special point $\widehat{x}_i \in (x_a, x_b)$, called **node**, a basis is Lagrangian if $\psi_i(\widehat{x}_j) = \delta_{ij}$, where δ_{ij} is Kronecker's symbol. As a result of that, for any $v_k = \sum_{i=1}^{N_k^r} c_i \psi_i$ the coefficients of the expansion satisfy $c_i = v_k(\widehat{x}_i)$.

Fig. 8.1 shows an example of functions $v_k^1 \in X_k^1(0,1)$ and $v_k^2 \in X_k^2(0,1)$ defined on a uniform partition of $(0,1)$ in $k = 4$ elements, while in Fig. 8.2 we depict some elements of a Lagrangian basis of $X_k^1(0,1)$ (left) and $X_k^2(0,1)$ (right). We observe that in the linear case, nodes and vertexes of the partition coincide. The number of degrees of freedom of the linear finite element space is thus $\dim(X_k^1(0,1)) = N_k^1 = k + 1$. Furthermore, each element of the Lagrangian basis is a piecewise linear function with unit value in one particular node and vanishing in all other nodes of the mesh.

For the case of quadratic elements, since a second order polynomial is uniquely defined by fixing it on 3 points, we observe that the set of nodes must be richer than the vertexes of the partition. In particular, it is sufficient to complement the vertexes with the midpoints of each interval in order to obtain a complete set of interpolation nodes that uniquely characterize quadratic finite element functions. As a result of that, we easily conclude that for the one-dimensional case $\dim(X_k^2(0,1)) = N_k^2 = 2k+1$ and in general $\dim(X_k^r(0,1)) = N_k^r = rk + 1$. Numbering nodes from left to right, Fig. 8.2 (right) shows the

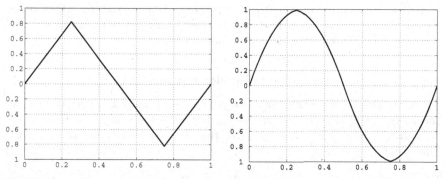

Fig. 8.1. Linear ($r = 1$, left) and quadratic ($r = 2$ right) finite element functions

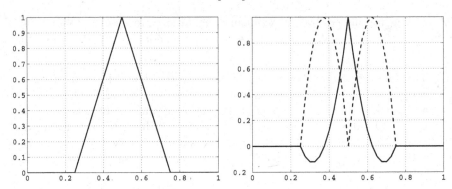

Fig. 8.2. Linear ($r = 1$, left) and quadratic ($r = 2$ right) Lagrangian basis functions for the finite element method

components ψ_4, ψ_5, ψ_6 of the Lagrangian basis for $X_k^2(0,1)$. In particular, ψ_4, ψ_6 are related to the element midpoints and are depicted by a dashed line, while ψ_5 refers to a node coinciding with a vertex and is depicted with continuous line.

The numerical approximation of boundary conditions with the finite element method is completely analogous with their treatment at the level of the abstract variational problem. Homogeneous Dirichlet boundary conditions are enforced into the definition of the discrete space V_k in order to satisfy $V_k \subset V = H_0^1(x_a, x_b)$. To this aim, we define a new discrete space

$$ X_{k,0}^r(x_a, x_b) = \{v_k \in X_k^r(x_a, x_b) \ : \ v_k(x_a) = v_k(x_b) = 0\} \subset H_0^1(x_a, x_b) $$

where we have removed the basis components related to nodes $x_0 = x_a$ and $x_{rk} = x_b$ from the original finite element space $X_k^r(x_a, x_b)$. Non homogeneous Dirichlet conditions are reformulated as homogeneous conditions plus a right hand side term by means of their lifting on the entire domain. This technique is straightforwardly extended at the discrete level, provided that a discrete lifting is applied, i.e. a lifting belonging to $X_k^r(x_a, x_b)$. Finally, the discretization of Neumann and Robin conditions do not require any particular care, because they are naturally embedded into the bilinear form.

8.8.2 Error analysis of the finite element method

The Céa Lemma shows that the Galerkin method is convergent provided that the family of approximation spaces V_k becomes dense in $V = H^1(x_a, x_b)$. Namely, given $u \in V$ the solution of the abstract variational problem, we know that

$$ \|u - u_k\|_{H^1} \leq \frac{M}{\alpha} \inf_{v_k \in V_k} \|u - v_k\|_{H^1}. \tag{8.79} $$

By consequence, we require that the discrete approximation spaces V_k satisfy the following approximation property

$$\lim_{k \to \infty} \inf_{v_k \in V_k} \|v - v_k\|_{H^1} = 0, \quad \forall v \in H^1(x_a, x_b). \tag{8.80}$$

In this section we analyze how, for a fixed value of the polynomial degree r, the finite element spaces $X_k^r(x_a, x_b)$ satisfy property (8.80) when k increases, or similarly, the partition of the domain (x_a, x_b) is refined by decreasing the characteristic mesh size h. Furthermore, we aim to determine the asymptotic **order of convergence**, that is the largest value of the exponent p such that

$$\lim_{k \to \infty} \|u - u_k\|_{H^1} = \lim_{k \to \infty} C((x_b - x_a)/k)^p = \lim_{h \to 0} C h^p$$

being C a generic positive constant independent on k, h. The following Theorem[5] is at the basis of the approximation properties of the finite element method.

Theorem 8.7. *For any value of the polynomial degree $r \geq 1$ there exists a positive constant C_r, uniformly independent on k, h such that, for any $v \in H^s(x_a, x_b)$ with $s \geq 2$ it holds*

$$\inf_{v_k \in X_k^r} \|v - v_k\|_{H^1} \leq C_r h^l \|v\|_{H^{l+1}} \quad where \quad l = \min[s-1, r], \quad h = \frac{x_b - x_a}{k}. \tag{8.81}$$

This shows that the finite element space $X_k^r(x_a, x_b)$ satisfy (8.80) provided that the solution to be approximated is regular enough. More precisely, an *a priori* error estimate for the finite element method is obtained by combining inequalities (8.79) and (8.81), leading to the following result.

Corollary 8.1. *Provided that problem $B(u, v) = Lv$ admits a regular solution $u \in H^s(x_a, x_b)$ with $s \geq 2$, then the finite element approximation of degree $r \geq 1$ satisfies,*

$$\|u - u_k\|_{H^1} \leq C(M, \alpha, r) h^l \|u\|_{H^{l+1}} \tag{8.82}$$

where

$$l = \min[s-1, r] \quad and \quad C(M, \alpha, r) = C_r \frac{M}{\alpha}.$$

Conversely, when no more than $u \in H^1(x_a, x_b)$ can be guaranteed, the asymptotic order of convergence with respect to the characteristic mesh size can be arbitrarily low.

---

[5] For a proof, we remand the interested reader to *Quarteroni* [42].

8.8.3 The finite element method for the approximation of advection, diffusion, reaction problems

We aim to use the finite element method for the approximation of problem (8.4), that for the sake of clarity is reported below, complemented with homogeneous Dirichlet boundary conditions,

$$ - (p(x)u')' + q(x)u' + r(x)u = f(x), \text{ in } (x_a, x_b), \qquad (8.83) $$

where $u(x_a) = u(x_b) = 0$, with $p(x), q(x) \in C^1([0,1])$ with $p(x) \geq a_0 > 0$ and $r(x), f(x) \in C^0([0,1])$, whose variational formulation consists to find $u \in H_0^1(x_a, x_b)$ such that

$$ B(u,v) = \int_{x_a}^{x_b} (p(x)u'v' + q(x)u'v + r(x)uv)dx = Lv = \int_{x_a}^{x_b} f(x)vdx, $$

$$ \forall v \in H_0^1(x_a, x_b). $$

Then, at the discrete level, we look for $u_k \in X_{k,0}^r(x_a, x_b)$ such that $B(u_k, \psi_i) = F\psi_i$ for any $i = 1, \ldots, rk + 1$.

We notice that Corollary 8.1 is applicable also to this case and in particular the values of continuity and coercivity constants of the bilinear form, M and α respectively, depend on p, q, r as follows,

$$ M = (\|p\|_{L^\infty} + C_P\|q\|_{L^\infty} + C_P^2\|r\|_{L^\infty}), \quad \alpha = a_0. $$

Accordingly, the a priori error estimate constant $C(M, \alpha, r)$ may become arbitrarily large. By consequence, either the mesh characteristic size is chosen to be be small enough, or the approximation is not satisfactory. In particular, the more the advection terms dominate over the diffusion one, the more the scheme becomes **unstable**. This effect is emphasized in Fig. 8.3 where we show the solution obtained approximating (8.83) and $u(0) = 0$, $u(1) = 1$ by means of linear finite elements. On the left, we show the comparison between approximations obtained when $p = 10^{-2}, q = 1, r = 0$, using uniform meshes characterized by $h = 0.1$ and $h = 0.01$ identified by dashed and continuous lines respectively. On the right, we perform similar numerical tests for a reaction dominated problem with $p = 10^{-2}, q = 0, r = 10$. When the mesh is not refined enough, spurious numerical oscillations clearly appear in both cases.

To stabilize the finite element method, we analyze the scheme at the level of the algebraic nodal equations. To this purpose, we first subdivide the bilinear form $B(u,v)$ into diffusion, transport and reaction terms,

$$ B_p(u,v) = \int_{x_a}^{x_b} p(x)u'v'dx, \quad B_q(u,v) = \int_{x_a}^{x_b} q(x)u'vdx, $$

$$ B_r(u,v) = \int_{x_a}^{x_b} r(x)uv. $$

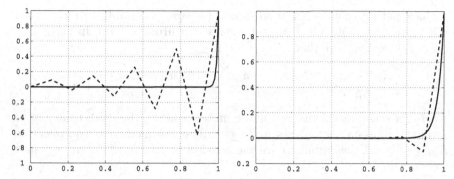

Fig. 8.3. Numerical instability of the finite element method for advection and reaction dominated problems, depicted on the left and right panels respectively. Dashed line corresponds to the approximation obtained using $h = 0.1$, while continuous line corresponds to the numerical solution with $h = 0.01$, which turns out to be very accurate

and in the case of linear finite elements the corresponding algebraic equations become,

$$B_p(u_k, \psi_i) = c_{i-1} \int_{x_{i-1}}^{x_i} p(x)\psi'_{i-1}\psi'_i + c_i \int_{x_{i-1}}^{x_{i+1}} p(x)(\psi'_i)^2 + c_{i+1} \int_{x_i}^{x_{i+1}} p(x)\psi'_{i+1}\psi'_i$$

$$= -\frac{p}{h}\left(c_{i-1} - 2c_i + c_{i+1}\right),$$

$$B_q(u_k, \psi_i) = c_{i-1} \int_{x_{i-1}}^{x_i} q(x)\psi'_{i-1}\psi_i + c_i \int_{x_{i-1}}^{x_{i+1}} q(x)\psi'_i\psi_i + c_{i+1} \int_{x_i}^{x_{i+1}} q(x)\psi'_{i+1}\psi_i$$

$$= \frac{q}{2}\left(c_{i+1} - c_{i-1}\right),$$

$$B_r(u_k, \psi_i) = c_{i-1} \int_{x_{i-1}}^{x_i} r(x)\psi_{i-1}\psi_i + c_i \int_{x_{i-1}}^{x_{i+1}} r(x)\psi_i^2 + c_{i+1} \int_{x_i}^{x_{i+1}} r(x)\psi_{i+1}\psi_i$$

$$= \frac{r}{6}h\left(c_{i+1} + 4c_i + c_{i-1}\right).$$

Since we are using a Lagrangian basis to represent the discrete space, the unknown coefficients c_i coincide with the nodal values of the finite element function, namely $c_i = u_k(x_i) = u_i$. By multiplying by h the above expressions, it emerges that, for uniform partitions, the one-dimensional linear finite element approximation of diffusion and transport terms coincide with the corresponding finite difference schemes, while the two families of methods differ in the approximation of reaction terms.

From Chapter 2, we know that for transport dominated problems it is more convenient to discretize qu' using a one sided difference ratio. This observation gives rise to the **upwind** scheme, where the one-sided approximation of the first order derivative is suitably chosen according to the sign of q. Shifting to one-sided schemes does not apply to the finite element framework. However, we have put into evidence that the upwind scheme is equivalent to a centered difference scheme modified with an **artificial diffusion** term. In particular, for $q > 0$ we easily see that

$$\frac{q(u_i - u_{i-1})}{h} = \frac{q(u_{i+1} - u_{i-1})}{2h} - \frac{qh}{2}\frac{(u_{i+1} - 2u_i + u_{i-1})}{h^2}.$$

Then, for one-dimensional and linear finite elements, the correct amount of artificial diffusion that allows us to **stabilize** the method is equivalent to $h\|q\|_{L^\infty}/2$ and the stabilized scheme consists to find $\widehat{u}_k \in X_k^1(x_a, x_b)$ such that

$$\widehat{B}(\widehat{u}_k, v_k) = Lv_k, \quad \forall v_k \in X_{k,0}^1(x_a, x_b)$$

where

$$\widehat{B}(u_k, v_k) = B(u_k, v_k) + \frac{h\|q\|_{L^\infty}}{2} \int_{x_a}^{x_b} u_k' v_k' dx.$$

The finite element method turn out to be unstable also for the approximation of reaction dominated problems. In this context, the **mass lumping technique** is often used to cure the instabilities. For linear finite elements, it corresponds to modify the nodal equations of the scheme, namely replacing $(u_{i+1} + 4u_i + u_{i-1})/6$ with the single value u_i. The error introduced by this modification can be thoroughly analyzed and, in conclusion, it does not substantially affect the convergence of the scheme[6].

8.8.4 The finite element method in two space dimensions

First, we address the problem of partitioning two-dimensional domains into elements. The simplest case, consists in subdividing a polygonal domain $\Omega \subset \mathbb{R}^2$ into triangles. However, not all possible partitions are admissible for the construction of a finite element space. The following requirements must be satisfied by any admissible **triangulation**:

a) all elements K of the partition are triangles. We assume K is an open set;
b) for any distinct K_1, K_2 we have $K_1 \cap K_2 = \emptyset$;
c) if $\overline{K}_1 \cap \overline{K}_2 = E \neq \emptyset$, then E is either an entire face or a vertex;
d) the diameter of each element K, i.e. $h_K = \text{diam}(K)$, is upper bounded, namely there exists $h > 0$ such that $h_k < h$.

The value h is called the characteristic mesh size. Furthermore, a triangulation is called quasi-uniform when it satisfies the following additional constraints:

e) there exists a constant $C > 1$ such that for any K the radii of the inner circle and outer circle with respect to K, i.e. ρ_K and h_K respectively, satisfy $1 < h_K/\rho_K < C$;
f) there exist two constants $C, c > 0$ such that for any K it holds $ch < h_K < Ch$.

---

[6] For further details, we refer to Section 5.5, Chapter 5, *Quarteroni* [42].

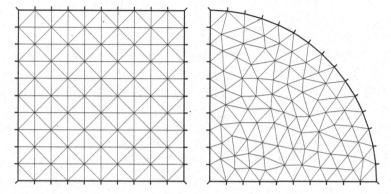

Fig. 8.4. A uniform (left) and quasi-uniform (right) partitions. We see that the finite element method easily adapts to generic configurations of the domain

The collection of all triangles of the partition is usually denoted as T_h. In Fig. 8.4 we show an example of uniform partition on the left and quasi uniform partition on the right.

To build up the finite element method in two space dimensions, we denote by $\mathbb{P}^r(K)$ the space of polynomial functions with maximum degree less or equal to r over the triangle K. Then the finite element space of degree r over the mesh T_h is given by

$$X_k^r = \{v_k \in C^0(\Omega) \ : \ v_k|_K \in \mathbb{P}^r(K), \ \forall K \in T_h\}$$

whose dimension is denoted as $N_k^r = \dim(X_k^r)$. As an instance, for linear finite elements in two space dimensions N_k^1 is equivalent to the number of vertexes of the computational mesh. This conclusion naturally emerges when looking at the construction of a Lagrangian basis for the linear finite element space. Reasoning as for the one-dimensional case, the Lagrangian basis is given by the collection of functions that vanish in all but one vertex of the mesh. A component of such basis is depicted in Fig. 8.5 and according to their shape, these are often called *hat functions*.

The fundamental properties of the finite element method that we have presented for the approximation of one-dimensional problems also apply to multiple space dimensions. In particular Theorem 8.7 and Corollary 8.1 still hold true, but particular care should be devoted to check the regularity requirement for the exact solution to be approximated, namely $u \in H^s(\Omega)$ with $s \geq 2$. Indeed, it is sufficient to consider a second order problem over a non convex polygonal domain to violate the previous regularity condition[7]. In the forthcoming section, we will investigate the behaviour of the finite element method in these sub-optimal regularity conditions[8].

---

[7] See *Salsa* [13].

[8] For a detailed analysis of the finite element method we refer the interested reader to *Quarteroni, Valli* [45].

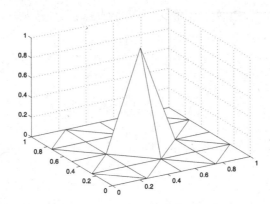

Fig. 8.5. A Lagrangian basis function of a two-dimensional linear finite element space

8.9 Exercises

8.1. Consider the following boundary value problem

$$\begin{cases} -u'' = 5x - 1 & \text{in } (-1, 2) \\ u(-1) = 1/2 \\ u'(2) = 2. \end{cases} \tag{8.84}$$

Write the associated variational formulation and determine existence, uniqueness and stability of the solution.

8.2. Analyze the following boundary value problem

$$\begin{cases} -(\mu(x)u')' + (\beta(x)u)' + \sigma(x)u = f(x) & a < x < b \\ u(a) = 0, \quad \mu(b)u'(b) = \beta(b)u(b) \end{cases} \tag{8.85}$$

where $a, b \in \mathbb{R}$ and μ, β, σ are given functions, with $\mu \geq \mu_0 > 0$. Introduce the functional spaces in order to write the variational formulation of the problem and prove that it is well-posed, under suitable assumptions on the data.

8.3. Consider the function

$$f(x) = \begin{cases} 2\pi^2 \sin \pi x & 0 \leq x < \dfrac{1}{2} \\ 2\pi^2 \sin \pi x - 6\left(\dfrac{1}{2} - x\right) + \left(\dfrac{1}{2} - x\right)^3 & \dfrac{1}{2} \leq x \leq 1. \end{cases}$$

Determine the functional spaces and write the variational formulation associated to the problem

$$\begin{cases} -u'' + \pi^2 u = f(x) & 0 < x < 1 \\ u(0) = 0, \ u(1) = -1/8. \end{cases}$$

Prove that the problem is well-posed and verify that

$$
u(x) = \begin{cases} \sin \pi x & 0 \leq x < \dfrac{1}{2} \\ \sin \pi x + \left(\dfrac{1}{2} - x \right)^3 & \dfrac{1}{2} \leq x \leq 1 \end{cases}
$$

is the variational solution. Deduce the regularity of the solution in terms of Sobolev spaces.

8.4. Analyze the following elliptic problem

$$
\begin{cases} -\Delta u + \sigma u = f & \text{in } \Omega \\ \alpha u + \nabla u \cdot \mathbf{n} = g & \text{on } \partial\Omega \end{cases} \tag{8.86}
$$

where Ω is an open, bonded, smooth domain in $\mathbb{R}^2$, $\mathbf{n}$ is the outward normal unit vector to $\partial\Omega$, $f = f(x)$ and $g = g(x)$ are assigned functions and α, σ are real constants, where $\sigma > 0$.
Write the variational formulation of the problem, introducing suitable assumptions on the data, and give sufficient conditions for the well-posedness.

8.5. Consider $\Omega = [0, \pi] \times [0, 1]$, where $\partial\Omega = \Gamma_D \cup \Gamma_N$ with

$$
\Gamma_N = \left\{ (x_1, x_2) \in \mathbb{R}^2 : x_1 = \pi, \, 0 < x_2 < 1 \right\}
$$
$$
\cup \left\{ (x_1, x_2) \in \mathbb{R}^2 : 0 < x_1 < \pi, x_2 = 1 \right\}
$$

and let $\mathbf{n}$ be the outward normal unit vector to $\partial\Omega$. The function $f \in L^2(\Omega)$ and the vector

$$
\mathbf{b} = \begin{bmatrix} x_2^2 \\ \sin x_1 \end{bmatrix}
$$

are assigned; consider the problem

$$
\begin{cases} -\Delta u + \mathbf{b} \cdot \nabla u = f(x,y) & \text{in } \Omega \\ \nabla u \cdot \mathbf{n} = 0 & \text{on } \Gamma_N \\ u = 0 & \text{on } \Gamma_D. \end{cases}
$$

Write the weak formulation and prove existence and uniqueness of the solution, verifying the hypothesis of the Lax-Milgram Lemma. Then, give an a priori estimate of the solution.

8.6 (A simple system). Consider the problem

$$
\begin{cases} -\alpha \Delta u + \beta v = f/2 & \text{in } \Omega \\ \alpha \Delta v + \beta u = -f/2 & \text{in } \Omega \\ u + v = 2 & \text{on } \Gamma_D \\ u - v = 0 & \text{on } \Gamma_D \\ \nabla(u + v) \cdot \mathbf{n} = 0 & \text{on } \Gamma_N \\ \nabla(u - v) \cdot \mathbf{n} = 0 & \text{on } \Gamma_N \end{cases}
$$

where $\Omega \subset \mathbb{R}^2$, $\partial\Omega = \Gamma_N \cup \Gamma_D$, f is a function defined on Ω, $\alpha > 0$ and $\beta \geq 0$ are constants ($\Gamma_N \cap \Gamma_D = \emptyset$). Find the variational formulation of the problem in the suitable functional spaces and find sufficient conditions on the data for the well-posedness.

8.7. Write the weak formulation of the following problem

$$\begin{cases} \cos x\, u'' - \sin x\, u' - xu = 1 & 0 < x < \pi/6 \\ u'(0) = -u(0),\ u(\pi/6) = 0. \end{cases}$$

Discuss existence and uniqueness and derive a stability estimate.

8.8 (Transmission conditions I). Consider the problem

$$\begin{cases} (p(x)\,u')' = f & \text{in } (a,b) \\ u(a) = u(b) = 0 \end{cases} \tag{8.87}$$

where $f \in L^2(a,b)$, $p(x) = p_1 > 0$ in (a,c) and $p(x) = p_2 > 0$ in (c,b), where $a < c < b$.

Show that problem (8.87) has a unique weak solution in $H^1(a,b)$, satisfying the conditions

$$\begin{cases} p_1 u'' = f & \text{in } (a,c) \\ p_2 u'' = f & \text{in } (c,b) \\ p_1 u'(c-) = p_2 u'(c+). \end{cases}$$

Observe the jump of the derivative of u at $x = c$.

8.9 (Transmission conditions II). Let Ω_1 and Ω be bounded, Lipschitz domains in $\mathbb{R}^n$ such that $\Omega_1 \subset\subset \Omega$. Let $\Omega_2 = \Omega \backslash \overline{\Omega}_1$. In Ω_1 and Ω_2 consider the following bilinear forms

$$a_k(u,v) = \int_{\Omega_k} \mathbf{A}^k(\mathbf{x})\nabla u \cdot \nabla v\ d\mathbf{x} \qquad (k = 1,2)$$

with $\mathbf{A}^k$ *uniformly elliptic*. Assume that the entries of A^k are continuous in $\overline{\Omega}_k$, but that the matrix

$$\mathbf{A}(\mathbf{x}) = \begin{cases} \mathbf{A}^1(\mathbf{x}) \text{ in } \overline{\Omega}_1 \\ \mathbf{A}^2(\mathbf{x}) \text{ in } \Omega_2 \end{cases}$$

may have *a jump across* $\Gamma = \partial\Omega_1$. Let $u \in H_0^1(\Omega)$ be the weak solution of the equation

$$a(u,v) = a_1(u,v) + a_2(u,v) = (f,v)_0 \qquad \forall v \in H_0^1(\Omega),$$

where $f \in L^2(\Omega)$.

a) Which boundary value problem does u satisfy?

b) Which conditions on Γ do express the coupling between u_1 and u_2?

8.10. Let $\Omega = (0, 1) \times (0, 1) \subset \mathbb{R}^2$. Prove that the functional

$$E\left(v\right) = \frac{1}{2}\int_{\Omega}\left\{|\nabla v|^2 - xv\right\}dxdy$$

has a unique minimizer $u \in H_0^1(\Omega)$. Write the Euler equation and find an explicit formula for u.

8.11 (Distributed observation and control, Neumann conditions). Let $\Omega \subset \mathbb{R}^n$ be a bounded, smooth domain and Ω_0 an open (non empty) subset of Ω. Set $V = H^1(\Omega)$, $H = L^2(\Omega)$ and consider the following control problem: *minimize the cost functional*

$$J\left(u, z\right) = \frac{1}{2}\int_{\Omega_0}\left(u - u_d\right)^2 dx + \frac{\beta}{2}\int_{\Omega}z^2 dx$$

over $(u, z) \in H^1(\Omega) \times L^2(\Omega)$, *with state system*

$$\begin{cases} \mathcal{L}u = -\Delta u + a_0 u = z & \text{in } \Omega \\ \partial_\nu u = g & \text{on } \partial\Omega \end{cases} \tag{8.88}$$

where a_0 is a positive constant, $g \in L^2(\partial\Omega)$ and $z \in L^2(\Omega)$.

a) Show that there exists a unique minimizer.

b) Write the optimality conditions: adjoint problem and Euler equations.

8.12 (Distributed observation and boundary control, Neumann conditions). Let $\Omega \subset \mathbb{R}^n$ be a bounded, smooth domain. Consider the following control problem: *minimize the cost functional*

$$J\left(u, z\right) = \frac{1}{2}\int_{\Omega}\left(u - u_d\right)^2 dx + \frac{\beta}{2}\int_{\partial\Omega}z^2 dx$$

over $(u, z) \in H^1(\Omega) \times L^2(\partial\Omega)$, *with state system*

$$\begin{cases} -\Delta u + a_0 u = f & \text{in } \Omega \\ \partial_\nu u = z & \text{on } \partial\Omega \end{cases}$$

where a_0 is a positive constant, $f \in L^2(\Omega)$ and $z \in L^2(\partial\Omega)$.

a) Show that there exists a unique minimizer.

b) Write the optimality conditions: adjoint problem and Euler equations.

8.9.1 Approximation of boundary conditions in the finite element method

Let us consider the variational formulation of the problem addressed in Exercise 8.2 with $(a, b) = (0, 1)$, that is

$$\int_0^1 (\mu u'v' - \beta uv' + \sigma uv)dx = \int_0^1 fvdx$$

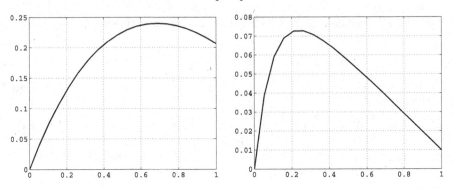

Fig. 8.6. Numerical solution of the problem described in 8.9.1 with $\beta = -1$ (left) and $\beta = -10$ (right)

arising from a constant coefficient an advection, diffusion, reaction equation complemented with a homogeneous Dirichlet condition on $x = 0$ and a Robin condition $-\mu u' + \beta u = 0$ on $x = 1$. Let us consider its approximation by means of piecewise linear finite elements. For the treatment of Dirichlet boundary conditions on $x_a = 0$ we introduce the discrete space,

$$X^1_{k,\{0\}}(0,1) = \{v_k \in X^1_k(0,1) \;:\; v_k(0) = 0\}$$

while $-\mu u' + \beta u = 0$ is enforced weakly, since it is the natural condition associated to the advection-diffusion operator. Then, the finite element approximation consists on finding $u_k \in X^1_{k,\{0\}}(0,1)$ such that

$$\int_0^1 (\mu u'_k v'_k - \beta u_k v'_k + \sigma u_k v_k)dx = \int_0^1 f v_k dx.$$

On the left panel of Fig. 8.6 we plot the solution obtained with coefficients $\mu = 1, \beta = -1, \sigma = 1$ and $f = 1$, while on the right the advective term is increased up to $\beta = -10$. We also notice that on $x_b = 1$, the solution features a positive value but a negative slope, in agreement with the condition $-\mu u' + \beta u = 0$.

8.9.2 Approximation of Robin boundary conditions

We address problem proposed in Exercise 8.4 on the unit circle Ω, with $\sigma = 0$. The Robin boundary condition $-\nabla u \cdot \mathbf{n} = \alpha u - g$ is treated by substituting $\alpha u - g$ into the boundary integral on $\partial \Omega$, after integration by parts. For this reason, additional terms appear on the left and right hand side of the variational formulation,

$$(\nabla u_k, \nabla v_k)_{L^2(\Omega)} + (\alpha u_k, v_k)_{L^2(\partial\Omega)} = (f, v_k)_{L^2(\Omega)} + (g, v_k)_{L^2(\partial\Omega)}.$$

We address two possible combinations of coefficients. In the case $\alpha = 1$, $g = y$, $f = 0$ the solution of the problem is $u(x,y) = \frac{1}{2}y$. Indeed, the Laplace

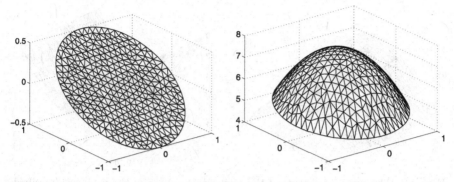

Fig. 8.7. Numerical solution of the problem described in Exercise 8.4 with $\alpha = 1$, $g = y$, $f = 0$ (left) and $\alpha = 1$, $g = y$, $f = 10$ (right)

equation is trivially satisfied, as well as the Robin condition since $\mathbf{n} = x\,\mathbf{i} + y\,\mathbf{j}$ on the unit circle and $\nabla u \cdot \mathbf{n} = \frac{1}{2}y = g - \alpha u$. As shown in Fig. 8.7 (left), the numerical solution, computed on a piecewise linear finite element space $X_h^1(\Omega)$, nicely fits to the exact one. More precisely, since $u(x,y) = \frac{1}{2}y \in X_h^1(\Omega)$, the numerical solution is exact up to machine precision in this case.

In the second test case we set $f = 10$. Fig. 8.7 (right) shows that, as expected, the solution of the problem is no longer linear.

8.9.3 Approximation of a system of equations

We address, as an example, the approximation of the system of equations of Exercise 8.6, restricted, for simplicity, to the case of homogeneous Dirichlet boundary conditions. The main difficulty in the application of the standard finite element method to this case consists in the treatment of the boundary conditions, because they combine values of the two unknown u and v. However, it is possible to reformulate the system at hand in a more convenient form by adding and subtracting the governing equations. By this way, we obtain a new system for the unknowns $z_1 = u + v$ and $z_2 = u - v$, more precisely

$$\begin{cases} -\alpha\Delta z_1 - \beta z_2 = f & \text{in } \Omega \\ -\alpha\Delta z_2 + \beta z_1 = 0 & \text{in } \Omega \\ z_1 = 0 & \text{on } \Gamma_D \\ z_2 = 0 & \text{on } \Gamma_D \\ \nabla z_1 \cdot \mathbf{n} = 0 & \text{on } \Gamma_N \\ \nabla z_2 \cdot \mathbf{n} = 0 & \text{on } \Gamma_N. \end{cases}$$

In this form the system can be more easily discretized by means of finite elements. To this purpose, we define Ω as the unit square, where on the vertical sides we set Dirichlet conditions, Γ_D, and on the horizontal sides we set Neumann conditions, Γ_N. Then, given the finite element space $X_{k,\Gamma_D}^1(\Omega)$,

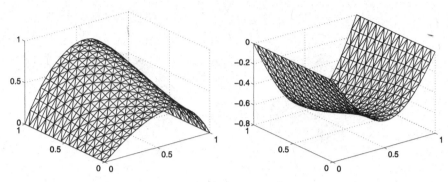

Fig. 8.8. The two components of the numerical solution of the problem described in 8.9.3 with $\alpha = 0.1$, $\beta = 1$, $f = x + y$

we aim to find $z_{k,1}$, $z_{k,2} \in X^1_{k,\Gamma_D}(\Omega)$ such that

$$\begin{cases} (\alpha \nabla z_{k,1}, \nabla v_{k,1}) - (\beta z_{k,2}, v_{k,1}) = (f, v_{k,1}), & \forall v_{k,1} \in X^1_{k,\Gamma_D}(\Omega), \\ (\alpha \nabla z_{k,2}, \nabla v_{k,2}) + (\beta z_{k,1}, v_{k,2}) = 0, & \forall v_{k,2} \in X^1_{k,\Gamma_D}(\Omega). \end{cases}$$

At the matrix level, this system is equivalent to

$$\begin{bmatrix} A_{11} & A_{12} \\ A_{12} & A_{22} \end{bmatrix} \cdot \begin{bmatrix} Z_1 \\ Z_2 \end{bmatrix} = \begin{bmatrix} F_1 \\ F_2 \end{bmatrix}$$

where the matrix blocks are $A_{11,ij} = A_{22,ij} = (\alpha \nabla \psi_j, \nabla \psi_i)$, $A_{12,ij} = -(\beta \psi_j, \psi_i)$, $A_{21} = -A^T_{12}$, the right hand side is $F_{1,i} = (f, \psi_i)$ and Z_1, Z_2 are the vectors of degrees of freedom to be determined by solving the system. Setting $\alpha = 0.1, \beta = 1$ and $f(x,y) = x + y$ the numerical solution of the system is reported in Fig. 8.8, where $z_{k,1}$ is depicted on the left and $z_{k,2}$ on the right.

8.9.4 Effect of problem regularity on the convergence of the finite element method

Let us consider the Poisson problem defined on a circular sector S_σ, with unit radius and angular coordinate $-\sigma/2 < \theta < \sigma/2$ where σ is a given value $0 < \sigma < 2\pi$. Denoting by Γ^1_σ the sides of the sector and by Γ^2_σ the circular perimeter, we aim to approximate u_σ, solution of the following problem,

$$\begin{cases} \Delta u_\sigma = 0 & \text{in } S_\sigma \\ u_\sigma = 0 & \text{on } \Gamma^1_\sigma \\ u_\sigma = \cos(\theta \pi / \sigma) & \text{on } \Gamma^2_\sigma \end{cases} \tag{8.89}$$

that is $u_\sigma = r^{\pi/\sigma} \cos(\theta\pi/\sigma)$ with $\theta \in (-\sigma/2, \sigma/2)$. Indeed, u_σ satisfies the boundary conditions and $\Delta u_\sigma = 0$, because u_σ is the real part of the olomorfic

Table 8.1. Convergence rates of the linear finite element method applied to the approximation of (8.89) for different values of σ

σ h	$\frac{1}{2}\pi$		$\frac{5}{4}\pi$		$\frac{3}{2}\pi$		$\frac{7}{4}\pi$	
	$\|\widehat{u}_\sigma - u_k\|_{H^1}$	p	$\|\widehat{u}_\sigma - u_k\|_{H^1}$	p	$\|\widehat{u}_\sigma - u_k\|_{H^1}$	p	$\|\widehat{u}_\sigma - u_k\|_{H^1}$	p
0.1250	9.434880e-02	–	5.240410e-02	–	1.106750e-01	–	1.822270e-01	–
0.0625	4.663180e-02	1.017	3.135850e-02	0.741	7.495930e-02	0.562	1.186980e-01	0.618
0.0312	2.343740e-02	0.993	1.824140e-02	0.782	4.797670e-02	0.644	8.455800e-02	0.489

function $z^{\frac{\pi}{\sigma}}$. Furthermore, owing to Lax-Milgram Lemma we assert that $u_\sigma \in H^1(S_\sigma)$, but $u_\sigma \notin H^2(S_\sigma)$ when $\sigma < \pi$. The latter conclusion can be proved by observing that in any neighborhood of the origin we have $\partial_{xx}u \simeq r^{\pi/\sigma-2}$ and thus

$$\int_{S_\sigma} |\partial_{xx}u_\sigma|^2 dxdy \simeq \int_{-\sigma/2}^{\sigma/2}\int_0^1 r^{2\pi/\sigma-4}rdrd\theta.$$

The integral on the right hand side is bounded provided that $2\pi/\sigma - 3 > -1$ namely $\sigma < \pi$. Conversely, when $\sigma > \pi$ we have $\partial_{xy}u_\sigma \notin L^2(S_\sigma)$ that is $u_\sigma \notin H^2(S_\sigma)$ and the more σ approaches 2π, the less regular is the solution.

In order to practically quantify the convergence rate of the finite element method, we solve the problem at hand over a sequence of progressively refined grids, characterized by parameters $h_1 > h_2$. Then, an estimate of the convergence rate is given by the value p such that

$$p = \left(\log\left(\frac{\|u_\sigma - u_{k,1}\|_{H^1}}{\|u_\sigma - u_{k,2}\|_{H^1}}\right)\right)\left(\log\left(\frac{h_1}{h_2}\right)\right)^{-1}.$$

Since the first order derivatives of u_σ are singular in the origin, the true error $\|u_\sigma - u_k\|_{H^1}$ can not be easily evaluated. We use the value $\|\widehat{u}_\sigma - u_k\|_{H^1}$ instead, where $\widehat{u}_\sigma$ is the piecewise linear interpolant of u_σ on a mesh that is four times more refined than the one used to compute u_k. In view of Theorem 8.7, we expect that when a piecewise linear approximation is sought, the numerical solution converges linearly to u_σ if $\sigma < \pi$, while the convergence rate progressively decreases when σ is larger than π, as confirmed by the results reported in Table 8.1.

9

Weak formulation of evolution problems

9.1 Parabolic Equations

In Chapter 3 we have considered the diffusion equation and some of its generalizations, as in the reaction-diffusion model (Section 3.3.4). This kind of equation belongs to the class of *parabolic equations,* that we have already classified in spatial dimension 1, in Section 6.5, and that we are going to define in a more general setting.

Let $\Omega \subset \mathbb{R}^n$ be a *bounded Lipschitz* domain, $T > 0$ and consider the space-time cylinder $Q_T = \Omega \times (0, T)$. Let $\mathbf{A} = \mathbf{A}(\mathbf{x})$ be a square matrix of order n, $\mathbf{b} = \mathbf{b}(\mathbf{x})$ a vector in[1] $\mathbb{R}^n$, $c = c(\mathbf{x})$ and $f = f(\mathbf{x},t)$ real functions. We consider equations of the type

$$u_t + \mathcal{L}u = f \tag{9.1}$$

where

$$\mathcal{L}u = -\operatorname{div}(\mathbf{A}\nabla u) + \mathbf{b} \cdot \nabla u + cu.$$

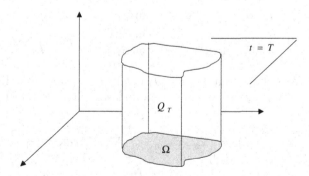

Fig. 9.1. Space-time cylinder

---

[1] For simplicity we consider time-independent coefficients.

Salsa S., Vegni F.M.G., Zaretti A., Zunino P.: *A Primer on PDEs. Models, Methods, Simulations.*
Unitext – La Matematica per il 3+2 65.
DOI 10.1007/978-88-470-2862-3_9, © Springer-Verlag Italia 2013

The operator $\mathcal{L}$ is called **parabolic** if

$$A(\mathbf{x})\,\boldsymbol{\xi}\cdot\boldsymbol{\xi} > 0 \qquad \text{a.e. in } \Omega, \ \forall \boldsymbol{\xi} \in \mathbb{R}^n, \ \boldsymbol{\xi} \neq \mathbf{0}.$$

For parabolic equations we may repeat the arguments concerning elliptic equations in Sections 8.1 and 8.2. Also in this case, different notions of solutions may be given, with the obvious corrections due to the evolutionary nature of (9.1). For identical reasons, we develop the theory for divergence form equations. Thus, given f in Q_T and $g \in L^2(\Omega)$, we want to determine a solution u, of the *parabolic* equation

$$u_t + \mathcal{L}u = f \text{ in } Q_T$$

satisfying an *initial (or Cauchy)* condition

$$u(\mathbf{x},0) = g(\mathbf{x}) \text{ in } \Omega$$

and one of the usual boundary conditions (*Dirichlet, Neumann, mixed or Robin*) on the lateral boundary $S_T = \partial\Omega \times [0,T]$.

To emphasize the ideas and to reduce the technical complexity, we will work with some degree of regularity of the data, which is actually not strictly necessary.

9.2 The Heat Equation

The *star* among parabolic equations is clearly the heat equation. We use this equation to introduce a possible *weak formulation* of the most common initial-boundary value problems.

9.2.1 The Cauchy-Dirichlet problem

Suppose we are given the problem

$$\begin{cases} u_t - \alpha\Delta u = f(\mathbf{x},t) & \text{in } Q_T \\ u = 0 & \text{in } \Omega \\ u(\mathbf{x},0) = g(\mathbf{x}) & \text{on } S_T = \partial\Omega \times (0,T) \end{cases} \qquad (9.2)$$

where $\alpha > 0$. We assume that $f \in L^2(Q_T)$ and[2] $g \in V = H_0^1(\Omega)$. Recall that, by Poincaré inequality, we can choose the following inner product and norm in V

$$(\varphi, \psi)_0 = (\nabla\varphi, \nabla\psi)_0 = \int_\Omega \nabla\varphi \cdot \nabla\psi\, d\mathbf{x}$$

$$\|\varphi\|_0 = \|\nabla\varphi\|_0 = \left(\int_\Omega |\nabla\varphi|^2\, d\mathbf{x}\right)^{1/2}.$$

---

[2] Also $g \in H^1(\Omega)$ is admissible.

To find a weak formulation we first proceed formally. We multiply the diffusion equation by a smooth function $v = v(\mathbf{x})$, vanishing at the boundary of Ω to take into account the homogeneous Dirichlet condition, and integrate over Ω. We find:

$$\int_\Omega u_t(\mathbf{x},t)\, v(\mathbf{x})\, dx - \alpha \int_\Omega \Delta u(\mathbf{x},t)\, v(\mathbf{x})\, dx = \int_\Omega f(\mathbf{x},t)\, v(\mathbf{x})\, dx.$$

Integrating by parts the second term, since $v = 0$ on $\partial\Omega$, we get, for $t \in (0,T)$

$$\int_\Omega u_t(\mathbf{x},t)\, v(\mathbf{x})\, dx + \alpha \int_\Omega \nabla u(\mathbf{x},t) \cdot \nabla v(\mathbf{x})\, dx = \int_\Omega f(\mathbf{x},t)\, v(\mathbf{x})\, dx. \quad (9.3)$$

This looks like what we did for elliptic equations, except for the presence of u_t. Moreover, here we will have somehow to take into account the initial condition. Which could be a correct functional setting for the solution u?

Due to the homogeneous Dirichlet conditions, it is natural to require that the function $\mathbf{x} \mapsto u(\mathbf{x},t)$ belongs to V for at least a.e. $t \in (0,T)$. Moreover since we will need later on to perform an integration in time of (9.3) over $(0,T)$ we end up to require that $u \in H^1_{0,S_T}(Q_T)$ where

$$H^1_{0,S_T}(Q_T) = \{u : u, u_t, |\nabla u| \in L^2(Q_T),\ u = 0 \text{ on } S_T\}.$$

Notice that by the Poincaré inequality,

$$\int_0^T \int_\Omega u^2 dx dt \le C_P^2 \int_0^T \int_\Omega |\nabla u|^2\, dx dt.$$

Thus $H^1_{0,S_T}(Q_T)$ is a Hilbert space with inner product and norm given by

$$(u,v)_{H^1_{0,S_T}(Q_T)} = \int_0^T \int_\Omega \nabla u \cdot \nabla v\, dx dt,$$

$$\|u\|_{H^1_{0,S_T}(Q_T)} = \left(\int_0^T \int_\Omega |\nabla u|^2\, dx dt \right)^{1/2}.$$

Accordingly, we choose our test functions $v \in V$.

Let us examine the initial conditions. Look at the *real* function

$$U : t \longmapsto \int_\Omega |\nabla u(\mathbf{x},t)|^2\, dx.$$

We will show that U is continuous in $[0,T]$. To express this fact we say that

$$u \in C([0,T];V).$$

Therefore, the initial condition $u(0) = g$ means that

$$\int_\Omega (\nabla u(\mathbf{x},t) - \nabla g(\mathbf{x}))^2\, dx \to 0 \text{ as } t \to 0. \quad (9.4)$$

It turns out that $C\left(\left[0,T\right];V\right)$ is a Banach space with the norm given by

$$\|u\|_{C\left(\left[0,T\right];V\right)} = \max_{t\in\left[0,T\right]} \left(\int_{\Omega} |\nabla u\left(\mathbf{x},t\right)|^2 \, d\mathbf{x}\right)^{1/2}.$$

Note that, by Poincaré inequality, we also have $u \in C\left(\left[0,T\right];L^2\left(\Omega\right)\right)$ and

$$\int_{\Omega} \left(u\left(\mathbf{x},t\right) - g\left(\mathbf{x}\right)\right)^2 d\mathbf{x} \to 0 \text{ as } t \to 0.$$

In conclusion, we are lead to the following definition.

Definition 9.1. *A weak solution to problem (9.2) is a function* $u \in H^1_{0,S_T}$ *$\left(Q_T\right)$ such that:*

i)   *For every* $v \in V$,

$$\int_{\Omega} u_t\left(\mathbf{x},t\right) v\left(\mathbf{x}\right) \, d\mathbf{x} + \alpha \int_{\Omega} \nabla u\left(\mathbf{x},t\right) \cdot \nabla v\left(\mathbf{x}\right) \, d\mathbf{x} = \int_{\Omega} f\left(\mathbf{x},t\right) v\left(\mathbf{x}\right) \, d\mathbf{x}$$

   for a.e. $t \in \left(0,T\right)$.

ii) $u\left(\mathbf{x},t\right) = g\left(\mathbf{x}\right)$ *in* Ω, *in the sense of (9.4).*

Remark 9.1. We leave it to the reader to check that if a *weak* solution u is smooth, i.e. $u \in C^{2,1}\left(\overline{Q}_T\right)$, then u is actually a classical solution.

We prove the following theorem.

Theorem 9.1. *There exists a unique weak solution u of (9.2). Moreover the following stability estimates hold:*

$$\max_{t\in\left[0,T\right]} \int_{\Omega} |\nabla u\left(\mathbf{x},t\right)|^2 \, d\mathbf{x} \le \|g\|_1^2 + \frac{1}{\alpha} \int_0^T \int_{\Omega} f^2\left(\mathbf{x},s\right) d\mathbf{x} ds \tag{9.5}$$

and

$$\int_0^T \int_{\Omega} u_t^2\left(\mathbf{x},s\right) d\mathbf{x} ds \le \alpha \|g\|_1^2 + \int_0^T \int_{\Omega} f^2\left(\mathbf{x},s\right) d\mathbf{x} ds. \tag{9.6}$$

Note in particular how the first estimate deteriorates as α approaches to zero, when f in not identically zero.

To prove Theorem 9.1, we shall use the so-called Faedo-Galerkin method, also convenient for the numerical approximations techniques. Let us explain briefly the general strategy, that consists in the following 4 steps.

1. *Galerkin approximations.* First discretize in space and construct a sequence of approximating functions u_m by solving a system of ordinary differential equations.

2. *Energy estimates.* Prove that suitable norms of $\{u_m\}$ are controlled by the norms of the data f and g in their respective spaces.

3. *Existence of a weak solution.* Prove that $\{u_m\}$, or at least a subsequence of it, converges in a suitable sense to a weak solution $u \in H^1_{0,S_T}(Q_T)$.

4. *Uniqueness and stability.* Show that u is the unique weak solution of problem (9.2) and that it depends continuously from the data f and g (*energy estimates for u*).

9.2.2 Galerkin approximations

Select a sequence of smooth functions $\{w_k\}_{k=1}^{\infty}$ constituting[3]

$$\text{an } orthogonal \text{ basis in } V = H^1_0(\Omega)$$

and

$$\text{an } orthonormal \text{ basis in } L^2(\Omega).$$

In particular, we can write

$$g = \sum_{k=1}^{\infty} \hat{g}_k w_k$$

where $\hat{g}_k = \int_{\Omega} g w_k \, d\mathbf{x}$ and the series converges in V.

Note that, since $(\nabla w_h, \nabla w_k)_0 = 0$ if $h \neq k$, letting $\lambda_k = \|\nabla w_k\|_0^2$, we have

$$\|g\|_1^2 = \sum_{k=1}^{\infty} \lambda_k \hat{g}_k^2. \tag{9.7}$$

Moreover

$$f(\mathbf{x}, t) = \sum_{k=1}^{\infty} \hat{f}_k(t) w_k(\mathbf{x}),$$

where $\hat{f}_k(t) = \int_{\Omega} f(\mathbf{x}, t) w_k(\mathbf{x}) \, d\mathbf{x}$ and the series converges in $L^2(\Omega)$ for a.e. $t \in [0, T]$. Since also $(w_h, w_k)_0 = 0$ if $h \neq k$ and $\|w_k\|_0 = 1$, we have

$$\int_{\Omega} f^2(\mathbf{x}, t) \, d\mathbf{x} = \sum_{k=1}^{\infty} \hat{f}_k^2(t). \tag{9.8}$$

Construct now the sequence of finite-dimensional subspaces

$$V_m = \text{span}\{w_1, w_2, \ldots, w_m\}.$$

Clearly

$$V_m \subset V_{m+1} \quad \text{and} \quad \overline{\cup V_m} = V.$$

For $m \geq 1$ fixed, let

$$f_m(\mathbf{x}, t) = \sum_{k=1}^{m} \hat{f}_k(t) w_k(\mathbf{x}), \quad g_m(\mathbf{x}) = \sum_{k=1}^{m} \hat{g}_k w_k(\mathbf{x}).$$

---

[3] We can choose as w_k the Dirichlet eigenfunctions of the Laplace operator, normalized with respect to the norm in $L^2(\Omega)$ (see Theorem 8.3).

and

$$u_m\left(\mathbf{x},t\right)=\sum_{k=1}^{m}c_{mk}\left(t\right)w_k\left(\mathbf{x}\right).$$

We can find the unknown coefficients $c_{mk}=c_{mk}\left(t\right)$ by solving the following approximating problem, obtained by projecting the differential equation onto the subspace V_m: determine $u_m\in H_{0,S_T}^1\left(Q_T\right)$, such that:

$i_m)$ For every $h=1,\ldots,m$,

$$\int_\Omega\partial_t u_m\left(\mathbf{x},t\right)w_h\left(\mathbf{x}\right)\,d\mathbf{x}+\alpha\int_\Omega\nabla u_m\left(\mathbf{x},t\right)\cdot\nabla w_h(\mathbf{x})\,d\mathbf{x}=\int_\Omega f_m\left(\mathbf{x},t\right)w_h\left(\mathbf{x}\right)d\mathbf{x}$$

for a.e. $t\in(0,T)$

$ii_m)$ $u_m\left(\mathbf{x},t\right)=g_m\left(\mathbf{x}\right)$ in Ω, in the sense of (9.4).

Note that since the integro/differential equation in $i_m)$ is true for each element of the basis w_h, $h=1,\ldots,m$, then it is true for every $v\in V_m$. We call u_m a *Galerkin approximation* of the solution u.

The following lemma holds.

Lemma 9.1. For all m, there exists a unique solution $u_m\in H_{0,S_T}^1\left(Q_T\right)$ of the approximating problem $i_m)$, $ii_m)$. Moreover $u_m\in C\left([0,T];V\right)$ and

$$\int_\Omega\left|\nabla u_m\left(\mathbf{x},t\right)-\nabla g_m\left(\mathbf{x}\right)\right|^2d\mathbf{x}\to 0\text{ as }t\to 0.\qquad(9.9)$$

Proof. Since $w_1,\ldots,w_m$ are mutually orthonormal in $L^2\left(\Omega\right)$, we have

$$\int_\Omega\partial_t u_m\left(\mathbf{x},t\right)w_h\left(\mathbf{x}\right)d\mathbf{x}=\sum_{h=1}^{m}\int_\Omega\dot{c}_{mk}\left(t\right)w_k\left(\mathbf{x}\right)w_h\left(\mathbf{x}\right)d\mathbf{x}=\dot{c}_{mh}\left(t\right).$$

Also, $w_1,\ldots,w_m$ is an orthogonal system in V_m, hence, for each $h=1,\ldots,m:$

$$\int_\Omega\nabla u_m\left(\mathbf{x},t\right)\cdot\nabla w_h\left(\mathbf{x}\right)d\mathbf{x}=\sum_{k=1}^{m}\int_\Omega c_{mk}\left(t\right)\nabla w_k\left(\mathbf{x}\right)\cdot\nabla w_h\left(\mathbf{x}\right)d\mathbf{x}=\lambda_h c_{mh}\left(t\right).$$

Thus $i_m)$ is equivalent to the following system of m *uncoupled* linear ordinary differential equations, with constant coefficients

$$\begin{cases}\dot{c}_{mh}\left(t\right)+\alpha\lambda_h c_{mh}\left(t\right)=\hat{f}_h\left(t\right)&\text{q.o. in }(0,T)\\c_{mh}\left(0\right)=\hat{g}_h\end{cases}\qquad h=1,\ldots,m.\quad(9.10)$$

In particular, it follows that the coefficients c_{mh} **do not depend** on m:

$$c_{mh}\left(t\right)=c_h\left(t\right).$$

From the o.d.e. theory, each equation has a unique solution $c_h \in C([0,T])$ with $\dot{c}_h \in L^2(0,T)$, given by the explicit formula[4]

$$c_h(t) = \hat{g}_h + \int_0^t e^{-\alpha\lambda_h(t-s)} \hat{f}_h(s)\, ds. \tag{9.12}$$

We deduce that $u_m \in C([0,T]; V)$ and that $\partial_t u_m \in L^2(Q_T)$. Moreover, since we have

$$c_h(t) \to \hat{g}_h \text{ as } t \to 0 \text{ for every } k = 1,\dots,m,$$

we can write

$$\int_\Omega |\nabla u_m(\mathbf{x},t) - \nabla g_m(\mathbf{x})|^2\, d\mathbf{x} = \sum_{h=1}^m |c_h(t) - \hat{g}_h|^2 \lambda_h \to 0$$

as $t \to 0$.   □

Remark 9.2. Notice that the fundamental calculus formula holds for c_h and c_h^2:

$$c_h(t) = \hat{g}_h + \int_0^t \dot{c}_h(s)\, ds \quad \text{for } t \in [0,T]. \tag{9.13}$$

and

$$c_h^2(t) = \hat{g}_h^2 + \int_0^t \frac{d}{ds} c_h^2(s)\, ds \quad \text{for } t \in [0,T]. \tag{9.14}$$

In fact, $c_h, c_h^2 \in C([0,T])$ and both $\dot{c}_h$, $\frac{d}{ds} c_h^2(s) = 2c_h \dot{c}_h$ belongs to $L^2(0,T)$.

Remark 9.3. We have chosen a basis $\{w_k\}$ orthonormal in $L^2(\Omega)$ and orthogonal in V because with respect to this basis, the Laplace operator becomes a *diagonal operator*, as it is reflected by the uncoupling in the approximate problem (9.10). The reader has realized that what we have performed is nothing but a variant of the method of *separation of variables*.

However, the method also works using any countable basis $\{w_k\}$ for both spaces, with no orthogonality property. This is particularly important in the

---

[4] Let h fixed. If $\hat{f}_h$ is continuous in $[0,T]$ the formula comes from elementary theory. In the general case, by the density of $C_0^\infty(\Omega)$ in $L^2(0,T)$, we can find a sequence $\{f_{jh}\} \subset C_0^\infty(\Omega)$, such that $f_{jh} \to f_h$ in $L^2(0,T)$, as $j \to \infty$. Then, formula (9.12) holds for f_{jh} :

$$c_h(t) = \hat{g}_h e^{-\lambda_h t} + \int_0^t e^{-\lambda_h(t-s)} \hat{f}_h(r)\, dr. \tag{9.11}$$

It is easy to check that $\{c_{jh}\}$ is a Cauchy sequence in $C([0,T])$. Thus it converges to c_h in this space. Passing to the limit in (9.11) we deduce that (9.12) holds for c_h. By direct differentiation, c_h is a solution of (9.10), with the desired properties. Uniqueness follows since the homogeneous problem $\dot{w} + \lambda_h w = 0$, $w(0) = 0$ has only the null solution.

numerical implementation of the method, where, in general, the elements of the basis in V_m are not mutually orthogonal. If

$$u_m\left(\mathbf{x},t\right) = \sum_{k=1}^{m} b_{mk}\left(t\right) w_k\left(\mathbf{x}\right),$$

then

$$\int_{\Omega} \partial_t u_m\left(\mathbf{x},t\right) w_h\left(\mathbf{x}\right) d\mathbf{x} = \int_{\Omega} \sum_{h=1}^{m} \dot{b}_{mk}\left(t\right) w_k\left(\mathbf{x}\right) w_h\left(\mathbf{x}\right) d\mathbf{x}$$

$$= \sum_{h=1}^{m} P_{hk} \dot{c}_{mk}\left(t\right)$$

where $P_{hk} = (w_h, w_k)_0$. Moreover

$$\int_{\Omega} \nabla u_m\left(\mathbf{x},t\right) \cdot \nabla w_h\left(\mathbf{x}\right) d\mathbf{x} = \sum_{k=1}^{m} \int_{\Omega} c_{mk}\left(t\right) \nabla w_k\left(\mathbf{x}\right) \cdot \nabla w_h\left(\mathbf{x}\right) d\mathbf{x}$$

$$= \sum_{h=1}^{m} W_{hk} c_{mk}\left(t\right)$$

where $W_{hk} = (\nabla w_k, \nabla w_h)_0$.

Letting

$$\hat{\mathbf{F}}_m\left(t\right) = (\hat{f}_1\left(t\right), \dots, \hat{f}_m\left(t\right))^{\mathsf{T}}$$

and

$$\mathbf{B}_m\left(t\right) = (b_{m1}\left(t\right), \dots, b_{mm}\left(t\right))^{\mathsf{T}}, \qquad \hat{\mathbf{g}}_m = (\hat{g}_1, \dots, \hat{g}_m)^{\mathsf{T}}$$

the approximating problem (9.10) becomes

$$\dot{\mathbf{B}}_m\left(t\right) = -\alpha \mathbf{P}^{-1} \mathbf{W} \mathbf{B}_m\left(t\right) + \hat{\mathbf{F}}_m\left(t\right), \qquad \text{a.e. } t \in [0, T] \qquad (9.15)$$

where[5]

$$\mathbf{P} = (P_{hk}), \qquad \mathbf{W} = (W_{hk})$$

with initial condition

$$\mathbf{B}_m\left(0\right) = \hat{\mathbf{g}}_m.$$

Also for problem (9.15) we could exhibit an explicit formula that allows to reach the same conclusion in Lemma 9.1. We skip the details.

9.2.3 Energy estimates

We want to show that the norms

$$\|\partial_t u_m\|_{L^2(Q_T)}, \|u_m\|_{C([0,T];V)},$$

_____

[5] Since $w_1, \dots, w_m$ is a basis in $L^2\left(\Omega\right)$, the matrix $\mathbf{P}$ is non singular.

hence also $\|u_m\|_{H^1_{0,S_T}(Q_T)}$, can be estimated by the appropriate norms of the data, **and that the estimates are independent of** m.

Thus, let $u_m(\mathbf{x},t) = \sum_{k=1}^m c_k(t) w_k(\mathbf{x})$ be the m−Galerkin approximation.

Lemma 9.2. For every $t \in [0,T]$, the following estimates holds:

$$\max_{[0,T]} \int_\Omega |\nabla u_m(\mathbf{x},t)|^2 \, d\mathbf{x} \le \|g\|_1^2 + \frac{1}{\alpha} \int_0^t \int_\Omega f^2 \, d\mathbf{x}ds \qquad (9.16)$$

and

$$\int_0^t \int_\Omega (\partial_t u_m)^2 d\mathbf{x}ds \le \alpha \|g\|_1^2 + \int_0^t \int_\Omega f^2 \, d\mathbf{x}ds. \qquad (9.17)$$

Proof. Observe that

$$\int_\Omega |\nabla u_m(\mathbf{x},t)|^2 \, d\mathbf{x} = \sum_{k=1}^m \lambda_k c_k^2(t) \qquad (9.18)$$

and

$$\int_\Omega |\partial_t u_m(\mathbf{x},t)|^2 \, d\mathbf{x} = \sum_{k=1}^m \dot{c}_k^2(t). \qquad (9.19)$$

Multiply the differential equation

$$\dot{c}_h(t) + \alpha \lambda_h c_h(t) = \hat{f}_h(t) \qquad (9.20)$$

by $\dot{c}_h(t)$. We get

$$\dot{c}_h^2(t) + \alpha \lambda_h c_h(t) \dot{c}_h(t) = \hat{f}_h(t) \dot{c}_h(t) \qquad (9.21)$$

for a.e. $t \in [0,T]$. From the elementary inequality

$$|ab| \le \frac{a^2}{2\varepsilon} + \frac{\varepsilon b^2}{2} \qquad \forall a,b \in \mathbb{R}, \, \varepsilon > 0, \qquad (9.22)$$

for $\varepsilon = 1$, we deduce

$$\dot{c}_h^2(t) + \frac{\alpha \lambda_h}{2} \frac{d}{dt} c_h^2(t) \le \frac{1}{2} \hat{f}_h^2(t) + \frac{1}{2} \dot{c}_h^2(t)$$

for a.e. $t \in [0,T]$. Thus, from (9.21) we obtain

$$\dot{c}_h^2(t) + \alpha \lambda_h \frac{d}{dt} c_h^2(t) \le \hat{f}_h^2(t).$$

We now integrate over $(0,t)$ and use (9.14) to find:

$$\int_0^t \dot{c}_h^2(s) \, ds + \alpha \lambda_h c_h^2(t) \le \alpha \lambda_h \hat{g}_h^2 + \int_0^t \hat{f}_h^2(s) \, ds. \qquad (9.23)$$

Summing for $h = 1, \ldots, m$ we find

$$\int_0^t \sum_{k=1}^m \dot{c}_h^2(s)\, ds + \alpha \sum_{k=1}^m \lambda_k c_k^2(t) \leq \alpha \sum_{k=1}^m \lambda_k \hat{g}_k^2 + \int_0^t \sum_{k=1}^m \hat{f}_h^2(s)\, ds$$

$$\leq \alpha \sum_{k=1}^\infty \lambda_k \hat{g}_k^2 + \int_0^t \sum_{k=1}^\infty \hat{f}_h^2(s)\, ds$$

which is equivalent to

$$\int_0^t \int_\Omega (\partial_t u_m)^2 dx\, ds + \alpha \int_\Omega |\nabla u_m(\mathbf{x},t)|^2\, dx \leq \alpha \|g\|_1^2 + \int_0^t \int_\Omega f^2 dx\, ds$$

by (9.7), (9.8), (9.18) and (9.19).

The inequalities (9.16) and (9.17) follow easily.   □

9.2.4 Existence and stability

Existence and stability estimates. Lemma 9.2 and (9.18), (9.19) show that the series

$$\sum_{k=1}^\infty \lambda_k c_k^2(t)$$

is uniformly convergent in $[0,T]$ (or in $C([0,T])$) while the series

$$\sum_{k=1}^\infty \dot{c}_k^2(t)$$

converges in $L^1(0,T)$. Hence we have

$$u_m(\mathbf{x},t) \to u(\mathbf{x},t) = \sum_{k=1}^\infty c_k(t) w_s(\mathbf{x})$$

in $C([0,T]\,; V)$ and

$$\max_{t \in [0,T]} \int_\Omega |\nabla u(\mathbf{x},t)|^2\, dx = \max_{t \in [0,T]} \sum_{k=1}^\infty \lambda_k c_k^2(t)$$

$$\leq \sum_{k=1}^\infty \lambda_k \hat{g}_k^2 + \frac{1}{\alpha} \int_0^T \sum_{k=1}^\infty \hat{f}_h^2(s)\, ds$$

$$= \|g\|_1^2 + \frac{1}{\alpha} \int_0^T \int_\Omega f^2 dx\, ds$$

which is (9.5). Analogously,

$$\partial_t u_m(\mathbf{x},t) \to v(\mathbf{x},t) = \sum_{k=1}^\infty \dot{c}_k(t) w_s(\mathbf{x})$$

in $L^2(Q_T)$ and

$$\int_0^T \int_\Omega v^2(\mathbf{x}, s)\, d\mathbf{x}ds = \int_0^T \sum_{k=1}^\infty \dot{c}_k^2(s)\, ds \tag{9.24}$$

$$\leq \alpha \sum_{k=1}^\infty \lambda_k \hat{g}_k^2 + \int_0^T \sum_{k=1}^\infty \hat{f}_k^2(s)\, ds$$

$$= \alpha \|g\|_1^2 + \int_0^T \int_\Omega f^2 d\mathbf{x}ds.$$

We want to show that $v = \partial_t u$ in the sense of distributions in Q_T. Take a function $\varphi \in C_0^\infty(Q_T)$. Then we have, since $\partial_t u_m \to v$ in $L^2(Q_T)$:

$$\int_0^T \int_\Omega v\varphi\, d\mathbf{x}ds = \lim_{m\to\infty} \int_0^T \int_\Omega \partial_t u_m \varphi\, d\mathbf{x}ds$$

$$= -\lim_{m\to\infty} \int_0^T \int_\Omega u_m \varphi_t\, d\mathbf{x}ds$$

$$= -\int_0^T \int_\Omega u\varphi_t\, d\mathbf{x}ds$$

whence $v = \partial_t u$ in the sense of distributions and (9.6) follows from (9.24).

To show that u is a weak solution, integrate equation $i_m)$ over $(0, t)$:

$$\int_0^t \int_\Omega \partial_t u_m w_h\, d\mathbf{x} + \alpha \int_0^t \int_\Omega \nabla u_m \cdot \nabla w_h\, d\mathbf{x} = \int_0^t \int_\Omega f_m w_h d\mathbf{x}.$$

Letting $m \to +\infty$, by the convergence properties of $\{u_m\}$ and $\{\partial_t u_m\}$, we get

$$\int_0^t \int_\Omega \partial_t u w_h\, d\mathbf{x} + \alpha \int_0^t \int_\Omega \nabla u \cdot \nabla w_h\, d\mathbf{x} = \int_0^t \int_\Omega f w_h d\mathbf{x} \tag{9.25}$$

which is now valid for all $h \geq 1$. But $\{w_k\}_{k\geq 1}$ is a basis in V and therefore (9.25) holds for every $v \in V$. Differentiating back with respect to t, we obtain

$$\int_\Omega \partial_t u v\, d\mathbf{x} + \alpha \int_\Omega \nabla u \cdot \nabla v\, d\mathbf{x} = \int_\Omega f v d\mathbf{x}$$

for all $v \in V$ and a.e. in $(0, T)$. Therefore u is a weak solution of problem (9.2).

For the initial condition, we have:

$$u(\mathbf{x}, 0) = \sum_{k=1}^\infty c_k(0) w_s(\mathbf{x}) = \sum_{k=1}^\infty \hat{g}_k(0) w_s(\mathbf{x}) = g(\mathbf{x}) \qquad \text{a.e. in } (0, T)$$

and (9.4) holds since $u \in C(0, T; V)$.

Uniqueness. Let

$$z\left(\mathbf{x},t\right)=\sum_{k=1}^{\infty}d_{k}\left(t\right)w_{k}\left(\mathbf{x}\right)$$

another weak solutions of the same problem. Then, $U=u-z$ is a weak solution of

$$\int_{\Omega}\partial_{t}U\ v\ d\mathbf{x}+\alpha\int_{\Omega}\nabla U\cdot\nabla v d\mathbf{x}=0$$

for all $v\in V$ and a.e. $t\in[0,T]$, with initial data $U\left(\mathbf{x},0\right)=0$.

Let $U_{k}\left(t\right)=c_{k}\left(t\right)-d_{k}\left(t\right)$. Then, U_{k} solves

$$\dot{U}_{k}\left(t\right)+\alpha\lambda_{k}U_{k}\left(t\right)=0$$

with $U_{k}\left(0\right)=0$. It follows that $U_{k}=0$ for all $k\geq1$ and therefore $u=z$ a.e. in Q_{T}.

Theorem 9.1 is now completely proved.

9.2.5 Neumann and mixed boundary conditions

The Faedo-Galerkin method works with the other common boundary conditions, with small adjustments. For instance, consider the following initial-Neumann value problem

$$\begin{cases} u_{t}-\alpha\Delta u=f\left(\mathbf{x},t\right) & \text{in } Q_{T} \\ \partial_{\nu}u=0 & \text{on } S_{T} \\ u\left(\mathbf{x},0\right)=g\left(\mathbf{x}\right) & \text{in } \Omega \end{cases} \tag{9.26}$$

where $f\in L^{2}\left(Q_{T}\right)$, $g\in V=H^{1}\left(\Omega\right)$. Recall that inner product and norm in V are given by

$$\left(\varphi,\psi\right)_{1,2}=\left(\varphi,\psi\right)_{0}+\left(\nabla\varphi,\nabla\psi\right)_{0}=\int_{\Omega}\varphi\psi d\mathbf{x}+\int_{\Omega}\nabla\varphi\cdot\nabla\psi d\mathbf{x}$$

$$\left\|\varphi\right\|_{1,2}=\left(\left\|\varphi\right\|_{1,2}+\left\|\nabla\varphi\right\|_{0}\right)^{1/2}=\left(\int_{\Omega}\varphi^{2}d\mathbf{x}+\int_{\Omega}\left|\nabla\varphi\right|^{2}d\mathbf{x}\right)^{1/2}.$$

For the weak formulation, as in the case of the Dirichlet conditions, we multiply the diffusion equation by a test function[6] $v\in V$ and integrate it over Ω. After an integration by parts, we find

$$\int_{\Omega}u_{t}\left(\mathbf{x},t\right)v\left(\mathbf{x}\right)\ d\mathbf{x}+\alpha\int_{\Omega}\nabla u\left(\mathbf{x},t\right)\cdot\nabla v\left(\mathbf{x}\right)d\mathbf{x}-\alpha\int_{\partial\Omega}\partial_{\nu}u\left(\boldsymbol{\sigma},t\right)\ v(\boldsymbol{\sigma})d\sigma$$

$$=\int_{\Omega}f\left(\mathbf{x},t\right)v\left(\mathbf{x}\right)d\mathbf{x}.$$

Taking into account the homogeneous Neumann condition, we are lead to the following definition.

---

[6] Here there is no reason to require that $v=0$ on S_{T}.

Definition 9.2. *A weak solution to problem (9.26) is a function $u \in H^1(Q_T)$ such that:*

i) *For every $v \in V$,*

$$\int_\Omega u_t(\mathbf{x},t)\, v(\mathbf{x})\ dx + \alpha \int_\Omega \nabla u(\mathbf{x},t) \cdot \nabla v(\mathbf{x})\ dx = \int_\Omega f(\mathbf{x},t)\, v(\mathbf{x})\ dx$$

   for a.e. $t \in (0,T)$.

ii) *$u(\mathbf{x},t) = g(\mathbf{x})$ in Ω, in the sense of (9.4).*

Remark 9.4. Again, we leave it to the reader to check that if a *weak* solution u is *smooth*, i.e. $u \in C^{2,1}(\overline{Q}_T)$, then u is actually a classical solution.

The following theorem holds.

Theorem 9.2. *There exists a unique weak solution u of (9.26). Moreover, for every $t \in [0,T]$, the following stability estimates hold*

$$\int_\Omega \left(u^2(\mathbf{x},t) + |\nabla u(\mathbf{x},t)|^2 \right) dx \leq e^{2\alpha t} \|g\|_{1,2}^2 + \frac{1}{\alpha} \int_0^t \int_\Omega e^{2\alpha(t-s)} f^2 dx ds \tag{9.27}$$

$$\int_0^t \int_\Omega u_t^2 dx ds \leq \alpha e^{2\alpha t} \|g\|_{1,2}^2 + \int_0^t \int_\Omega e^{2\alpha(t-s)} f^2 dx ds. \tag{9.28}$$

Proof. We follow the proof of Theorem 9.1, emphasizing the differences that arise from the Dirichlet conditions.

Since for a.e. $t \in (0,T)$ we have that the function $\mathbf{x} \to u(\mathbf{x},t)$ belongs to V, to discretize in space, we select a sequence of smooth functions $\{w_k\}_{k=1}^\infty$ constituting[7]

an orthogonal basis in $V = H^1(\Omega)$

and

an orthonormal basis in $L^2(\Omega)$.

In particular, we can write

$$g = \sum_{k=1}^\infty \hat{g}_k w_k$$

where $\hat{g}_k = \int_\Omega g w_k\ dx$ and the series converges in V.

Note that, since $(w_h, w_k)_{1,2} = 0$ if $h \neq k$, letting $\mu_k = \|w_k\|_{1,2}^2$, we have

$$\|g\|_{1,2}^2 = \sum_{k=1}^\infty \mu_k \hat{g}_k^2. \tag{9.29}$$

---

[7] We can choose as w_k the Neumann eigenfunctions of the Laplace operator, normalized with respect to the norm in $L^2(\Omega)$ (see Theorem 8.3). These functions are also orthogonal in $H^1(\Omega)$.

Moreover

$$f(\mathbf{x},t) = \sum_{k=1}^{\infty} \hat{f}_k(t)\, w_k(\mathbf{x}).$$

Moreover

$$f(\mathbf{x},t) = \sum_{k=1}^{\infty} \hat{f}_k(t)\, w_k(\mathbf{x}),$$

where $\hat{f}_k(t) = \int_\Omega f(\mathbf{x},t)\, w_k(\mathbf{x})\, d\mathbf{x}$ and the series converges in $L^2(\Omega)$ for a.e. $t \in [0,T]$. Since also $(w_h, w_k)_0 = 0$ if $h \neq k$ and $\|w_k\|_0 = 1$, we have

$$\int_\Omega f^2(\mathbf{x},t)\, d\mathbf{x} = \sum_{k=1}^{\infty} \hat{f}_k^2(t). \tag{9.30}$$

The Galerkin approximations

$$u_m(\mathbf{x},t) = \sum_{k=1}^{m} c_{mk}(t)\, w_k(\mathbf{x})$$

are solutions of the problem

$$\begin{cases} \int_\Omega \partial_t u_m w_h\, d\mathbf{x} + \alpha \int_\Omega \nabla u_m \cdot \nabla w_h\, d\mathbf{x} = \int_\Omega f_m w_h\, d\mathbf{x}, & \text{a.e } t \in [0,T] \\ u_m(0) = g_m \end{cases}$$

for every $h = 1, \dots, m$.

The main difference comes from the fact that for the Cauchy-Dirichlet problem the bilinear form in the $\mathbf{x}$−space $H_0^1(\Omega)$ given by

$$B(\varphi, \psi) = \alpha \int_\Omega \nabla\varphi(\mathbf{x}) \cdot \nabla\psi(\mathbf{x})\, d\mathbf{x}$$

is $H_0^1(\Omega)$−coercive. This is crucial for deriving the energy inequalities for u_m. Here, due to the lack of a Poincaré inequality, the bilinear form B is not V−coercive.

However, B is **weakly coercive**, i.e. that there exist $\gamma > 0$, $\lambda \geq 0$ such that

$$\tilde{B}(\varphi, \varphi) = a(\varphi, \varphi) + \lambda \|\varphi\|_0^2 \geq \gamma \|\varphi\|_{1,2}^2 \qquad \forall \varphi \in V \tag{9.31}$$

that is

$$\tilde{B}(\varphi, \psi) = B(\varphi, \psi) + \lambda\alpha(\varphi, \psi)_0$$

is V−coercive. In fact, the simple choice (not unique, of course) $\lambda = \alpha$ gives

$$B(\varphi, \varphi) + \alpha \|\varphi\|_0^2 = \alpha \|\varphi\|_{1,2}^2 \qquad \forall \varphi \in V.$$

But then, going back to our approximating problem, performing the simple change of variable

$$z_m(\mathbf{x},t) = e^{-\alpha t} u_m(\mathbf{x},t)$$

we see that

$$\partial_t z_m\left(\mathbf{x},t\right) = e^{-\alpha t}\partial_t u_m\left(\mathbf{x},t\right) - \alpha e^{-\alpha t}u_m\left(\mathbf{x},t\right) = e^{-\alpha t}\partial_t u_m\left(\mathbf{x},t\right) - \alpha z_m\left(\mathbf{x},t\right)$$

so that, z_m solves

$$\begin{cases} \int_\Omega \partial_t z_m w_s d\mathbf{x} + \alpha \int_\Omega \left(z_m w_s + \nabla z_m \cdot \nabla w_s\right)d\mathbf{x} = \int_\Omega e^{-\alpha t} f_m\ w_s d\mathbf{x}, \quad \text{a.e in } [0,T] \\ z_m\left(0\right) = g_m. \end{cases}$$

The o.d.e. system for the coefficients of $z_m\left(t\right)$, $d_{mh}\left(t\right) = e^{-\alpha t}c_{mh}\left(t\right)$, takes the (coupled) form

$$\begin{cases} \dot{d}_{mh}\left(t\right) + \alpha\mu_h d_{mh}\left(t\right) = \hat{f}_h\left(t\right) \quad \text{q.o. in } \left(0,T\right) \\ d_{mh}\left(0\right) = \hat{g}\cdot_h. \end{cases} \tag{9.32}$$

With small adjustments, the technique used in the proof of Theorem 9.1 yields for z_m the stability estimate

$$\int_0^t\int_\Omega z_m^2 d\mathbf{x}ds + \alpha\int_\Omega \left\{z_m^2\left(\mathbf{x},t\right) + \left|\nabla z_m\left(\mathbf{x},t\right)\right|^2\right\}d\mathbf{x}$$
$$\leq \|g\|_{1,2}^2 + \int_0^t\int_\Omega e^{-2\alpha s} f_m^2\ d\mathbf{x}ds.$$

for all $t \in [0,T]$. The estimates for $u_m = e^{\alpha t}z_m$ reads

$$\int_0^t\int_\Omega u_m^2 d\mathbf{x}ds + \alpha\int_\Omega \left\{u_m^2\left(\mathbf{x},t\right) + \left|\nabla u_m\left(\mathbf{x},t\right)\right|^2\right\}d\mathbf{x}$$
$$\leq e^{2\alpha t}\|g\|_{1,2}^2 + \int_0^t\int_\Omega e^{2\alpha(t-s)} f_m^2\ d\mathbf{x}ds.$$

Now the proof proceeds as in Theorem 9.1; we skip the details.   □

Mixed boundary conditions. Let Γ_D be a smooth subset of $\partial\Omega$ and $\Gamma_N = \partial\Omega\backslash\Gamma_D$. Set $D_T = \Gamma_D \times \left(0,T\right)$ and $N_T = \Gamma_N \times \left(0,T\right)$. Consider the problem

$$\begin{cases} u_t - \alpha\Delta u = f & \text{in } Q_T \\ u\left(\mathbf{x},0\right) = g\left(\mathbf{x}\right) & \text{in } \Omega \\ u = 0 & \text{on } D_T \\ \partial_\nu u = 0 & \text{on } N_T. \end{cases} \tag{9.33}$$

A correct functional setting for u is $H_{0,D_T}^1\left(Q_T\right)$ where

$$H_{0,D_T}^1\left(Q_T\right) = \left\{u : u, u_t, \left|\nabla u\right| \in L^2\left(Q_T\right), u = 0 \text{ on } D_T\right\}.$$

Note that for a.e. $t \in \left(0,T\right)$, the function $\mathbf{x} \to u\left(\mathbf{x},t\right)$ belongs to $V = H_{0,\Gamma_D}^1\left(\Omega\right)$. Recall that in V the Poincaré inequality holds

$$\|v\|_0 \leq C_P \|v\|_1$$

so that we choose the inner product $\left(u,v\right)_1 = \left(\nabla u, \nabla v\right)_0$ with norm $\|\cdot\|_1$. This is the space of test functions.

Note that the bilinear form

$$B\left(w,v\right) = \alpha\left(\nabla w, \nabla v\right)_0$$

is continuous and V−coercive.

As a consequence, the Poincaré inequality

$$\int_0^T \int_\Omega u^2 \, d\mathbf{x}dt \leq C_P^2 \int_0^T \int_\Omega |\nabla u|^2 \, d\mathbf{x}dt$$

holds in $H^1_{0,D_T}\left(Q_T\right)$ so that $H^1_{0,D_T}\left(Q_T\right)$ is a Hilbert space with inner product and norm given by

$$(u,v)_{H^1_{0,D_T}(Q_T)} = \int_0^T \int_\Omega \nabla u \cdot \nabla v \, d\mathbf{x}dt,$$

$$\|u\|_{H^1_{0,D_T}(Q_T)} = \left(\int_0^T \int_\Omega |\nabla u|^2 \, d\mathbf{x}dt\right)^{1/2}.$$

Here is a weak formulation, that takes into account the homogeneous boundary conditions

Definition 9.3. *A weak solution to problem (9.33) is a function* $u \in H^1_{0,D_T}$ *(Q_T) such that:*

i) *For every* $v \in V$,

$$\int_\Omega u_t\left(\mathbf{x},t\right)v\left(\mathbf{x}\right)\, d\mathbf{x} + \alpha \int_\Omega \nabla u\left(\mathbf{x},t\right)\cdot\nabla v\left(\mathbf{x}\right)\, d\mathbf{x} = \int_\Omega f\left(\mathbf{x},t\right)v\left(\mathbf{x}\right)\, d\mathbf{x}$$

for a.e. $t \in (0,T)$.

ii) *$u\left(\mathbf{x},t\right) = g\left(\mathbf{x}\right)$ in Ω, in the sense of (9.4).*

If we assume that $f \in L^2\left(Q_T\right)$ and $g \in V$, there exists exactly one weak solution. Indeed, a theorem perfectly analogous to Theorem 9.1 holds with the same proof, except for small adjustments. We leave the details to the reader.

9.3 General Equations

9.3.1 Weak formulation of initial-boundary value problems

We now consider divergence form operators with drift and reaction

$$\mathcal{L}u = -\mathrm{div}\mathbf{A}\nabla u + \mathbf{b}\cdot\nabla u + cu.$$

The matrix $\mathbf{A} = (a_{i,j}(\mathbf{x}))$, in general different from a multiple of the identity matrix, encodes the anisotropy of the medium with respect to diffusion. For instance, a matrix of the type

$$\begin{pmatrix} \alpha & 0 & 0 \\ 0 & \varepsilon & 0 \\ 0 & 0 & \varepsilon \end{pmatrix}$$

with $\alpha \gg \varepsilon > 0$, denotes higher propensity of the medium towards diffusion along the x_1-axis, than along the other directions. As in the stationary case, for the control of the stability of numerical algorithms, it is important to compare the effects of the drift, reaction and diffusion terms. We make the following hypotheses:

(a) the coefficients a_{ij}, b_{ij}, c are bounded with

$$|a_{ij}| \le K, \ |\mathbf{b}| \le \gamma, \ |c| \le \gamma_0, \qquad \text{a.e. in } \Omega$$

and the matrix $\mathbf{A}$ is symmetric: $a_{ij} = a_{ij}$, $i,j = 1,\dots,n$;

(b) $\mathcal{L}$ is *uniformly elliptic*:

$$\alpha |\boldsymbol{\xi}|^2 \le \mathbf{A}(\mathbf{x})\boldsymbol{\xi}\cdot\boldsymbol{\xi} \le K|\boldsymbol{\xi}|^2 \qquad \text{for all } \boldsymbol{\xi} \in \mathbb{R}^n, \ \boldsymbol{\xi} \ne \mathbf{0}, \text{ a.e. in } \Omega.$$

We consider initial value problems of the form:

$$\begin{cases} u_t + \mathcal{L}u = f & \text{in } Q_T \\ \mathcal{B}u = 0 & \text{on } S_T \\ u(\mathbf{x},0) = g(\mathbf{x}) & \text{in } \Omega \end{cases} \tag{9.34}$$

where $\mathcal{B}u$ stands for one of the usual *homogeneous* boundary conditions. For instance, $\mathcal{B}u = \partial_\nu u + hu$ for the Robin condition.

The weak formulation of problem (9.34) follows the pattern of the previous sections. Let us briefly review the main ingredients.

Functional setting

Test functions, data and solution. We choose the test functions in a Hilbert space V, $H_0^1(\Omega) \subseteq V \subseteq H^1(\Omega)$, determined by the type of boundary condition we are dealing with.

We assume $f \in L^2(Q_T)$ and $g \in V$.

The familiar choices are $V = H_0^1(\Omega)$ for the homogeneous Dirichlet condition, $V = H^1(\Omega)$ for the Neumann or Robin condition, $V = H_{0,\Gamma_D}^1(\Omega)$ in the case of mixed Neumann-Dirichlet or Robin-Dirichlet conditions[8].

---

[8] There is no problem in dealing with non homogeneous Robin or Neumann conditions $\mathcal{B}u = \partial_\nu u + hu = q$, with $q \in L^2(\partial\Omega)$. To treat non homogeneous Dirichlet conditions we should develop the theory with a more general right hand side f (see for instance *Evans* [2]).

Correspondingly, guided by the examples in the previous sections, we look for $u \in C\left([0,T];V\right)$ such that $u_t \in L^2\left(Q_T\right)$. Observe that

$$\|u\|_{C([0,T];V)} = \max_{t\in[0,T]} \|u\left(\cdot,t\right)\|_V.$$

The bilinear form on V

We define

$$B\left(\varphi,\psi\right) = \int_\Omega \left\{\mathbf{A}\nabla\varphi\cdot\nabla\psi + \left(\mathbf{b}\cdot\nabla\varphi\right)\psi + c\varphi\psi\right\}d\mathbf{x}$$

and, in the case of Robin condition,

$$B\left(\varphi,\psi\right) = \int_\Omega \left\{\mathbf{A}\nabla\varphi\cdot\nabla\psi + \left(\mathbf{b}\cdot\nabla\varphi\right)\psi + c\varphi\psi\right\}d\mathbf{x} + \int_{\partial\Omega} h\varphi\psi\,d\sigma$$

where we require $h \in L^\infty\left(S_T\right)$, $h \geq 0$ a.e. on S_T.

Recall the trace inequality:

$$\|\varphi\|_{L^2(\partial\Omega)} \leq C_{tr}\|\varphi\|_V.$$

Under the stated hypotheses, it is not difficult to show that

$$|B\left(\varphi,\psi\right)| \leq M\|\varphi\|_V\|\psi\|_V \tag{9.35}$$

so that B is *continuous* in V. The constant M depends on K, γ, γ_0 that is, on the size of the coefficients a_{ij}, b_j, c (and on $\|h\|_{L^\infty(\partial\Omega)}$ in the case of Robin condition).

Also, B is *weakly coercive*. In fact, we have, for every[9] $\varepsilon > 0$

$$\int_\Omega \left(\mathbf{b}\cdot\nabla\varphi\right)\varphi\,d\mathbf{x} \geq -\gamma\|\nabla\varphi\|_0\|\varphi\|_0$$

$$\geq -\frac{\gamma}{2}\left[\varepsilon\|\nabla\varphi\|_0^2 + \frac{1}{\varepsilon}\|\varphi\|_0^2\right]$$

and

$$\int_\Omega c\varphi^2 d\mathbf{x} \geq -\gamma_0\|\varphi\|_0^2$$

whence, as $h \geq 0$ a.e. on $\partial\Omega$,

$$B\left(\varphi,\varphi\right) \geq \left[\alpha - \frac{\gamma_0\varepsilon}{2}\right]\|\nabla\varphi\|_0^2 - \left[\frac{\gamma}{2\varepsilon} + \gamma_0\right]\|\varphi\|_0^2. \tag{9.36}$$

We distinguish three cases.

---

[9] Using once more the elementary inequality

$$2ab \leq \varepsilon a^2 + \frac{1}{\varepsilon}b^2$$

which holds $\forall a,b \in \mathbb{R}$, $\forall \varepsilon > 0$.

If $\gamma = 0$ and $\gamma_0 = 0$ the bilinear form is $V-$coercive when $V = H_0^1(\Omega)$. If $H_0^1(\Omega) \subseteq V \subseteq H^1(\Omega)$,

$$\tilde{B}(\varphi, \psi) = B(\varphi, \psi) + \lambda_0 (\varphi, \psi)_0 \qquad (9.37)$$

is $V-$coercive for any $\lambda_0 > 0$.

If $\gamma = 0$ and $\gamma_0 > 0$, (9.37) is $V-$coercive for any $\lambda_0 > \gamma_0$.

If $\gamma > 0$, choose in (9.36)

$$\varepsilon = \frac{\alpha}{\gamma} \quad \text{and} \quad \lambda_0 = 2\left[\frac{\gamma}{2\varepsilon} + \gamma_0\right] = 2\left[\frac{\gamma^2}{2\alpha} + \gamma_0\right].$$

Then

$$\tilde{B}(\varphi, \psi) \geq \frac{\alpha}{2} \|\nabla\varphi\|_0^2 + \frac{\lambda_0}{2} \|\varphi\|_0^2 \geq \alpha_0 \|u\|_{1,2}^2 \qquad (9.38)$$

where $\alpha_0 = \min\{\alpha/2, \lambda_0/2\}$, so that B is *weakly coercive.*

The weak formulation

To obtain a weak formulation, we multiply the differential equation $u_t + \mathcal{L}u = 0$ by a test function and we integrate it over Ω. After an integration by parts of the divergence term, taking into account the homogeneous boundary conditions, we are lead to the following definition of weak solution of the initial-boundary value problem (9.34).

Definition 9.4. *A weak solution of problem* (9.34) *is a function* $u \in C([0,T]; V)$ *such that* $u_t \in L^2(Q_T)$ *and:*

i)   *For every* $v \in V$,

$$\int_\Omega u_t(\mathbf{x}, t) v(\mathbf{x})\ d\mathbf{x} + B(u, v) = \int_\Omega f(\mathbf{x}, t) v(\mathbf{x})\ d\mathbf{x}$$

   for a.e. $t \in (0, T)$.

ii) $u(\mathbf{x}, t) = g(\mathbf{x})$ *in* Ω, *in the sense of* (9.4).

The following theorem holds.

Theorem 9.3. *There exists a unique weak solution* u *of* (9.34). *Moreover the following stability estimate holds*

$$\int_0^T \int_\Omega u_t^2\ d\mathbf{x}ds + \max_{t\in[0,T]} \|u(\cdot, t)\|_V^2 \leq C\left\{\|g\|_V^2 + \int_0^T \int_\Omega f^2 d\mathbf{x}dt\right\} \qquad (9.39)$$

where C *depends only on* $\Omega, n, \alpha, K, \gamma_0, \gamma, T$.

The method of Faedo-Galerkin may be used also in this case, as in the previous sections. However, its implementation requires some deep results of

Functional Analysis, out of the aims of this brief introduction. Therefore, we show the construction of the Galerkin approximations and the proof of the energy estimates, only sketching the existence part of the proof.

We choose an orthonormal basis $\{w_k\}$ in H, orthogonal in V, and let

$$V_m = \text{span}\{w_1, w_2, \ldots, w_m\}.$$

• *Galerkin approximations.* Keeping the same notations of Remark 9.3, we set

$$u_m(\mathbf{x}, t) = \sum_{k=1}^{m} c_{mk}(t) w_k(\mathbf{x})$$

and look at the approximating problem

$$\int_\Omega \partial_t u_m(\mathbf{x}, t) w_h(\mathbf{x}) \, d\mathbf{x} + B(u_m, w_h) = \int_\Omega f_m(\mathbf{x}, t) w_h(\mathbf{x}) \, d\mathbf{x} \qquad (9.40)$$

for every $h = 1, \ldots, m$, with initial condition

$$u_m(\mathbf{x}, 0) = g_m(\mathbf{x}) \text{ in } \Omega. \qquad (9.41)$$

Problem (9.40), (9.41) leads to the following linear system of ordinary differential equations for the coefficients c_{mk}

$$\begin{cases} \dot{\mathbf{C}}_m(t) = -\mathbf{W}\mathbf{C}_m(t) + \hat{\mathbf{F}}_m(t), & \text{a.e. } t \in [0, T], \\ \mathbf{C}_m(0) = \hat{\mathbf{g}}_m \end{cases} \qquad (9.42)$$

where $\mathbf{C}_m(t) = (c_{m1}(t), \ldots, c_{mm}(t))^\top$, $\hat{\mathbf{F}}_m(t) = ((\hat{f}_1(t), \ldots, \hat{f}_m(t))^\top$ and the entries of the matrix $\mathbf{W}$ are

$$W_{hk} = B(w_k, w_h).$$

Since $\hat{\mathbf{F}}_m \in L^2(0, T; \mathbb{R}^m)$, for every $m \geq 1$ there exists a unique solution $\mathbf{C}_m \in H^1(0, T; \mathbb{R}^m)$ of problem (9.42).

Consequently, there exists a unique solution $u_m \in H^1(Q_T)$ of problem (9.40), (9.41). Moreover, $u_m \in C([0, T]; V)$ and $\partial_t u_m(\cdot, t) \in V$ for a.e. $t \in [0, T]$.

• *Energy estimates.* We have the following lemma.

Lemma 9.3. *Let u_m be the solution of problem (9.40), (9.41). Then, for every $t \in [0, T]$:*

$$\int_0^t \int_\Omega (\partial_t u_m)^2 d\mathbf{x} ds + \|u_m(\cdot, t)\|_V^2 \leq C \left\{ \|g_m\|_V^2 + \int_0^t \int_\Omega f_m^2 d\mathbf{x} ds \right\} \qquad (9.43)$$

where C depends only on $\Omega, n, \alpha, K, \gamma_0, \gamma, T$.

Proof. First, since B is in general only weakly coercive, we make the change of variable

$$z_m(t) = e^{-\lambda_0 t} u_m(t)$$

and set $d_{mk}(t) = e^{-\lambda_0 t} c_{mk}(t)$. Then (9.40) becomes

$$\int_\Omega \partial_t z_m(\mathbf{x}, t) w_h(\mathbf{x}) \, dx + \tilde{B}(z_m, w_h) = \int_\Omega f_m(\mathbf{x}, t) w_h(\mathbf{x}) \, dx \qquad (9.44)$$

with the coercive bilinear form $\tilde{B}$ in (9.37). Since the system of o.d.e. is *coupled*, we cannot deduce the *energy estimates* for z_m by analogous estimates on the single coefficients $d_{mk}(t)$ as we did in the previous section. We have to use directly z_m. Hence, multiply equation (9.44) by $\dot{d}_{mh}$ and sum for $h = 1, \dots, m$. We find:

$$\int_\Omega (\partial_t z_m(\mathbf{x}, t))^2 \, dx + \tilde{B}(z_m, \partial_t z_m) = \int_\Omega f_m(\mathbf{x}, t) \partial_t z_m(\mathbf{x}) \, dx. \qquad (9.45)$$

We have, referring for instance to the Robin problem:

$$\tilde{B}(z_m, \partial_t z_m) = \int_\Omega \{\mathbf{A}\nabla z_m \cdot \nabla \partial_t z_m + (\mathbf{b} \cdot \nabla z_m)\partial_t z_m + (c + \lambda_0) z_m \partial_t z_m\} \, dx$$

$$+ \int_{\partial\Omega} h z_m \partial_t z_m \, d\sigma.$$

Since the matrix $\mathbf{A}$ is symmetric and all the coefficients, a_{ij}, b_j, c, h are independent of time, we can write:

$$\tilde{B}(z_m, \partial_t z_m) = \frac{1}{2}\frac{d}{dt} \int_\Omega \{\mathbf{A}\nabla z_m \cdot \nabla z_m + (c + \lambda_0) z_m^2\} \, dx + \frac{1}{2}\frac{d}{dt} \int_{\partial\Omega} h z_m^2 \, d\sigma$$

$$+ \int_\Omega (\mathbf{b} \cdot \nabla z_m)\partial_t z_m.$$

Now

$$\left| \int_\Omega (\mathbf{b} \cdot \nabla z_m)\partial_t z_m \right| \leq \gamma \|z_m(\cdot, t)\|_V \|\partial_t z_m(\cdot, t)\|_0 \qquad (9.46)$$

$$\leq 4\gamma^2 \|z_m(\cdot, t)\|_V^2 + \frac{1}{4}\|\partial_t z_m(\cdot, t)\|_0^2.$$

As before:

$$\left| \int_\Omega f_m \partial_t z_m \, dx \right| \leq \frac{1}{2}\int_\Omega (f_m)^2 \, dx + \frac{1}{2}\int_\Omega (\partial_t z_m)^2 \, dx. \qquad (9.47)$$

Integrating (9.45) over $(0, t)$, taking into account (9.38), (9.35), (9.46) and (9.47), we obtain:

$$\int_0^t \int_\Omega (\partial_t z_m)^2 \, dx ds + 2\alpha_0 \|z_m(\cdot, t)\|_V^2 \qquad (9.48)$$

$$\leq 2M \|g_m\|_V^2 + 2\int_0^t \int_\Omega (f_m)^2 \, dx ds + 4\gamma^2 \int_0^t \|z_m(\cdot, s)\|_V^2 \, ds.$$

380   9 Weak formulation of evolution problems

We have now to bound the last integral in terms of f_m and g_m. To do it, this time multiply equation (9.44) by d_{mh} and sum for $h = 1, \ldots, m$.

We find, after some adjustments:

$$\frac{d}{dt} \int_\Omega z_m^2 (\mathbf{x}, t) \, d\mathbf{x} + \tilde{B} (z_m, z_m) = \int_\Omega f_m (\mathbf{x}, t) z_m (\mathbf{x}) \, d\mathbf{x}. \qquad (9.49)$$

Now, $\tilde{B} (z_m, z_m) \geq \alpha_0 \| z_m (\cdot, t) \|_V^2$ for all $t \in [0, T]$ and, by the usual elementary inequality:

$$\left| \int_\Omega f_m z_m d\mathbf{x} \right| \leq \frac{2}{\alpha_0} \int_\Omega (f_m)^2 \, d\mathbf{x} + \frac{\alpha_0}{2} \int_\Omega (z_m)^2 \, d\mathbf{x}$$

$$\leq \frac{2}{\alpha_0} \int_\Omega (f_m)^2 \, d\mathbf{x} + \frac{\alpha_0}{2} \| z_m (\cdot, t) \|_V^2.$$

Then ,integrating (9.49) over $(0, t)$ we get

$$\| z_m (\cdot, t) \|_0^2 + \frac{\alpha_0}{2} \int_0^t \| z_m (\cdot, s) \|_V^2 \, ds \leq \| g_m \|_0^2 + \frac{2}{\alpha_0} \int_0^t \int_\Omega (f_m)^2 \, d\mathbf{x} ds$$

and in particular:

$$\int_0^t \| z_m (\cdot, s) \|_V^2 \, ds \leq \frac{2}{\alpha_0} \| g_m \|_0^2 + \int_0^t \int_\Omega (f_m)^2 \, d\mathbf{x} ds.$$

Substituting into (9.48) we find

$$\int_0^t \int_\Omega (\partial_t z_m)^2 \, d\mathbf{x} ds + 2\alpha_0 \| z_m (\cdot, t) \|_V^2$$

$$\leq \left(2M + \frac{4\gamma^2}{\alpha_0} \right) \| g_m \|_V^2 + (2 + 4\gamma^2) \int_0^t \int_\Omega (f_m)^2 \, d\mathbf{x} ds.$$

Going back to z_m we finally arrive to (9.43).

• *Existence and uniqueness.* From (9.43) we deduce that $\{u_m\}$ is bounded in both $L^2 (0, T; V)$ and $C ([0, T], V)$ while $\{\partial_t u_m\}$ is bounded in $L^2 (Q_T)$. As in subsection 9.2.4, we can extract from $\{u_m\}$ a subsequence converging in a suitable sense to a weak solution u, for $m \to \infty$. It turns out that u is the unique weak solution and satisfies the estimate (9.39). □

Example 9.1. Fig. 9.2 shows the graph of the solution of the Cauchy-Dirichlet problem

$$\begin{cases} u_t - u_{xx} + 2u_x = 0.2tx & 0 < x < 5, t > 0 \\ u (x,0) = \max (2 - 2x, 0) & 0 < x < 5 \\ u (0, t) = 2 - t/6, u (5, t) = 0 & t > 0. \end{cases} \qquad (9.50)$$

Note the tendency of the drift term $2u_x$, to "transport to the right" initial data and the effect of the source term $0.2tx$ to increase the solution near $x = 5$, more and more with time.

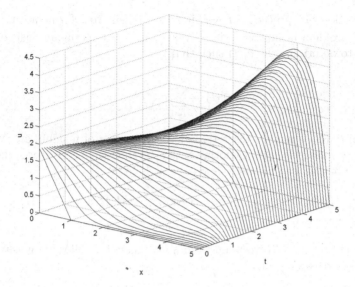

Fig. 9.2. The solution of problem (9.50) in Example 9.1

9.4 Numerical methods

9.4.1 A Faedo-Galerkin/finite element method for the heat equation

Let us consider the variational formulation of the Cauchy-Dirichlet problem for the heat equation in multiple space dimensions. Precisely, we aim to find $\mathbf{u} \in L^2(0, T; H_0^1(\Omega)) \cap C^0([0, T]; L^2(\Omega))$ such that,

$$\begin{cases} (\mathbf{u}'(t), v) + B\left[\mathbf{u}(t), v\right] = (\mathbf{f}(t), v), & \forall v \in H_0^1(\Omega) \\ \mathbf{u}(0) = u_0 \end{cases} \tag{9.51}$$

where $\mathbf{f}(t) \in L^2(\Omega \times (0, T])$, $u_0 \in L^2(\Omega)$, $B\left[u, v\right]$ is a continuous and coercive bilinear form and lower case bold symbols denote functions from $(0, T)$ to a function space V.

Given a family of discrete spaces $V_k \subset H_0^1(\Omega)$, the Faedo-Galerkin method consists on selecting a suitable $\mathbf{u}_k(t) \in V_k$ to approximate $\mathbf{u}(t)$, solution of (9.51). A possible choice for the spatial approximation is to use finite elements, namely $V_k = X_{k,0}^r$. Then, for any $t \in (0, T]$ we aim to find $\mathbf{u}_k(t) \in X_{k,0}^r$ such that

$$\begin{cases} (\mathbf{u}_k'(t), v_k) + B\left[\mathbf{u}_k(t), v_k\right] = (\mathbf{f}(t), v_k), & \forall v_k \in X_{k,0}^r \\ \mathbf{u}_k(t = 0) = u_{k,0} \end{cases} \tag{9.52}$$

where $u_{k,0}$ is the projection of $u_0 \in L^2(\Omega)$ on the finite element space. Problem (9.52) is said to be **semi-discrete**, because we have just performed the space discretization. To proceed further, it is convenient to reformulate the semi-discrete problem in matrix form. Given $\{\psi_i\}_{i=1}^{N_k^r}$ a Lagrangian basis of $X_{k,0}^r$, we aim to determine the $c_j(t)$ such that

$$\mathbf{u}_k(t) = \sum_{j=1}^{N_k^r} c_j(t)\psi_j.$$

Problem (9.52) is not yet discrete, indeed its degrees of freedom are N_k^r time dependent functions $c_j(t)$. Denoting with $c_j'(t)$ their time derivatives, the semi-discrete problem can be easily rewritten as follows,

$$\begin{cases} \sum_{j=1}^{k} c_j'(t)(\psi_j, \psi_i) + \sum_{j=1}^{N_k^r} c_j(t)B\left[\psi_j, \psi_i\right] = (\mathbf{f}(t), \psi_i), \quad i = 1, \ldots, N_k^r \\ c_j(0) = c_{j,0} \end{cases}$$

where the values $c_{j,0}$ correspond to $u_{k,0}$ and given the following notations for matrices and vectors

$$\mathbf{M}_{kr,ij} = (\psi_j, \psi_i), \quad \mathbf{A}_{kr,ij} = B\left[\psi_j, \psi_i\right], \quad \mathbf{F}_{kr,i}(t) = (\mathbf{f}(t), \psi_i)$$

we see that (9.52) is equivalent to solve a linear system of ordinary differential equations in the unknowns $\mathbf{c}(t) = \{c_j(t)\}_{j=1}^{N_k^r}$,

$$\mathbf{M}_{kr}\mathbf{c}'(t) + \mathbf{A}_{kr}\mathbf{c}(t) = \mathbf{F}_{kr}(t), \quad \mathbf{c}(t=0) = \mathbf{c}_0.$$

In case of a constant coefficient heat equation, matrices $\mathbf{M}_{kr} \in \mathbb{R}^{N_k^r \times N_k^r}$ and $\mathbf{A}_{kr} \in \mathbb{R}^{N_k^r \times N_k^r}$, respectively called **mass** and **stiffness** matrices, are constant.

For the time discretization, we restrict to one step schemes for ordinary differential equations and in particular on the $\theta-$**method**. Given a sequence of uniform time steps t^n, namely $t^n = n \cdot \tau$ where $\tau > 0$, and given an initial state $\mathbf{c}_0$, we aim to discretize the continuous evolution of $\mathbf{c}(t)$ by means of a sequence of vectors $\mathbf{c}_n$ such that

$$\frac{1}{\tau}\mathbf{M}_{kr}\left(\mathbf{c}_n - \mathbf{c}_{n-1}\right) + \theta\mathbf{A}_{kr}\mathbf{c}_n + (1 - \theta)\mathbf{A}_{kr}\mathbf{c}_{n-1} = \theta\mathbf{F}_{kr}(t^n) + (1 - \theta)\mathbf{F}_{kr}(t^{n-1}),$$
$$(9.53)$$

where θ is a free parameter to be suitably chosen in $[0, 1]$. Using the change of variable $\mathbf{c}_n^\theta = \theta\mathbf{c}_n + (1 - \theta)\mathbf{c}_{n-1}$, for any $\theta \neq 0$ equation (9.53) can be reformulated as follows,

$$\mathbf{C}_{kr}^\tau\mathbf{c}_n^\theta = \frac{1}{\theta\tau}\mathbf{M}_{kr}\mathbf{c}_{n-1} + \theta\mathbf{F}_{kr}(t^n) + (1 - \theta)\mathbf{F}_{kr}(t^{n-1}),$$

$$\text{where} \quad \mathbf{C}_{kr}^\tau = \frac{1}{\theta\tau}\mathbf{M}_{kr} + \mathbf{A}_{kr}.$$

The properties of the scheme, stability in particular, are affected by the choice of θ. Setting $\theta = 0$ leads to the forward Euler scheme, while $\theta = 1$

corresponds to backward Euler. Both schemes have already been studied in Chapter 3. Although the scheme obtained using $\theta = 0$ is usually said to be explicit, we observe that in the case of finite elements it is not possible to explicitly determine the coefficients $\mathbf{c}_n$ once $\mathbf{c}_{n-1}$ is given, because matrix $\mathbf{M}_{kr}$ is not diagonal. On the one hand, it is anyway computationally convenient to perform one step of forward Euler method rather than backward Euler, because the mass matrix, $\mathbf{M}_{kr}$, is better conditioned than the stiffness one, $\mathbf{A}_{kr}$. On the other hand, the stability constraints may reverse this balance. Taking for simplicity $\mathbf{F}_{kr}(t) = 0$, we recall from Chapter 3 that (9.53) is stable if for any $\mathbf{c}_0$ it satisfies

$$\lim_{n \to \infty} \|\mathbf{c}_n\|_\infty = 0. \tag{9.54}$$

The previous requirement can be enforced by looking at the spectrum of $\mathbf{C}_{kr}^\tau$. In the framework of the finite element method it can be proved that the following stability conditions hold true[10].

Theorem 9.4. *The scheme* (9.53) *with* $\theta \in [\frac{1}{2}, 1)$ *is unconditionally stable, namely condition* (9.54) *is satisfied for any possible value of* h, τ. *When* $\theta \in [0, \frac{1}{2})$, θ−*method is stable under the condition* $\tau \leq \frac{C_r h^2}{(1-2\theta)}$, *where* C_r *is a positive constant possibly dependent on* r, *but not on* k *or* h.

Theorem 9.4 highlights that the choice $\theta = \frac{1}{2}$ might be particularly interesting. Indeed, in this case (9.53) is simultaneously unconditionally stable and second order accurate. The resulting scheme is often called **Crank-Nicholson method**.

9.5 Exercises

9.1. The potassium concentration $c(x, y, z, t)$ in a cell of spherical shape Ω and radius R, with boundary Γ, satisfies the evolution problem

$$\begin{cases} c_t - \operatorname{div}(\mu \nabla c) - \sigma c = 0 & \text{in } \Omega \times (0, T) \\ \mu \nabla c \cdot \mathbf{n} + \chi c = \chi c_{ext} & \text{on } \Gamma \times (0, T) \\ c(x, y, z, 0) = c_0(x, y, z) & \text{on } \Omega \end{cases} \tag{9.55}$$

where c_{ext} is the given external concentration which is constant, σ and χ positive scalars and μ is a strictly positive function. Write the weak formulation and analyze the well-posedness, providing suitable assumptions on the coefficients and on the data.

---

[10] We refer to [42], Chapter 6 for a detailed proof.

9.2. Consider the following parabolic problem

$$
\begin{cases}
\dfrac{\partial u}{\partial t} - \Delta u + xy\dfrac{\partial u}{\partial x} + x^2 y^2 \dfrac{\partial u}{\partial y} = f & \text{in } \Omega \times (0,T) \\[2mm]
u = 0 & \text{on } \Gamma_D \times (0,T) \\[2mm]
\dfrac{\partial u}{\partial n} = 0 & \text{on } \Gamma_N \times (0,T) \\[2mm]
u(\mathbf{x},0) = u_0(\mathbf{x}) & \text{on } \Omega
\end{cases}
$$

where $\Omega = B_1$ in $\mathbb{R}^2$, whose boundary is $\partial\Omega = \Gamma_D \cup \Gamma_N$ with $\Gamma_D \cap \Gamma_N = \emptyset$. Write the variational formulation. Deduce the existence and the uniqueness of the solution, under sufficient conditions on the data.

9.3. We introduce the domain

$$
\Omega = \left\{ \mathbf{x} \in \mathbb{R}^2 : x_1^2 + 4x_2^2 < 4 \right\},
$$
$$
\Gamma_D = \partial\Omega \cap \{x_1 \geq 0\}, \qquad \Gamma_N = \partial\Omega \cap \{x_1 < 0\}.
$$

Consider the problem

$$
\begin{cases}
u_t - \operatorname{div}\left[A_\alpha \nabla u\right] + \mathbf{b} \cdot \nabla u - \alpha u = x_2 & \text{in } \Omega \times (0,T) \\
u(\mathbf{x},t) = 0 & \text{on } \Gamma_D \times (0,T) \\
-A_\alpha \nabla u \cdot \mathbf{n} = \cos x_1 & \text{on } \Gamma_N \times (0,T) \\
u(\mathbf{x},0) = \mathcal{H}(x_1) & \text{on } \Omega
\end{cases}
\qquad (9.56)
$$

where

$$
A_\alpha = \begin{bmatrix} 1 & 0 \\ 0 & \alpha\, e^{x_1^2 + x_2^2} \end{bmatrix}, \qquad
\mathbf{b} = \begin{bmatrix} \sin(x_1 + x_2) \\ x_1^2 + x_2^2 \end{bmatrix}
$$

and $\mathcal{H}$ is the Heavyside function. For which values of the parameter $\alpha \in \mathbb{R}$ the problem is parabolic? Give a weak formulation of the problem and deduce existence and uniqueness of the solution, computing the values of the continuity and coercivity constants.

9.4. Consider the problem

$$
\begin{cases}
u_t - (a(x)\, u_x)_x + b(x)\, u_x + c(x)u = f(x,t) & 0 < x < 1, 0 < t < T \\
u(x,0) = g(x), & 0 \leq x \leq 1 \\
u(0,t) = 0,\ u(1,t) = k(t). & 0 \leq t \leq T.
\end{cases}
$$

1) Modifying u suitably, reduce the problem to homogeneous Dirichlet conditions.

2) Write a weak formulation for the new problem.

3) Prove the well-posedness of the problem, indicating the hypotheses on the coefficients a, b, c and the data f, g. Write a stability estimate for the original u.

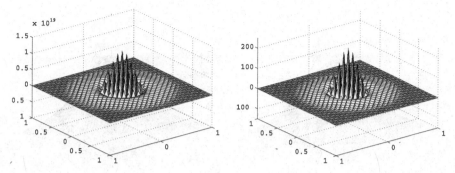

Fig. 9.3. Solution of the Cauchy-Dirichlet problem at time $t = 0.1$ approximated by the θ−method with $\theta = 0$ (left) and $\theta = \frac{1}{4}$ (right). We notice that the two plots do not share the same scale

9.5. Consider the Neumann problem (9.26) with non-homogeneous boundary condition $\partial_\nu u = h$, with $h \in L^2(\partial\Omega)$:

a)   Give a weak formulation of the problem and derive the main estimates for the Galerkin approximations.

b)   Deduce existence and uniqueness of the solution.

9.5.1 Verification of Euler methods stability properties

We address the Cauchy-Dirichlet problem (9.51) on $\Omega = (-1,1) \times (-1,1)$ defined by $B[u,v] = (\nabla u, \nabla v)$ and $u_0 = \exp(-10(x^2 + y^2))$, which is discretized by means of (9.53) using linear finite elements, namely $r = 1$. Given $h = 0.05$ and $\tau = 0.01$, we compare the numerical approximations obtained with $\theta = 0, \frac{1}{4}, \frac{1}{2}, 1$.

   Since the selected values for h and τ do not satisfy the condition $\tau \leq \frac{C_r h^2}{(1-2\theta)}$ for $\theta = 0, \frac{1}{4}$, we expect instabilities to appear in those cases. Fig. 9.3 confirms this behavior, with decreasing magnitude of the oscillations passing from $\theta = 0$ to $\theta = \frac{1}{4}$, in agreement with Theorem 9.4, which shows that the stability condition becomes less restrictive when θ increases. Finally, above the threshold $\theta = \frac{1}{2}$, unconditional stability is verified. However, a sufficiently refined grid in space and time is still mandatory to obtain an accurate solution.

9.5.2 Numerical simulation of mass transfer

Let us consider the problem defined in Exercise 9.1, describing the transfer of chemicals, such as potassium, through the cell membrane. For simplicity, we restrict to two space dimensions and we model the cell with a unit circle. According to equation $c_t - \mathrm{div}(\mu\nabla c) - \sigma c = 0$, the concentration evolves by diffusion and reaction. We notice that a positive value of the reaction coefficient corresponds to mass production. A simple model to describe the

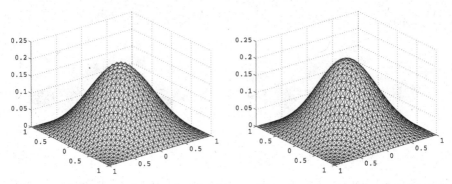

Fig. 9.4. Solution of the Cauchy-Dirichlet problem at time $t = 0.1$ approximated by the θ–method with $\theta = \frac{1}{2}$ (left) and $\theta = 1$ (right)

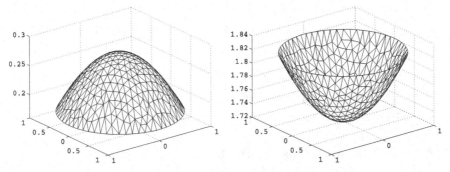

Fig. 9.5. Numerical solution of the chemical transfer problem at time $t = 1$ for $1 = c_0 > c_{ext} = 0$ (left) and for $1 = c_0 < c_{ext} = 2$ (right)

behaviour of the cell membrane is $-\mu \nabla c \cdot \mathbf{n} = \chi(c - c_{ext})$, which can be seen as a constitutive law for the mass flux through the membrane where c_{ext} is the outer concentration and χ is the membrane permeability.

The concentration inside the cell increases or decreases according to the sign and magnitude of the concentration jump $c - c_{ext}$. When it is positive, an outgoing flux is prescribed, and vice versa, when $c_{ext} > c$, chemicals penetrate into the cell. This is confirmed by the results reported in Fig. 9.5, relative to $\mu = 1$, $\sigma = 0$. Indeed, given $c_0 = 1$ and $c_{ext} = 0$, on the left panel we see that at time $t = 1$ the concentration inside the cell reaches nearly c_{ext}, approaching this value from above. On the right panel, we observe that starting from $c_0 = 1$ with $c_{ext} = 2$ the concentration in the cell rises to c_{ext} from below.

Part III
Solutions

10

Solutions of selected exercises

10.1 Section 2.8

2.1. The problem is analyzed in Section 2.3.3. Its solution is given by (2.34):

$$\rho(x,t) = \begin{cases} \rho_m & \text{for } x \leq -v_m t \\ \dfrac{\rho_m}{2}\left(1 - \dfrac{x}{v_m t}\right) & \text{for } -v_m t < x < v_m t \\ 0 & \text{for } x \geq v_m t \end{cases} .$$

Hence, the car density at the light (which is in $x = 0$) is $\rho(0,t) = \rho_m/2$ and is constant.

Furthermore, according to the model of traffic dynamics, the velocity of a car depends on the car density, according to the law

$$v(\rho) = v_m \left(1 - \frac{\rho}{\rho_m}\right).$$

While reaching the light, the car which at time t_0 is in the position $x_0 = -v_m t_0$ moves in the area $-v_m t_0 < x < v_m t_0$; denoting by $x = x(t)$ the position of the car with respect to time, using the solution ρ for the car density, we deduce:

$$v(\rho(x)) = \dot{x} = v_m \left(\frac{1}{2} - \frac{x}{2v_m t}\right).$$

The motion of the car is therefore described by the initial value problem:

$$\begin{cases} \dot{x} = \dfrac{v_m}{2} + \dfrac{x}{2t} \\ x(t_0) = -v_m t_0. \end{cases}$$

Salsa S., Vegni F.M.G., Zaretti A., Zunino P.: *A Primer on PDEs. Models, Methods, Simulations.* Unitext – La Matematica per il 3+2 65.
DOI 10.1007/978-88-470-2862-3_10, © Springer-Verlag Italia 2013

The equation is linear[1] and for $t > 0$ the general solution is $x(t) = \sqrt{t}\,(C + v_m\sqrt{t})$; from the initial condition we deduce

$$x(t) = v_m(t - 2\sqrt{t_0 t}).$$

Hence, we find $x(t) = 0$ for $t = 4t_0$. The car takes the time $4t_0$ to reach the light.

2.2. In the model introduced in Section 2.3 we assume that the velocity v of cars attains the maximum value when the concentration ρ is null and linearly decreases to zero, corresponding to the maximum density. The flux of cars (velocity times density) depends on the concentration according to the constitutive law

$$q(\rho) = \rho v(\rho) = v_m \rho \left(1 - \frac{\rho}{\rho_m}\right). \tag{10.1}$$

We find the characteristics based at $(x_0, 0)$. They are straight lines of equation (2.30); since

$$q'(\rho) = v_m \left(1 - \frac{2\rho}{\rho_m}\right)$$

we have

$$x = x_0 + v_m \left(1 - \frac{2\rho(x_0, 0)}{\rho_m}\right) t.$$

We consider the initial value problem: for $x_0 < 0$ the characteristics display according to the value of the parameter a:

$$x = x_0 + v_m (1 - 2a)\, t.$$

Besides, for $x_0 > 0$ the characteristics are vertical straight lines $x = x_0$.

Different situations may arise. If the characteristics that depart from the points of negative abscissa have positive slope, then characteristics carrying different data intersect at a finite time and we will have to determine the equation of the shock curve along which the discontinuity of the initial value problem propagates in the first quadrant of the plane (x, t). Conversely, if the characteristics that originate at the points of abscissa negative have negative slope, in the origin will be based a rarefaction area. Then, three cases have to be discussed according to the values of the parameter a in the given range. For $v_m(1 - 2a) > 0$, namely $0 \le a < 1/2$, we witness the propagation of a discontinuity in the first quadrant; for $(1 - 2a) < 0$, i.e. $1/2 < a \le 1$, there is a rarefaction wave; for $a = 1/2$, the characteristics are all parallel, and traffic moves at same speed along the road described by the model.

---

[1] The general solution for the linear first order equation $\dot{y} = \alpha(t)y(t) + \beta(t)$ is

$$y(t) = e^{\int \alpha\, dt} \left(C + \int \beta e^{-\int \alpha\, dt}\, dt\right).$$

In the case $0 \le a < 1/2$ we determine the equation of the line $s = s(t)$ along which the discontinuity propagates using the Rankine-Hugoniot condition (2.38). Denote by q^+ and q^-, respectively, the flow to the right and to the left of the shock curve, separating areas of the plane (x,t) reached by the characteristics that start at $t = 0$ from positive x-axis points, and carrying the data $\rho^+ = \rho_m/2$ to those that start at points of negative abscissa and transport the data $\rho^- = a\rho_m$. We have

$$q^+ = \frac{v_m \rho_m}{4} \quad \text{and} \quad q^- = a v_m \rho_m (1-a);$$

the condition (2.38) becomes:

$$\dot{s} = \frac{q^+ - q^-}{\rho^+ - \rho^-} = \frac{v_m \rho_m/4 - v_m a \rho_m (1-a)}{\rho_m/2 - a\rho_m} = v_m \left(\frac{1}{2} - a \right)$$

which can be easily integrated. Hence, for $0 \le a < 1/2$, a line of discontinuity of the solution is propagated from from the origin with equation $s(t) = v_m (1/2 - a) t$.

We consider the case $1/2 < a \le 1$; the characteristics of the rarefaction area

$$v_m(1 - 2a)t < x < 0$$

are fanlike distributed from the origin

$$x = ht \quad \text{with } v_m(1 - 2a) < h < 0$$

and along each of them the density is constant. At time $\bar{t}$ the car density occupying the position between $v_m(1 - 2a)\bar{t}$ and 0 linearly decreases from $a\rho_m$ to $\rho_m/2$.

The car density in the rarefaction area can be formally obtained from the characteristics equation $x = v_m (1 - 2\rho/\rho_m)$, with $x_0 = 0$, expliciting ρ:

$$\rho(x,t) = \frac{(v_m t - x)\rho_m}{2 v_m t}.$$

2.3. The Burgers equation, introduced in Section 2.6, is a special case of scalar conservation law $u_t + q(u)_x = 0$, where the flux q of the concentration u is represented by the function $q(u) = u^2/2$. In general, the characteristics based at the point $(x_0, 0)$ have equation $x = x_0 + q'(g(x_0))t$; for the Burgers equation we find

$$x = x_0 + g(x_0)t.$$

We analyze the three cases assigned.

a) Considering the data $g(x)$, the characteristics are vertical lines $x = x_0$ if $x_0 < 0$ or if $x_0 > 1$, and are the straight lines of equation $x = x_0 + t$ if $0 < x < 1$.

The area $S = \{0 < x < t\}$ is not reached by characteristics. In this sector we define u as a rarefaction wave that connects with continuity the values from 0 to 1. Proceeding formally as in Section 2.3 for the traffic dynamics, we deduce the equation of the rarefaction wave based on the point (x_0, t_0) using the formula

$$u(x, t) = r \left(\frac{x - x_0}{t - t_0} \right),$$

where r is the inverse function of q'. For the Burgers equation, since $q'(u) = u$, the rarefaction wave with vertex in the origin is $u = x/t$.

Let us determine the shock curve $s = s(t)$ which is generated from the encounter of the vertical characteristics which carry the initial condition $u^- = 0$, with the characteristics $x = x_0 + t$ (for $0 < x_0 < 1$) carrying the data $u^+ = 1$; respectively, on one side and the other part of the shock curve, we have $q^- = q(u^-) = 0$ and $q^+ = q(u^+) = 1/2$. The shock curve satisfies the Rankine-Hugoniot condition (2.38), therefore

$$\dot{s} = \frac{q^+ - q^-}{u^+ - u^-} = \frac{1}{2}$$

and the initial condition for the shock curve is $s(0) = 1$. Denoting x the space variable, the shock curve has equation $x(t) = t/2 + 1$ (Fig. 10.1).

What we have found so far is valid until time $t = 2$ (when the characteristic $x = t$, the shock curve and the characteristic $x = 2$ which is carrying the initial condition $u = 0$ cross). From this moment on, the shock curve is due to the discontinuity between the characteristics carrying the data $u^- = 0$ and the characteristics carrying the rarefaction solution u, namely $u^+(s, t) = s/t$ (and, consequently, $q^+ = q(u^+) = s^2/2t^2$). Using again the condition (2.38), the differential equation for the shock curve becomes

$$\dot{s} = \frac{s}{2t}$$

with the initial condition $s(2) = 2$. Separation of the variables gives the solution $s = \sqrt{2t}$.

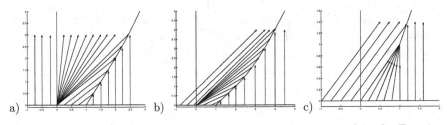

Fig. 10.1. The characteristics for the Burgers equation, assigned in the Exercise 2.3 and respectively referred to the the initial condition given in the three cases

In summary, the pattern of the characteristics and of the shock curve is shown in Fig. 10.1; it corresponds to the solution

$$u(x,t) = \begin{cases} 0 & \text{if } x \leq 0 \\ x/t & \text{if } 0 \leq x \leq t, \text{ with } t \leq 2 \text{ or } 0 \leq x \leq \sqrt{2t} \\ 1 & \text{if } t \leq x < t/2 + 1 \text{ with } t < 2 \\ 0 & \text{if } x > t/2 + 1, \text{ with } t \leq 2 \text{ or } x > \sqrt{2t}, \text{ with } t \geq 2. \end{cases}$$

b) For the assigned initial condition $g(x_0)$, the characteristics have equation

$$\begin{cases} x = x_0 + t & \text{if } x_0 < 0 \\ x = x_0 + 2t & \text{if } 0 < x_0 < 1 \\ x = x_0 & \text{if } x_0 > 1 \end{cases}$$

and are represented in Fig. 10.1.

In the area $S = \{t < x < 2t\}$ there are no characteristics; proceeding as in the traffic model we can define the solution u as a rarefaction wave that connects with continuity (for fixed values of t), the values between $u = 1$ and $u = 2$, respectively, transported by the characteristics $x = t$ and $x = 2t$. As for the case a), the rarefaction wave based at the point (x_0, t_0) is given by the formula

$$u(x,t) = r\left(\frac{x - x_0}{t - t_0}\right),$$

where r is the inverse function of q'. For the Burgers equation, the rarefaction wave with vertex in the origin is $u = x/t$. Along the straight lines $x = ht$, with $1 < h < 2$, u is constant.

The collision of characteristics creates a discontinuity $x = x(t)$ that propagates following the Rankine-Hugoniot condition (2.38). At least for small times t, the shock curve is due to discontinuity created by the impact between the value $u^+ = 2$ (transported along the characteristics $x = x_0 + 2t$) and the value $u^- = 0$ (transported along the vertical characteristics). Taking into account that $q = u^2/2$, those data are respectively associated to the fluxes $q^+ = q(u^+) = 2$ and $q^- = q(u^-) = 0$. Therefore, the shock curve is given by the solution of the problem

$$\begin{cases} \dot{x} = 1 \\ x(0) = 1 \end{cases} \qquad \text{i.e.} \qquad x = t + 1.$$

For times $t > 1$, namely from the point $(2, 1)$, the shock curve is diverted because the vertical characteristics $x = x_0$, with $x_0 > 2$, run up against the rarefaction wave $u = x/t$. In this situation, the Rankine-Hugoniot condition (2.38) has to be solved using $u^+ = x/t$ and $q^+ = x^2/2t^2$; hence, the shock curve is determined by the Cauchy problem

$$\begin{cases} \dot{x} = x/2t \\ x(1) = 2. \end{cases}$$

The differential equation is separable and the solution of the Cauchy problem is $x = 2\sqrt{t}$.

For times $t > 4$, namely from the point $(4, 4)$, the shock curve crosses the characteristics $x = x_0 + t$, with $x_0 < 0$, carrying the data $u = 1$. Therefore it is deviated according to (2.38); in this case we have $u^+ = 1$ and $q^+ = q(u^+) = 1/2$. The shock curve $x = x(t)$ is found solving the problem

$$\begin{cases} \dot{x} = 1/2 \\ x(4) = 4 \end{cases} \quad \text{i.e.} \quad x = \frac{t}{2} + 2.$$

The shock curve is made up of three parts connected at the points $(2, 1)$ and $(4, 4)$ and is represented in Fig. 10.1. The solution is

$$u(x, t) = \begin{cases} 1 & \text{if } x < t < 4 \text{ or } x < t/2 + 2 \text{ with } t > 4 \\ x/t & \text{if } t \le x \le 2t, \text{ with } x \le 2\sqrt{t} \\ 2 & \text{if } 2t \le x < t+1 \\ 0 & \text{if } t+1 < x < 2 \text{ or } t < x^2/4 \text{ with } 2 \le x < 4 \\ & \text{or } t < 2x - 4 \text{ with } x \ge 4. \end{cases}$$

The solution is visualized in three dimension in Fig. 10.2.

c) With the given data we deduce that the characteristics have equation

$$\begin{cases} x = x_0 + t & \text{if } x_0 \le 0 \\ x = x_0 + (1 - x_0)t & \text{if } 0 < x_0 < 1 \\ x = x_0 & \text{if } x_0 \ge 1. \end{cases}$$

All the characteristics based on the segment $0 < x_0 < 1$ cross the point $(1, 1)$; therefore, for $0 < t < 1$ they don't collide. This fact means that in this time interval there is no discontinuity of the solution. A discontinuity originates

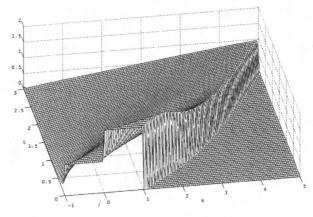

Fig. 10.2. The solution of the Exercise 2.3b, numerically calculated with upwind scheme

in the point $(1,1)$ because of the collision between the vertical characteristics (carrying the data $u^- = 0$) and the characteristics $x = x_0 + t$ (carrying the data $u^+ = 1$). Denoting with $q^- = 0$ and $q^+ = 1/2$ the flux from side to side of the discontinuity, according to the condition (2.38), the shock curve solves the problem

$$\begin{cases} \dot{x} = 1/2 \\ x(1) = 1 \end{cases}$$

namely, it is $x(t) = (t+1)/2$ (Fig. 10.1).

In the area $S = \{0 \le x < 1, 0 \le t < x\}$ the solution is implicitly defined by the equation $u = g(x - q'(u)t)$. In this case, given that $g(x) = 1 - x$, we deduce

$$u(x,t) = \frac{1-x}{1-t}.$$

The solution of the problem is

$$u(x,t) = \begin{cases} 1 & \text{if } x < t < 1 \text{ or } x < (t+1)/2 \text{ with } t > 1 \\ (1-x)/(1-t) & \text{if } 0 < t < x < 1 \\ 0 & \text{if } t < 2x - 1 \text{ with } x > 1. \end{cases}$$

2.4. The flux function associated to this case is $q(u) = u^4/4$ $(q' = u^3)$. Since we have a conservation low, the characteristics are straight lines of equation $x = x_0 + q'(g(x_0))t$. Taking into account the initial data, we deduce

$$\begin{cases} x = x_0 & \text{if } x_0 < 0 \\ x = x_0 + t & \text{if } 0 < x_0 < 1 \\ x = x_0 & \text{if } x_0 > 1. \end{cases}$$

The characteristics are represented in Fig. 10.3. The characteristics based at points $x_0 \le 0$ or $x_0 \ge 1$ are vertical and carry the data $u = 0$; the characteristics carrying the data $u = 1$ are based on the segment $0 < x_0 < 1$. The point $(0,0)$ is the vertex of a fanlike of rarefaction solution, that we explicitly compute: from that point we invert the relation $x = q'(u)t$, that implicitly defines the solution. Since $q' = u^3$, we deduce that the rarefaction solution is $u = \sqrt[3]{x/t}$.

The collision between characteristics carrying different data causes a discontinuity $s = s(t)$, which propagates according to the Rankine-Hugoniot condition (2.38); here, the point where the shock curve is based is $(1,0)$, therefore, for small times, $s = s(t)$ solves the following initial value problem

$$\begin{cases} \dot{s} = 1/4 \\ s(0) = 1 \end{cases} \quad \text{deducing} \quad s = \frac{t}{4} + 1.$$

The considerations made so far are valid until the shock curve meets the characteristic $x = t$, in the point $(4/3, 4/3)$. For times $t > 4/3$, the discontinuity is generated by the collision between the vertical characteristics that originate at the point of abscissa $x_0 > 1$ and carry the data $u^- = 0$ and

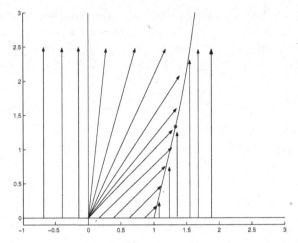

Fig. 10.3. The characteristics of Exercise 2.4. For $0 < t < 4/3$ the shock curve has equation $x = t/4 + 1$, while for $t > 4/3$ it has equation $x = \sqrt[4]{4^3 t/3^3}$. The solution in the rarefaction area based in the origin is $u = \sqrt[3]{x/t}$

the rarefaction solution $u^+ = \sqrt[3]{s/t}$. Using the Rankine-Hugoniot condition (2.38) with the corresponding values for the flux functions $q^+ = 4^{-1} (s/t)^{4/3}$ and $q^- = 0$, we find that the shock curves propagates according to the Cauchy value problem

$$\begin{cases} \dot{s} = \dfrac{1}{4} \dfrac{s}{t} \\ s(4/3) = 4/3. \end{cases}$$

The equation can be integrated by separation of the variables and we deduce the solution $s = \sqrt[4]{4^3 t/3^3}$. The progression of the shock curve and the set of characteristics are represented in Fig. 10.3. Summarizing, the solution of the problem is

$$u(x,t) = \begin{cases} 1 & \text{if } 0 < t < x < 1 + t/4 \\ 0 & \text{if } x \leq 0 \text{ or } 1 < x < t/4 + 1 \text{ with } t < 4/3 \\ & \text{or } x > \sqrt[4]{4^3 t/3^3} \text{ with } t > 4/3 \\ \sqrt[3]{x/t} & \text{if } 0 \leq x \leq t \text{ with } t < 4/3 \text{ or } x < \sqrt[4]{4^3 t/3^3} \text{ with } t \geq 4/3. \end{cases}$$

2.5. For the Burgers equation $q' = u$: the characteristics are then $x = x_0 + g(x_0)t$. On account of the assigned initial condition, the characteristics are the straight lines of equation $x = x_0 + \sin x_0\, t$ if $0 < x_0 < \pi$, elsewhere they are vertical straight lines.

Because of the convexity of the flux function q, the characteristics based at the points $(x_0, 0)$ where the initial condition is increasing (namely, in this exercise, for $0 < x_0 < \pi/2$) spread wide and originate a rarefaction wave. On the other hand, the characteristics based at the points $(x_0, 0)$ where the initial

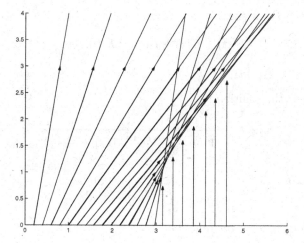

Fig. 10.4. The set of the characteristics associated to a conservation law with convex flux function and initial condition $g(x) = \sin x$ as $0 < x < \pi$ and zero elsewhere. The envelope of the family of the characteristics is clearly visible

condition is decreasing (in this exercise for $\pi/2 < x_0 < \pi$) evolve closer and closer and originate a compression wave, therefore a shock curve arises. The starting point of the shock curve coincide with the point with the minimum time on the envelope of the characteristics starting form the points of the interval $[\pi/2, \pi]$. In order to determine the equation of the envelope of the characteristics, we consider the system

$$\begin{cases} x = x_0 + \sin x_0 \, t \\ 0 = 1 + \cos x_0 \, t \end{cases}$$

where the second equation is obtained from the first, after differentiation with respect to x_0. Then, the parametric equations of the envelope can be written as

$$\begin{cases} x = x_0 - \tan x_0 \\ t = -1/\cos x_0. \end{cases}$$

We deduce that the first (positive) time where the shock curve originates is $t = 1$; it corresponds to the point $x = \pi$ (Fig. 10.4).

The solution is shown in Fig. 2.25. It has been numerically calculated either using an upwind scheme either with a more accurate method, which is necessary to capture its strong discontinuities.

2.6. Hint: Multiply by u the equation. Use $a > 0$ and the inequality

$$2fu \leq f^2 + u^2$$

to obtain

$$\frac{d}{dt} \int_0^R u^2(x,t)dx \leq \int_0^R f^2(x,t)dx + \int_0^R u^2(x,t)dx.$$

Prove that if $E(t)$ satisfies

$$E'(t) \le G(t) + E(t), \quad E(0) = 0$$

then $E(t) \le e^t \int_0^t G(s)ds$.

2.10. Answer: the solution is

$$u(x,t) = \cfrac{1}{1 + \cfrac{\text{erfc}(-x/\sqrt{4\epsilon t})}{\text{erfc}((x-t)/\sqrt{4\epsilon t})} \exp\left(\frac{x-t/2}{2\epsilon}\right)}$$

where

$$\text{erfc}(s) = \int_s^{+\infty} \exp(-z^2)dz$$

is the *complementary error function*.

10.2 Section 3.7

3.1. We use the one-dimensional model described in Section 3.1.3, with Cauchy-Dirichlet conditions on the parabolic boundary:

$$\begin{cases} u_t - Du_{xx} = 0 & (0,1) \times (0,+\infty) \\ u(x,0) = g(x) & 0 \le x \le 1 \\ u(0,t) = u_0, \ u(1,t) = u_1 & t > 0. \end{cases}$$

In this case, we solve the problem:

$$\begin{cases} u_t - Du_{xx} = 0 & (0,1) \times (0,+\infty) \\ u(x,0) = x & 0 \le x \le 1 \\ u(0,t) = 1, \ u(1,t) = 0 & t > 0. \end{cases}$$

We try to conjecture what could happen. Since the initial data is increasing in x, heat initially flows along the bar from right to left. On the other hand, since $u_0 > u_1$, from the left (the hotter) end point heat will begin to flow to the right, which causes the increase of the temperature inside. These two opposite fluxes will tend to stabilize as time increases to eventually reach a steady state.

First, we determine the steady state, u^{St}; since $u_{xx}^{St} = 0$ and it satisfies the Dirichlet conditions $u^{St}(0) = 1$, $u^{St}(1) = 0$, we deduce $u^{St} = 1 - x$.

Now we introduce the transient $v = u - u^{St}$. The conjecture is that it tends to zero as $t \to +\infty$, if so the rate of convergence could also be determined. The function v solves the Dirichlet homogeneous boundary problem

$$\begin{cases} v_t - Dv_{xx} = 0 & (0,1) \times (0,+\infty) \\ v(x,0) = 2x - 1 & 0 < x < 1 \\ v(0,t) = 0, \ v(1,t) = 0 & t > 0. \end{cases} \tag{10.2}$$

Let us attempt to find a solution which is not identically zero satisfying the boundary conditions but with the following property: v is a product in which the dependence of v on x, t is separated, that is:

$$v(x,t) = y(x)\,w(t).$$

A substitution into the first of (10.2) gives $w'(t)\,y(x) - D\,w(t)\,y''(x) = 0$. Rearranging terms, we deduce:

$$\frac{w'(t)}{D\,w(t)} = \frac{y''(x)}{y(x)} \tag{10.3}$$

this equality must hold for every $x \in (0,1)$ and every $t > 0$, hence both sides are equal to a common constant λ.

In particular we deduce that y solves the eigenvalue problem $y'' - \lambda y = 0$ with $y(0) = y(1) = 0$. If $\lambda \geq 0$, the only solution which is compatible with the zero-boundary conditions is $y = 0$. This fact is immediate to be verified, indeed if $\lambda = 0$ the solutions are straight lines. If $\lambda > 0$ we have exponential solutions like

$$y(x) = Ae^{\sqrt{\lambda}t} + Be^{-\sqrt{\lambda}t}$$

and the boundary conditions $y(0) = y(1) = 0$ lead to

$$\begin{cases} A + B = 0 \\ Ae^{\sqrt{\lambda}} + Be^{-\sqrt{\lambda}} = 0 \end{cases}$$

giving $A = B = 0$ since the coefficient matrix is non singular. The case $\lambda = -\mu^2 < 0$ gives solutions like

$$y(x) = A\sin\mu t + B\cos\mu t.$$

The boundary conditions $y(0) = y(1) = 0$ require

$$\begin{cases} B = 0 \\ A\sin\mu + B\cos\mu = 0 \end{cases}$$

and we choose

$$A \text{ free}, \ B = 0, \ \mu = k\pi, \text{ with } k = 1,2,3,\cdots.$$

Therefore, the differential problem for y has non-trivial solutions like $y_k(x) = \sin k\pi x$ (eigenfunctions) only for $\lambda = -k^2\pi^2$ (eigenvalues). Correspondingly to such eigenvalues, w solves the linear equation with constant coefficients $w' + Dk^2\pi^2 w = 0$, whose general solution is

$$w_k = b_k e^{-D\pi^2 k^2 t}.$$

Thus, we have a family of non trivial solutions for the transient like

$$v_k(x,y) = b_k e^{-D\pi^2 k^2 t} \sin k\pi x$$

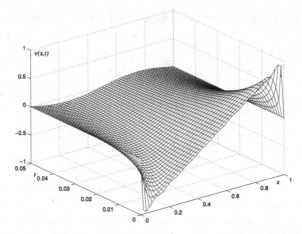

Fig. 10.5. The transient $v(x,t)$ in (10.4) represented with MatLab using the first 40 terms of the Fourier expansion

but none of them satisfies the initial condition. Exploiting the linear nature of the problem, we use the superposition principle and consider

$$v(x,t) = \sum_{k=1}^{\infty} v_k(x,t) = \sum_{k=1}^{\infty} b_k e^{-D\pi^2 k^2 t} \sin k\pi x.$$

For the initial condition $v(x,0) = 2x - 1$ we have:

$$v(x,0) = \sum_{k=1}^{\infty} b_k \sin k\pi x = 2x - 1.$$

The coefficients b_k coincide with the coefficient of the sine Fourier series[2] of $2x - 1$ in $(-1,1)$; from (A.5):

$$b_k = 2 \int_0^1 (2x - 1) \sin k\pi x \, dx.$$

Given the symmetry of the integrand function with respect to the medium point of the interval of integration, we find that

$$b_k = \begin{cases} 0 & \text{if } k \text{ is odd} \\ 4 \displaystyle\int_0^{1/2} (2x - 1) \sin k\pi x \, dx & \text{if } k \text{ is even} \end{cases}$$

$$= \begin{cases} 0 & \text{if } k \text{ is odd} \\ -\dfrac{4}{k\pi} & \text{if } k \text{ is even.} \end{cases}$$

---

[2] Appendix A.

Denoting $k = 2n$, we have then

$$v(x,t) = -\frac{2}{\pi} \sum_{n=1}^{\infty} \frac{1}{n} e^{-4D\pi^2 n^2 t} \sin 2n\pi x. \tag{10.4}$$

Let us analyze the behavior of the transient v as $t \to +\infty$ (Fig. 10.5): each term of the series is exponential and the leading one corresponds to $n = 1$, therefore

$$v(x,t) \sim -\frac{2}{\pi} e^{-4D\pi^2 t} \sin 2\pi x \longrightarrow 0,$$

and we deduce that, for increasing t, the solution of the problem

$$u(x,t) = u^{St} + v = 1 - x - \frac{2}{\pi} \sum_{n=1}^{\infty} \frac{1}{n} e^{-4D\pi^2 n^2 t} \sin 2n\pi x$$

tends to the stationary solution u^{St}.

3.2. We analyze a Cauchy-Neuman problem with homogeneous boundary conditions, hence the separation of variables can be immediately used; we search for a solution like

$$u(x,t) = y(x)w(t).$$

We are yield to the equation (10.3), studied in the Exercise 3.1. In this case the thermal response D is 1 and different boundary conditions are assigned. In particular, y solves $y'' - \lambda y = 0$ with $y'(0) = y'(L) = 0$. As usual, we distinguish 3 cases.

First case, we consider $\lambda = \mu^2 > 0$. The general solution is

$$y(x) = Ae^{\mu x} + Be^{-\mu x}$$

and we deduce that $y'(x) = \mu \left(Ae^{\mu x} - Be^{-\mu x} \right)$. The boundary conditions give

$$\begin{cases} A - B = 0 \\ e^{\mu L} A - e^{-\mu L} B = 0. \end{cases}$$

The system has the unique zero solution since the coefficient matrix is non-singular.

Second case, we consider $\lambda = 0$. We obtain $y(x) = Ax + B$. The boundary conditions are fulfilled with $A = 0$ and every B. Therefore, $\lambda = 0$ is an eigenvalue of the problem, and the constants are the corresponding eigenfunctions.

Final case, we consider $\lambda = -\mu^2$. The general solution is

$$y(x) = A \cos \mu x + B \sin \mu x$$

and we deduce that $y'(x) = \mu \left(-A \sin \mu x + B \cos \mu x \right)$. Since $y'(0) = y'(L) = 0$, we obtain

$$A \text{ free}, \ B = 0, \ \mu L = k\pi, \ \text{with } k = 1, 2, 3, \cdots .$$

Summing up the situations that generate non-zero solutions, the eigenvalues of the problem are $\lambda_k = -k^2\pi^2 L^{-2}$, corresponding to the eigenfunctions $y_k = \cos k\pi x L^{-1}$, with $k \in \mathbb{N}$.

The equation for the unknown w becomes

$$w'(t) + \frac{k^2\pi^2}{L^2}w(t) = 0$$

whose solution is

$$w_k = a_k e^{-\frac{k^2\pi^2}{L^2}t}.$$

We have found the family of solutions

$$u_k(x,t) = y(x)w(t) = a_k e^{-\frac{k^2\pi^2}{L^2}t} \cos \frac{k\pi}{L}x$$

with $k \in \mathbb{N}$, satisfying the homogeneous conditions at the boundaries. The initial condition $u(x,0) = x$ as $0 < x < L$ has to be fulfilled as well. We use the superposition principle and introduce

$$u(x,t) = \sum_{k=0}^{\infty} a_k e^{-\frac{k^2\pi^2}{L^2}t} \cos \frac{k\pi}{L}x$$

therefore, the initial condition becomes

$$u(x,0) = \sum_{k=0}^{\infty} a_k \cos \frac{k\pi}{L}x = x.$$

Hence, the coefficients a_k coincide with the coefficients of the cosine Fourier series of the function $g(x) = x$ in the interval $[-L, L]$:

$$g(x) = \frac{a_0}{2} + \sum_{k=1}^{\infty} a_k \cos \frac{k\pi x}{L}.$$

From the formulas (A.3) and (A.4), as $g(x) = x$, we deduce

$$a_0 = \frac{2}{L} \int_0^L x\,dx = L$$

$$a_k = \frac{2}{L} \int_0^L x \cos \frac{k\pi x}{L}\,dx$$

$$= \frac{2L}{k^2\pi^2}\left[(-1)^k - 1\right].$$

We deduce that the solution is

$$u(x,t) = \frac{L}{2} + \frac{2L}{\pi^2} \sum_{k=1}^{\infty} \frac{(-1)^k - 1}{k^2} e^{-\frac{k^2\pi^2}{L^2}t} \cos \frac{k\pi}{L}x$$

$$= \frac{L}{2} - \frac{4L}{\pi^2} \sum_{m=0}^{\infty} \frac{1}{(2m+1)^2} e^{-\frac{(2m+1)^2\pi^2}{L^2}t} \cos \frac{(2m+1)\pi}{L}x.$$

Finally, we note that as $t \to +\infty$ the solution u converges to the value $L/2$, that is the mean value of the data. In fact, the boundary conditions of Neumann type correspond to a problem where no interaction with the environment occurs[3]. The equation governs the evolution of the concentration of a substance subject to diffusion, whose total mass, if no exchanges with the outside are possible, is preserved and it tends to distributed uniformly on the segment $[0, L]$, reaching the stationary configuration.

3.3. We deal with a non homogeneous Cauchy-Neumann problem, with homogeneous boundary conditions. As the equation shows a exogenous heat source proportional to x and t, we expect the solution grows with respect to these variables.

Formally, we write the candidate solution as

$$u(x,t) = \sum_{k=0}^{\infty} c_k(t) \, v_k(x)$$

where v_k are the eigenfunctions of the eigenvalue problem associated to the homogeneous equation:

$$\begin{cases} v''(x) - \lambda v(x) = 0 & 0 < x < \pi \\ v'(0) = v'(\pi) = 0 \end{cases}$$

solved in the Exercise 3.2, where $L = \pi$. The eigenvalues are $\lambda_k = -k^2$ and the corresponding eigenfunctions are $v_k(x) = \cos kx$, with $k \in \mathbb{N}$.

We have then

$$u(x,t) = \sum_{k=0}^{\infty} c_k(t) \cos kx;$$

and the following conditions have to be fulfilled

$$\begin{cases} u_t(x,t) - u_{xx}(x,t) = \\ \quad = \sum_{k=0}^{\infty} \left(c_k'(t) + k^2 c_k(t) \right) \cos kx = t\,x \quad 0 < x < \pi, \, t > 0 \\ u(x,0) = \sum_{k=0}^{\infty} c_k(0) \cos kx = 1 \qquad\qquad 0 \le x \le \pi. \end{cases} \tag{10.5}$$

We write the function $g(x) = x$ in cosine Fourier series in the interval $(-\pi, \pi)$

$$x = \frac{a_0}{2} + \sum_{n=1}^{\infty} a_n \cos nx$$

where the coefficients a_0 and a_k are defined in (A.3) and (A.4) in Appendix A. We have

$$x = \frac{\pi}{2} + \sum_{n=1}^{\infty} \frac{-4}{\pi(2n+1)^2} \cos(2n+1)x.$$

---

[3] Adiabatic extremes, considering the heat conduction in the bar.

Comparing the first of (10.5) with the equation above, we deduce that the coefficients $c_k(t)$ must satisfy the following ordinary differential equations

$$\begin{cases} c'_0(t) = \pi t/2 \\ c'_{2n}(t) + 4n^2 c_{2n}(t) = 0 & n > 0 \\ c'_{2n+1}(t) + (2n+1)^2 c_{2n+1}(t) = \dfrac{-4t}{\pi(2n+1)^2} & n \geq 0. \end{cases}$$

These equations are respectively associated with the initial values that we get comparing the second of (10.5) with the cosine Fourier series of the constant 1:

$$\begin{cases} c_0(0) = 1 \\ c_{2n}(0) = 0 & n > 0 \\ c_{2n+1}(0) = 0 & n \geq 0. \end{cases}$$

We have a set of differential equations with constant coefficients whose solutions can be written using the superposition principle by adding a particular solution to the general solution of the homogeneous equation. Thus, their solutions are, respectively:

$$\begin{cases} c_0(t) = \dfrac{\pi t^2}{4} + 1 \\ c_{2n}(0) = 0 & n > 0 \\ c_{2n+1}(t) = \dfrac{-4}{\pi(2n+1)^4}\left[t + \dfrac{e^{-(2n+1)^2 t} - 1}{(2n+1)^2}\right] & n \geq 0. \end{cases}$$

Therefore, the solution of the problem is

$$u(x,t) = 1 + \frac{t^2\pi}{4} + \sum_{n=0}^{\infty} \frac{-4}{\pi(2n+1)^4}\left[t + \frac{e^{-(2n+1)^2 t} - 1}{(2n+1)^2}\right]\cos(2n+1)x \quad (10.6)$$

and it is represented in Fig. 10.6.

3.4. The concentration c satisfies the equation

$$c_t = Dc_{xx} \qquad 0 < x < L,\, t > 0.$$

If we denote by $\mathbf{i}$ the versor of the x-axis, according to the *Fick's law*, the flux entering in $x = 0$ is given by

$$\int_A \mathbf{q}\left(c\left(0,t\right)\right)\cdot\mathbf{i}\,dxdy = \int_A -Dc_x\left(0,t\right)dxdy = -DAc_x\left(0,t\right) = C_0 R_0$$

while the outgoing flux in $x = L$ is

$$\int_A \mathbf{q}\left(c\left(L,t\right)\right)\cdot\mathbf{i}\,dxdy = \int_A -Dc_x\left(L,t\right)dxdy = -DAc_x\left(L,t\right) = c\left(L,t\right) R_0.$$

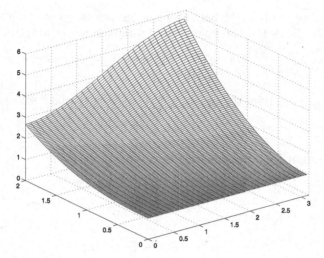

Fig. 10.6. The function (10.6), represented in the interval $0 \le t \le 2$ using MatLab and the first 15 non zero terms of the Fourier series

Therefore, we deduce the following Neumann-Robin conditions

$$c_x(0,t) = -B \quad \text{and} \quad c_x(L,t) + Ec(L,t) = 0$$

where we denoted

$$B = \frac{C_0 R_0}{DA} \quad \text{and} \quad E = \frac{R_0}{DA};$$

the problem is also associated to the initial condition

$$c(x,0) = c_0(x).$$

We determine, first, the stationary solution c^{St}, which satisfies the conditions

$$c_{xx}^{St} = 0 \qquad\qquad\qquad 0 < x < L, t > 0$$
$$c_x^{St}(0,t) = -B, \ c_x^{St}(L,t) + Ec^{St}(L,t) = 0 \quad t > 0.$$

We find

$$c^{St}(x) = B(L-x) + \frac{B}{E}.$$

Now, we analyze the transient $u(x,t) = c(x,t) - c^{St}(x)$. Then u solves the following problem

$$\begin{aligned}
&u_t = Du_{xx} && 0 < x < L, t > 0 \\
&u_x(0,t) = 0, \ u_x(L,t) + Eu(L,t) = 0 && t > 0 \\
&u(x,0) = c_0(x) - c^{St}(x) && 0 < x < L.
\end{aligned}$$

In this way, we are brought back to homogeneous boundary conditions, essential to use the method of separation of variables; we set $u(x,t) = y(x)w(t)$ and deduce

$$\frac{w'(t)}{Dw'(t)} = \frac{y''(x)}{y(x)} = \lambda.$$

In particular, w satisfies the equation $w'(t) = \lambda Dw(t)$ whose solution is

$$V(t) = e^{\lambda Dt}$$

while y is a solution of the following eigenvalue problem:

$$y''(x) - \lambda y(x) = 0$$

associated to the conditions

$$y'(0,t) = 0, \; y'(L,t) + Ey(L,t) = 0.$$

If $\lambda > 0$, the general solution is $y(x) = c_1 e^{-\sqrt{\lambda}x} + c_1 e^{\sqrt{\lambda}x}$. The boundary conditions give

$$\begin{cases} -c_1 + c_2 = 0 \\ c_1(E - \sqrt{\lambda})e^{-\sqrt{\lambda}L} + c_2(E + \sqrt{\lambda})e^{\sqrt{\lambda}L} = 0. \end{cases} \tag{10.7}$$

Now, we have that

$$\det \begin{pmatrix} -1 & 1 \\ (E - \sqrt{\lambda})e^{-\sqrt{\lambda}L} & (E + \sqrt{\lambda})e^{\sqrt{\lambda}L} \end{pmatrix}$$

$$= -(E + \sqrt{\lambda})e^{\sqrt{\lambda}L} - (E - \sqrt{\lambda})e^{-\sqrt{\lambda}L}$$

$$= (E + \sqrt{\lambda})e^{-\sqrt{\lambda}L}\left(\frac{\sqrt{\lambda} - E}{\sqrt{\lambda} + E} - e^{2\sqrt{\lambda}L} \right)$$

is negative, since $e^{2\sqrt{\lambda}L} > 1$ and $(\sqrt{\lambda} - E)/(\sqrt{\lambda} + E) < 1$. The system (10.7) has the only solution $c_1 = c_2 = 0$.

W can make the same deduction even in the case $\lambda = 0$.

If $\lambda < 0$, we find the conditions

$$\begin{cases} U'(0) = \sqrt{-\lambda}c_2 = 0 \\ U'(L,t) + EU(L,t) = c_1\left[\cos\left(\sqrt{-\lambda}L\right) - \sqrt{-\lambda}\sin\left(\sqrt{-\lambda}L\right)\right] = 0. \end{cases}$$

So, λ satisfies the equation

$$\cot\left(\sqrt{-\lambda}L\right) = \sqrt{-\lambda}.$$

Examining the graphs of functions $f_1(x) = \cot Lx$ and $f_2(x) = x$ (Fig. 10.7) we note that there are infinite points k_m, with $0 < k_m < m\pi/L$ and

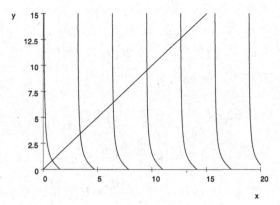

Fig. 10.7. Intersections of $y = x$ and $y = \cot x$

$m > 0$, where the two functions f_1 and f_2 intersect. Then, $\lambda_m = k_m^2$ and the corresponding eigenfunction is $\cos(k_m x)$. Therefore, the solution is

$$u\left(x, t\right) = \sum_{m=1}^{\infty} u_m e^{-Dk_m^2 t} \cos(k_m x)$$

where u_m is the coefficient of the Fourier series[4] of $u\left(x, 0\right)$ expanded with respect to the eigenfunction $\cos(k_m x)$, namely

$$u_m = \frac{1}{\alpha_m} \int_0^L u\left(x, 0\right) \cos\left(k_m x\right) dx \qquad \alpha_m = \int_0^L \cos^2\left(k_m x\right) dx.$$

Regarding the concentration c, we finally deduce the formula

$$c\left(x, t\right) = \frac{B}{E} + B\left(L - x\right) + \sum_{m=1}^{\infty} u_m e^{-Dk_m^2 t} \cos(k_m x).$$

Since $k_m > 0$ for every m, as t goes to $+\infty$ every term of the series converges to zero exponentially and therefore c settles to the steady solution c^{St}

$$c\left(x, t\right) \to C_0 + \frac{C_0 R_0}{DA}\left(L - x\right).$$

3.5. We proceed as in Section 3.3.2 and substitute into the equation $u_t - u_{xx} = 0$. Denoting

$$\xi = \frac{x}{\sqrt{t}}.$$

---

[4] It can be proved that $\int_0^L \cos(k_n x) \cos\left(k_m x\right) dx = 0$ if $m \neq n$. Furthermore, the functions $\varphi_m\left(x\right) = \cos\left(k_m x\right)/\sqrt{\alpha_m}$ are a base of the space $L^2\left(0, L\right)$ of square-integrable functions (Chapter 7).

we obtain:

$$\frac{\partial \xi}{\partial x} = \frac{1}{\sqrt{t}} \qquad \frac{\partial \xi}{\partial t} = -\frac{x}{2t\sqrt{t}} \qquad \frac{\partial^2 \xi}{\partial x^2} = 0.$$

From $u(x,t) = U(x/\sqrt{t})$, we have

$$u_t(x,t) = U'(\xi)\frac{\partial \xi}{\partial t} = -U'(\xi)\frac{x}{2t\sqrt{t}}$$

$$u_x(x,t) = U'(\xi)\frac{\partial \xi}{\partial t} = U'(\xi)\frac{1}{\sqrt{t}}$$

$$u_{xx}(x,t) = \frac{1}{\sqrt{t}}U''(\xi)\frac{\partial \xi}{\partial x} = U''(\xi)\frac{1}{t}.$$

The equation $u_t - u_{xx} = 0$ becomes $-U'(\xi)\frac{x}{2t\sqrt{t}} - \frac{1}{t}U''(\xi) = 0$ and it can be rewritten as

$$U'(\xi)\xi + 2U''(\xi) = 0.$$

It is a linear first order equation with respect to U', whose general solution is

$$U'(\xi) = ce^{-\xi^2/4}.$$

Integration with respect to ξ gives

$$U(\xi) = c_1 + c_2 \int e^{-\xi^2/4}\, d\xi = c_1 + c_2 \int_0^{\xi/2} e^{-z^2}\, dz.$$

If we use the error function (whose graph is in Fig. 10.8), we finally obtain

$$u(x,t) = U\left(\frac{x}{\sqrt{t}}\right) = c_1 + c_2 \mathrm{erf}\left(\frac{x}{\sqrt{t}}\right).$$

In order to fulfill the boundary conditions

$$u(0,t) = C, \qquad \lim_{x \to +\infty} u(x,t) = 0 \qquad t > 0$$

we set $c_1 = C$ and $c_2 = -2C/\sqrt{\pi}$, and hence

$$u(x,t) = C\left(1 - \mathrm{erf}\left(\frac{x}{2\sqrt{t}}\right)\right). \tag{10.8}$$

3.6. We have to solve the following diffusion problem:

$$\begin{cases} u_t - \Delta u = 0 & \mathbf{x} \in B_R, t > 0 \\ u(\mathbf{x},0) = U & \mathbf{x} \in B_R \\ u(\sigma,t) = 0 & \sigma \in \partial B_R, t > 0. \end{cases}$$

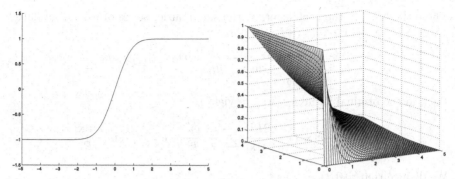

Fig. 10.8. On the left, the error function $\mathrm{erf}(x) = \frac{2}{\sqrt{\pi}} \int_0^x e^{-z^2}\,dz$. On the right, the function (10.8), obtained with MatLab; it represents the diffusion on the half straight line $x > 0$ of a source $C = 1$ concentrated in the origin and constant with time

Since the problem is radial, it seems convenient to search for a radial solution, namely $u(\mathbf{x}, t) = u(r, t)$, with $r = |\mathbf{x}|$. We write the laplacian in polar coordinates[5], and remark that

$$\Delta u = u_{rr} + \frac{2}{r} u_r = \frac{1}{r}(ru)_{rr}.$$

The problem becomes, then

$$\begin{cases} u_t - r^{-1}(ru)_{rr} = 0 & 0 < r < R, t > 0 \\ u(r, 0) = U & 0 \le r < R \\ u(R, t) = 0 & t > 0. \end{cases}$$

Introducing $v = ru$, the new unknown v satisfies the following one-dimensional problem

$$\begin{cases} v_t - v_{rr} = 0 & 0 < r < R, t > 0 \\ v(r, 0) = ru(r, 0) = rU & 0 \le r < R \\ v(0, t) = v(R, t) = 0 & t > 0. \end{cases}$$

The one-dimensional problem with homogeneous Dirichlet boundary conditions has been solved in the Exercise 3.1 with the technique of the separation of variables[6], and has eigenvalues $-k^2\pi^2/R^2$ and corresponding eigenfunctions $\sin k\pi r/R$, with $k = 1, 2, \cdots$. Therefore, we deduce

$$v(r, t) = \sum_{k=1}^{\infty} b_k e^{-\frac{k^2\pi^2}{R^2}t} \sin \frac{k\pi r}{R}$$

---

[5] Appendx D.
[6] See equation (10.2).

where the b_k's are the coefficients of the sine Fourier series of the initial data $g(r) = rU$ on the interval $(-R, R)$, namely

$$b_k = \frac{2U}{R} \int_0^R r \sin \frac{k\pi r}{R} \, dr = \frac{2RU}{k\pi}(-1)^{k+1}.$$

Back in the original variables, we are yield to

$$u(\mathbf{x}, t) = \frac{1}{|\mathbf{x}|} v(\mathbf{x}, t) = \frac{2RU}{\pi |\mathbf{x}|} \sum_{k=1}^{\infty} \frac{(-1)^{k+1}}{k} e^{-\frac{k^2 \pi^2}{R^2} t} \sin \frac{k\pi |\mathbf{x}|}{R}.$$

We deduce that $u(\mathbf{0}, t) \to 0$ as $t \to +\infty$.

3.8. Hint. Choose h, k such that $v(x, t) = u(x, t) e^{hx + kt}$ is a solution of $v_t = Dv_{xx}$.

3.9. Hint. Extend g to $x < 0$ by odd reflection: $g(-x) = -g(x)$. Solve the corresponding global Cauchy problem and write the result as an integral on $(0, +\infty)$.

3.10. Hint. Assume first that $c(\mathbf{x}, t) \le a < 0$. Then reduce to this case by setting $u = ve^{kt}$ with a suitable $k > 0$.

3.11. Hint. The solution is radial so that $u = u(r, t)$, $r = |\mathbf{x}|$. Observe that $\Delta u = u_{rr} + \frac{2}{r} u_r = \frac{1}{r}(ru)_{rr}$. Let $v = ru$, reduce to homogeneous Dirichlet condition and use separation of variables.

3.14. Answer. The solution is

$$u(r, t) = \frac{2}{r} \sum_{n=1}^{\infty} \frac{(-1)^n}{\lambda_n} \sin(\lambda_n r) \left\{ \frac{q}{1 - \lambda_n^2} \left(e^{-t} - e^{-\lambda_n^2 t} \right) - U e^{-\lambda_n^2 t} \right\}$$

where $\lambda_n = n\pi$.

10.3 Section 4.7

4.1. We use the superposition principle and split the problem into four simpler problems. Each problem has different boundary conditions: it has homogeneous boundary conditions on three sides of the rectangle out of four, while on the other side is assigned the Dirichlet condition corresponding to the given function. For instance, we find u_1 solving

$$\begin{cases} \Delta u_1 = 0 & \text{in } R \\ u_1(0, y) = g_1(y), u_1(L, y) = 0 & 0 < y < H \\ u_1(x, 0) = 0, u_1(x, H) = 0 & 0 < x < L. \end{cases}$$

We use the method of separation of the variables and consider $u_1(x, y) = X(x)Y(y)$; assuming that X and Y are nonzero, from the Laplace equation we deduce

$$\frac{X''(x)}{X(x)} = -\frac{Y''(y)}{Y(y)} = \lambda$$

where λ is constant. Hence Y satisfies the eigenvalue problem

$$\begin{cases} Y''(y) + \lambda Y(y) = 0, & 0 < y < H \\ Y(0) = Y(H) = 0. \end{cases}$$

If $\lambda = 0$ we have the zero solution. For $\lambda = -\mu^2 < 0$, we have

$$Y(y) = c_1 \sinh \mu y + c_2 \cosh \mu y$$

and the boundary conditions require that $c_1 = c_2 = 0$. On the other hand, we find a nonzero solution in the case $\lambda = \mu^2 > 0$, obtaining

$$Y(y) = c_1 \sin \mu y + c_2 \cos \mu y.$$

Indeed, the boundary conditions yield to $c_2 = 0$ and $\mu H = k\pi$, where $k \in \mathbb{N} \setminus \{0\}$; consequently

$$Y_k(y) = \sin \frac{k\pi y}{H}$$

are the eigenfunctions and

$$\lambda_k = \frac{k^2 \pi^2}{H^2}$$

are the eigenvalues of the problem. Correspondingly, we consider the equation

$$X''(x) - \lambda_k X(x) = 0$$

with the condition $X(L) = 0$. It is convenient to introduce a shift in the x direction and we write its solution as

$$X_k(x) = c_1 \sinh \frac{k\pi(x - L)}{H} + c_2 \cosh \frac{k\pi(x - L)}{H}.$$

Hence, on account of the boundary zero condition, we deduce $c_2 = 0$ and

$$X_k(x) = c_1 \sinh \frac{k\pi(x - L)}{H}.$$

Therefore, by linearity, a candidate solution, satisfying the zero condition on the three sides of the rectangle is

$$u_1(x, y) = \sum_{k=1}^{\infty} c_k \sinh \frac{k\pi(x - L)}{H} \sin \frac{k\pi y}{H}.$$

We choose c_k in order to fulfill the non zero boundary condition, i.e.

$$u_1(0, y) = -\sum_{k=1}^{\infty} c_k \sinh \frac{k\pi L}{H} \sin \frac{k\pi y}{H} = g_1(y).$$

Hence, denoting with a_k the coefficients of the Fourier sine series associated to g_1 on the interval $(-H, H)$, namely

$$a_k = \frac{2}{H} \int_0^H g_1(y) \sin \frac{k\pi y}{H} dy$$

we have that

$$c_k = \frac{a_k}{- \sinh \frac{k\pi L}{H}}.$$

Finally, we deduce that

$$u_1(x, y) = -\sum_{k=1}^\infty \frac{a_k}{\sinh \frac{k\pi L}{H}} \sinh \frac{k\pi(x - L)}{H} \sin \frac{k\pi y}{H}.$$

With analogous steps, the solution of the problems with nonzero data on the other sides of the rectangle can be found.

4.2. A function u is harmonic in Ω if it has continuous second derivatives and also $\Delta u = 0$ in Ω. Furthermore, it has the mean value property and from the Theorem 4.5, u is $C^\infty(\Omega)$. Then, we just have to prove that if u is harmonic, then

$$w = \frac{\partial u}{\partial x_i}$$

is harmonic as well. In fact, since the Schwarz's theorem holds, we have

$$\Delta w = \Delta \frac{\partial u}{\partial x_i} = \frac{\partial \Delta u}{\partial x_i} = 0.$$

4.3. Given the symmetry of the domain, it is convenient to use the polar coordinates in the plane. We assume that $f = f(r, \theta)$ can be expanded in sine Fourier series with respect to θ, in $[0, 2\pi]$:

$$f(r, \theta) = \sum_{m=1}^\infty f_m(r) \sin m\theta.$$

We write the candidate solution in the form

$$u(r, \theta) = \sum_{m=1}^\infty u_m(r) \sin m\theta$$

where the coefficients $u_m(r)$ have to be determined. Substituting into the Laplace equation (4.16), written in polar coordinates[7], we deduce:

$$\sum_{m=1}^\infty \left\{ u_m'' \sin m\theta(r) + \frac{1}{r} u'(r) \sin m\theta - \frac{m^2}{r^2} u_m(r) \frac{\partial^2 \sin m\theta}{\partial \theta^2} \right\}$$

$$= \sum_{m=1}^\infty f_m(r) \sin m\theta.$$

---

[7] Appendix D.

Then, the coefficients u_m solve the set of ordinary differential equations

$$u_m''(r) + \frac{1}{r}u'(r) - \frac{m^2}{r^2}u_m(r) = f_m(r) \qquad m \geq 1 \qquad (10.9)$$

(called Euler equations) with the conditions

$$u_m(R) = 0 \text{ and } u_m \text{ bounded in } [0,1].$$

4.4. We use the linearity to split the problem into the following two:

$$\begin{cases} \Delta w(x,y) = y & \text{in } B_1 \\ w = 0 & \text{on } \partial B_1 \end{cases} \qquad \begin{cases} \Delta z(x,y) = 0 & \text{in } B_1 \\ z = 1 & \text{on } \partial B_1 \end{cases}$$

and $u = w+z$. It is evident that we have $z = 1$ in B_1. In order to solve the first problem, we use the solution of the Exercise 4.3. Note that writing $y = r\sin\theta$ the right hand side function is expanded in Fourier series with respect to θ, in $[0, 2\pi]$. In particular, we search for a solution like:

$$w(r,\theta) = \sum_{m=1}^{\infty} w_m(r)\sin m\theta$$

satisfying the equations in (10.9). We deduce the following set of ordinary differential equations with variable coefficients:

$$\begin{cases} w_1'' + w_1'/r - w_1/r^2 = r \\ w_m'' + w_m'/r - m^2 w_m/r^2 = 0 \end{cases} \qquad \text{for } m > 1 \qquad (10.10)$$

associated to the conditions

$$w_m(1) = 0 \text{ and } w_m \text{ bounded in } (0,1)$$

for every $m \geq 1$.

Using the substitution[8] $t = \log r$ in the first of (10.10), we are yield to the constant coefficients equation $v'' - v = e^{3t}$, whose general solution is

$$v(t) = c_1 e^t + c_2 e^{-t} + \frac{1}{8}e^{3t}.$$

Going back to the variable r we obtain

$$w_1(r) = c_1 r + c_2 \frac{1}{r} + \frac{1}{8}r^3.$$

We choose $c_2 = 0$ to obtain a bounded function and setting $w_1(1) = 0$ we deduce $c_1 = -1/8$. The same procedure can be used to show that the equations for w_m, with $m > 1$, have the zero solution. Hence, we have

$$w(r,\theta) = \frac{r}{8}(r^2 - 1)\sin\theta.$$

---

[8] It is a standard substitution for Euler equations.

The solution of the initial problem, in cartesian coordinates, is therefore

$$u(x, y) = \frac{y}{8}(x^2 + y^2 - 1) + 1.$$

4.5. We generalize the technique used in Exercise 4.3 applied to the ring $B_{1,R}$. We exploit the symmetry of the domain and search for a solution in polar coordinates like

$$u(r, \theta) = \alpha_0(r) + \sum_{n=1}^{\infty} [\alpha_n(r) \cos n\theta + \beta_n(r) \sin n\theta]. \tag{10.11}$$

In order to substitute into the laplacian in polar coordinates

$$\Delta u = u_{rr} + \frac{1}{r}u_r + \frac{1}{r^2}u_{\theta\theta}, \tag{10.12}$$

we calculate

$$u_r(r, \theta) = \alpha_0'(r) + \sum_{n=1}^{\infty} [\alpha_n'(r) \cos n\theta + \beta_n'(r) \sin n\theta],$$

$$u_{rr}(r, \theta) = \alpha_0''(r) + \sum_{n=1}^{\infty} [\alpha_n''(r) \cos n\theta + \beta_n''(r) \sin n\theta],$$

$$u_{\theta\theta}(r, \theta) = -\sum_{n=1}^{\infty} n^2 [\alpha_n(r) \cos n\theta + \beta_n(r) \sin n\theta].$$

The equation (10.12) then becomes

$$\alpha_0''(r) + \frac{\alpha_0'(r)}{r} + \sum_{n=1}^{\infty} \left[\left(\alpha_n''(r) + \frac{\alpha_n'(r)}{r} - \frac{n^2}{r^2}\alpha_n(r) \right) \cos n\theta \right.$$

$$\left. + \left(\beta_n''(r) + \frac{\beta_n'(r)}{r} - \frac{n^2}{r^2}\beta_n(r) \right) \sin n\theta \right] = 0.$$

Let us expand in Fourier series the smooth, periodic data g and h

$$g(\theta) = \frac{a_0}{2} + \sum_{n=1}^{\infty} (a_n \cos n\theta + b_n \sin n\theta),$$

$$h(\theta) = \frac{A_0}{2} + \sum_{n=1}^{\infty} (A_n \cos n\theta + B_n \sin n\theta).$$

The boundary conditions imply that the values of the coefficients of the solutions $u(1, \theta)$ and $u(R, \theta)$ coincide with the coefficients of the Fourier series associated to g and h, respectively. We are yield to study the family of boundary value problems

$$\begin{cases} \alpha_0''(r) + \dfrac{1}{r}\alpha_0'(r) = 0 \\[2mm] \alpha_0(1) = \dfrac{a_0}{2}, \quad \alpha_0(R) = \dfrac{A_0}{2} \end{cases} \tag{10.13}$$

$$\begin{cases} \alpha_n''(r) + \dfrac{1}{r}\alpha_n'(r) - \dfrac{n^2}{r^2}\alpha_n(r) = 0 \\[2mm] \alpha_n(1) = a_n, \quad \alpha_n(R) = A_n \end{cases} \tag{10.14}$$

$$\begin{cases} \beta_n''(r) + \dfrac{1}{r}\beta_n'(r) - \dfrac{n^2}{r^2}\beta_n(r) = 0 \\[2mm] \beta_n(1) = b_n, \quad \beta_n(R) = B_n. \end{cases} \tag{10.15}$$

In (10.13) we have a first order equation with respect to α_0', whose general solution is $\alpha_0(r) = c_1 + c_2 \ln r$; using the boundary conditions we deduce

$$\alpha_0(r) = \frac{a_0}{2} + \frac{A_0 - a_0}{2\ln R}\ln r.$$

In order to find the general solutions of the (Euler) equations in (10.14) and in (10.15) we search for a particular solution like r^k, where k has to be determined (see Exercise 4.4). A substitution gives

$$\left[k(k-1) + k - n^2\right]r^{k-2} = 0$$

and we deduce that $k = \pm n$. Hence, the general solution of the problems (10.14) and (10.15) is like $c_1 r^n + c_2 r^{-n}$. Using the boundary conditions and rearranging terms, we obtain

$$\alpha_n(r) = a_n K_n(r)r^{-n} + A_n H_n(r)\left(\frac{r}{R}\right)^n$$

$$\beta_n(r) = b_n K_n(r)r^{-n} + B_n H_n(r)\left(\frac{r}{R}\right)^n$$

where

$$H_n(r) = \frac{1 - r^{-2n}}{1 - R^{-2n}} \quad \text{and} \quad K_n(r) = \frac{1 - R^{-2n}r^{2n}}{1 - R^{-2n}}.$$

Finally, a substitution into (10.11) gives the solution of the problem.

Since the functions g and h are regular and periodic, the series associated to their Fourier expansions

$$\sum_{n=1}^{\infty}(|a_n| + |b_n|), \quad \sum_{n=1}^{\infty}(|A_n| + |B_n|)$$

are convergent; furthermore, we have that

$$H_n(r) \le \frac{1}{1 - R^{-1}}, \qquad K_n(r) \le \frac{1}{1 - R^{-1}}, \qquad \frac{r}{R} < 1, \qquad \text{and} \quad \frac{1}{r} < 1.$$

Hence, the series obtained in (10.11) is absolutely uniformly-convergent, and it can be differentiated under the series; then, (10.11) is the only solution of the problem.

Regarding the case $g = \sin\theta$ and $h = 1$, we point out that the functions f and g are already expanded in Fourier series, with the nonzero coefficients corresponding to $b_1 = 1$ and $A_0 = 2$, only. We deduce

$$a_0(r) = \frac{\ln r}{\ln R}, \qquad \beta_1(r) = \frac{R^2 - r^2}{(R^2 - 1)r}$$

while $\alpha_n(r) \equiv 0$ for $n \ge 1$ and $\beta_m(r) \equiv 0$ for $m \ge 2$. Then, the solution is

$$u(r, \theta) = \frac{\ln r}{\ln R} + \frac{R^2 - r^2}{(R^2 - 1)r} \sin\theta.$$

4.6. Let v be the solution of the problem $\Delta v = 0$ in B_1 such that $v = U$ on ∂B_1. This function exists, it is unique and can be explicitly written using the Poisson's formula (Section 4.3.3).

Consider the function $v(x, -y)$, which is harmonic in B_1 and on ∂B_1 attains the values of v, changed in sign.

Now, denote $w(x, y) = v(x, y) + v(x, -y)$; for the maximum principle $w \equiv 0$, indeed w solves the problem

$$\begin{cases} \Delta w(x, y) = 0 & \text{in } B_1 \\ w = 0 & \text{on } \partial B_1. \end{cases}$$

Hence, $v(x, y) = -v(x, -y)$, i.e. v is odd with respect to y, and, in particular $v(x, 0) = 0$; then, v solves the problem

$$\begin{cases} \Delta v(x, y) = 0 & \text{in } B_1^+ \\ v = u & \text{on } \partial B_1^+. \end{cases}$$

Since u is a solution of the same problem, uniqueness gives us that $v \equiv u \equiv U$ in B_1^+. Furthermore, since both v and U are odd functions with respect to y, we have that $v \equiv U$ in B_1 and then $\Delta U = 0$ in B_1.

4.7. a) For $r \le R$, let us denote

$$g(r) = \frac{1}{2\pi r} \int_{\partial B_r(\mathbf{x})} u(\boldsymbol{\sigma}) \, d\sigma.$$

Changing the variables and then differentiating, we can reproduce the first steps of the proof of the Theorem 4.2; we obtain

$$g'(r) = \frac{r}{2\pi} \int_{B_1(\mathbf{0})} \Delta u(\mathbf{x} + r\mathbf{y}) \, d\mathbf{y} \ge 0.$$

Hence g is a non decreasing function, and since $g(r) \to u(\mathbf{x})$ as $r \to 0$ the first inequality is proved. The second formula can be deduced from the first one with $R = r$ mutiplying by r and integrating both sides from 0 to R; we have

$$\frac{R^2}{2} u(\mathbf{x}) \geq \frac{1}{2\pi} \int_0^R dr \int_{\partial_r(\mathbf{x})} u(\boldsymbol{\sigma}) d\sigma = \frac{1}{2\pi} \int_{B_R(\mathbf{x})} u(\mathbf{y}) d\mathbf{y}$$

which implies the thesis.

b) We prove the statement for subharmonic functions. Suppose that for some $\mathbf{x}_0 \in \Omega$ we have

$$u(\mathbf{x}_0) = \sup_{x \in \Omega} u(\mathbf{x}).$$

Using the inequality proved in the point a) for a ball of radius r centered at $\mathbf{x}_0$ small enough to be all inside Ω, we have

$$u(\mathbf{x}_0) \leq \frac{1}{\pi r^2} \int_{B_r(\mathbf{x}_0)} u(\mathbf{y}) d\mathbf{y}$$

implying

$$\int_{B_r(\mathbf{x}_0)} (u(\mathbf{y}) - u(\mathbf{x}_0)) d\mathbf{y} \geq 0.$$

The integrand is a continuous function (by assumption) and it is not positive; hence,

$$u(\mathbf{y}) - u(\mathbf{x}_0) \equiv 0,$$

namely u is constant in $B_r(\mathbf{x}_0)$. This procedure can be repeated substituting to $\mathbf{x}_0$ any point of $B_r(\mathbf{x}_0)$. Now, since Ω is a domain, given any other point $\mathbf{y} \in \Omega$ we can determine a finite sequence of balls $B(\mathbf{x}_j) \subset\subset \Omega$ with $j = 0, \cdots, m$ such that $\mathbf{x}_i \in B(\mathbf{x}_{i-1})$ and $x_m = \mathbf{y}$. Using the same procedure, we show that u is constant in every ball, and the thesis follows from the arbitrariness of the point $\mathbf{y}$.

c) The functions u and u^2 are $C^\infty(\Omega)$. Using the formula:

$$\Delta(uv) = v\Delta u + u\Delta v + 2\nabla u \cdot \nabla v$$

with $u = v$ we deduce

$$\Delta(u^2) = 2u\Delta u + 2|\nabla u|^2 = 2|\nabla u|^2 \geq 0$$

hence u^2 is subharmonic.

d) Denote $w = F(u)$. We have that

$$w_x = F'(u)u_x, \quad w_{xx} = F''(u)u_x^2 + F'(u)u_{xx}$$

and the same rules hold for the y variable. We deduce that

$$\Delta u = F''(u)|\nabla u|^2 + F'(u)\Delta u.$$

If F is an increasing and convex function then the composition function w is subharmonic.

4.10. Hint. Write the mean formula in a ball $B_R(0)$ for u. Use the Schwarz inequality and let $R \to +\infty$.

4.12. Answer. We obtain

$$G(\mathbf{x}, \mathbf{y}) = -\frac{1}{2\pi}[\log|\mathbf{x} - \mathbf{y}| - \log(\frac{|\mathbf{x}|}{R}|\mathbf{x}^* - \mathbf{y}|)],$$

where $\mathbf{x}^* = R^2 \mathbf{x} |\mathbf{x}|^{-2}$, $\mathbf{x} \neq 0$.

4.13. Hint. (a) Let $B_r(\mathbf{x}) \subset \Omega$ and let w be the harmonic function in $\Omega \backslash \overline{B}_r(\mathbf{x})$ such that $w = 0$ on $\partial\Omega$ and $w = 1$ on $\partial B_r(\mathbf{x})$. Show that, for every r small enough,

$$G(\mathbf{x}, \cdot) > w(\cdot)$$

in $\Omega \backslash \overline{B}_r(\mathbf{x})$.

(b) For fixed $\mathbf{x} \in \Omega$, define $w_1(\mathbf{y}) = G(\mathbf{x}, \mathbf{y})$ and $w_2(\mathbf{y}) = G(\mathbf{y}, \mathbf{x})$. Apply Green's identity (4.32) in $\Omega \backslash B_r(\mathbf{x})$ to w_1 and w_2. Let $r \to 0$.

10.4 Section 5.6

5.1. This is a two-dimensional diffusion problem with linear reaction term. a) Denoting δ the Dirac distribution and $\rho = \sqrt{x^2 + y^2}$ the distance from the origin, the population density $P = P(x, y, t)$ solves the problem

$$\begin{cases} P_t - D\Delta P = aP & (x, y) \in \mathbb{R}^2,\ t > 0 \\ P(x, y, 0) = M\delta & (x, y) \in \mathbb{R}^2 \\ \lim_{\rho \to +\infty} P(x, y, t) = 0 & t > 0. \end{cases}$$

Using the substitution $P(x, y, t) = e^{at}u(x, y, t)$; we deduce:

$$P_t = ae^{at}u + e^{at}u_t \qquad \text{and} \qquad \Delta P = e^{at}\Delta u.$$

Therefore, u solves the problem

$$\begin{cases} u_t - D\Delta u = 0 & (x, y) \in \mathbb{R}^2,\ t > 0 \\ u(x, y, 0) = P(x, y, 0) = M\delta & (x, y) \in \mathbb{R}^2 \\ \lim_{\rho \to +\infty} u(x, y, t) = 0 & t > 0 \end{cases}$$

whose (positive autosimilar) solution is the fundamental solution (see Section 3.3.5) with $n = 2$:

$$u(x, y, t) = \frac{M}{4\pi Dt}e^{-(x^2+y^2)/4Dt}.$$

Going back to the original unknown, we deduce

$$P(x, y, t) = \frac{M}{4\pi Dt} e^{at - (x^2 + y^2)/4Dt}.$$

b) The evolution of the total number of individuals can be found solving the integral:

$$
\begin{aligned}
M(t) &= \int_{\mathbb{R}^2} P(x, y, t)\, dx\, dy \\
&= \frac{M}{4\pi Dt} e^{at} \int_{\mathbb{R}^2} e^{-(x^2 + y^2)/4Dt}\, dx\, dy \\
&= \frac{M}{4\pi Dt} e^{at} \int_0^{2\pi} d\theta \int_0^\infty e^{-\rho^2/4Dt}\, \rho\, d\rho = M e^{at}
\end{aligned}
$$

hence the population increases exponentially.

c) Individuals who establish themselves in the rural area are equal to that of the original core. We have that

$$
\begin{aligned}
M &= \int_{\mathbb{R}^2 \setminus B_{R(t)}} P(x, y, t)\, dx\, dy \\
&= \frac{M}{4\pi Dt} e^{at} \int_{\mathbb{R}^2 \setminus B_{R(t)}} e^{-(x^2 + y^2)/4Dt}\, dx\, dy \\
&= \frac{M}{4\pi Dt} e^{at} \int_0^{2\pi} d\theta \int_{R(t)}^\infty e^{-\rho^2/4Dt}\, \rho\, d\rho = M e^{at - R^2(t)/4Dt}
\end{aligned}
$$

and we deduce

$$R(t) = 2t\sqrt{aD}.$$

The metropolitan front, therefore, moves with constant speed equal to $2\sqrt{aD}$.

5.2. a) We use the change of variables

$$u = \omega e^{at}$$

therefore

$$u_t = \omega_t e^{at} + \omega a e^{at} \quad \text{and} \quad u_{xx} = \omega_{xx} e^{at}.$$

Substituting into the equation we find that ω fulfills the problem

$$
\begin{cases}
\omega_t - \omega_{xx} = 0 & 0 < x < 1,\ t > 0 \\
\omega(x, 0) = g(x) & 0 \le x \le 1 \\
\omega_x(0, t) = \omega_x(1, t) = 0 & t > 0.
\end{cases}
$$

Using the separation of the variables we have $\omega(x, t) = X(x)T(t)$ and

$$\frac{T'}{T} = \frac{X''}{X} = \lambda.$$

The equation $X'' - \lambda X = 0$ is associated to the homogeneous Neumann problem and has nontrivial solution if and only if $\lambda \leq 0$. The eigenvalues of the problem are $\lambda_k = -k^2\pi^2$ with $k \in \mathbb{N}$ whose corresponding eigenfunctions are

$$X_k = a_k \cos k\pi x.$$

The equation for the variable t is $T' + k^2\pi^2 T = 0$ And we deduce that its solution is $T = e^{-k^2\pi^2 t}$. We obtain

$$\omega = \sum_{k=0}^{\infty} a_k e^{-k^2\pi^2 t} \cos k\pi x$$

where a_k has to be determined using the initial condition. In particular

$$x^2 - \frac{1}{3} = \sum_{k=0}^{\infty} a_k \cos k\pi x$$

and a_k are the (even) Fourier coefficients of the function $x^2 - 1/3$. We deduce

$$a_0 = 0$$

$$a_k = 2 \int_0^1 \left(x^2 - \frac{1}{3} \right) \cos k\pi x \, dx = \frac{4(-1)^k}{k^2\pi^2}.$$

The solution of the initial problem is

$$u(x,t) = \frac{4}{\pi^2} \sum_{k=1}^{\infty} \frac{(-1)^k}{k^2} e^{(a-k^2\pi^2)t} \cos k\pi x.$$

b) As $t \to +\infty$, we have

$$u \sim -\frac{4}{\pi^2} e^{(a-\pi^2)t} \cos \pi x,$$

and we deduce that

$$\begin{cases} \text{if } a < \pi^2 \ u \to 0 \\ \text{if } a = \pi^2 \ u \sim -\dfrac{4}{\pi^2} \cos \pi x \\ \text{if } a > \pi^2 \ u \to +\infty. \end{cases}$$

10.5 Section 6.9

6.1. The Cauchy-Dirichlet problem has homogeneous boundary conditions so the separation of variables technique can be used. We search for solutions of the form

$$U(x,t) = w(t) v(x).$$

Substituting into the equation we find

$$0 = U_{tt} - c^2 U_{xx} = w''(t) v(x) - c^2 w(t) v''(x)$$

hence, we get to an equality among two functions of a different variable

$$\frac{1}{c^2} \frac{w''(t)}{w(t)} = \frac{v''(x)}{v(x)} \qquad (10.16)$$

therefore both sides are equal to the same constant λ. We deduce two ordinary differential equations

$$v''(x) - \lambda v(x) = 0 \qquad \text{and} \qquad w''(t) - \lambda c^2 w(t) = 0. \qquad (10.17)$$

The problem for v is associated to the boundary conditions

$$v(0) = v(1) = 0 \qquad (10.18)$$

and the general solution of the first of (10.17) depends on the sign of λ. Indeed, if $\lambda = 0$, the general solution is $v(x) = A + Bx$ and the conditions (10.18) imply $A = B = 0$. If $\lambda = \mu^2 > 0$, the general solution is $v(x) = Ae^{-\mu x} + Be^{\mu x}$ and, again, the conditions (10.18) imply $A = B = 0$. On the other hand, we deduce nontrivial solution if, finally, $\lambda = -\mu^2 < 0$; the general solution is $v(x) = A \sin \mu x + B \cos \mu x$. The conditions (10.18) become

$$v(0) = B = 0$$
$$v(1) = A \sin \mu L + B \cos \mu L = 0$$

and we deduce

$$A \text{ free, } B = 0, \ \mu L = m\pi, \ m = 1, 2, \dots .$$

Therefore nontrivial solutions are

$$v_m(x) = A_m \sin \mu_m x, \qquad \mu_m = \frac{m\pi}{L}.$$

Using the eigenvalues $\lambda = -\mu_m^2 = -m^2\pi^2/L^2$, the second of (10.17) has the general solution

$$w_m(t) = C_m \cos(\mu_m ct) + D_m \sin(\mu_m ct).$$

Therefore, we obtain solutions like

$$U_m(x,t) = [a_m \cos(\mu_m ct) + b_m \sin(\mu_m ct)] \sin \mu_m x$$

where a_m and b_m are free. Each of these functions represents a possible movement of the string, known as m^{th}−component of the vibration or m^{th}−harmonic, corresponding to the vibration of frequency $mc/2L$. The lowest frequency, corresponding to $m = 1$ is called fundamental, while the other frequencies are integer multiples of the fundamental one: it is for this property that the string is capable of producing tones of good musical quality (this is not a vibrant membrane, see Section 6.6.3).

If the datum g is exactly

$$g(x) = a_m \sin \mu_m x$$

then the solution of the problem coincide with U_m and the string vibrates in its mth mode of oscillation. In the general case, the idea is to build a solution by superposition of the infinite harmonic U_m by the formula

$$u(x,t) = \sum_{m=1}^{\infty} [a_m \cos(\mu_m ct) + b_m \sin(\mu_m ct)] \sin \mu_m x. \tag{10.19}$$

The initial conditions require

$$u(x,0) = \sum_{m=1}^{\infty} a_m \sin \mu_m x = g(x) \tag{10.20}$$

$$u_t(x,0) = \sum_{m=1}^{\infty} c\mu_m b_m \sin \mu_m x = 0 \tag{10.21}$$

for $0 \le x \le L$.

We deduce $b_m = 0$ for every $m \ge 1$. Assuming that g can be expanded in sine Fourier series in the interval $[0, L]$, we consider

$$\hat{g}_m = \frac{2}{L} \int_0^L g(x) \sin\left(\frac{m\pi}{L}x\right) dx \qquad m \ge 1$$

the Fourier coefficients associated to g. In the formula (10.19) we choose

$$a_m = \hat{g}_m, \text{ and } b_m = 0, \tag{10.22}$$

hence (10.19) fulfill the conditions (10.20), (10.21) and it is the candidate to solve the problem assigned.

Note that U_m is a standing wave

$$\hat{g}_m \cos(\mu_m ct) \sin \mu_m x$$

and u is a superposition of such sinusoidal vibrations with increasing frequency.

If the coefficients a_m tend to zero fast enough as $m \to \infty$, for instance if

$$|\hat{g}_m| \le \frac{C}{m^4}, \tag{10.23}$$

is not difficult to check that you can differentiate twice term by term, and deduce that u is indeed the (classical) solution of the our problem.

The formulas (10.19) and (10.22) indicate that the vibration of the string consists of the superposition of those harmonics whose amplitude corresponds to the Fourier coefficients associated to the initial data. The presence (or absence) of various harmonics gives to the sound emitted by a string a particular "timbre", In contrast to the "pure tone" produced by an electronic instrument, corresponding to a single frequency.

6.2. We associate to the Tricomi equation the quadratic form (6.36) in Section 6.5.1

$$H(p,q) = p^2 - tq^2$$

and we deduce that the equation is hyperbolic in the half plane $t > 0$ (parabolic at $t = 0$), and elliptic in the half plane $t < 0$, analogously with the types of conic section $H(p,q) = 1$.

For $t < 0$ the characteristics are not real. For $t = 0$ we have the family of characteristics $\phi(x,t) = k$ solution of the ordinary differential equation

$$\frac{dx}{dt} = 0.$$

We obtain $\phi(x,t) = x$, therefore the characteristics and the set $t = 0$ intersect at a single point, and the set $t = 0$ has no interior points and is not, indeed, characteristic.

Considering the case $t > 0$, we have two families of characteristics $\phi(x,t) = k$ and $\psi(x,t) = k$ solutions of the ordinary differential equation

$$\left(\frac{dx}{dt}\right)^2 - t = 0,$$

that is $\frac{dx}{dt} = \pm\sqrt{t}$. We find immediately that the characteristics correspond to level lines of the surfaces

$$\xi = \phi(x,t) = 3x + t^{3/2} \quad \text{and} \quad \eta = \psi(x,t) = 3x - t^{3/2}$$

in the half plane $t > 0$.

6.3. The assigned equation is elliptic in the half plane $y < 0$, hyperbolic in the half plane $y > 0$ and parabolic for $y = 0$.

The characteristics are real only if $y \leq 0$; in the parabolic case the characteristics are the constants $y = k$, and, among them, just the straight line $y = 0$ belongs to the characteristic set.

In the half plane $y > 0$ we search for two functions $\phi(x,y) = k$ and $\psi(x,y) = k$ solving the characteristic equation (6.45), in this case it is

$$\left(\frac{dy}{dx}\right)^2 - y = 0$$

namely $y' = \pm\sqrt{y}$. Using the separation of variables we obtain

$$2\sqrt{y} = \pm x + \text{constant}.$$

Then, the characteristics correspond to the level lines of the surfaces

$$\begin{cases} \xi = \phi(x,y) = 2\sqrt{y} + x \\ \eta = \psi(x,y) = 2\sqrt{y} - x. \end{cases}$$

In order to write the equation in canonical form we change the variables as written above and exploit the conditions:

$$\xi_x = 1 \qquad \xi_y = y^{-1/2} \qquad \eta_x = -1 \qquad \eta_y = y^{-1/2}.$$

For the solution $u(x,y) = U(\xi(x,y), \eta(x,y))$, we deduce

$$u_x = U_\xi \xi_x + U_\eta \eta_x = U_\xi - U_\eta$$

$$u_y = U_\xi \xi_y + U_\eta \eta_y = \frac{U_\xi + U_\eta}{\sqrt{y}}$$

$$u_{xx} = U_{\xi\xi}\xi_x + U_{\xi\eta}\eta_x - U_{\eta\xi}\xi_x - U_{\eta\eta}\eta_x = U_{\xi\xi} - 2U_{\xi\eta} + U_{\eta\eta}$$

$$u_{yy} = -\frac{1}{2}y^{-3/2}(U_\xi + U_\eta) + \frac{1}{\sqrt{y}}(U_{\xi\xi}\xi_y + U_{\xi\eta}\eta_y + U_{\eta\xi}\xi_y + U_{\eta\eta}\eta_y)$$

$$= -\frac{U_\xi + U_\eta}{2y^{3/2}} + \frac{1}{y}(U_{\xi\xi} + 2U_{\xi\eta} + U_{\eta\eta}).$$

A substitution into the equation $u_{xx}x - yu_{yy} - \frac{1}{2}u_y = 0$ leads to the canonical form:

$$-4U_{\eta\xi} = 0$$

and we deduce

$$U(\xi, \eta) = f(\xi) + g(\eta)$$

where f and g are arbitrary functions. Back to the original variables, the general solution of the equation is

$$u(x,y) = f(2\sqrt{y} + x) + g(2\sqrt{y} - x).$$

6.4. The equation assigned is parabolic. Then, there exists a family of characteristics $\phi(x,t) = k$ satisfying the differential condition

$$t^2\left(\frac{dx}{dt}\right)^2 + 2t\frac{dx}{dt} + 1 = 0$$

namely

$$t\frac{dx}{dt} - 1 = 0.$$

Integrating, we find $x = \log|t| + \text{costant}$, i.e. $te^{-x} = k$, $k \in \mathbb{R}$. Hence, $\phi(x,t) = te^{-x}$.

Now that the function ϕ has been found, in order to change the variables to rewrite the equation in canonical form, we consider a smooth function ψ, that we will choose later, in a way that $\nabla\phi$ and $\nabla\psi$ are independent and that

$$t^2\psi_t^2 + 2t\psi_t\psi_x + \psi_x^2 = A \neq 0.$$

For the sake of simplicity, we consider $\psi = \psi(x)$, with positive first derivative. We change the coordinates[9]

$$\begin{cases} \xi = te^{-x} \\ \eta = \psi(x) \end{cases}$$

and deduce that

$$\xi_x = -te^{-x} \qquad \eta_x = \psi' \qquad \xi_t = e^{-x} \qquad \eta_t = 0.$$

Consider $U(\xi, \eta)$ the function such that $u(x,t) = U(te^{-x}, \psi(x))$; we obtain

$$u_x = -te^{-x}U_\xi + U_\eta\psi'$$
$$u_t = e^{-x}U_\xi$$
$$u_{xx} = te^{-x}U_\xi + t^2e^{-2x}U_{\xi\xi} - 2te^{-x}\psi'U_{\eta\xi} + (\psi')^2U_{\eta\eta} + \psi''U_\eta$$
$$u_{x,t} = -e^{-x}U_\xi - te^{-2x}U_{\xi\xi} + e^{-x}\psi'U_{\eta\xi}$$
$$u_{tt} = e^{-2x}U_{\xi\xi}.$$

Substituting into the equation $t^2u_{tt} + 2tu_{xt} + u_{xx} - u_x = 0$ gives

$$(\psi')^2U_{\eta\eta} + (\psi'' - \psi')U_\eta = 0$$

the canonical form of the equation. We choose

$$\psi(x) = e^x$$

hence the second coefficient vanishes and we have $U_{\eta\eta} = 0$; an integration gives

$$U = f(\xi) + \eta g(\xi)$$

where f and g are arbitrary functions. Back to the original variables we obtain the general solution of the equation

$$u(x,t) = f(te^{-x}) + e^x g(te^{-x}).$$

6.6. Answer: we find

$$u(x,t) = \frac{2}{\pi^2}\sum_{k=1}^{\infty}\frac{(-1)^k}{k^2}\sin k\pi t \sin k\pi x.$$

Using the Weierstrass criterion the series defining u is uniformly convergent and therefore u is continuous.

The energy associated to the string is

$$E(t) = \frac{1}{2}\left(\int_0^1 (|u_t(x,t)|^2 + |u_x(x,t)|^2)dx\right.$$

---

[9] The condition $\psi' > 0$ ensures the local invertibility of the transformation.

It is constant and the value can be determined using the initial conditions.

6.7. Answer. We find

$$u(x,t) = \sin x \int_0^t g(t-\tau)\sin \tau d\tau.$$

6.8. Answer. The solution is

$$u(r,t) = \sum_{n=1}^{\infty} a_n J_0(\lambda_n r)\cos \lambda_n t$$

where J_0 is the Bessel function of order zero, $\lambda_1, \lambda_2, \cdots$ are the zeroes of J_0 and the coefficients a_n are given by

$$a_n = \frac{2}{c_n^2}\int_0^1 s\,g(s)J_0(\lambda_n s)ds$$

where

$$c_n = \sum_{n=1}^{\infty}\frac{(-1)^k}{k!(k+1)!}\left(\frac{\lambda_n}{2}\right)^{2k+1}.$$

6.10. Answer. a) we have $\alpha = k/2$ and therefore

$$v_{tt} - \frac{k^2}{4}v = c^2\Delta v.$$

b) For $\beta = k/2c$ we have

$$w_{tt} = c^2\Delta_{\mathbf{x},x_3}w.$$

10.6 Section 7.7

7.1. We have

$$\|\mathbf{Ax}\|^2 = \langle \mathbf{Ax}, \mathbf{Ax}\rangle = \langle \mathbf{A}^\top\mathbf{Ax}, \mathbf{x}\rangle.$$

The dimension of the matrix $\mathbf{A}^\top\mathbf{A}$ is (n,n); furthermore A is symmetric and non negative. Now,

$$\sup_{\|\mathbf{x}\|\leq 1}\langle \mathbf{A}^\top\mathbf{Ax}, \mathbf{x}\rangle = \sup_{\|\mathbf{x}\|=1}\langle \mathbf{A}^\top\mathbf{Ax}, \mathbf{x}\rangle = \Lambda_M$$

where Λ_M is the maximum eigenvalue of $\mathbf{A}^\top\mathbf{A}$. Therefore $\|L\| = \sqrt{\Lambda_M}$.

7.2. Schwarz's inequality yields

$$|L_g f| = \left|\int_\Omega fg\right| \leq \left(\int_\Omega |f|^2\right)^{1/2}\left(\int_\Omega |g|^2\right)^{1/2} = \|g\|_0\,\|f\|_0$$

so that $L_g \in L^2(\Omega)^*$ and $\|L_g\| \leq \|g\|_0$. Actually $\|L_g\| = \|g\|_0$ since, choosing $f = g$, we have

$$\|g\|_0^2 = L_g(g) \leq \|L_g\| \|g\|_0$$

whence also $\|L_g\| \geq \|g\|_0$.

7.3. Schwarz's inequality yields

$$|(x,y)| \leq \|x\| \|y\|,$$

whence $L_1 \in H^*$ and $\|L_1\| \leq \|y\|$. Actually $\|L_1\| = \|y\|$ since, choosing $x = y$, we have

$$\|y\|^2 = |L_1 y| \leq \|L_1\| \|y\|,$$

or $\|L_1\| \geq \|y\|$. Observe that this argument provides the following alternative definition of the norm of an element $y \in H$:

$$\|y\| = \sup_{\|x\|=1} (x,y). \qquad (10.24)$$

7.4. Since $u \in L^2(0,1)$, necessarily $u \in L^2(0,1/2)$. The Schwarz's inequality gives

$$\left| \int_0^{1/2} u(t)\, dt \right| \leq \int_0^{1/2} |u(t)|\, dt \leq \frac{\sqrt{2}}{2} \left(\int_0^1 |u(t)|^2 dt \right)^{1/2} = \frac{\sqrt{2}}{2} \|u(t)\|_{L^2(0,1)}$$

and we deduce that the functional is well-defined. F is bounded, hence continuous. Then, since the Riesz theorem, there exists a unique function $f \in H$ such that

$$F(u) = \int_0^1 f(t)u(t)dt$$

and $\|F\|_{H'} = \|f\|_H$. We have that

$$f = \begin{cases} 1 & \text{if } 0 < x < 1/2 \\ 0 & \text{if } 1/2 < x < 1. \end{cases}$$

Finally, we have

$$\|F\|_{H'} = \sup_{\|u\|_{L^2} \leq 1} |F(u)| = \frac{\sqrt{2}}{2}.$$

7.5. We show three sequences converging to δ in $D'(\mathbb{R}^n)$.

a) Denote with $\chi_E(x)$ the characteristic function of the set $E \subset \mathbb{R}^n$. This function is identically 1 if $x \in E$ and is zero if $x \notin E$. We consider B_r the ball of radius r centered at the origin, we prove that

$$\lim_{r \to 0} \frac{1}{|B_r|} \chi_{B_r} = \delta \qquad \text{in } D'(\mathbb{R}^n).$$

We have to show that for every φ in $\mathcal{D}(\mathbb{R}^n)$,

$$\int_{\mathbb{R}^n} \frac{1}{|B_r|} \chi_{B_r}(\mathbf{x}) \varphi(\mathbf{x}) \, d\mathbf{x} \to \varphi(\mathbf{0}), \qquad \text{as } r \to 0.$$

In fact, using the Integral Mean Value Theorem, we obtain

$$\int_{\mathbb{R}^n} \frac{1}{|B_r|} \chi_{B_r}(\mathbf{x}) \varphi(\mathbf{x}) \, d\mathbf{x} = \frac{1}{|B_r|} \int_{B_r} \varphi(\mathbf{x}) \, d\mathbf{x} = \varphi(\mathbf{x}_r)$$

where $\mathbf{x}_r$ is a point in B_r. Since φ is continuous, $\varphi(\mathbf{x}_r) \to \varphi(\mathbf{0})$ as $\mathbf{x}_r \to 0$.

b) Consider the function η_ε

$$\eta_\varepsilon = \frac{1}{\varepsilon^n} \eta \left(\frac{|\mathbf{x}|}{\varepsilon} \right)$$

where

$$\eta(\mathbf{x}) = \begin{cases} c e^{\frac{1}{|\mathbf{x}|^2 - 1}} & 0 \le |\mathbf{x}| < 1 \\ 0 & |\mathbf{x}| \ge 1. \end{cases}$$

We prove that

$$\lim_{\varepsilon \to 0} \eta_\varepsilon = \delta \qquad \text{in } \mathcal{D}'(\mathbb{R}^n).$$

In fact, we take φ in $\mathcal{D}(\mathbb{R}^n)$ and deduce

$$\int_{\mathbb{R}^n} \eta_\varepsilon(\mathbf{x}) \varphi(\mathbf{x}) \, d\mathbf{x} = \frac{1}{\varepsilon^n} \int_{B_\varepsilon} \eta \left(\frac{\mathbf{x}}{\varepsilon} \right) \varphi(\mathbf{x}) \, d\mathbf{x}$$

$$\underset{\mathbf{x}=\varepsilon\mathbf{y}}{=} \int_{B_1} \eta(\mathbf{y}) \varphi(\varepsilon\mathbf{y}) \, d\mathbf{y}.$$

Using the Lebesgue's Dominated Convergence Theorem (Section 7.1.4) we deduce

$$\lim_{\varepsilon \to 0} \int_{B_1} \eta(\mathbf{y}) \varphi(\varepsilon\mathbf{y}) \, d\mathbf{y} = \varphi(\mathbf{0}).$$

c) Consider $\Gamma_D(\mathbf{x}, t)$ the fundamental solution of the heat equation; we have already seen that for $\varphi \in \mathcal{D}(\mathbb{R}^n)$, then

$$\lim_{t \to 0^+} \int_{\mathbb{R}^n} \Gamma_D(\mathbf{y} - \mathbf{x}, t) \varphi(\mathbf{x}) \, d\mathbf{x} = \varphi(\mathbf{y}).$$

In terms of distributions, this fact means that for $\mathbf{y}$ fixed,

$$\Gamma_D(\mathbf{y} - \cdot, t) \to \delta_{\mathbf{y}} \qquad \text{in } \mathcal{D}'(\mathbb{R}^n)$$

as $t \to 0^+$.

7.6. Take $\varphi \in \mathcal{D}(\mathbb{R})$. We deduce

$$\langle u', \varphi \rangle = -\langle u, \varphi' \rangle = -\int_{\mathbb{R}} |x| \, \varphi'(x) \, dx = \int_{-\infty}^{0} x\varphi'(x) \, dx - \int_{0}^{\infty} x\varphi'(x) \, dx$$

$$= -\int_{-\infty}^{0} \varphi(x) \, dx + \int_{0}^{\infty} \varphi(x) \, dx = \int_{\mathbb{R}} \mathrm{sign}(x)\varphi(x) \, dx = \langle S, \varphi \rangle.$$

Furthermore,

$$\langle \frac{d}{dt} \,\mathrm{sign}\, t, \varphi \rangle = -\langle \mathrm{sign}\, t, \varphi' \rangle$$

$$= -\int_{\mathbb{R}} \mathrm{sign}\, t\varphi'(t) \, dt = -\int_{-a}^{a} \mathrm{sign}\, t\varphi'(t) \, dt \quad \text{with } \mathrm{supp}\varphi \subset [-a, a]$$

$$= \int_{-a}^{0} \varphi'(t) \, dt - \int_{0}^{a} \varphi'(t) \, dt = 2\phi(0) = \langle 2\delta, \varphi \rangle.$$

7.7. We have that

$$\langle \frac{d}{dx} \arctan \frac{1}{x}, \varphi \rangle = -\langle \arctan \frac{1}{x}, \varphi' \rangle = -\int_{\mathbb{R}} \arctan \frac{1}{x} \varphi'(x) \, dx$$

$$= -\int_{-a}^{a} \arctan \frac{1}{x} \varphi'(x) \, dx \quad \text{con } \mathrm{supp}\varphi \subset [-a, a]$$

$$= -\int_{-a}^{0} \arctan \frac{1}{x} \varphi'(x) \, dx - \int_{0}^{a} \arctan \frac{1}{x} \varphi'(x) \, dx$$

$$= \left[\arctan \frac{1}{x} \varphi \right]_{-a}^{0^-} - \left[\arctan \frac{1}{x} \varphi \right]_{0^+}^{a} + \int_{-a}^{a} \frac{\varphi(x)}{1 + x^2} dx$$

$$= \pi\varphi(0) - \int_{\mathbb{R}} \frac{\varphi(x)}{1 + x^2} dx = \langle \pi\delta - \frac{1}{1 + x^2}, \varphi \rangle.$$

7.8. Using the definition of derivative in the distributional sense, with $\varphi \in \mathcal{D}(\mathbb{R}^2)$ we find

$$\langle \mathcal{H}_{xy}, \varphi \rangle = \langle \mathcal{H}, \varphi_{xy} \rangle = \int_{\mathbb{R}^2} \mathcal{H}(x, y)\varphi_{xy} \, dx \, dy$$

$$= \int_{[-a,a] \times [-a,a]} \mathcal{H}(x, y)\varphi_{xy} \, dx \, dy \quad \text{with } \mathrm{supp}\varphi \subset [-a, a] \times [-a, a]$$

$$= \int_{0}^{a} \int_{0}^{a} \varphi_{xy} \, dx \, dy = \int_{0}^{a} [\varphi_x]_{0}^{a} \, dx$$

$$= -\int_{0}^{a} \varphi_x(x, 0) \, dx = -[\varphi(x,0)]_{0}^{a} = \varphi(0,0) = \langle \delta, \varphi \rangle.$$

7.9. First, we search for an orthonormal basis $\{u_1, u_2\}$ of the subspace $V \subset L^2(0, 1)$, which has dimension 2. As a first vector u_1 we take v_1 itself. In order to find the second vector u_2, we choose $\alpha, \beta \in \mathbb{R}$ such that $v = \alpha + \beta x$ has the following properties:

$$\begin{cases} v \perp v_1 \\ \|v\| = 1 \end{cases}$$

namely, in terms of inner product in L^2

$$\begin{cases} \int_0^1 (\alpha + x\beta)\, dx = 0 \\ \int_0^1 (\alpha + x\beta)^2\, dx = 1. \end{cases}$$

We find

$$\begin{cases} \alpha + \dfrac{\beta}{2} = 0 \\ \alpha^2 + \alpha\beta + \dfrac{\beta^2}{3} = 1 \end{cases}$$

we deduce two pairs of solutions $(\alpha, \beta) = \pm(\sqrt{3}, -2\sqrt{3})$. We consider $u_2 = \sqrt{3}(1 - 2x)$.

With an orthonormal basis of V, the projection of x^2 on V is exactly

$$P_V x^2 = (x^2, u_1)u_1 + (x^2, u_2)u_2.$$

We deduce that

$$\begin{cases} (x^2, u_1) = \displaystyle\int_0^1 x^2\, dx = \dfrac{1}{3} \\ (x^2, u_2) = \sqrt{3} \displaystyle\int_0^1 x^2(1 - 2x)\, dx = -\dfrac{1}{2\sqrt{3}} \end{cases}$$

and we immediately deduce that $P_v x^2 = -1/6 + x$.

Changing point of view, if we search for the function in V with the minimum distance form x^2 we have to solve

$$\inf_{\lambda_1, \lambda_2} \sqrt{\int_0^1 (x^2 - \lambda_1 - \lambda_2 x)^2 dx}.$$

This problem is equivalent to find the infimum of the corresponding square value. We have to minimize the function

$$\mathcal{U} = \lambda_1^2 + \frac{\lambda_2^2}{3} - \frac{2}{3}\lambda_1 - \frac{\lambda_2}{2} + \lambda_1\lambda_2 + \frac{1}{5}$$

whose stationary points solve the system

$$\begin{cases} 2\lambda_1 + \lambda_2 = \dfrac{2}{3} \\ \lambda_1 + \dfrac{2}{3}\lambda_2 = \dfrac{1}{2} \end{cases}$$

we deduce $\lambda_1 = -1/6$ and $\lambda_2 = 1$. The infimum coincide with function $P_V x^2$.

We can change again our point of view. In order to find the projection of x^2 on the vector space of the straight lines we can exploit the second thesis of Theorem 7.8, and proceed searching for the orthogonal complement of the projection

$$x^2 = P_V x^2 + u \qquad \text{where } u \perp V$$

from which $u = x^2 - P_V x^2 = x^2 - (a + bx) \perp v$ for every $v \in V$. The orthogonality condition in $L^2(0,1)$ becomes

$$\int_0^1 [x^2 - (a + bx)](\lambda_1 + \lambda_2 x)\, dx = 0 \qquad \forall \lambda_1, \lambda_2 \in \mathbb{R}.$$

Solving the integral we are yield to

$$\lambda_1 \left[\frac{1}{3} - a - \frac{b}{2}\right] + \lambda_2 \left[\frac{1}{4} - \frac{a}{2} - \frac{b}{3}\right] = 0 \qquad \forall \lambda_1, \lambda_2 \in \mathbb{R}$$

and, one again, we find the solution $a = -1/6$ and $b = 1$.

7.10. Note that since t and t^2 are odd and even respectively, they are orthogonal in $L^2(-1,1)$. We have

$$\|t\|^2_{L^2(-1,1)} = \int_{-1}^1 t^2\, dt = \frac{2}{3}$$

$$\|t^2\|^2_{L^2(-1,1)} = \int_{-1}^1 t^4\, dt = \frac{2}{5}.$$

The functions

$$e_1 = \sqrt{\frac{3}{2}}\, t \quad \text{and} \quad e_2 = \sqrt{\frac{5}{2}}\, t^2$$

are therefore an orthonormal basis of V. The projection of e^t can be found using the Fourier series:

$$P_V e^t = (e^t, e_1)e_1 + (e^t, e_2)e_2.$$

Integration by parts gives

$$\int te^t\, dt = e^t(t - 1)$$

$$\int t^2 e^t\, dt = e^t(t^2 - 2t + 2)$$

and then

$$(e^t, e_1) = \sqrt{\frac{3}{2}}\, [e^t(t - 1)]_{-1}^1 = e^{-1}\sqrt{6}$$

$$(e^t, e_2) = \sqrt{\frac{5}{2}}\, [e^t(t^2 - 2t + 2)]_{-1}^1 = \sqrt{\frac{5}{2}}(e - 5e^{-1}).$$

Hence,

$$P_V e^t = 3e^{-1}t + \frac{5}{2}(e - 5e^{-1})t^2.$$

7.11. We recall that the product of a distribution $u \in \mathcal{D}'(\mathbb{R})$ by a function $\psi \in C^\infty(\mathbb{R})$ is

$$\langle u\psi, \varphi \rangle = \langle u, \psi\varphi \rangle \qquad \forall \varphi \in \mathcal{D}(\mathbb{R}).$$

In this case the Leibniz rule for the derivative of the product can be used, namely

$$\langle (u\psi)', \varphi \rangle = \langle u'\psi, \varphi, \rangle + \langle u\psi', \varphi \rangle.$$

We deduce that $x\mathcal{H}(x)$ is a primitive of $\mathcal{H}(x)$, indeed, $(x\mathcal{H}(x))' = \mathcal{H}(x) + x\delta$ and we know that $x\delta = 0$; hence $(x - a)\mathcal{H}(x - a)$ is a primitive of $\mathcal{H}(x - a)$, for every $a \in \mathbb{R}$. Therefore, integrating once our equation we have

$$u' = C - \mathcal{H}(x - 1/2) - \mathcal{H}(x + 1/2),$$

and integrating twice we are yield to

$$u = Cx + B - (x - 1/2)\mathcal{H}(x - 1/2) - (x + 1/2)\mathcal{H}(x + 1/2).$$

The boundary conditions require $C = B = 1$. The solution is represented in Fig. 10.9.

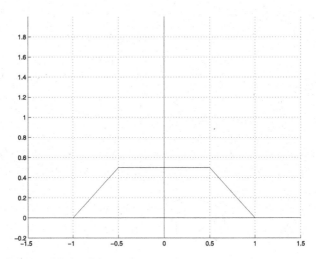

Fig. 10.9. Solution of the Exercise 7.11

7.12. The function fulfill the boundary conditions. Calculations of its derivatives (in distributional sense) gives

$$y' = -\frac{\cosh \sqrt{k}(1 - |x|)}{2 \cosh \sqrt{k}} \operatorname{sign} x$$

$$y'' = \frac{\sinh \sqrt{k}(1 - |x|)}{2 \cosh \sqrt{k}} \operatorname{sign}^2 x - \frac{\cosh \sqrt{k}(1 - |x|)}{\cosh \sqrt{k}} \delta$$

$$= k\frac{\sinh \sqrt{k}(1 - |x|)}{2\sqrt{k} \cosh \sqrt{k}} - \delta = ky - \delta$$

using the property $v\delta = v(0)\delta$ with $v \in C^\infty(\mathbb{R})$. Furthermore, if k tends to 0, we find that

$$y \longrightarrow \frac{1 - |x|}{2}$$

as we expect with vanishing elastic force.

7.15. Answer. $p(x) = a_0 L_0(x) + a_1 L_1(x) + \ldots + a_n L_n(x)$, where L_n is the $n - th$ Legendre polynomial and $a_n = (n + 1/2)(f, L_n)_{L^2(-1,1)}$.

7.17. Answer. $u(r, \varphi) = \sum_{n=0}^\infty a_n r^n L_n(\cos \varphi)$, where L_n is the $n - th$ Legendre polynomial and

$$a_n = \frac{2n + 1}{2} \int_{-1}^1 g\left(\cos^{-1} x\right) L_n(x)\, dx.$$

At a certain point, the change of variable $x = \cos \varphi$ is required.

7.18. Hint.

$$u(r, \theta, t) = \sum_{p,j=0}^\infty J_p(\alpha_{pj} r)\{A_{pj} \cos p\theta + B_{pj} \sin p\theta\} \cos(\sqrt{\alpha_{pj}} t)$$

where the coefficients A_{pj} and B_{pj} are determined by the expansion of the initial condition $u(r, \theta, 0) = g(r, \theta)$.

10.7 Section 8.9

8.1. It is a mixed problem with non homogeneous Dirichlet boundary condition. First, we consider an extension of the data at the point $x = -1$; the simplest extension is the constant. Now, we consider the problem for the function $w = u - 1/2$, we have

$$\begin{cases} -w'' = -u'' = 5x - 1 & -1 < x < 2 \\ w(-1) = 0 \\ w'(2) = 2. \end{cases} \tag{10.25}$$

The correct functional setting is

$$V = H^1_{0,-1}(-1,2)$$

i.e. the set of functions in $H^1(-1,2)$ (which are continuous) vanishing in -1; we exploit also the Poincaré inequality and take

$$\|v\|_V = \|v'\|_{L^2(-1,2)}.$$

We multiply the first equation of the system by $v \in V$ and integrate by parts, we find

$$\int_{-1}^2 w'v' \, dx - [w'v]^2_{-1} = \int_{-1}^2 (5x - 1)v \, dx$$

that is, taking into account the boundary conditions

$$\int_{-1}^2 w'v' \, dx = \int_{-1}^2 (5x - 1)v \, dx + 2v(2). \qquad (10.26)$$

We introduce the functional $F : V \longrightarrow \mathbb{R}$:

$$Fv = \int_{-1}^2 (5x - 1)v \, dx + 2v(2)$$

where $v \in V$. From (10.26) we deduce the variational formulation

$$\textit{find } w \in V \textit{ such that } \langle w, v \rangle = Fv \textit{ for every } v \in V. \qquad (10.27)$$

Since the left hand side of (10.27) is an inner product, we can use the Riesz theorem.

In order to prove the continuity of the functional F we use a "trace" inequality for the (continuous) functions of $H^1_{0,x_1}(x_1, x_2)$, namely

$$|v(y)| \le (x_2 - x_1)^{1/2}\|v'\|_{L^2(x_1,x_2)} \qquad \forall y \in [x_1, x_2]. \qquad (10.28)$$

In fact, for the fundamental theorem if integral calculus, we have

$$|v(y)| = \left| \int_{x_1}^y v'(s) \, ds \right| \le \int_{x_1}^y |v'(s)| \, ds \le \int_{x_1}^{x_2} |v'(s)| \, ds$$

$$\le (x_2 - x_1)^{1/2} \left(\int_{x_1}^{x_2} |v'(s)|^2 \, ds \right)^{1/2} = (x_2 - x_1)^{1/2}\|v'\|_{L^2(x_1,x_2)}.$$

The continuity of F, therefore, arises from

$$|Fv| = \left| \int_{-1}^2 (5x - 1)v \, dx + 2v(2) \right| \le \int_{-1}^2 |5x - 1||v| \, dx + 2|v(2)|$$

$$\le 9 \int_{-1}^2 |v| \, dx + 2\sqrt{3}\|v'\|_{L^2(-1,2)} \le 9\sqrt{3}\|v\|_{L^2(-1,2)} + 2\sqrt{3}\|v'\|_{L^2(-1,2)}$$

$$\le \sqrt{3}(9C_P + 2)\|v\|_V$$

where C_P is the Poincaré constant. We conclude that there exists a unique solution $w \in V$ of the problem (10.27) and that

$$\|w\|_V \le \sqrt{3}(9C_P + 2).$$

From the stability estimate on w we deduce an analogous estimate for the solution $u = w + 1/2$ of the problem (8.84).

8.2. This is a problem with drift term and Dirichlet-Robin boundary conditions. Since the Dirichlet problem is homogeneous, using the Poincaré inequality we consider

$$V = H_{0,a}^1(a, b) \text{ using } \|v\|_V = \|v'\|_{L^2(a,b)}.$$

Multiplying the equation by $v \in V$ and integrating we find

$$\int_a^b \mu u'v' \, dx - [\mu u'v]_a^b - \int_a^b \beta uv' \, dx + [\beta uv]_a^b + \int_a^b \sigma uv \, dx = \int_a^b fv \, dx.$$

Taking into account the boundary conditions we are yield to consider the bilinear form B and the functional L, which are

$$B(u, v) = \int_a^b \mu u'v' \, dx - \int_a^b \beta uv' \, dx + \int_a^b \sigma uv \, dx$$

$$Lv = \int_a^b fv \, dx$$

and the problem can be stated in abstract form as follows

find $u \in V$ such that $B(u, v) = Lv$ for every $v \in V$.

We introduce sufficient conditions on the data in order to apply the Lax-Milgram Theorem. Assuming $f \in L^2(a, b)$, the functional L is continuous in V, thanks to Schwarz's and Poincaré's inequalities

$$|Lv| \le \|f\|_{L^2(a,b)} \|v\|_{L^2(a,b)} \le C_P \|f\|_{L^2(a,b)} \|v\|_V.$$

On the other hand, assuming $\mu, \beta, \sigma \in L^\infty(a, b)$, the bilinear form B is continuous, in fact we obtain

$$|B(u, v)| \le \|\mu\|_{L^\infty} \|u\|_V \|v\|_V + \|\beta\|_{L^\infty} \|u\|_{L^2(a,b)} \|v\|_V$$
$$+ \|\sigma\|_{L^\infty} \|u\|_{L^2(a,b)} \|v\|_{L^2(a,b)}$$
$$\le \left(\|\mu\|_{L^\infty} + C_P \|\beta\|_{L^\infty} + C_P^2 \|\sigma\|_{L^\infty} \right) \|u\|_V \|v\|_V.$$

Under suitable assumption the bilinear form B is coercive. Indeed an integration by parts gives

$$B(u, u) \ge \mu_0 \|u\|_V^2 - \frac{1}{2} \int_a^b \beta(x)(u^2)' dx + \int_a^b \sigma(x)u^2 dx$$

$$= \mu_0 \|u\|_V + \int_a^b \left(\sigma + \frac{1}{2}\beta' \right) u^2 dx - \frac{1}{2}[\beta u^2]_a^b.$$

Assuming that

$$\beta \in H^1(a,b), \ \beta(b) \leq 0 \ \text{and} \ \left(\sigma + \frac{1}{2}\beta'\right) \geq 0 \ \text{a.e. in } (a,b) \qquad (10.29)$$

then, we deduce that

$$B(u,u) \geq \mu_0 \|u\|_V^2.$$

Summing up, if $f \in L^2(a,b)$, $\mu \in L^\infty(a,b)$, $\mu \geq \mu_0 > 0$ and if (10.29) hold, the the problem (8.85) has one unique solution in V, such that the following estimate holds

$$\|u'\|_{L^2(a,b)} \leq \frac{C_P}{\mu_0} \|f\|_{L^2(a,b)}.$$

8.3. We introduce, as an extension of the Dirichlet data, $g(x) = -x/8$ and consider the function $w = u - g$ submitted to the following homogeneous Dirichlet problem

$$\begin{cases} -w'' + \pi^2 w = f - g & 0 < x < 1 \\ w(0) = 0, \ w(1) = 0. \end{cases} \qquad (10.30)$$

We use the space $V = H_0^1(0,1)$ with inner product

$$\langle w, v \rangle_V = (w', v')_{L^2} + \pi^2 (w, v)_{L^2}$$

and introduce the norm

$$\|w\|_V = \sqrt{\|w'\|_{L^2}^2 + \pi^2 \|w\|_{L^2}^2}$$

and consider the functional $F : V \longrightarrow \mathbb{R}$ so defined

$$Fv = \int_0^1 (f - g)v \, dx = (f - g, v)_{L^2}.$$

The problem (10.30) can be rewritten in abstract form

$$\text{find } w \in V \text{ such that } \langle w, v \rangle_V = Fv \text{ for every } v \in V.$$

The problem has a unique solution in V thanks to the projection theorem. Since the problem for w has unique solution, the problem for u has unique solution too, being $u = w + g$.

We calculate the derivatives in distributional sense of the solution assigned in the exercise, using the Heavyside function

$$u = \sin \pi x + \left(\frac{1}{2} - x\right)^3 \mathcal{H}\left(x - \frac{1}{2}\right)$$

$$u' = \pi \cos \pi x - 3\left(\frac{1}{2} - x\right)^2 \mathcal{H}\left(x - \frac{1}{2}\right)$$

$$u'' = -\pi^2 \sin \pi x + 6\left(\frac{1}{2} - x\right) \mathcal{H}\left(x - \frac{1}{2}\right);$$

therefore u is a distributional solution of the equation. Differentiating twice more, we get

$$u''' = -\pi^3 \cos \pi x - 6\mathcal{H}\left(x - \frac{1}{2}\right)$$

$$u^{(iv)} = \pi^4 \sin \pi x - 6\delta\left(x - \frac{1}{2}\right)$$

where $\delta(x - 1/2)$ is the distribution $\langle \delta(x - 1/2), \phi \rangle = \phi(1/2)$ for every $\phi \in \mathcal{D}(\mathbb{R})$. The latter generalized function does not belong to $H^1(0,1)$, and we conclude that $u \in H^3(0,1)$, but $u \notin H^4(0,1)$.

8.4. Denote with Γ the boundary of Ω. We have a Robin problem that we embed in the Sobolev space

$$V = H^1(\Omega), \text{ with } \|v\|_V = \left(\|u\|_{L^2(\Omega)}^2 + \|\nabla u\|_{L^2(\Omega)}^2\right)^{1/2}$$

and the following trace inequality holds

$$\|v\|_{L^2(\Gamma)} \leq C_T \|v\|_V. \tag{10.31}$$

We consider a function $v \in V$; multiply by v the equation and integrate by parts on Ω; on account of the Robin boundary condition we obtain

$$\int_\Omega \nabla u \cdot \nabla v \, d\mathbf{x} + \alpha \int_\Gamma uv \, d\sigma + \sigma \int_\Omega uv \, d\mathbf{x} = \int_\Omega fv \, d\mathbf{x} + \int_\Gamma gv \, d\sigma.$$

We introduce the bilinear form B and the functional F:

$$B(u,v) = \int_\Omega \nabla u \cdot \nabla v \, d\mathbf{x} + \alpha \int_\Gamma uv \, d\sigma + \sigma \int_\Omega uv \, d\mathbf{x}$$

$$Fv = \int_\Omega fv \, d\mathbf{x} + \int_\Gamma gv \, d\sigma.$$

The problem (8.86) can be rewritten in the variational form:

$$\textit{find } u \in V \textit{ such that } B(u,v) = Fv \textit{ for every } v \in V.$$

We apply the Lax-Milgram Theorem.

On account of the trace inequality (10.31) we prove the continuity of B and F. In fact:

$$|B(u,v)| \leq \|\nabla u\|_{L^2(\Omega)}\|\nabla v\|_{L^2(\Omega)} + \sigma\|u\|_{L^2(\Omega)}\|v\|_{L^2(\Omega)} + \alpha\|u\|_{L^2(\Gamma)}\|v\|_{L^2(\Gamma)}$$

$$\leq (1 + \sigma)\|u\|_V\|v\|_V + \alpha C_T^2\|u\|_V\|v\|_V$$

$$\leq (1 + \sigma + \alpha C_T^2)\|u\|_V\|v\|_V$$

$$Fv \leq \|f\|_{L^2(\Omega)}\|v\|_{L^2(\Omega)} + \|g\|_{L^2(\Gamma)}\|v\|_{L^2(\Gamma)}$$

$$\leq \left(\|f\|_{L^2(\Omega)} + C_T\|g\|_{L^2(\Gamma)}\right)\|v\|_V.$$

The coercivity of B is immediate:

$$B(u,u) = \|\nabla u\|^2_{L^2(\Omega)} + \sigma\|u\|^2_{L^2(\Omega)} + \alpha\int_\Gamma u^2 d\sigma$$
$$\geq \min(1,\sigma)\|u\|^2_V.$$

The Lax-Milgram Theorem gives existence, uniqueness of the solution u and the following estimate holds

$$\|u\|_{H^1(\Omega)} \leq \frac{\|f\|_{L^2(\Omega)} + C_T\|g\|_{L^2(\Gamma)}}{\min(1,\sigma)}.$$

8.5. We consider the space $V = H^1_{0,\Gamma_D}(\Omega)$ and $v \in V$; taking into account the Poincaré inequality, as norm in V we use $\|v\|_V = \|\nabla v\|_{L^2}$. We have that

$$\|\mathbf{b}\|_{L^\infty(\Omega)} = 1.$$

Testing the equation with the function v and using the Green's formula we find

$$\int_\Omega \nabla u \cdot \nabla v \, d\mathbf{x} - \int_{\Gamma_N\cup\Gamma_D} \nabla u \cdot \mathbf{n} v \, d\sigma + \int_\Omega \mathbf{b}\cdot\nabla u \, v \, d\mathbf{x} = \int_\Omega f v \, d\mathbf{x} \qquad \forall v \in V.$$

On account of the boundary conditions, we consider

$$B(u,v) = \int_\Omega \nabla u \cdot \nabla v + \int_\Omega \mathbf{b}\cdot\nabla u \, v$$
$$Fv = \int_\Omega f v$$

and the problem can be rewritten in variational form

find $u \in V$ such that $B(u,v) = Fv$ for every $v \in V$.

We use the Lax-Milgram Theorem. The linearity of B and of the functional F is obvious. The continuity of the functional and of the bilinear form are a consequence of the Schwarz's inequality:

$$|Fv| \leq \|f\|_{L^2}\|v\|_{L^2} \leq C_P\|f\|_{L^2}\|v\|_V$$
$$|B(u,v)| \leq \|u\|_V\|v\|_V + \|\mathbf{b}\|_\infty\|u\|_V\|v\|^2_L$$
$$\leq (1+C_P)\|u\|_V\|v\|_V.$$

In order to prove the coercivity of B, we exploit the Green's formula

$$B(u, u) = \int_\Omega |\nabla u|^2 dx + \int_\Omega (\mathbf{b} \cdot \nabla u) u \, dx$$

$$= \int_\Omega |\nabla u|^2 dx + \int_\Omega \mathbf{b} \cdot \frac{1}{2} \nabla u^2 dx$$

$$= \int_\Omega |\nabla u|^2 dx - \frac{1}{2} \int_\Omega \operatorname{divb} u^2 dx + \frac{1}{2} \int_{\Gamma_N \cup \Gamma_D} \mathbf{b} \cdot \mathbf{n} \, u^2 d\sigma$$

$$= \int_\Omega |\nabla u|^2 dx + \frac{1}{2} \int_{\Gamma_N} \mathbf{b} \cdot \mathbf{n} \, u^2 d\sigma$$

$$= \int_\Omega |\nabla u|^2 dx + \frac{1}{2} \int_0^\pi \sin x_1 \, u^2(x_1, 1) \, dx_1 + \frac{1}{2} \int_0^1 x_2^2 \, u^2(\pi, x_2) \, dx_2$$

$$\geq \|u\|_V^2.$$

Since the coercivity constant is 1, the solution fulfill the following stability estimate

$$\|u\|_V \leq \|F\|_{V'} \leq C_P \|f\|_{L^2}.$$

8.6. We rewrite the problem adding and subtracting, respectively, the equations assigned. Introducing the new unknowns

$$w = u + v$$
$$w_2 = u - v$$

we find

$$\begin{cases} -\alpha \Delta w - \beta w_2 = f & \text{in } \Omega \\ -\alpha \Delta w_2 + \beta w = 0 & \text{in } \Omega \\ w = 2 & \text{on } \Gamma_D \\ w_2 = 0 & \text{on } \Gamma_D \\ \nabla w \cdot \mathbf{n} = 0 & \text{on } \Gamma_N \\ \nabla w_2 \cdot \mathbf{n} = 0 & \text{on } \Gamma_N. \end{cases}$$

We consider an extension of the Dirichlet datum; for the sake of simplicity we consider the constant 2 and rewrite the problem for the unknown $w_1 = w - 2$:

$$\begin{cases} -\alpha \Delta w_1 - \beta w_2 = f & \text{in } \Omega \\ -\alpha \Delta w_2 + \beta w_1 = -2\beta & \text{in } \Omega \\ w_1 = 0 & \text{on } \Gamma_D \\ w_2 = 0 & \text{on } \Gamma_D \\ \nabla w_1 \cdot \mathbf{n} = 0 & \text{on } \Gamma_N \\ \nabla w_2 \cdot \mathbf{n} = 0 & \text{on } \Gamma_N. \end{cases}$$

We consider $V = H_{0,\Gamma_D}^1(\Omega)$ and two functions $v_i \in V$, $i = 1, 2$; the first two equations can be respectively tested with v_1 and v_2. On account of the

Neumann boundary conditions we obtain

$$\begin{cases} \alpha \int_\Omega \nabla w_1 \cdot \nabla v_1 \, d\mathbf{x} - \beta \int_\Omega w_2 v_1 \, d\mathbf{x} = \int_\Omega f v_1 \, d\mathbf{x} \\ \alpha \int_\Omega \nabla w_2 \cdot \nabla v_2 \, d\mathbf{x} + \beta \int_\Omega w_1 v_2 \, d\mathbf{x} = -2\beta \int_\Omega v_2 \, d\mathbf{x}. \end{cases}$$

Denoting

$$\mathbf{w} = \begin{bmatrix} w_1 \\ w_2 \end{bmatrix} \qquad \mathbf{v} = \begin{bmatrix} v_1 \\ v_2 \end{bmatrix} \qquad \mathbf{f}(\mathbf{v}) = \begin{bmatrix} \int_\Omega f v_1 \, d\mathbf{x} \\ -2\beta \int_\Omega v_2 \, d\mathbf{x} \end{bmatrix}$$

we consider the bilinear form

$$a : V^2 \times V^2 \longrightarrow \mathbb{R}$$
$$(\mathbf{w}, \mathbf{v}) \quad \longmapsto \quad a(\mathbf{w}, \mathbf{v})$$

where

$$a(\mathbf{w}, \mathbf{v}) = \alpha \int_\Omega (\nabla w_1 \cdot \nabla v_1 + \nabla w_2 \cdot \nabla v_2) \, d\mathbf{x} - \beta \int_\Omega (w_2 v_1 - w_1 v_2) \, d\mathbf{x}.$$

In the space V^2 we use the norm

$$\|\mathbf{v}\|_{V^2} = \left(\|v_1\|^2_{H^1(\Omega)} + \|v_2\|^2_{H^1(\Omega)} \right)^{1/2}$$

and we also denote

$$\|\mathbf{v}\|_{L^2} = \left(\|v_1\|^2_{L^2(\Omega)} + \|v_2\|^2_{L^2(\Omega)} \right)^{1/2}.$$

The problem can be rewritten in the following variational form

find $\mathbf{w} \in V^2$ *such that* $a(\mathbf{w}, \mathbf{v}) = \mathbf{f}(\mathbf{v})$ *for every* $\mathbf{v} \in V^2$.

The bilinear form a is continuous, indeed we have

$$|a(\mathbf{w}, \mathbf{v})| \le \alpha(\|\nabla w_1\|_{L^2}\|\nabla v_1\|_{L^2} + \|\nabla w_2\|_{L^2}\|\nabla v_2\|_{L^2})$$
$$+ \beta(\|w_2\|_{L^2}\|v_1\|_{L^2} + \|w_1\|_{L^2}\|v_2\|_{L^2})$$
$$\le \alpha(\|w_1\|_{H^1}\|v_1\|_{H^1} + \|w_2\|_{H^1}\|v_2\|_{H^1})$$
$$+ \beta(\|w_2\|_{H^1}\|v_1\|_{H^1} + \|w_1\|_{H^1}\|v_2\|_{H^1})$$
$$\le 2(\alpha + \beta)\|\mathbf{w}\|_{H^1}\|\mathbf{v}\|_{H^1}$$

and we obtain the continuity constant $M = 2(\alpha + \beta)$. On the other hand, denoting by $|\Omega|$ the measure of Ω, the continuity of the functional can be proved as follows

$$|\mathbf{f}(\mathbf{v})| \le \|f\|_{L^2}\|v_1\| + 2|\Omega|\|v_2\|_{L^2} \le (\|f\|_{L^2} + 2|\Omega|)\|\mathbf{v}\|_{L^2}.$$

The coercivity of a is immediate

$$a(\mathbf{w}, \mathbf{w}) = \alpha \int_\Omega \left(|\nabla w_1|^2 + |\nabla w_2|^2\right) dx = \alpha\|\nabla \mathbf{w}\|_{L^2}^2$$

with coercivity constant α. We use the Lax-Milgram Theorem to deduce existence and uniqueness of the solution $\mathbf{w}$ and, furthermore, the following a priori estimate

$$\|\mathbf{w}\|_{V^2} \le \frac{\|f\|_{L^2} + 2|\Omega|}{\alpha}.$$

8.9. Hint. b) $u_1{}_{|\Gamma} = u_2{}_{|\Gamma}$ and $\mathbf{A}^1\nabla u_1 \cdot \boldsymbol{\nu} = \mathbf{A}^2\nabla u_2 \cdot \boldsymbol{\nu}$, where $\boldsymbol{\nu}$ points outward with respect to Ω_1.

8.11. Hint. a) Observe that, if $u[z]$ is the solution of (8.88) the map $z \longmapsto u[z] - u[0]$ is linear. Then write

$$\tilde{J}(z) = \tfrac{1}{2}\int_{\Omega_0} \left(u[z] - u[0] + u[0] - u_d\right)^2 dx + \tfrac{\beta}{2}\int_\Omega z^2 dx$$

and adjust the bilinear form (8.71) accordingly.

b) Answer. The adjoint problem is $(\mathcal{L} = \mathcal{L}^*)$

$$\begin{cases} -\Delta p + a_0 p = (u - z_d)\chi_{\Omega_0} & \text{in } \Omega \\ \partial_\nu p = 0 & \text{on } \partial\Omega. \end{cases}$$

Where χ_{Ω_0} is the characteristic function of Ω_0. The Euler equation is: $p + \beta z = 0$ in $L^2(\Omega)$.

8.12. Hint. a) See Exercise 8.11.

b) Answer. The adjoint problem is

$$\begin{cases} -\Delta p + a_0 p = u - z_d & \text{in } \Omega \\ \partial_\nu p = 0 & \text{on } \partial\Omega. \end{cases}$$

The Euler equation is: $p + \beta z = 0$ in $L^2(\partial\Omega)$.

10.8 Section 9.5

9.1. The model for an exchange of substances through the walls of a cell is also numerically solved in the last section of the chapter.

A Robin problem is assigned. Since μ is a positive function, we write

$$\mu(x) \ge \mu_0 > 0 \quad \text{a.e. } (\mathbf{x}, t) \text{ in } Q_T = \Omega \times [0, T]$$

and this inequality ensures the uniform ellipticity of the divergence term.

We consider $V = H^1(\Omega)$ and $v \in V$, with the usual norm. We recall that in $H^1(\Omega)$ the following trace inequality holds

$$\|v\|_{L^2(\Gamma)} \leq C_T \|v\|_V.$$

We denote $\mathbf{c}(t) = c(\cdot, t)$ and consider $c(\mathbf{x}, t)$ as a function of the variable t with values in V and write $\mathbf{c}'$ instead of $\mathbf{c}_t$. Let us proceed formally and multiply the first equation of the system (9.55) by v; integrating on Ω, we find

$$\langle \mathbf{c}', v \rangle_* + \int_\Omega \mu \nabla \mathbf{c} \cdot \nabla v \, d\mathbf{x} - \int_\Gamma \mu \nabla \mathbf{c} \cdot \mathbf{n} \, v \, d\sigma - \int_\Omega \sigma c v \, d\mathbf{x} = 0$$

namely

$$\langle \mathbf{c}', v \rangle_* + \int_\Omega \mu \nabla \mathbf{c} \cdot \nabla v \, d\mathbf{x} + \int_\Gamma \chi c v \, d\sigma - \int_\Omega \sigma c v \, d\mathbf{x} = \int_\Gamma \chi c_{est} v \, d\sigma.$$

Let us introduce the bilinear form $\overline{B}$ and the functional F

$$\overline{B}(w, v; t) = \int_\Omega \mu \nabla w \cdot \nabla v \, d\mathbf{x} + \int_\Gamma \chi w v \, d\sigma - \int_\Omega \sigma w v \, d\mathbf{x}$$

$$Fv = \int_\Gamma \chi c_{est} v \, d\sigma.$$

The problem can be written in weak form as follows. *Find* $\mathbf{c} \in L^2(0, T; V)$ *such that* $\mathbf{c}' \in L^2(0, T; V')$ *and*

1. *for every* $v \in V$ *and for a.e.* $t \in [0, T]$

$$\langle \mathbf{c}'(t), v \rangle_* + \overline{B}(\mathbf{c}(t), v) = Fv;$$

2. $\mathbf{c}(0) = c_0$ *in* Ω.

In order to use the theory for the well-posedness of the problem we have to show that $\overline{B}$ is continuous and weakly coercive and that F is continuous. Using the trace and the Schwarz inequalities we find that the functional F is continuous

$$|Fv| \leq \chi c_{est} |\Gamma|^{1/2} \|v\|_{L^2(\Gamma)}$$
$$\leq 2R\sqrt{\pi} \chi c_{est} C_T \|v\|_V.$$

On the other hand, since $\mu \in L^\infty(\Omega)$, the bilinear form $\overline{B}$ is $V-$continuous, in fact

$$|\overline{B}(c, v)| \leq \|\mu\|_{L^\infty} \|\nabla c\|_{L^2} \|\nabla v\|_{L^2} + \sigma \|c\|_{L^2} \|v\|_{L^2} + \chi C_T^2 \|c\|_V \|v\|_V$$
$$\leq \left(\|\mu\|_{L^\infty} + \sigma + \chi C_T^2 \right) \|c\|_V \|v\|_V.$$

The weak coercivity of $\overline{B}$ can be proved in the following way

$$\overline{B}(v,v) + \lambda\|v\|_{L^2} = \int_\Omega \mu|\nabla c|^2 dx + \int_\Gamma v^2 d\sigma + \int_\Omega (\lambda - \sigma)v^2 dx$$
$$\geq \min(\mu_0, \lambda - \sigma)\|v\|_V^2.$$

We can choose the constant λ in a way that, for instance, $\lambda - \sigma = \mu_0/2$. The constant of weak coercivity is then $\mu_0/2$.

9.2. We have a mixed problem with homogeneous Dirichlet and Neumann conditions. We use the Poincaré inequality and consider

$$V = H^1_{0,\Gamma_D}(\Omega) \text{ with } \|v\|_V = \|\nabla v\|_{L^2}.$$

We analyze first the drift term $\beta \cdot \nabla u$ where

$$\beta = \begin{pmatrix} xy \\ x^2 y^2 \end{pmatrix}$$

since the domain in the circle of radius 1 we have $|xy| \leq 1$ and $x^2 y^2 \leq 1$. We multiply the drift term by v and integrate on Ω, exploiting the Schwarz inequality we deduce that:

$$|\int_\Omega \beta \cdot \nabla u\, v\, dx| \leq \int_\Omega (|u_x| + |u_y|)|v|\, dx$$
$$\leq \|u_x\|_{L^2}\|v\|_{L^2} + \|u_y\|_{L^2}\|v\|_{L^2}$$
$$\leq 2\|\nabla u\|_{L^2}\|v\|_{L^2}$$
$$\leq 2C_P\|u\|_V\|v\|_V. \tag{10.32}$$

On the other hand, using the elementary (Young) inequality we get

$$|ab| \leq \varepsilon a^2 + \frac{1}{4\varepsilon}b^2 \tag{10.33}$$

and therefore we deduce

$$|\int_\Omega \beta \cdot \nabla u\, v\, dx| \leq \int_\Omega (|u_x| + |u_y|)|v|\, dx$$
$$\leq \varepsilon(\|u_x\|^2_{L^2} + \|u_y\|^2_{L^2}) + \frac{1}{2\varepsilon}\|v\|_{L^2}. \tag{10.34}$$

Denote $\mathbf{u}(t) = u(\cdot, t)$ and read $u(\mathbf{x}, t)$ as a function of t with values in V and write $\mathbf{u}'$ instead of $\mathbf{u}_t$. If we multiply the equation by $v \in V$ and integrate, using the Green formula, we obtain the weak problem

$$\textit{find } u \in V \textit{ such that } ; \langle \mathbf{u}', v \rangle + a(\mathbf{u}, v) = Fv \qquad \forall v \in V$$

where the bilinear form a and the functional F have been introduced

$$a(u,v) = \int_\Omega \nabla u \cdot \nabla v \, d\mathbf{x} + \int_\Omega \boldsymbol{\beta} \cdot \nabla u v \, d\mathbf{x}$$

$$Fv = \int_\Omega fv \, d\mathbf{x}.$$

The continuity of F is immediate using the Schwarz inequality with continuity constant $C = C_P \|f\|_{L^2}$; the continuity of a is a consequence of (10.32), where the continuity constant is $M = 1 + 2C_P$. In order to deduce the weak coercivity of a we use (10.34):

$$a(u,u) + \lambda \int_\Omega u^2 \, d\mathbf{x} \geq (1-\varepsilon)\|u\|_V + \left(\lambda - \frac{1}{2\varepsilon}\right)\|u\|_{L^2}^2$$

choosing $\varepsilon = 1/2$ and, as a consequence, $\lambda > 1$. Then, it is possible to use the Faedo Galerkin theory to prove the well-posedness of the weak problem

find $\mathbf{u} \in L^2(0,T;V)$ such that $\mathbf{u}' \in L^2(0,T;V')$, $\mathbf{u}(0) = u_0$ in Ω
and for every $v \in V$ a.e. t in $[0,T]$ $\langle \mathbf{u}'(t), v\rangle_* + a(\mathbf{u}(t), v) = Fv$.

9.3. Primarily we verify that A_α, with $\alpha > 0$, satisfies the uniform ellipticity condition. In fact,

$$A_\alpha(x_1, x_2)\boldsymbol{\xi} \cdot \boldsymbol{\xi} \geq \xi_1^2 + \alpha\xi_2^2 \geq \min(1,\alpha)(\xi_1^2 + \xi_2^2) \qquad (10.35)$$

for every $\boldsymbol{\xi} \in \mathbb{R}^2$. Therefore, the problem is parabolic for every $\alpha > 0$.

We have a mixed problem with homogeneous Dirichlet conditions on the boundary Γ_D and, then, we are allowed to use the Poincaré inequality; we consider

$$V = H^1_{0,\Gamma_D}(\Omega), \text{ with } \|v\|_V^2 = \|\nabla v\|_{L^2(\Omega)}^2.$$

Using the notation $\mathbf{u}(t) = u(\cdot, t)$, we read $u(\mathbf{x}, t)$ as a function of the variable t with values in V and write $\mathbf{u}'$ instead of $\mathbf{u}_t$. Let us proceed formally multiplying the first equation of the problem (9.56) by $v \in V$ and integrating

$$\langle \mathbf{u}', v\rangle_* + \int_\Omega A_\alpha \nabla \mathbf{u} \cdot \nabla v \, d\mathbf{x} - \int_{\Gamma_N} A_\alpha \nabla \mathbf{u} \cdot \mathbf{n} \, v \, d\sigma + \int_\Omega \mathbf{b} \cdot \nabla \mathbf{u} \, v \, d\mathbf{x}$$

$$- \alpha \int_\Omega \mathbf{u} \, v \, d\mathbf{x} = \int_\Omega x_2 v \, d\mathbf{x}.$$

On account of the third raw of (9.56), we find

$$\langle \mathbf{u}', v\rangle_* + \int_\Omega A_\alpha \nabla \mathbf{u} \cdot \nabla v \, d\mathbf{x} + \int_\Omega \mathbf{b} \cdot \nabla \mathbf{u} \, v \, d\mathbf{x} - \alpha \int_\Omega \mathbf{u} \, v \, d\mathbf{x}$$

$$= \int_\Omega x_2 v \, d\mathbf{x} - \int_{\Gamma_N} \cos x_1 \, v \, d\sigma.$$

We introduce the bilinear form B and the functional F:

$$B(u,v) = \int_\Omega A_\alpha \nabla \mathbf{u} \cdot \nabla v \, dx + \int_\Omega \mathbf{b} \cdot \nabla u \, v \, dx - \alpha \int_\Omega \mathbf{u} \, v \, dx$$

$$Fv = \int_\Omega x_2 v \, dx - \int_{\Gamma_N} \cos x_1 \, v \, d\sigma$$

in order to use the Faedo Galerkin theory to prove the well-posedness of the problem (9.56) restated in a weak form as

> find $\mathbf{u} \in L^2(0,T;V)$ such that $\mathbf{u}' \in L^2(0,T;V')$, $\mathbf{u}(0) = u_0$ in Ω and for every $v \in V$ a.e. t in $[0,T]$ $\langle \mathbf{u}'(t), v \rangle_* + B(\mathbf{u}(t), v) = Fv$.

Now, we prove the continuity of B and F and the weak coercivity of B.

We recall that for the functions in V a trace inequality

$$\|v\|_{\Gamma_N} \le C_T \|v\|_V$$

holds; we exploit it to prove the continuity of F with the Schwarz inequality

$$|Fv| \le \left(\int_\Omega x_2^2 dx \right)^{1/2} \|v\|_{L^2(\Omega)} + \left(\int_{\Gamma_N} \cos^2 x_1 d\sigma \right)^{1/2} \|v\|_{\Gamma_N}$$

$$\le \left(|\Omega|^{1/2} + |\Gamma_N|^{1/2} C_T \right) \|v\|_V = (\sqrt{\pi} + \sqrt{3} C_T) \|v\|_V$$

where $|\Omega|$ and $|\Gamma_N|$ denote the measure of the sets Ω and Γ_N.

On account of the fact that in the domain Ω we have $x_1^2 + x_2^2 \le 4$, with the Schwarz inequality we can prove the continuity of the bilinear form B

$$|B(u,v)| \le \int |u_{x_1} v_{x_1} + \alpha e^{x_1^2 + x_2^2} u_{x_2} v_{x_2}| dx + \alpha \int_\Omega |uv| \, dx$$

$$+ \int_\Omega |\sin(x_1 + x_2) u_{x_1} + (x_1^2 + x_2^2) u_{x_2}||v| dx$$

$$\le \max\{1, \alpha e^4\}$$

$$\int_\Omega |\nabla u \cdot \nabla v| dx + \alpha \|u\|_{L^2(\Omega)} \|v\|_{L^2(\Omega)} + 5 \|\nabla u\|_{L^2(\Omega)} \|v\|_{L^2(\Omega)}$$

$$\le \left(\max\{1, \alpha e^4\} + \alpha C_P^2 + 5 C_P \right) \|u\|_V \|v\|_V.$$

In order to prove the weak coercivity of the bilinear form B, we use the Young inequality (10.33) to give an estimate of the drift term

$$\left| \int_\Omega \beta \cdot \nabla u \, v \, dx \right| \le \int_\Omega (|\sin(x_1^2 + x_2^2)||u_{x_1}||v| + (x_1^2 + x_2^2)|u_y||v|) dx$$

$$\le \int_\Omega (|u_{x_1}||v| + 4|u_y||v|) dx$$

$$\le \varepsilon \|u_{x_1}\|_{L^2(\Omega)}^2 + \frac{1}{4\varepsilon} \|v\|_{L^2(\Omega)}^2 + \varepsilon \|u_{x_2}\|_{L^2(\Omega)}^2 + \frac{4}{\varepsilon} \|v\|_{L^2(\Omega)}^2$$

$$\le \varepsilon \|u\|_V^2 + \frac{17}{4\varepsilon} \|v\|_{L^2(\Omega)}^2. \tag{10.36}$$

Thanks to the uniform ellipticity of A_α proved in (10.35) and to (10.36), we deduce that

$$B(u, u) + \lambda \|u\|_{L^2(\Omega} \geq (\min(1, \alpha) - \epsilon) \|u\|_V^2 + \left(\lambda - \frac{17}{4\varepsilon} - \alpha \right) \|u\|_{L^2(\Omega)}^2.$$

We chose ε and, consequently, λ in order that the two coefficients appearing in the right hand side are positive. Hence, the weak coercivity of the bilinear form B is proved.

Part IV
Appendices

A

Fourier Series

A.1 Fourier coefficients

Let u be a $2T-$periodic function in $\mathbb{R}$ and assume that u can be expanded in a trigonometric series as follows:

$$u(x) = U + \sum_{k=1}^{\infty} \{a_k \cos k\omega x + b_k \sin k\omega x\} \qquad (A.1)$$

where $\omega = \pi/T$.

First question: how u and the coefficients U, a_k and b_k are related to each other? To answer, we use the following so called *orthogonality relations*, whose proof is elementary

$$\int_{-T}^{T} \cos k\omega x \, \cos m\omega x \, dx = \int_{-T}^{T} \sin k\omega x \, \sin m\omega x \, dx = 0 \qquad \text{if } k \neq m$$

$$\int_{-T}^{T} \cos k\omega x \, \sin m\omega x \, dx = 0 \quad \text{for all } k, m \geq 0.$$

Moreover

$$\int_{-T}^{T} \cos^2 k\omega x \, dx = \int_{-T}^{T} \sin^2 k\omega x \, dx = T. \qquad (A.2)$$

Now, suppose that the series (A.1) converges *uniformly* in $\mathbb{R}$. Multiplying (A.1) by $\cos n\omega x$ and integrating term by term over $(-T,T)$, the orthogonality relations and (A.2) yield, for $n \geq 1$,

$$\int_{-T}^{T} u(x) \cos n\omega x \, dx = T a_n$$

or

$$a_n = \frac{1}{T} \int_{-T}^{T} u(x) \cos n\omega x \, dx. \qquad (A.3)$$

Salsa S., Vegni F.M.G., Zaretti A., Zunino P.: *A Primer on PDEs. Models, Methods, Simulations.*
Unitext – La Matematica per il 3+2 65.
DOI 10.1007/978-88-470-2862-3_11, © Springer-Verlag Italia 2013

For $n = 0$ we get

$$\int_{-T}^{T} u(x)\ dx = 2UT$$

or, setting $U = a_0/2$,

$$a_0 = \frac{1}{T} \int_{-T}^{T} u(x)\ dx \tag{A.4}$$

which is coherent with (A.3) as $n = 0$.

Similarly, we find

$$b_n = \frac{1}{T} \int_{-T}^{T} u(x) \sin n\omega x\ dx. \tag{A.5}$$

Thus, if u has the uniformly convergent expansion (A.1), the coefficients a_n, b_n (with $a_0 = 2U$) must be given by the formulas (A.3) and (A.5). In this case we say that the trigonometric series

$$\frac{a_0}{2} + \sum_{k=1}^{\infty} \{a_k \cos k\omega x + b_k \sin k\omega x\} \tag{A.6}$$

is the *Fourier series of* u and the coefficients (A.3), (A.4) and (A.5) are called the *Fourier coefficients* of u.

• *Odd and even functions.* If u is an *odd* function, i.e. $u(-x) = -u(x)$, we have $a_k = 0$ for every $k \geq 0$, while

$$b_k = \frac{2}{T} \int_{0}^{T} u(x) \sin k\omega x\ dx.$$

Thus, if u is odd, its Fourier series is a *sine* Fouries series

$$u(x) = \sum_{k=1}^{\infty} b_k \sin k\omega x.$$

Similarly, if u is *even*, i.e. $u(-x) = u(x)$, we have $b_k = 0$ for every $k \geq 1$, while

$$a_k = \frac{2}{T} \int_{0}^{T} u(x) \cos k\omega x\ dx.$$

Thus, if u is even, its Fourier series is a *cosine* Fouries series

$$u(x) = \frac{a_0}{2} + \sum_{k=1}^{\infty} a_k \cos k\omega x.$$

• *Fourier coefficients of a derivative.* Let $u \in C^1(\mathbb{R})$ be $2T-$periodic. Then we may compute the Fourier coefficients a_k' and b_k' of u'. We have, integrating by parts, for $k \geq 1$

$$a'_k = \frac{1}{T} \int_{-T}^{T} u'(x) \cos k\omega x \, dx$$

$$= \frac{1}{T} \left[u(x) \cos k\omega x \right]_{-T}^{T} + \frac{k\omega}{T} \int_{-T}^{T} u(x) \sin k\omega x \, dx$$

$$= \frac{k\omega}{T} \int_{-T}^{T} u(x) \sin k\omega x \, dx$$

$$= k\omega b_k$$

and

$$b'_k = \frac{1}{T} \int_{-T}^{T} u'(x) \sin k\omega x \, dx$$

$$= \frac{1}{T} \left[u(x) \sin k\omega x \right]_{-T}^{T} - \frac{k\omega}{T} \int_{-T}^{T} u(x) \cos k\omega x \, dx$$

$$= -\frac{k\omega}{T} \int_{-T}^{T} u(x) \cos k\omega x \, dx$$

$$= -k\omega a_k.$$

Thus, the Fourier coefficients a'_k and b'_k are related to a_k and b_k by the following formulas

$$a'_k = k\omega b_k, \quad b'_k = -k\omega a_k. \tag{A.7}$$

• *Complex form of a Fourier series.* Using the Euler identities

$$e^{\pm ik\omega x} = \cos k\omega x \pm i \sin k\omega x$$

the Fourier series (A.6) can be expressed in the complex form

$$\sum_{k=-\infty}^{\infty} c_k e^{ik\omega x},$$

where the complex Fourier coefficients c_k are given by

$$c_k = \frac{1}{2T} \int_{-T}^{T} u(z) \, e^{-ik\omega z} dz.$$

The relations among the real and the complex Fourier coefficients are

$$c_0 = \frac{1}{2} a_0$$

and

$$c_k = \frac{1}{2} (a_k - b_k), \quad c_{-k} = \bar{c}_k \quad \text{for } k > 0.$$

A.2 Expansion in Fourier series

In the above computations we started from a function u admitting a uniform convergent expansion in Fourier series. Adopting a different point of view, let u be a $2T-$periodic function and assume we can compute its Fourier coefficients, given by formulas (A.3) and (A.5). Thus, we can *associate* with u its Fourier series and write

$$u \sim \frac{a_0}{2} + \sum_{k=1}^{\infty} \{a_k \cos k\omega x + b_k \sin k\omega x\}.$$

The main questions are now the following:

1. Which conditions on u do assure "the convergence" of its Fourier series? Of course there are several notions of convergence (e.g pointwise, uniform, least squares).

2. If the Fourier series is convergent in some sense, does it always have sum u?

A complete answer to the above questions is not elementary. The convergence of a Fourier series is a rather delicate matter. We indicate some basic results: for the proofs, see e.g. *Rudin* [33] and [34], *Royden* [32], or *Wheeden and Zygmund* [36].

• *Least squares or L^2 convergence.* This is perhaps the most natural type of convergence for Fourier series. Let

$$S_N(x) = \frac{a_0}{2} + \sum_{k=1}^{N} \{a_k \cos k\omega x + b_k \sin k\omega x\}$$

be the $N-$partial sum of the Fourier series of u. We have:

Theorem A.1. *Let u be a square integrable function[1] on $(-T, T)$. Then*

$$\lim_{N \to +\infty} \int_{-T}^{T} [S_N(x) - u(x)]^2 \, dx = 0.$$

Moreover, the following Parseval relation holds

$$\frac{1}{T} \int_{-T}^{T} u^2 = \frac{a_0^2}{2} + \sum_{k=1}^{\infty} \left(a_k^2 + b_k^2 \right). \tag{A.8}$$

Since the numerical series in the right hand side of (A.8) is convergent, we deduce the following important consequence.

---

[1] That is $\int_{-T}^{T} u^2 < \infty$.

Corollary A.1 (Riemann-Lebesgue).

$$\lim_{k \to +\infty} a_k = \lim_{k \to +\infty} b_k = 0.$$

• *Pointwise convergence.* We say that u satisfies the *Dirichlet conditions* in $[-T, T]$ if it is continuous in $[-T, T]$ except possibly at a finite number of points of jump discontinuity and moreover if the interval $[-T, T]$ can be partitioned in a finite numbers of subintervals such that u is monotone in each one of them.

The following theorem holds.

Theorem A.2. *If u satisfies the Dirichlet conditions in $[-T, T]$ then the Fourier series of u converges at each point of $[-T, T]$. Moreover*[2]:

$$\frac{a_0}{2} + \sum_{k=1}^{\infty} \{a_k \cos k\omega x + b_k \sin k\omega x\} = \begin{cases} \dfrac{u(x+) + u(x-)}{2} & x \in (-T, T) \\ \dfrac{u(T-) + u(-T+)}{2} & x = \pm T. \end{cases}$$

In particular, under the hypotheses of Theorem A.2, at every point x of continuity of u the Fourier series converges to $u(x)$.

• *Uniform convergence.* A simple criterion of uniform convergence is provided by the Weierstrass test (see Section 1.4). Since

$$|a_k \cos k\omega x + b_k \sin k\omega x| \le |a_k| + |b_k|$$

we deduce: *If the numerical series*

$$\sum_{k=1}^{\infty} |a_k| \quad \text{and} \quad \sum_{k=1}^{\infty} |b_k|$$

are convergent, then the Fourier series of u is uniformly convergent in $\mathbb{R}$, with sum u.

This is the case, for instance, if $u \in C^1(\mathbb{R})$ and is $2T$ periodic. In fact, from (A.7) we have for every $k \ge 1$,

$$a_k = -\frac{1}{\omega k} b_k' \quad \text{and} \quad b_k = \frac{1}{\omega k} a_k'.$$

Therefore

$$|a_k| \le \frac{1}{\omega k^2} + (b_k')^2$$

and

$$|b_k| \le \frac{1}{\omega k^2} + (a_k')^2.$$

---

[2] We set $f(x\pm) = \lim_{y \to \pm x} f(y)$.

Now, the series $\sum \frac{1}{k^2}$ is convergent. On the other hand, also the series

$$\sum_{k=1}^{\infty}(a_k')^2 \quad \text{and} \quad \sum_{k=1}^{\infty}(b_k')^2$$

are convergent, by Parseval's relation (A.8) applied to u' in place of u. The conclusion is that *if* $u \in C^1(\mathbb{R})$ *and* $2T$ *periodic, its Fourier series is uniformly convergent in* $\mathbb{R}$ *with sum* u.

Another useful result is a refinement of Theorem A.2:

Theorem A.3. *Assume* u *satisfies the Dirichlet conditions in* $[-T, T]$. *Then*

a) *If* u *is continuous in* $[a, b] \subset (-T, T)$, *then its Fourier series converges uniformly in* $[a, b]$.

b) *If* u *is continuous in* $[-T, T]$ *and* $u(-T) = u(T)$, *then its Fourier series converges uniformly in* $[-T, T]$ *(and therefore in* $\mathbb{R}$*).*

B

Notes on ordinary differential equations

B.1 Bidimensional autonomous systems

Consider the system

$$\begin{cases} \dot{x} = f(x,y) \\ \dot{y} = g(x,y) \end{cases} \tag{B.1}$$

where f and g are C^1 functions in the open set $D \subseteq \mathbb{R}^2$. This hypothesis guarantees both existence and uniqueness for the solution of the Cauchy problem with any initial data

$$x(0) = x_0, \quad y(0) = y_0, \quad (x_0, y_0) \in D.$$

A solution of the system (B.1) is a function $t \mapsto \mathbf{r}(t) = (x(t), y(t))$ whose graph is a subset of $\mathbb{R}^3$, given by the points of coordinates $(t, x(t), y(t))$ and, in general, it represents a curve in the three-dimensional space. The fact that the system (B.1) is autonomous allows to study the solutions using their projections on the state plane (x, y), called *phase plane*. Equivalently, we may think of t as a parameter and $(x(t), y(t))$ as parametric equations of a curve in the plane, called *orbit* or *trajectory* of the system. The vector $\dot{\mathbf{r}}(t) = (\dot{x}(t), \dot{y}(t))$ represents the *velocity vector* along the trajectory and it is tangent to the trajectory itself (Fig. B.1).

The possibility of operating in dimension two rather than three is strictly tied to the fact that the system is autonomous and it entails a significant reduction of complexity. Actually, autonomous systems enjoy some properties which are listed below and which allow us to read the behavior of the solutions from the behavior of the corresponding orbits.

a) *Time translation invariance.* Let $\mathbf{r}(t) = (x(t), y(t))$ be a solution of the system defined in (a, b), then any $\tau-$time shifted function $\mathbf{s}(t) = \mathbf{r}(t + \tau) = (x(t + \tau), y(t + \tau))$ is a solution of the *same* system, defined in $(a - \tau, b - \tau)$.

b) *Each point* $\mathbf{p}^0 = (x_0, y_0)$ *of the domain* D *belongs to exactly one orbit*; as a consequence, the orbits do not cross each other.

Salsa S., Vegni F.M.G., Zaretti A., Zunino P.: *A Primer on PDEs. Models, Methods, Simulations.*
Unitext – La Matematica per il 3+2 65.
DOI 10.1007/978-88-470-2862-3_12, © Springer-Verlag Italia 2013

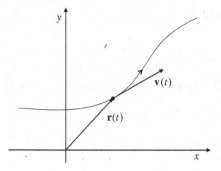

Fig. B.1. Position and velocity vector along the trajectory

Particularly important orbits are the *steady states* (also called *equilibria*, *critical* points or *singular* points), corresponding to constant solutions, and the *cycles*.

Definition B.1. *A point* (x^*, y^*) *is a* **steady state** *for the system* (B.1) *if* $f(x^*, y^*) = g(x^*, y^*) = 0$.

The point (x^*, y^*) is the orbit corresponding to the constant solution $x(t) \equiv x^*$, $y(t) \equiv y^*$ whose graph is a straight line parallel to the t−axis. In other words, whenever we start from an equilibrium point, we remain there forever.

Some solution may tend to an equilibrium point as $t \to \pm\infty$; no solution can reach an equilibrium in finite time otherwise we would have two different orbits passing through the same point, in contradiction with property (b). On the other hand if a solution $(x(t), y(t))$ tends to the point (x^*, y^*) as $t \to \pm\infty$, then (x^*, y^*) must be an equilibrium point. Indeed, in this case, $(\dot{x}(t), \dot{y}(t))$ tends to $(0, 0)$ and, passing to the limit in both equations of the system, we obtain

$$\begin{cases} 0 = f(x^*, y^*) \\ 0 = g(x^*, y^*) \end{cases}$$

this means that (x^*, y^*) is a steady state.

Example B.1. The Lotka-Volterra model describes the evolution of a population of preys $x = x(t)$ and predators $y = y(t)$

$$\begin{cases} \dot{x} = x(a - by) \\ \dot{y} = y(-c + dx) \end{cases} \qquad a, b, c, d > 0. \qquad (B.2)$$

The equilibria are the origin and the point $(c/d, a/b)$.

Example B.2. A variant of the Lotka-Volterra model introduces a competition among individuals of the same species, still assuming that the influence of the environment may be neglected

$$\begin{cases} \dot{x} = x(a - by - ex) \\ \dot{y} = y(-c + dx - fy) \end{cases} \qquad a, b, c, d, e, f > 0 \qquad \text{(B.3)}$$

the equilibrium points can be found solving the system

$$\begin{cases} x(a - by - ex) = 0 \\ y(-c + dx - fy) = 0. \end{cases}$$

Besides the origin, they are

$$P_1 = \left(0, -\frac{c}{f}\right) \quad P_2 = \left(\frac{a}{e}, 0\right), \quad P_3 = \left(\frac{bc + af}{bd + ef}, \frac{ad - ce}{bd + ef}\right).$$

The following definitions are generalizations from the scalar case. We use the symbol $\varphi\,(t; \mathbf{q})$ to denote the solution of the system starting from the point $\mathbf{q}$.

Definition B.2. *The steady state* $\mathbf{p}^* = (x^*, y^*)$ *is:*

a) **stable** *(or* **neutrally stable***) if, for every* $\varepsilon > 0$, *there exists* $\delta = \delta_\varepsilon$ *such that, if* $|\mathbf{q} - \mathbf{p}^*| < \delta$, *the solution* $\varphi\,(t; \mathbf{q})$ *exists for every* $t \geq 0$ *and* $|\varphi\,(t; \mathbf{q}) - \mathbf{p}^*| < \varepsilon$ *for every* $t \geq 0$. *Intuitively: "an orbit starting close enough to* $\mathbf{p}^*$ *always remains close enough to it".*

b) **asymptotically stable** *if it is stable and, moreover, there exists* δ_1 *such that if* $|\mathbf{q} - \mathbf{p}^*| < \delta_1$, *then* $\varphi\,(t; \mathbf{q}) \to \mathbf{p}^*$ *as* $t \to +\infty$. *Intuitively: "any solution starting close enough to* $\mathbf{p}^*$ *not only remains always close enough to it but also converges to* $\mathbf{p}^*$*".*

c) **unstable** *if it is not stable (that is, if condition (a) does not hold).*

A complete analysis of a bidimensional autonomous system requires the description of the *global phase portrait* (for linear systems we can also classify the trajectories; when nonlinearities occur the analysis may be sometimes hard). In general the first step is to find the steady states, solving the system $f(x, y) = g(x, y) = 0$. In the applications, the phase portrait in the neighborhood of each equilibrium point is studied, in order to determine whether the steady state is or not; on the other hand, a *global* point of view can also be assumed. Some common global techniques are briefly described below.

• *The differential equation of the trajectory.* Writing the system in the form

$$\begin{cases} \dfrac{dx}{dt} = f\,(x, y) \\[2mm] \dfrac{dy}{dt} = g\,(x, y) \end{cases}$$

and formally dividing side by side[1], we obtain

$$\frac{dy}{dx} = \frac{g(x, y)}{f\,(x, y)} \qquad \text{(B.4)}$$

---

[1] Rigorously, we should use the inverse function theorem.

when $f(x, y) \neq 0$. In this case, the trajectories can be locally represented by functions of the form $y = y(x)$ and the family of solutions of (B.4) coincides (locally, at least) with the family of trajectories of the system. Thus, whenever we are able to determine the general solution of (B.4), we easily deduce the phase portrait.

• *Vertical/horizontal-slope isoclines and velocity field.* One of the first thing to draw when studying the phase portrait are the curves of equation $dy = 0$

$$g(x, y) = 0 \quad (horizontal\text{-}slope\ isocline)$$

and $dx = 0$

$$f(x, y) = 0 \quad (vertical\text{-}slope\ isocline)$$

whose intersections are the steady states.

The name isocline is due to the fact that any trajectory (but for the equilibria) which crosses the line $g(x, y) = 0$ has a horizontal tangent at the intersection point, since the vertical displacement at the intersection point is zero $(dy = 0)$. Analogously, any trajectory crossing the curve $f(x, y) = 0$, at a point which is not a steady state, has a vertical tangent in the intersection point, since the horizontal displacement is zero at that point $(dx = 0)$.

Once the two isoclines are determined, a study of the the signs of the functions f and g leads to a partition of the phase plane in different regions, where x and y are increasing or decreasing, determining at the same time the orientation of the trajectories.

B.2 Linear systems

General solution

For linear systems with constant coefficients it is possible to write a formula for the general solution. We consider the homogeneous system

$$\begin{cases} \dot{x} = ax + by \\ \dot{y} = cx + dy \end{cases} \tag{B.5}$$

where $a, b, c, d \in \mathbb{R}$. We rewrite it in the form

$$\dot{\mathbf{r}}(t) = \mathbf{A}\mathbf{r}(t) \tag{B.6}$$

with

$$\mathbf{r}(t) = \begin{pmatrix} x \\ y \end{pmatrix} \quad \text{and} \quad \mathbf{A} = \begin{pmatrix} a & b \\ c & d \end{pmatrix}.$$

In order to determine the general solution of the system (B.6), we search for a solution like

$$\mathbf{r}(t) = \mathbf{v}e^{\lambda t} \tag{B.7}$$

where $\mathbf{v} \in \mathbb{R}^2$ is an appropriate vector. By substitution in (B.6), since $\mathbf{r}'(t) = \lambda \mathbf{v} e^{\lambda t}$, we find $\lambda \mathbf{v} e^{\lambda t} = \mathbf{A} \mathbf{v} e^{\lambda t}$ and therefore $\mathbf{A} \mathbf{v} = \lambda \mathbf{v}$. Thus, the system has solutions like (B.7), if λ and $\mathbf{v}$ are, respectively, an *eigenvalue* and an *eigenvector* associated to the matrix $\mathbf{A}$. The eigenvalues are the solutions of the characteristic equation

$$\lambda^2 - (\mathrm{tr}\mathbf{A})\lambda + \det \mathbf{A} = 0. \tag{B.8}$$

Denoted $\Delta = (\mathrm{tr}\mathbf{A})^2 - 4\det \mathbf{A}$, we distinguish three cases according to the sign of Δ.

• Case of **real and distinct eigenvalues** ($\Delta > 0$). There exists two real and distinct eigenvalues λ_1 and λ_2, with corresponding linearly independent eigenvectors $\mathbf{h}^1$ and $\mathbf{h}^2$. The system has then two linearly independent solutions $\mathbf{h}^1 e^{\lambda_1 t}$ and $\mathbf{h}^2 e^{\lambda_2 t}$. The general solution of the system is

$$\mathbf{r}(t) = c_1 \mathbf{h}^1 e^{\lambda_1 t} + c_2 \mathbf{h}^2 e^{\lambda_2 t}, \qquad c_1, c_2 \in \mathbb{R}. \tag{B.9}$$

• Case of **one multiple eigenvalue** ($\Delta = 0$). We have one single real eigenvalue $\lambda = \mathrm{tr}\mathbf{A}/2$. We need to distinguish two situations, if λ is regular or not.

If λ is regular, that is that its algebraic multiplicity is equal to the geometric multiplicity, we have two linearly independent eigenvectors $\mathbf{h}^1$ and $\mathbf{h}^2$ associated to λ and the general solution of the system is

$$\mathbf{r}(t) = c_1 \mathbf{h}^1 e^{\lambda t} + c_2 \mathbf{h}^2 e^{\lambda t}, \qquad c_1, c_2 \in \mathbb{R}.$$

For bidimensional systems this situation occurs only if the matrix $\mathbf{A}$ is diagonal.

If the eigenvalue λ is not regular, the matrix $\mathbf{A}$ cannot be diagonalized. We have an eigenvector $\mathbf{h}$ corresponding to the solution $\mathbf{h} e^{\lambda t}$; thus, we search for another solution of the form $\mathbf{r}(t) = \mathbf{v}^1 e^{\lambda t} + \mathbf{v}^2 t e^{\lambda t}$. A substitution in (B.6) gives

$$(\mathbf{A} - \lambda \mathbf{I}) \mathbf{v}^1 = \mathbf{v}^2 \tag{B.10}$$
$$\mathbf{A} \mathbf{v}^2 = \lambda \mathbf{v}^2.$$

As a consequence, $\mathbf{v}^2$ is an eigenvector and we choose $\mathbf{v}^2 = \mathbf{h}$. Furthermore, we deduce from (B.10), i.e. $(\mathbf{A} - \lambda \mathbf{I}) \mathbf{v}^1 = \mathbf{h}$, that $\mathbf{v}^1$ is a *generalized eigenvector*. Another solution of the system is thus $(\mathbf{h}^1 + \mathbf{h}t)e^{\lambda t}$ and its general solution is

$$\mathbf{r}(t) = c_1 \mathbf{h} e^{\lambda t} + c_2 \left(\mathbf{h}^1 + \mathbf{h}t \right) e^{\lambda t}, \qquad c_1, c_2 \in \mathbb{R}. \tag{B.11}$$

• Case of **complex and conjugate eigenvalues** ($\Delta < 0$): $\lambda = \alpha + i\beta$ and $\bar{\lambda} = \alpha - i\beta$, with $\alpha, \beta \in \mathbb{R}$. Two corresponding eigenvectors are complex and conjugate as well. $\mathbf{h} = \mathbf{h}^1 + i\mathbf{h}^2$, $\overline{\mathbf{h}} = \mathbf{h}^1 - i\mathbf{h}^2$ with $\mathbf{h}^1$ and $\mathbf{h}^2$ real vectors. Using Euler's formula, we find the two linearly independent solutions

$$\varphi(t) = \mathbf{h}e^{(\alpha+i\beta)t} = \mathbf{h}e^{\alpha t}\left(\cos\beta t + i\sin\beta t\right)$$

$$\overline{\varphi}(t) = \overline{\mathbf{h}}e^{(\alpha-i\beta)t} = \overline{\mathbf{h}}e^{\alpha t}\left(\cos\beta t - i\sin\beta t\right).$$

It is convenient to replace these solutions with a pair of real solutions using two appropriate linear combinations of them[2]

$$\psi^1(t) = e^{\alpha t}\left(\mathbf{h}^1\cos\beta t - \mathbf{h}^2\sin\beta t\right) \text{ and } \psi^2(t) = e^{\alpha t}\left(\mathbf{h}^2\cos\beta t + \mathbf{h}^1\sin\beta t\right).$$

Therefore, the general solution of the system is

$$\mathbf{r}(t) = e^{\alpha t}\left[\left(c_1\mathbf{h}^1 + c_2\mathbf{h}^2\right)\cos\beta t + \left(c_2\mathbf{h}^1 - c_1\mathbf{h}^2\right)\sin\beta t\right] \qquad c_1, c_2 \in \mathbb{R}.$$
$$(\text{B.12})$$

Stability of the zero solution

The homogeneous system has the zero solution $\mathbf{x}(t) \equiv \mathbf{0}$ and often, in applications, it is important to determine whether other solutions converge to $\mathbf{0}$ as $t \to +\infty$. If this case, the zero solution is an *asymptotically stable* equilibrium point. In the bidimensional case, the formulas for the general solution indicate that $\mathbf{0}$ is asymptotically stable *if and only if the eigenvalues, or their real part in case of complex eigenvalues, are negative*.

Actually, in order to conclude about the stability of the zero solution it is not necessary to write the general solution. Since it is sufficient to check that the solutions of (B.8) are negative or have negative real part, from $\lambda_1 + \lambda_2 = \text{tr }\mathbf{A}$ and $\lambda_1\lambda_2 = \det\mathbf{A}$ we deduce that $\mathbf{0}$ is asymptotically stable if and only if

$$\text{tr}\mathbf{A} < 0 \qquad \text{and} \qquad \det\mathbf{A} > 0.$$

Classification of the steady states

We assume that the coefficient matrix $\mathbf{A}$ is *non singular*, that is $\det\mathbf{A} = ad - bc \neq 0$; this fact entails that the origin is the only steady state. We discover that there are six possible types of behavior for the orbits near $(0,0)$ so that, accordingly, we are led to a classification of the origin into six different types of equilibria. We proceed with our analysis, by exploiting the formulas for the solution that we have discovered in the previous section and, again, we distinguish three cases according to the sign of $\Delta = (\text{tr}\mathbf{A})^2 - 4\det\mathbf{A}$.

• Case of **real and distinct eigenvalues.** The general solution is given by (B.9). Let us examine (B.9) in the phase plane, splitting the analysis into two

---

[2] Indeed, we have $\psi^1(t) = \dfrac{1}{2}\left(\varphi^1(t) + \varphi^2(t)\right)$ and $\psi^2(t) = \dfrac{1}{2i}\left(\varphi^1(t) - \varphi^2(t)\right)$.

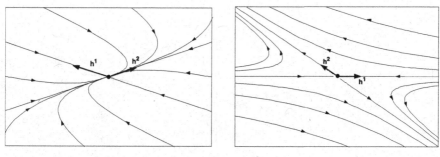

Fig. B.2. A proper node (left) and a saddle (right)

sub cases, according to the fact that the eigenvalues have the same sign or not.

We assume that the eigenvalues have the **same sign**; consider, for instance, $\lambda_1 < \lambda_2 < 0$. The typical configuration in the case of negative eigenvalues is illustrated in Fig. B.2, left. If the eigenvalues were both positive ($\lambda_1 > \lambda_2 > 0$) the direction of the arrows is reversed to take into account the fact that what happened as $t \to +\infty$ for the forward problem now happend as $t \to -\infty$. In this case the origin is called **proper node**; it is asymptotically stable if the eigenvalues are both negative, unstable if they are both positive. The straight lines in the direction of $\mathbf{h}^1$ and $\mathbf{h}^2$ are called *linear manifolds* (both manifolds are either stable or unstable, according to the sign of the eigenvalue).

Assume now that the eigenvalues have **opposite sign**; for instance, $\lambda_1 < 0 < \lambda_2$. The corresponding trajectories are represented in Fig. B.2. In this case the origin is called **saddle point**. Clearly, a saddle point is always unstable. The straight lines in the directions $\mathbf{h}^1$ and $\mathbf{h}^2$ are respectively called *stable* and *unstable manifolds*.

• Case of **one real eigenvalue** λ, with multiplicity 2. We consider two sub-cases, whether the eigenvalue is regular or not.

If the eigenvalue λ is regular the origin is a **star node**; a star is asymptotically stable if $\lambda < 0$, unstable if $\lambda > 0$. The phase portrait for the case $\lambda < 0$ is represented in Fig. B.3.

If, on the other hand, the eigenvalue λ is not regular the solution of the system is given by (B.11) and the origin is called **improper node**. The improper node is asymptotically stable if $\lambda < 0$, unstable if $\lambda > 0$ and the straight line in the direction $\mathbf{h}$ is, respectively, the *stable* or *unstable manifold*. The phase portrait is in Fig. B.3 in the case $\lambda < 0$. If $\lambda > 0$ (unstable node) the directions of the orbits are inverted.

• For **complex and conjugate eigenvalues**, the general solution is given by (B.12). Let us analyze the trajectories, according to the sign of the real part α of the eigenvalues, and in particular, we consider the case $\alpha = 0$.

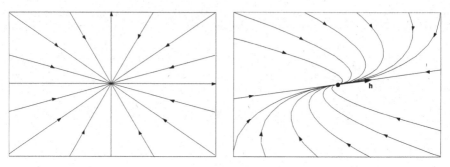

Fig. B.3. Asymptotically stable star (left), stable improper node (right)

If the eigenvalues are pure imaginary numbers, the (B.12) becomes

$$\mathbf{r}(t) = \left(c_1\mathbf{h}^1 + c_2\mathbf{h}^2\right)\cos\beta t + \left(c_2\mathbf{h}^1 - c_1\mathbf{h}^2\right)\sin\beta t \qquad c_1, c_2 \in \mathbb{R}.$$

The solutions are therefore periodic of period $2\pi/\beta$; the corresponding orbits are simple closed curves: in this case they are ellipses centered at the origin. Alternatively, observe that $\operatorname{tr}\mathbf{A} = 0$ namely $d = -a$, and the differential equation of the trajectories is exact. A potential function is $E(x,y) = cx^2 - 2axy - by^2$ so that the general solution is given by the level lines of the surface $z = E(x,y)$, meaning the family of ellipses of equation $E(x,y) = cx^2 - 2axy - by^2 = k$ with $k \in \mathbb{R}$. In this case the origin is a **center**, and it is neutrally stable, not asymptotically stable. The phase portrait is in Fig. B.4.

Consider now, for instance, the case $\alpha < 0$. In (B.12) the function $e^{\alpha t}$ rapidly vanishes as $t \to +\infty$, while the second factor in the same formula is bounded and determines a rotation of the vector $\mathbf{r}(t)$. Hence, every orbits spirals toward the origin as $t \to +\infty$. In this case the origin is called **focus** (or **vortex** or **spiral**), asymptotically stable if $\alpha < 0$, unstable if $\alpha > 0$ (Fig. B.4; for $\alpha > 0$ the arrows are reversed).

Lastly, we note that in the singular case when $\det \mathbf{A} = 0$, with $\mathbf{A} \neq \mathbf{0}$, we have a straight line of equilibrium points, and trajectories are half straight

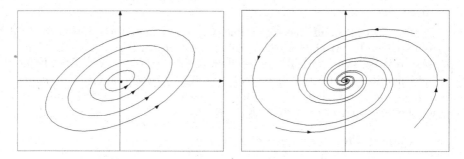

Fig. B.4. Center (left), asymptotically stable focus (right)

lines or segments, diverging or converging to an equilibrium point, respectively if the other eigenvalue λ (besides the eigenvalue 0) is positive or negative. In this case the general solution is

$$\mathbf{r}\,(t) = c_1 \mathbf{h}^1 + c_2 \mathbf{h}^2 e^{\lambda t}, \qquad c_1, c_2 \in \mathbb{R}.$$

We summarize the above results in the following table.

$\Delta > 0$	$\det \mathbf{A} > 0$ proper node	tr $\mathbf{A} < 0$ asymptotically stable tr $\mathbf{A} > 0$ unstable
	$\det \mathbf{A} < 0$ saddle (unstable)	
$\Delta = 0$	$b = c = 0$ star	tr $\mathbf{A} < 0$ asymptotically stable tr $\mathbf{A} > 0$ unstable
	$b \neq 0 \circ c \neq 0$ improper node	tr $\mathbf{A} < 0$ asymptotically stable tr $\mathbf{A} > 0$ unstable
$\Delta < 0$	tr $\mathbf{A} = 0$ center (neutrally stable)	
	tr $\mathbf{A} \neq 0$ fucus	tr $\mathbf{A} < 0$ asymptotically stable tr $\mathbf{A} > 0$ unstable

B.3 Non-linear systems

The linearization method

The results for linear systems can be generalized to a certain extent, under suitable hypotheses, to nonlinear systems. Let (x^*, y^*) be an equilibrium point for the system

$$\begin{cases} \dot{x} = f\,(x, y) \\ \dot{y} = g(x, y). \end{cases} \tag{B.13}$$

We assume that (x^*, y^*) is a *non degenerate* equilibrium, that is the Jacobian matrix

$$\mathbf{J}(x^*, y^*) = \begin{pmatrix} f_x(x^*, y^*) & f_y(x^*, y^*) \\ g_x(x^*, y^*) & g_y(x^*, y^*) \end{pmatrix},$$

is *non singular*; i.e.

$$\det \mathbf{J}(x^*, y^*) \neq 0.$$

This fact implies that (x^*, y^*) is an *isolated* steady state, that is, there exists a neighborhood of (x^*, y^*) contamining no other steady states.

In order to study the phase portrait in a neighborhood of (x^*, y^*) we apply the linearization method; we divide it in three steps.

1. We substitute the system (B.13) with the best linear approximation in a neighborhood of (x^*, y^*). To do this, we use the differentiability of f and g, recalling that

$$f(x^*, y^*) = g(x^*, y^*) = 0$$

and setting $\rho = \sqrt{(x-x^*)^2 + (y-y^*)^2}$ we have

$$\begin{cases} \dot{x} = f(x,y) = f_x(x^*,y^*)(x-x^*) + f_y(x^*,y^*)(y-y^*) + o(\rho) \\ \dot{y} = g(x,y) = g_x(x^*,y^*)(x-x^*) + g_y(x^*,y^*)(y-y^*) + o(\rho). \end{cases} \quad \text{(B.14)}$$

If we assume to be close enough to (x^*, y^*), we may consider the approximation error $o(\rho)$ negligible; furthermore, let us translate (x^*, y^*) to the origin, letting

$$u = x - x^*, \quad v = y - y^*.$$

Since $\dot{u} = \dot{x}$ and $\dot{v} = \dot{y}$ we obtain

$$\begin{cases} \dot{u} = f_x(x^*,y^*)u + f_y(x^*,y^*)v \\ \dot{v} = g_x(x^*,y^*)u + g_y(x^*,y^*)v. \end{cases} \quad \text{(B.15)}$$

We call *linearized system* in (x^*, y^*) the system (B.15).

2. We apply the linear theory. The fundamental condition that the origin is the only critical point is guaranteed by $\mathbf{J}(x^*, y^*)$. Thus, for systems (B.15) the origin may be classified according to the six categories introduced previously.

3. We have to transfer the classification of the origin for the linarized system (at the step 2), into some information on the phase portrait of the original system near (x^*, y^*).

If we want to transfer the conclusions about the stability or instability, then, the following important theorem holds.

Theorem B.1 (Stability via linearization). *If the origin is asymptotically stable (or unstable) for system (B.15), then (x^*, y^*) is locally asymptotically stable (or unstable) for (B.13).*

Note that the Theorem B.1 does not work when the origin is neutrally stable for the linearized system (B.15). Indeed, in this case (x^*, y^*) can be either stable, or unstable or asymptotically stable for the system (B.13). More powerful methods have to be applied when the linearization method fails.

If we want more precise information about the phase portrait in a neighborhood of an equilibrium point, and not just its stability, we need to extend to nonlinear systems the classification of equilibria valid for linear systems. We will describe an intuitive approach.

Let us change point of view and consider the nonlinear system (B.14) as a *perturbation* of the linear system (B.15), with $o(\rho)$ encoding the perturbation error. Our basic question is: how would a saddle point, a node, a focus or a centre transform under a perturbation of order $o(\rho)$?

In order to answer this question, we remark that two orbits' configurations are topologically equivalent if one is a continuous deformation of the other one (see Fig. B.5). Then, the following results hold.

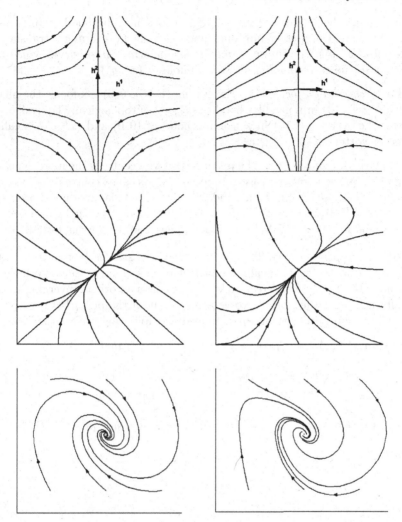

Fig. B.5. From above to below: perturbation of a saddle, a node and a focus

(a) *Saddle point perturbation.* Suppose that $(0,0)$ is a saddle for (B.15). Then the phase portrait for (B.13) in a neighborhood of (x^*, y^*) is topologically equivalent to the phase portrait of the system (B.15) in a neighborhood of $(0,0)$ and (x^*, y^*) is also called saddle point. Furthermore, after a translation of $(0,0)$ into (x^*, y^*), *the stable* (resp. *unstable*) *manifold is deformed into a curve, tangent to the stable* (resp. *unstable*) *manifold*, but it is not a straight line any more.

(b) *Perturbation of a proper or improper node.* The same arguments of the previous case hold. If for (B.15) $(0,0)$ is a proper or improper node, then the phase portrait for (B.13) in a neighborhood of (x^*, y^*) is topologically equivalent to the phase portrait for (B.15) in a neighborhood of $(0,0)$ and (x^*, y^*) is called proper or improper node as well. Furthermore, translating $(0,0)$ into (x^*, y^*), *the linear manifolds are deformed into curves which are tangents to the corresponding linear manifold at* (x^*, y^*).

(c) *Perturbation of a focus.* If for the system (B.15) $(0,0)$ is a focus, the phase portrait of (B.13) in a neighborhood of (x^*, y^*) is topologically equivalent to that of the system (B.15) in a neighborhood of $(0,0)$ and (x^*, y^*) is called focus (or vortex or spiral) as well.

(d) *Perturbation of a center.* The perturbation of a centre is the more delicate. First of all, in the nonlinear case, a point (x^*, y^*) is called centre if there exists a sequence of closed orbits Γ_n around (x^*, y^*), whose diameter goes to zero as $n \to +\infty$. If $(0,0)$ is a centre for (B.15), then the point (x^*, y^*) can be either a centre (as for the Lotka-Volterra model, page 456) or a focus for (B.13).

(e) *Star node perturbation.* The perturbation of a star node is also delicate. In order to have a topologically equivalent orbit's configuration, we have to guarantee that given any direction $\mathbf{w}$, there exists an orbit converging to the equilibrium point tangentially to $\mathbf{w}$. This is true if the perturbation error is of "slightly lower order than $o(\rho)$". Indeed, the following theorem holds:

Theorem B.2. *Assume that there exists* $\varepsilon > 0$ *such that*

$$f(x, y) = f(x^*, y^*) + f_x(x^*, y^*)(x - x^*) + f_y(x^*, y^*)(y - y^*) + o\left(\rho^{1+\varepsilon}\right)$$
$$g(x, y) = g(x^*, y^*) + g_x(x^*, y^*)(x - x^*) + g_y(x^*, y^*)(y - y^*) + o\left(\rho^{1+\varepsilon}\right)$$

then, if $(0,0)$ *is a star node for* (B.15), (x^*, y^*) *is a star node for* (B.13).

C

Finite difference approximation of time dependent problems

We address here the main steps to pursue the error analysis of finite difference methods for time dependent problems. Some examples of such methods have been addressed for the approximation of scalar conservation laws and for the heat equation. For simplicity, we restrict to problems in one space dimension. We aim to approximate the solution u of the following problem

$$\begin{cases} u_t + Lu = 0 & 0 < x < 1,\ 0 < t < T \\ Bu = 0 & x = 0,\ x = 1,\ 0 < t < T \\ u(x,0) = u_0(x) & 0 \leq x \leq 1, \end{cases} \tag{C.1}$$

where Lu is a generic linear differential operator that stands for $Lu = -\partial_{xx}u$ as in the heat equation or $Lu = a\partial_x u$ as for scalar conservation laws and where each case must be complemented by suitable boundary conditions, summarized by $Bu = 0$.

Let us consider an uniformly spaced computational mesh, such as the one of Fig. 3.9, whose nodes are given by

$$x_i = i\,h \text{ with } 0 \leq x_i \leq 1,\ x_0 = 0,\ x_N = 1, \quad t^n = n\tau \text{ with } 0 \leq t^n \leq T.$$

Let u_i^n be the numerical approximation of $u(x,t)$ in the node (x_i, t^n), namely $u_i^n \simeq u(x_i, t^n)$, and let $\mathbf{U}_n = \{u_i^n\}_i$ be the collection of degrees of freedom at time t^n. We observe that the precise range of variation of index i is affected by the choice of boundary conditions in (C.1), which has not been made precise yet. Anyway, there exists a suitable integer N such that $\mathbf{U}_n \in \mathbb{R}^N$.

In the finite difference context, the discretization of both the heat equation (3.96) and scalar conservations laws by the upwind scheme, such as (2.74), can be reformulated as the following general framework:

$$\text{given } \mathbf{U}_0 \in \mathbb{R}^N,\ \text{find a sequence } \mathbf{U}_{n+1} = \mathbf{C}_h^\tau \mathbf{U}_n \tag{C.2}$$

Salsa S., Vegni F.M.G., Zaretti A., Zunino P.: *A Primer on PDEs. Models, Methods, Simulations*.
Unitext – La Matematica per il 3+2 65.
DOI 10.1007/978-88-470-2862-3_13, © Springer-Verlag Italia 2013

where $\mathbf{C}_h^\tau \in \mathbb{R}^{N \times N}$ is the iteration matrix. In the forthcoming section, we aim to compare (C.2) with (C.1) and draw the general lines for the error analysis.[1]

C.1 Discrete scheme and the equivalence principle

In order to compare discrete and exact solutions, we introduce the restriction operator $\mathbf{R}u(t) = \{u(x_i, t)\}_i \in \mathbb{R}^N$ that collects the restrictions of the exact solution in the nodes of the finite difference scheme at any time $0 \le t \le T$. Then $\|\mathbf{R}u(t^n) - \mathbf{U}_n\|$ quantifies the global error, where $\|\cdot\|$ is a discrete norm in $\mathbb{R}^N$.

We aim to provide here a simple example of the so called **equivalence principle**, also known as Lax-Richtmyer Theorem, a general property which ensures that the error analysis, that is the verification of **convergence**, can be equivalently decomposed in two simpler steps, namely the study of **consistency** and **stability** of the scheme, that will be properly defined in what follows.

Definition C.1. *The scheme* (C.2) *is consistent with* (C.1) *provided that*

$$\lim_{\tau, h \to 0} \sup_{t \in (0, T)} \tau^{-1} \|\mathbf{R}u(t + \tau) - \mathbf{C}_h^\tau \mathbf{R}u(t)\| = 0. \tag{C.3}$$

The scheme turns out to be of order p and q in h and τ respectively, if $p, q > 0$ are th largest exponents such that

$$\lim_{\tau, h \to 0} \sup_{t \in (0, T)} \tau^{-1} \|\mathbf{R}u(t + \tau) - \mathbf{C}_h^\tau \mathbf{R}u(t)\| = \lim_{\tau, h \to 0} C(h^p + \tau^q) \tag{C.4}$$

where C is a positive constant independent on h and τ.

We notice that $\tau^{-1} \|\mathbf{R}u(t + \tau) - \mathbf{C}_h^\tau \mathbf{R}u(t)\|$ is the residual obtained when replacing the exact solution $u(x_i, t)$ into the scheme (C.2). In other words, it is equivalent to the **local truncation error** already addressed in the previous chapters. Its supremum over the time interval, precisely $\tau^{-1} \|\mathbf{R}u(t + \tau) - \mathbf{C}_h^\tau \mathbf{R}u(t)\|$ is called **global truncation error**. To sum up, the scheme turns out to be consistent if the global truncation error vanishes along with the discretization steps τ, h.

The stability of the scheme (C.2) depends on the iteration matrix $\mathbf{C}_h^\tau$. For a generic matrix $\mathbf{A} \in \mathbb{R}^{N \times N}$ we denote with $\|\mathbf{A}\|$ the matrix norm induced by $\|\cdot\|$ in $\mathbb{R}^N$ and we assert that (C.2) is stable if the following statement is verified.

---

[1] For a more detailed illustration of the equivalence principle, complemented with several examples, including for instance the error analysis of leapfrog scheme for the approximation of the wave equation, we refer the interested reader to *Dautray and Lions* [39], Chapter XX, Volume 6.

Definition C.2. *The scheme* (C.2) *turns out to be* **stable** *if there exists a positive constant* K, *independent on* h *and* τ *such that,*

$$\|(\mathbf{C}_h^\tau)^n\| \le K, \quad \forall\, n \text{ with } \tau n \le T. \tag{C.5}$$

Furthermore, the scheme is called **unconditionally stable** when (C.5) is satisfied for any admissible value of h and τ. Conversely, when (C.5) holds true for a restricted choice of h and τ, we have a **conditionally stable scheme**. Theorem 3.6 suggests that the satisfaction of (C.5) is closely related to the spectral properties of $\mathbf{C}_h^\tau$. For the particular case of forward and backward Euler schemes, this is put into evidence by Corollaries 3.4, 3.5. In the case of explicit methods for the approximation of conservation laws, such as the upwind scheme, we remind that CFL condition is necessary to prove stability. As a result of that, this family of schemes turns out to be only conditionally stable.

We address finally a precise definition of convergence.

Definition C.3. *Provided that the initial state of the discrete scheme converges to the exact one,*

$$\lim_{h \to 0} \|\mathbf{R}u_0 - \mathbf{U}_0\| = 0$$

for any $\tau, h \to 0$ *and* $n \to \infty$ *with* $n\tau = t$, *the solution of* (C.2) *converges to the one of* (C.1) *when*

$$\lim_{\tau,h \to 0} \|\mathbf{R}u(t) - \mathbf{U}_n\| = 0, \quad \forall t \in (0, T). \tag{C.6}$$

As highlighted by the following fundamental principle, consistency, stability and convergence are the pillars for the analysis of any numerical method.

Theorem C.1. *Let problem* (C.1) *be well posed and let us consider its approximation by means of* (C.2), *which is assumed to be consistent. Then, the scheme* (C.2) *is convergent if and only if it is stable.*

Proof. The direct implication, namely that stability implies convergence for any consistent scheme, is the most relevant to the purpose of error analysis. Then, let us assume that consistency and stability hold true for (C.2) and prove convergence. We start by looking at the error at time t^{n+1},

$$\mathbf{R}u(t^{n+1}) - \mathbf{U}_{n+1} = \mathbf{R}u(t^{n+1}) - \mathbf{C}_h^\tau \mathbf{U}_n$$
$$= \left(\mathbf{R}u(t^{n+1}) - \mathbf{C}_h^\tau \mathbf{R}u(t^n)\right) + \mathbf{C}_h^\tau\left(\mathbf{R}u(t^n) - \mathbf{U}_n\right).$$

We notice that the first term on the right hand side is proportional to the truncation error, while the second term is the error at time t^n. By propagating this expression recursively backward in time we obtain,

$$\mathbf{R}u(t^{n+1}) - \mathbf{U}_{n+1}$$
$$= \sum_{k=0}^{n} (\mathbf{C}_h^\tau)^k \left(\mathbf{R}u(t^{n+1-k}) - \mathbf{C}_h^\tau \mathbf{R}u(t^{n-k})\right) + (\mathbf{C}_h^\tau)^{n+1}\left(\mathbf{R}u_0 - \mathbf{U}_0\right)$$

which can be easily rearranged as follows,

$$\|\mathbf{R}u(t^{n+1}) - \mathbf{U}_{n+1}\|$$

$$\leq \sum_{k=0}^{n} \|(\mathbf{C}_h^\tau)^k\| \, \|\mathbf{R}u(t^{n+1-k}) - \mathbf{C}_h^\tau \mathbf{R}u(t^{n-k})\| + \|(\mathbf{C}_h^\tau)^{n+1}\| \, \|\mathbf{R}u_0 - \mathbf{U}_0\|.$$

Owing to the stability of (C.2) and assuming, without loss of generality, that $K \geq 1$ in (C.5) we have,

$$\|\mathbf{R}u(t^{n+1}) - \mathbf{U}_{n+1}\|$$

$$\leq \tau(n+1)K \sup_{t\in(0,T)} \tau^{-1}\|\mathbf{R}u(t+\tau) - \mathbf{C}_h^\tau \mathbf{R}u(t)\| + K\|\mathbf{R}u_0 - \mathbf{U}_0\|.$$

Observing that $\tau(n+1) = t^{n+1} \leq T$ and combining the consistency of the scheme with the convergence of the discrete initial state, we conclude that the previous inequality implies convergence. Furthermore, if (C.2) is consistent with order $\mathcal{O}(h^p + \tau^q)$ and if the same spatial accuracy holds for the approximation of the initial state, precisely $\|\mathbf{R}u_0 - \mathbf{U}_0\| = \mathcal{O}(h^p)$, then the error shares the same infinitesimal order with respect to h and τ than the truncation error. In other words, there exists a positive constant C, independent on h, τ such that

$$\lim_{\tau,h\to 0} \|\mathbf{R}u(t) - \mathbf{U}_n\| = \lim_{\tau,h\to 0} C(h^p + \tau^q).$$

The reverse statement, namely that convergence implies stability is more easily proved. First, we observe that convergence immediately leads to

$$\lim_{\tau,h\to 0} \|\mathbf{R}u(t) - (\mathbf{C}_h^\tau)^n \mathbf{U}_0\| = 0,$$

that is

$$\lim_{\tau,h\to 0} (\mathbf{C}_h^\tau)^n \mathbf{U}_0 = \mathbf{R}u(t), \quad \forall n\tau = t \in (0,T)$$

or equivalently that $\|(\mathbf{C}_h^\tau)^n \mathbf{U}_0\| \leq K$ for any $\mathbf{U}_0 \in \mathbf{R}^N$. Then, the uniform boundedness Theorem (also known as Banach-Steinhaus Theorem), ensures that $\|(\mathbf{C}_h^\tau)^n\| \leq K$ for any n such that $n\tau = t \leq T$, that is indeed the stability property. $\square$

D

Identities and Formulas

D.1 Gradient, Divergence, Curl, Laplacian

Let $\mathbf{F}$ be a smooth vector field and f a smooth real function, in $\mathbb{R}^3$.

Orthogonal cartesian coordinates

1. *gradient*
$$\nabla f = \frac{\partial f}{\partial x}\mathbf{i} + \frac{\partial f}{\partial y}\mathbf{j} + \frac{\partial f}{\partial z}\mathbf{k};$$

2. *divergence* ($\mathbf{F} = F_1\mathbf{i} + F_1\mathbf{j} + F_3\mathbf{k}$)
$$\operatorname{div}\mathbf{F} = \frac{\partial}{\partial x}F_1 + \frac{\partial}{\partial y}F_2 + \frac{\partial}{\partial z}F_3;$$

3. *laplacian*
$$\Delta f = \frac{\partial^2 f}{\partial x^2} + \frac{\partial^2 f}{\partial y^2} + \frac{\partial^2 f}{\partial z^2};$$

4. *curl*
$$\operatorname{curl}\mathbf{F} = \begin{vmatrix} \mathbf{i} & \mathbf{j} & \mathbf{k} \\ \partial_x & \partial_y & \partial_z \\ F_1 & F_2 & F_3 \end{vmatrix}.$$

Cylindrical coordinates
$$x = r\cos\theta, \ y = r\sin\theta, \ z = z \qquad (r > 0, \ 0 \le \theta \le 2\pi)$$
$$\mathbf{e}_r = \cos\theta\mathbf{i} + \sin\theta\mathbf{j}, \quad \mathbf{e}_\theta = -\sin\theta\mathbf{i} + \cos\theta\mathbf{j}, \quad \mathbf{e}_z = \mathbf{k}.$$

1. *gradient*
$$\nabla f = \frac{\partial f}{\partial r}\mathbf{e}_r + \frac{1}{r}\frac{\partial f}{\partial \theta}\mathbf{e}_\theta + \frac{\partial f}{\partial z}\mathbf{e}_z;$$

2. *divergence* ($\mathbf{F} = F_r\mathbf{e}_r + F_\theta\mathbf{e}_\theta + F_z\mathbf{k}$)
$$\operatorname{div}\mathbf{F} = \frac{1}{r}\frac{\partial}{\partial r}(rF_r) + \frac{1}{r}\frac{\partial}{\partial \theta}F_\theta + \frac{\partial}{\partial z}F_z;$$

Salsa S., Vegni F.M.G., Zaretti A., Zunino P.: *A Primer on PDEs. Models, Methods, Simulations.*
Unitext – La Matematica per il 3+2 65.
DOI 10.1007/978-88-470-2862-3_14, © Springer-Verlag Italia 2013

3. *laplacian*

$$\Delta f = \frac{1}{r}\frac{\partial}{\partial r}\left(r\frac{\partial f}{\partial r}\right) + \frac{1}{r^2}\frac{\partial^2 f}{\partial\theta^2} + \frac{\partial^2 f}{\partial z^2} = \frac{\partial^2 f}{\partial r^2} + \frac{1}{r}\frac{\partial f}{\partial r} + \frac{1}{r^2}\frac{\partial^2 f}{\partial\theta^2} + \frac{\partial^2 f}{\partial z^2};$$

4. *curl*

$$\text{curl } \mathbf{F} = \frac{1}{r}\begin{vmatrix} \mathbf{e}_r & r\mathbf{e}_\theta & \mathbf{e}_z \\ \partial_r & \partial_\theta & \partial_z \\ F_r & rF_\theta & F_z \end{vmatrix}.$$

Spherical coordinates

$$x = r\cos\theta\sin\psi, \; y = r\sin\theta\sin\psi, \; z = r\cos\psi$$
$$backslash \; (r > 0, \, 0 \le \theta \le 2\pi, \, 0 \le \psi \le \pi)$$

$$\mathbf{e}_r = \cos\theta\sin\psi\mathbf{i} + \sin\theta\sin\psi\mathbf{j} + \cos\psi\mathbf{k}$$
$$\mathbf{e}_\theta = -\sin\theta\mathbf{i} + \cos\theta\mathbf{j}$$
$$\mathbf{e}_\psi = \cos\theta\cos\psi\mathbf{i} + \sin\theta\cos\psi\mathbf{j} - \sin\psi\mathbf{k}.$$

1. *gradient*

$$\nabla f = \frac{\partial f}{\partial r}\mathbf{e}_r + \frac{1}{r\sin\psi}\frac{\partial f}{\partial\theta}\mathbf{e}_\theta + \frac{1}{r}\frac{\partial f}{\partial\psi}\mathbf{e}_\psi;$$

2. *divergence* $(\mathbf{F} = F_r\mathbf{e}_r + F_\theta\mathbf{e}_\theta + F_\psi\mathbf{e}_\psi)$

$$\text{div } \mathbf{F} = \underbrace{\frac{\partial}{\partial r}F_r + \frac{2}{r}F_r}_{\text{radial part}} + \underbrace{\frac{1}{r}\left[\frac{1}{\sin\psi}\frac{\partial}{\partial\theta}F_\theta + \frac{\partial}{\partial\psi}F_\psi + \cot\psi F_\psi\right]}_{\text{spherical part}};$$

3. *laplacian*

$$\Delta f = \underbrace{\frac{\partial^2 f}{\partial r^2} + \frac{2}{r}\frac{\partial f}{\partial r}}_{\text{radial part}} + \underbrace{\frac{1}{r^2}\left\{\frac{1}{(\sin\psi)^2}\frac{\partial^2 f}{\partial\theta^2} + \frac{\partial^2 f}{\partial\psi^2} + \cot\psi\frac{\partial f}{\partial\psi}\right\}}_{\text{spherical part (Laplace-Beltrami operator)}};$$

4. *curl*

$$\text{rot } \mathbf{F} = \frac{1}{r^2\sin\psi}\begin{vmatrix} \mathbf{e}_r & r\mathbf{e}_\psi & r\sin\psi\mathbf{e}_\theta \\ \partial_r & \partial_\psi & \partial_\theta \\ F_r & rF_\psi & r\sin\psi F_z \end{vmatrix}.$$

D.2 Formulas

Gauss' formulas

In $\mathbb{R}^n$, $n \geq 2$, let:

- Ω be a bounded smooth domain and and ν the outward unit normal on $\partial\Omega$;
- $\mathbf{u}, \mathbf{v}$ be vector fields of class $C^1\left(\overline{\Omega}\right)$;
- φ, ψ be real functions of class $C^1\left(\overline{\Omega}\right)$;
- $d\sigma$ be the area element on $\partial\Omega$.

1. $\int_\Omega \text{div } \mathbf{u} \, d\mathbf{x} = \int_{\partial\Omega} \mathbf{u} \cdot \nu \, d\sigma$ (Divergence Theorem);
2. $\int_\Omega \nabla\varphi \, d\mathbf{x} = \int_{\partial\Omega} \varphi\nu \, d\sigma$;
3. $\int_\Omega \Delta\varphi \, d\mathbf{x} = \int_{\partial\Omega} \nabla\varphi \cdot \nu \, d\sigma = \int_{\partial\Omega} \partial_\nu\varphi \, d\sigma$;
4. $\int_\Omega \psi \text{ div} \mathbf{F} \, d\mathbf{x} = \int_{\partial\Omega} \psi\mathbf{F} \cdot \nu \, d\sigma - \int_\Omega \nabla\psi \cdot \mathbf{F} \, d\mathbf{x}$ (Integration by parts);
5. $\int_\Omega \psi\Delta\varphi \, d\mathbf{x} = \int_{\partial\Omega} \psi\partial_\nu\varphi \, d\sigma - \int_\Omega \nabla\varphi \cdot \nabla\psi \, d\mathbf{x}$ (Green's identity I);
6. $\int_\Omega (\psi\Delta\varphi - \varphi\Delta\psi) \, d\mathbf{x} = \int_{\partial\Omega}(\psi\partial_\nu\varphi - \varphi\partial_\nu\psi) \, d\sigma$ (Green's identity II);
7. $\int_\Omega \text{curl } \mathbf{u} \, d\mathbf{x} = -\int_{\partial\Omega} \mathbf{u} \times \nu \, d\sigma$;
8. $\int_\Omega \mathbf{u} \cdot \text{curl } \mathbf{v} \, d\mathbf{x} = \int_\Omega \mathbf{v} \cdot \text{curl } \mathbf{u} \, d\mathbf{x} - \int_{\partial\Omega}(\mathbf{u} \times \mathbf{v}) \cdot \nu \, d\sigma$.

Identities

1. div curl $\mathbf{u} = 0$;
2. curl grad $\varphi = \mathbf{0}$;
3. div $(\varphi\mathbf{u}) = \varphi$ div $\mathbf{u} + \nabla\varphi \cdot \mathbf{u}$;
4. curl $(\varphi\mathbf{u}) = \varphi$ curl $\mathbf{u} + \nabla\varphi \times \mathbf{u}$;
5. curl $(\mathbf{u} \times \mathbf{v}) = (\mathbf{v}\cdot\nabla)\mathbf{u} - (\mathbf{u}\cdot\nabla)\mathbf{v} + (\text{div } \mathbf{v})\mathbf{u} - (\text{div } \mathbf{u})\mathbf{v}$;
6. div $(\mathbf{u} \times \mathbf{v}) = \text{curl}\mathbf{u} \cdot \mathbf{v} - \text{curl}\mathbf{v} \cdot \mathbf{u}$;
7. $\nabla(\mathbf{u} \cdot \mathbf{v}) = \mathbf{u} \times \text{curl } \mathbf{v} + \mathbf{v} \times \text{curl } \mathbf{u} + (\mathbf{u}\cdot\nabla)\mathbf{v} + (\mathbf{v}\cdot\nabla)\mathbf{u}$;
8. $(\mathbf{u}\cdot\nabla)\mathbf{u} = \text{curl}\mathbf{u} \times \mathbf{u} + \frac{1}{2}\nabla|\mathbf{u}|^2$;
9. curl curl $\mathbf{u} = \nabla(\text{div } \mathbf{u}) - \Delta\mathbf{u}$.

References

Partial Differential Equations

1. DiBenedetto E.: Partial Differential Equations. Birkhäuser, Basel (1995).

2. Evans L.C.: Partial Differential Equations. A.M.S., Graduate Studies in Mathematics, Providence (1998).

3. Friedman A.: Partial Differential Equations of Parabolic Type. Prentice-Hall, Englewood Cliffs (1964).

4. Gilbarg D., Trudinger N.: Elliptic Partial Differential Equations of Second Order. II edizione. Springer-Verlag, Berlin Heidelberg (1998).

5. John F.: Partial Differential Equations. 4th ed. Springer-Verlag, New York (1982).

6. Kellog O.: Foundations of Potential Theory. Springer-Verlag, New York (1967).

7. Lieberman G.M.: Second Order Parabolic Partial Differential Equations. World Scientific, Singapore (1996).

8. Lions J.L., Magenes E.: Nonhomogeneous Boundary Value Problems and Applications. Springer-Verlag, New York (1972).

9. McOwen R.: Partial Differential Equations: Methods and Applications. Prentice-Hall, New Jersey (1996).

10. Protter M., Weinberger H.: Maximum Principles in Differential Equations. Prentice-Hall, Englewood Cliffs (1984).

11. Renardy M., Rogers R.C.: An Introduction to Partial Differential Equations. Springer-Verlag, New York (1993).

12. Rauch J.: Partial Differential Equations. Springer-Verlag, Berlin Heidelberg (1992).

13. Salsa S.: Equazioni a derivate parziali. Metodi, modelli e applicazioni. 2nd ed. Springer-Verlag, Milan (2010).

14. Salsa S., Verzini G.: Equazioni a derivate parziali. Complementi ed esercizi. Springer-Verlag, Milan (2005).

Salsa S., Vegni F.M.G., Zaretti A., Zunino P.: *A Primer on PDEs. Models, Methods, Simulations.*
Unitext – La Matematica per il 3+2 65.
DOI 10.1007/978-88-470-2862-3, © Springer-Verlag Italia 2013

15. Salsa S.: Partial differential equations in action. From modelling to theory. Springer-Verlag, Milan (2008).
16. Smoller J.: Shock Waves and Reaction-Diffusion Equations. Springer-Verlag, New York (1983).
17. Strauss W.: Partial Differential Equation: An Introduction. Wiley, New York (1992).
18. Widder D.V.: The Heat Equation. Academic Press, New York (1975).

Mathematical Modelling

19. Acheson A.J.: Elementary Fluid Dynamics. Clarendon Press, Oxford (1990).
20. Billingham J., King A.C.: Wave Motion. Cambridge University Press, Cambridge (2000).
21. Courant R., Hilbert D.: Methods of Mathematical Phisics. Vol. 1 and 2. Wiley, New York (1953).
22. Dautray R., Lions J.L.: Mathematical Analysis and Numerical Methods for Science and Technology. Vol. 1-5. Springer-Verlag, Berlin Heidelberg (1985).
23. Lin C.C., Segel L.A.: Mathematics Applied to Deterministic Problems in the Natural Sciences. SIAM Classics in Applied Mathematics, 4th ed. Philadelphia (1995).
24. Murray J.D.: Mathematical Biology. 3rd edn. Springer-Verlag, Berlin Heidelberg (2003).
25. Segel L.A.: Mathematics Applied to Continuum Mechanics. Dover Publications, Inc., New York (1987).
26. Thomas D., Kernevez J.P. (eds): Analysis and Control of Immobilized Enzyme Systems. Springer-Verlag, Berlin Heidelberg New York (1975).
27. Whitham G.B.: Linear and Nonlinear Waves. Wiley-Interscience, New York (1974).

Analysis and Functional Analysis

28. Adams R.: Sobolev Spaces. Academic Press, New York (1975).
29. Brezis H.: Analyse Fonctionnelle. Masson, Paris (1983).
30. Evans L.C., Gariepy R.F.: Measure Theory and Fine properties of Functions. CRC Press, Boca Raton (1992).
31. Maz'ya V.G.: Sobolev Spaces. Springer-Verlag, Berlin Heidelberg (1985).
32. Royden H.L.: Real Analysis. McMillan, London (1988).
33. Rudin W.: Principles of Mathematical Analysis. 3rd ed. McGraw-Hill, New York (1976).
34. Rudin W.: Real and Complex Analysis. 2nd ed. McGraw-Hill, New York (1974).
35. Schwartz L.: Théorie des Distributions. Hermann, Paris (1966).
36. Wheeden R., Zygmund A.: Measure and Integral. Marcel Dekker, New York (1977).
37. Yoshida K.: Functional Analysis. Springer-Verlag, Berlin Heidelberg (1965).

Numerical Analysis

38. Comincioli V.: Analisi Numerica: Metodi Modelli Applicazioni. McGraw-Hill, Milano (1995).

39. Dautray R., Lions J.L.: Mathematical Analysis and Numerical Methods for Science and Technology. Vol. 6. Springer-Verlag, Berlin Heidelberg (2000).

40. Le Veque R.J.: Numerical methods for conservation laws. Birkhäuser, Basel (1992).

41. Le Veque R.J.: Finite difference methods for ordinary and partial differential equations. Society for Industrial and Applied Mathematics (SIAM), Philadelphia (2007).

42. Quarteroni A.: Numerical Models for Differential Problems. Springer-Verlag, Milan (2009).

43. Quarteroni A., Sacco R., Saleri F.: Numerical Mathematics. 2nd ed. Springer-Verlag, Berlin Heidelberg (2007).

44. Quarteroni A., Saleri F.: Calcolo Scientifico. 5th ed. Springer-Verlag, Milan (2012).

45. Quarteroni A., Valli A.: Numerical Approximation of Partial Differential Equations. 2nd ed. Springer-Verlag, Berlin Heidelberg (1997).

Index

Salsa S., Vegni F.M.G., Zaretti A., Zunino P.: *A Primer on PDEs. Models, Methods, Simulations.*
Unitext – La Matematica per il 3+2 65.
DOI 10.1007/978-88-470-2862-3, © Springer-Verlag Italia 2013

Collana Unitext – La Matematica per il 3+2

Series Editors:
A. Quarteroni (Editor-in-Chief)
L. Ambrosio
P. Biscari
C. Ciliberto
G. van der Geer
G. Rinaldi
W.J. Runggaldier

Editor at Springer:
F. Bonadei
francesca.bonadei@springer.com

As of 2004, the books published in the series have been given a volume number. Titles in grey indicate editions out of print.
As of 2011, the series also publishes books in English.

A. Bernasconi, B. Codenotti
Introduzione alla complessità computazionale
1998, X+260 pp, ISBN 88-470-0020-3

A. Bernasconi, B. Codenotti, G. Resta
Metodi matematici in complessità computazionale
1999, X+364 pp, ISBN 88-470-0060-2

E. Salinelli, F. Tomarelli
Modelli dinamici discreti
2002, XII+354 pp, ISBN 88-470-0187-0

S. Bosch
Algebra
2003, VIII+380 pp, ISBN 88-470-0221-4

S. Graffi, M. Degli Esposti
Fisica matematica discreta
2003, X+248 pp, ISBN 88-470-0212-5

S. Margarita, E. Salinelli
MultiMath – Matematica Multimediale per l'Università
2004, XX+270 pp, ISBN 88-470-0228-1

A. Quarteroni, R. Sacco, F.Saleri
Matematica numerica (2a Ed.)
2000, XIV+448 pp, ISBN 88-470-0077-7
2002, 2004 ristampa riveduta e corretta
(1a edizione 1998, ISBN 88-470-0010-6)

13. A. Quarteroni, F. Saleri
Introduzione al Calcolo Scientifico (2a Ed.)
2004, X+262 pp, ISBN 88-470-0256-7
(1a edizione 2002, ISBN 88-470-0149-8)

14. S. Salsa
Equazioni a derivate parziali - Metodi, modelli e applicazioni
2004, XII+426 pp, ISBN 88-470-0259-1

15. G. Riccardi
Calcolo differenziale ed integrale
2004, XII+314 pp, ISBN 88-470-0285-0

16. M. Impedovo
Matematica generale con il calcolatore
2005, X+526 pp, ISBN 88-470-0258-3

17. L. Formaggia, F. Saleri, A. Veneziani
Applicazioni ed esercizi di modellistica numerica
per problemi differenziali
2005, VIII+396 pp, ISBN 88-470-0257-5

18. S. Salsa, G. Verzini
Equazioni a derivate parziali – Complementi ed esercizi
2005, VIII+406 pp, ISBN 88-470-0260-5
2007, ristampa con modifiche

19. C. Canuto, A. Tabacco
Analisi Matematica I (2a Ed.)
2005, XII+448 pp, ISBN 88-470-0337-7
(1a edizione, 2003, XII+376 pp, ISBN 88-470-0220-6)

20. F. Biagini, M. Campanino
Elementi di Probabilità e Statistica
2006, XII+236 pp, ISBN 88-470-0330-X

21. S. Leonesi, C. Toffalori
Numeri e Crittografia
2006, VIII+178 pp, ISBN 88-470-0331-8

22. A. Quarteroni, F. Saleri
Introduzione al Calcolo Scientifico (3a Ed.)
2006, X+306 pp, ISBN 88-470-0480-2

23. S. Leonesi, C. Toffalori
Un invito all'Algebra
2006, XVII+432 pp, ISBN 88-470-0313-X

24. W.M. Baldoni, C. Ciliberto, G.M. Piacentini Cattaneo
Aritmetica, Crittografia e Codici
2006, XVI+518 pp, ISBN 88-470-0455-1

25. A. Quarteroni
Modellistica numerica per problemi differenziali (3a Ed.)
2006, XIV+452 pp, ISBN 88-470-0493-4
(1a edizione 2000, ISBN 88-470-0108-0)
(2a edizione 2003, ISBN 88-470-0203-6)

26. M. Abate, F. Tovena
Curve e superfici
2006, XIV+394 pp, ISBN 88-470-0535-3

27. L. Giuzzi
Codici correttori
2006, XVI+402 pp, ISBN 88-470-0539-6

28. L. Robbiano
Algebra lineare
2007, XVI+210 pp, ISBN 88-470-0446-2

29. E. Rosazza Gianin, C. Sgarra
Esercizi di finanza matematica
2007, X+184 pp, ISBN 978-88-470-0610-2

30. A. Machì
Gruppi – Una introduzione a idee e metodi della Teoria dei Gruppi
2007, XII+350 pp, ISBN 978-88-470-0622-5
2010, ristampa con modifiche

31 Y. Biollay, A. Chaabouni, J. Stubbe
Matematica si parte!
A cura di A. Quarteroni
2007, XII+196 pp, ISBN 978-88-470-0675-1

32. M. Manetti
Topologia
2008, XII+298 pp, ISBN 978-88-470-0756-7

33. A. Pascucci
Calcolo stocastico per la finanza
2008, XVI+518 pp, ISBN 978-88-470-0600-3

34. A. Quarteroni, R. Sacco, F. Saleri
Matematica numerica (3a Ed.)
2008, XVI+510 pp, ISBN 978-88-470-0782-6

35. P. Cannarsa, T. D'Aprile
Introduzione alla teoria della misura e all'analisi funzionale
2008, XII+268 pp, ISBN 978-88-470-0701-7

36. A. Quarteroni, F. Saleri
Calcolo scientifico (4a Ed.)
2008, XIV+358 pp, ISBN 978-88-470-0837-3

37. C. Canuto, A. Tabacco
Analisi Matematica I (3a Ed.)
2008, XIV+452 pp, ISBN 978-88-470-0871-3

38. S. Gabelli
Teoria delle Equazioni e Teoria di Galois
2008, XVI+410 pp, ISBN 978-88-470-0618-8

39. A. Quarteroni
Modellistica numerica per problemi differenziali (4a Ed.)
2008, XVI+560 pp, ISBN 978-88-470-0841-0

40. C. Canuto, A. Tabacco
Analisi Matematica II
2008, XVI+536 pp, ISBN 978-88-470-0873-1
2010, ristampa con modifiche

41. E. Salinelli, F. Tomarelli
   Modelli Dinamici Discreti (2a Ed.)
   2009, XIV+382 pp, ISBN 978-88-470-1075-8

42. S. Salsa, F.M.G. Vegni, A. Zaretti, P. Zunino
   Invito alle equazioni a derivate parziali
   2009, XIV+440 pp, ISBN 978-88-470-1179-3

43. S. Dulli, S. Furini, E. Peron
   Data mining
   2009, XIV+178 pp, ISBN 978-88-470-1162-5

44. A. Pascucci, W.J. Runggaldier
   Finanza Matematica
   2009, X+264 pp, ISBN 978-88-470-1441-1

45. S. Salsa
   Equazioni a derivate parziali – Metodi, modelli e applicazioni (2a Ed.)
   2010, XVI+614 pp, ISBN 978-88-470-1645-3

46. C. D'Angelo, A. Quarteroni
   Matematica Numerica – Esercizi, Laboratori e Progetti
   2010, VIII+374 pp, ISBN 978-88-470-1639-2

47. V. Moretti
   Teoria Spettrale e Meccanica Quantistica – Operatori in spazi di Hilbert
   2010, XVI+704 pp, ISBN 978-88-470-1610-1

48. C. Parenti, A. Parmeggiani
   Algebra lineare ed equazioni differenziali ordinarie
   2010, VIII+208 pp, ISBN 978-88-470-1787-0

49. B. Korte, J. Vygen
   Ottimizzazione Combinatoria. Teoria e Algoritmi
   2010, XVI+662 pp, ISBN 978-88-470-1522-7

50. D. Mundici
   Logica: Metodo Breve
   2011, XII+126 pp, ISBN 978-88-470-1883-9

51. E. Fortuna, R. Frigerio, R. Pardini
   Geometria proiettiva. Problemi risolti e richiami di teoria
   2011, VIII+274 pp, ISBN 978-88-470-1746-7

52. C. Presilla
Elementi di Analisi Complessa. Funzioni di una variabile
2011, XII+324 pp, ISBN 978-88-470-1829-7

53. L. Grippo, M. Sciandrone
Metodi di ottimizzazione non vincolata
2011, XIV+614 pp, ISBN 978-88-470-1793-1

54. M. Abate, F. Tovena
Geometria Differenziale
2011, XIV+466 pp, ISBN 978-88-470-1919-5

55. M. Abate, F. Tovena
Curves and Surfaces
2011, XIV+390 pp, ISBN 978-88-470-1940-9

56. A. Ambrosetti
Appunti sulle equazioni differenziali ordinarie
2011, X+114 pp, ISBN 978-88-470-2393-2

57. L. Formaggia, F. Saleri, A. Veneziani
Solving Numerical PDEs: Problems, Applications, Exercises
2011, X+434 pp, ISBN 978-88-470-2411-3

58. A. Machì
Groups. An Introduction to Ideas and Methods of the Theory of Groups
2011, XIV+372 pp, ISBN 978-88-470-2420-5

59. A. Pascucci, W.J. Runggaldier
Financial Mathematics. Theory and Problems for Multi-period Models
2011, X+288 pp, ISBN 978-88-470-2537-0

60. D. Mundici
Logic: a Brief Course
2012, XII+124 pp, ISBN 978-88-470-2360-4

61. A. Machì
Algebra for Symbolic Computation
2012, VIII+174 pp, ISBN 978-88-470-2396-3

62. A. Quarteroni, F. Saleri, P. Gervasio
Calcolo Scientifico (5a ed.)
2012, XVIII+450 pp, ISBN 978-88-470-2744-2

The online version of the books published in this series is available at SpringerLink.
For further information, please visit the following link:
http://www.springer.com/series/5418